Nuclear Waste Management and Sustainability

Nuclear Waste Management and Sustainability

Editors

Michael I. Ojovan
Vladislav A. Petrov
Sergey V. Yudintsev

MDPI

Basel • Beijing • Wuhan • Barcelona • Belgrade • Novi Sad • Cluj • Manchester

Editors
Michael I. Ojovan
South Kensington Campus
London, UK

Vladislav A. Petrov
Institute of Geology of Ore Deposits, Petrography, Mineralogy and Geochemistry of Russian Academy of Sciences (IGEM RAS)
Moscow, Russia

Sergey V. Yudintsev
Institute of Geology of Ore Deposits, Petrography, Mineralogy and Geochemistry of Russian Academy of Sciences (IGEM RAS)
Moscow, Russia

Editorial Office
MDPI
St. Alban-Anlage 66
4052 Basel, Switzerland

This is a reprint of articles from the Special Issue published online in the open access journal *Sustainability* (ISSN 2071-1050) (available at: https://www.mdpi.com/journal/sustainability/special_issues/Nuclear_Waste_Management_and_Sustainability).

For citation purposes, cite each article independently as indicated on the article page online and as indicated below:

Lastname, A.A.; Lastname, B.B. Article Title. *Journal Name* **Year**, *Volume Number*, Page Range.

ISBN 978-3-0365-8606-9 (Hbk)
ISBN 978-3-0365-8607-6 (PDF)
doi.org/10.3390/books978-3-0365-8607-6

Cover image courtesy of Michael I. Ojovan, Vladislav A. Petrov and Sergey V. Yudintsev

Contents

About the Editors vii

Vladislav A. Petrov, Michael I. Ojovan and Sergey V. Yudintsev
Material Aspect of Sustainable Nuclear Waste Management
Reprinted from: *Sustainability* **2023**, *15*, 11934, doi:10.3390/su151511934 1

Vitaly Milyutin, Natalya Nekrasova, Pavel Kozlov, Arseni Slobodyuk, Darya Markova, Sergey Shaidullin, et al.
Removal of Cs-137 from Liquid Alkaline High-Level Radwaste Simulated Solution by Sorbents of Various Classes
Reprinted from: *Sustainability* **2023**, *15*, 8734, doi:10.3390/su15118734 9

Bella Zubekhina, Boris Burakov, Andrei Shiryaev, Xiaodong Liu and Yury Petrov
Long-Term Chemical Alteration of ^{238}Pu-Doped Borosilicate Glass in a Simulated Geological Environment with Bentonite Buffer
Reprinted from: *Sustainability* **2023**, *15*, 6306, doi:10.3390/su15076306 27

Erin Holland, Yannick Verbelen, Dean Connor, Tomas Martin, Matthew Higginson and Tom Scott
An Introduction to Nuclear Industrial Archaeology
Reprinted from: *Sustainability* **2023**, *15*, 6178, doi:10.3390/su15076178 39

Shengheng Tan, Jiong Chang, Xiao Liu, Shikuan Sun, Liang Xian and Shengdong Zhang
Influence of Radioactive Sludge Content on Vitrification of High-Level Liquid Waste
Reprinted from: *Sustainability* **2023**, *15*, 4937, doi:10.3390/su15064937 61

Marina Palamarchuk, Maxim Chervonetskiy, Natalya Polkanova and Svetlana Bratskaya
Toward Deep Decontamination of Intermediate-Level-Activity Spent Ion-Exchange Resins Containing Poorly Soluble Inorganic Deposits
Reprinted from: *Sustainability* **2023**, *15*, 3990, doi:10.3390/su15053990 75

Victor I. Malkovsky, Vladislav A. Petrov, Sergey V. Yudintsev, Michael I. Ojovan and Valeri V. Poluektov
Influence of Rock Structure on Migration of Radioactive Colloids from an Underground Repository of High-Level Radioactive Waste
Reprinted from: *Sustainability* **2023**, *15*, 882, doi:10.3390/su15010882 93

Norbert Clauer, Miroslav Honty, Lander Frederickx and Christophe Nussbaum
Evaluation of a Long-Term Thermal Load on the Sealing Characteristics of Potential Sediments for a Deep Radioactive Waste Disposal
Reprinted from: *Sustainability* **2022**, *14*, 14004, doi:10.3390/su142114004 103

Anna V. Matveenko, Andrey P. Varlakov, Alexander A. Zherebtsov, Alexander V. Germanov, Ivan V. Mikheev, Stepan N. Kalmykov and Vladimir G. Petrov
Natural Clay Minerals as a Starting Material for Matrices for the Immobilization of Radioactive Waste from Pyrochemical Processing of SNF
Reprinted from: *Sustainability* **2021**, *13*, 10780, doi:10.3390/su131910780 127

Victoria Krupskaya, Anatoliy Boguslavskiy, Sergey Zakusin, Olga Shemelina, Mikhail Chernov, Olga Dorzhieva and Ivan Morozov
The Influence of Liquid Low-Radioactive Waste Repositories on the Mineral Composition of Surrounding Soils
Reprinted from: *Sustainability* **2020**, *12*, 8259, doi:10.3390/su12198259 141

Ga Hyun Chun, Jin-ho Park and Jae Hak Cheong
Calculation of Potential Radiation Doses Associated with Predisposal Management of Dismantled Steam Generators from Nuclear Power Plants
Reprinted from: *Sustainability* **2020**, *12*, 5149, doi:10.3390/su12125149 . . . . **153**

Alexey V. Luzhetsky, Vladislav A. Petrov, Sergey V. Yudintsev, Viktor I. Malkovsky, Michael I. Ojovan, Maximilian S. Nickolsky, et al.
Effect of Gamma Irradiation on Structural Features and Dissolution of Nuclear Waste Na–Al–P Glasses in Water
Reprinted from: *Sustainability* **2020**, *12*, 4137, doi:10.3390/su12104137 . . . . **179**

Esther Phillip, Thye Foo Choo, Nurul Wahida Ahmad Khairuddin and Rehab O. Abdel Rahman
On the Sustainable Utilization of Geopolymers for Safe Management of Radioactive Waste: A Review
Reprinted from: *Sustainability* **2023**, *15*, 1117, doi:10.3390/su15021117 . . . . **203**

Nailia Rakhimova
Recent Advances in Alternative Cementitious Materials for Nuclear Waste Immobilization: A Review
Reprinted from: *Sustainability* **2023**, *15*, 689, doi:10.3390/su15010689 . . . . **233**

Rehab O. Abdel Rahman and Michael I. Ojovan
Toward Sustainable Cementitious Radioactive Waste Forms: Immobilization of Problematic Operational Wastes
Reprinted from: *Sustainability* **2021**, *13*, 11992, doi:10.3390/su132111992 . . . . **259**

Michael I. Ojovan, Vladislav A. Petrov and Sergey V. Yudintsev
Glass Crystalline Materials as Advanced Nuclear Wasteforms
Reprinted from: *Sustainability* **2021**, *13*, 4117, doi:10.3390/su13084117 . . . . **285**

About the Editors

Michael I. Ojovan

Michael I. Ojovan has been Professor of Imperial College London, Nuclear Engineer of International Atomic Energy Agency (IAEA), Associate Reader at the University of Sheffield, UK, Leading Scientist of Lomonosov Moscow State University and of the Institute of Geology of Ore Deposits, Petrography, Mineralogy and Geochemistry (IGEM) of Russian Academy of Sciences, and Chief Editor of journal "Science and Technology of Nuclear Installations". Michael has published 15 monographs including the "Handbook of Advanced Radioactive Waste Conditioning Technologies" and the three editions of "An Introduction to Nuclear Waste Immobilisation" by Elsevier. He has founded and led the IAEA International Predisposal Network (IPN) and the IAEA International Project on Irradiated Graphite Processing (GRAPA). Michael is known for the connectivity-percolation theory (CPT) of glass transition, Sheffield model of viscosity of glasses and melts, theoretical bases of condensed Rydberg matter, metallic and glass-composite materials for nuclear waste immobilisation, and self-sinking capsules to investigate Earth' deep interior.

Vladislav A. Petrov

Vladislav A. Petrov is Corresponding Member of the Russian academy of sciences, Professor of the Mendeleev University of Chemical Technology, Doctor of Sciences in Geology and Mineralogy, and Director of the Institute of Geology of Ore Deposits, Petrography, Mineralogy and Geochemistry (IGEM) of Russian Academy of Sciences. Vladislav graduated as mining engineer-geologist with specialization on deposits of rare earth and radioactive elements in Moscow Geology Prospecting Institute. He has experience in survey of uranium, gold, lead-zinc and REE deposits in Siberia as well as in site selection for radioactive waste disposal in Chelyabinsk, Krasnoyarsk and Eastern Regions of Russia.Vladislav is co-author of 10 monographs including "Petrophysical Properties of Crystalline Rocks" (Geological Society of London. 2005), "Isolation of spent nuclear fuel: geological-geochemical basis" (Moscow. 2008), "Internationalization of the Nuclear Fuel Cycle: Goals, Strategies, and Challenges" (The National Academies Press. 2009), and "Uranium and molybdenum-uranium deposits in regions of continental intracore magmatism development: geology, geodynamics and physical-chemical conditions of formation" (Moscow. 2012). Vladislav is known as expert in the fields of radiogeology and radiogeoecology.

Sergey V. Yudintsev

Sergey V. Yudintsev is Head of the Laboratory of Radiogeology and Radiogeoecology at the Institute of Geology of Ore Deposits, Petrography, Mineralogy, and Geochemistry of the Russian Academy of Sciences (IGEM RAS) where he has been involved in research on radioactive waste disposal problems, searching for new matrices for isolation of long-lived radionuclides of actinides and fission products (technetium). In his career at RAS he has worked as a researcher and principal scientist studying the issues of safe and sustainable development of nuclear energy, including problems of minor actinides transmutation and the thorium-fuel cycle. Dr. Yudintsev has published more than 300 papers in Russian and international journals. He received his undergraduate degree in geochemistry from Moscow State University in 1981, Ph. D. in 1989 and a D. Sc. in 2009, both from the IGEM RAS. In 2011 he was elected as a corresponding member of the Russian Academy of Sciences.

sustainability

Editorial

Material Aspect of Sustainable Nuclear Waste Management

Vladislav A. Petrov, Michael I. Ojovan * and Sergey V. Yudintsev

Institute of Geology of Ore Deposits, Petrography, Mineralogy, and Geochemistry, Russian Academy of Sciences, Staromonetny Lane, 35, 119017 Moscow, Russia; vlad243@igem.ru (V.A.P.); syud@igem.ru (S.V.Y.)
* Correspondence: m.i.ojovan@gmail.com

Citation: Petrov, V.A.; Ojovan, M.I.; Yudintsev, S.V. Material Aspect of Sustainable Nuclear Waste Management. *Sustainability* **2023**, *15*, 11934. https://doi.org/10.3390/su151511934

Received: 21 July 2023
Accepted: 1 August 2023
Published: 3 August 2023

1. Introduction

Effectively managing nuclear waste is crucial to ensure the safe sustainable usage of nuclear energy, which ranges from large-scale applications in power generation to numerous smaller-scale applications in medicine, industry, and agriculture, and scientific research is needed at the current state of development. The waste, which is generated as a by-product of nuclear energy use, is termed nuclear or radioactive waste since it contains levels of hazardous radionuclides that are above the clearance/exemption levels set by national authorities in each country. Depending on the radionuclide contents, the nuclear waste is classified as very-low-level, low-level, intermediate-level, or high-level waste (VLLW, LLW, ILW, and HLW, correspondingly). The lifecycle of nuclear waste management (NWM) contains the following stages: pretreatment, treatment, conditioning, and disposal. The characterization of these stages is mandatory and often crucial, and transportation is also a necessary linkage between the different players of NWM. Nuclear waste is first treated and conditioned for its safe handling, transportation, storage, and ultimate placement into a disposal facility. Finally, the conditioned nuclear waste is disposed at the back end of its lifecycle with the aim of permanently protecting humans and the biosphere from dangerous radioactive materials [1].

The largest source of nuclear waste by volume is the nuclear fuel cycle (NFC), which has nuclear power plants that produce electricity at its core. Figure 1 schematically shows the NFC activities and associated nuclear waste streams, emphasizing that hazardous radioactive materials (uranium ores) are initially mined from deep crystalline or shallow sedimentary formations; these materials, along with newly formed radiotoxic and artificial nuclides, are finally taken to geological disposal facilities (GDF) that are located deep underground within geological formations (crystalline rocks, salt, clays, etc.) that have been stable for millions of years (see, e.g., [2]).

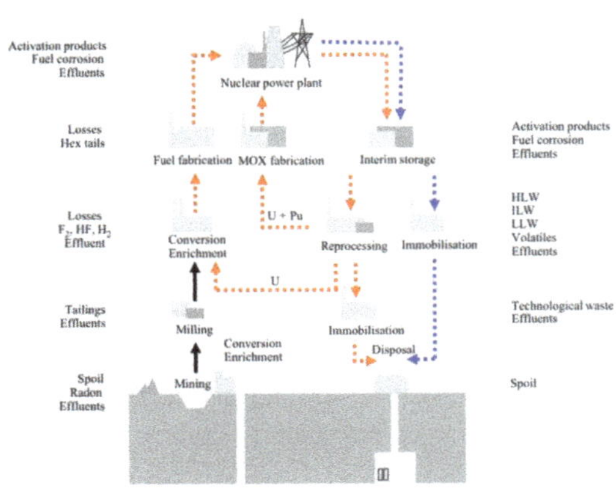

Figure 1. Closed (shown in orange) and open NFC activities with associated nuclear wastes (see [2], publicly available).

2. Materials for Nuclear Waste Management

The safety of nuclear waste relies on the materials that are used, and this aspect is especially important in its disposal. Nuclear waste processing, storage, and disposal activities encompass the utilization of advanced technologies and materials, aiming to ensure the reliability of long-term waste isolation. Considering the material aspect is hence one of the keys to ensure sustainable and safe nuclear waste management. Cements, geopolymers, glasses, glass composite materials, ceramics, and metals are generically the basic materials that are used and systematically analyzed for their expected long-term performance in the disposal environment. The selection of materials within the nuclear waste management problem is a complex task accounting for technology availability and waste characteristics, and typically involves a compromise between various constraints including flexibility, ease of processing, waste loading, characteristics of the waste, required lifetime performance affected by self-irradiation from decaying radionuclides, and water alteration accounting for the mineralogical–geochemical aspects [3–6]. In addition to laboratory tests, the available data on natural analogue materials that have been proven for their long-term stability and durability are used to ensure confidence in the multi-scale approaches that are currently used to predict the behavior of waste disposal systems on geological timescales. The generic rule is that the higher the nuclear waste hazard, the more durable materials are being used to immobilize and ultimately dispose of the radionuclides (Figure 2).

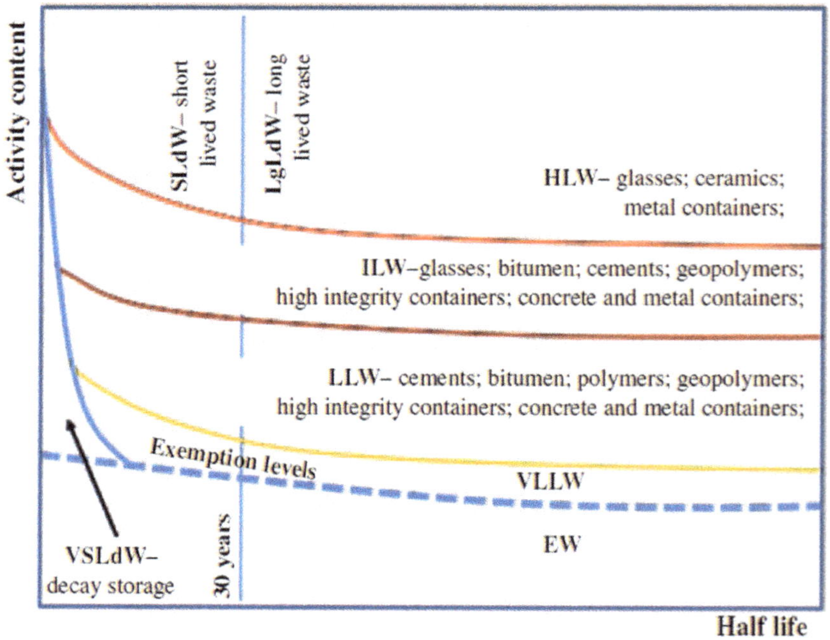

Figure 2. Materials and disposal routes of nuclear waste following IAEA classification scheme. Reproduced from [3] with permission from Elsevier. EW—waste exempt of regulatory control.

At the core of the multibarrier system preventing radionuclide migration into the biosphere is the wasteform [4]—the product of immobilizing waste into a matrix (Figure 3).

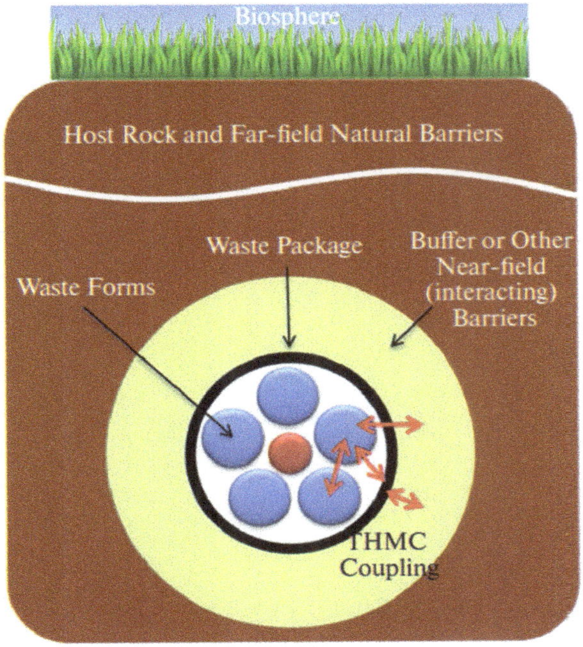

Figure 3. Schematic representation a of multibarrier system (see [4], publicly available). THMC—coupling of thermo–hydro–mechanical and chemical processes.

Immobilization is defined as the conversion of waste into a wasteform via solidification, embedding, or encapsulation aiming to facilitate handling, transportation, storage, and disposal. The immobilization of waste is achieved by its chemical incorporation into the structure of a suitable matrix (e.g., cement, glass, or ceramic) so it is captured and unable to escape. The management of HLW is the most challenging because the times required for radionuclides to decay to background levels are of geological timescales. Chemical immobilization (binding of radionuclides at atomic/molecular level) is typically applied for HLW, while encapsulation is used for ILW, and the immobilization of LLW is achieved by physically surrounding the waste constituents with materials such as bitumen or cement so it is isolated and so the radionuclides are retained. Physical encapsulation is often applied to ILW and can also be used for HLW, especially where the chemical incorporation of radionuclides in the surrounding matrix is also possible. Choosing the wasteforms and packages for higher activity radioactive wastes such as ILW and HLW is challenging, as the crucial decisions largely depend on the nature of the waste streams and are based on the proven technologies used worldwide [4–6].

VLLW does not require immobilization, although some forming packaging is used to assist with handling and transportation. LLW is typically immobilized using cement, although more durable materials such as bitumen or glass can provide a much better retention of contaminants and reduce the final waste disposal volume. Since ILW is more hazardous than LLW, it can be immobilized using glasses, bitumen, cements, and geopolymers. LLW and ILW can be packed in high-integrity metal and concrete containers. The most hazardous waste is HLW, typically resulting from spent fuel reprocessing, which requires the most durable and stable wasteforms. These are represented by durable crystalline ceramics that are currently under the deployment stage, and by glasses that have been used industrially for many decades. The vitrified HLW is typically encased in stainless steel canisters, while corrosion-resistant copper containers are used to dispose of spent nuclear fuel. Table 1 indicates the various nuclear waste classes from the papers that are published within this Special Issue (i.e., A, B, C. . .P are chapters of this book).

Table 1. Relevant nuclear waste classes within publications of current *Sustainability* Special Issue.

Chapter	Title of Contribution, Web Reference	Waste Class
A	"Material Aspect of Sustainable Nuclear Waste Management" (Editorial)	All
B	"Removal of Cs-137 from Liquid Alkaline High-Level Radwaste Simulate Solution by Sorbents of Various Classes" (Article), https://www.mdpi.com/2071-1050/15/11/8734 (accessed on 20 July 2023).	All
C	"Long-Term Chemical Alteration of ^{238}Pu-Doped Borosilicate Glass in a Simulated Geological Environment with Bentonite Buffer" (Article), https://www.mdpi.com/2071-1050/15/7/6306 (accessed on 20 July 2023).	HLW
D	"An Introduction to Nuclear Industrial Archaeology" (Article), https://www.mdpi.com/2071-1050/15/7/6178 (accessed on 20 July 2023).	All
E	"Influence of Radioactive Sludge Content on Vitrification of High-Level Liquid Waste" (Article), https://www.mdpi.com/2071-1050/15/6/4937 (accessed on 20 July 2023).	HLW
F	"Toward Deep Decontamination of Intermediate-Level-Activity Spent Ion-Exchange Resins Containing Poorly Soluble Inorganic Deposits" (Article), https://www.mdpi.com/2071-1050/15/5/3990 (accessed on 20 July 2023).	ILW
G	"Influence of Rock Structure on Migration of Radioactive Colloids from an Underground Repository of High-Level Radioactive Waste" (Article), https://www.mdpi.com/2071-1050/15/1/882 (accessed on 20 July 2023).	HLW
H	"Evaluation of a Long-Term Thermal Load on the Sealing Characteristics of Potential Sediments for a Deep Radioactive Waste Disposal" (Article), https://www.mdpi.com/2071-1050/14/21/14004 (accessed on 20 July 2023).	HLW
I	"Natural Clay Minerals as a Starting Material for Matrices for the Immobilization of Radioactive Waste from Pyrochemical Processing of SNF" (Article), https://www.mdpi.com/2071-1050/13/19/10780 (accessed on 20 July 2023).	HLW, ILW
J	"The Influence of Liquid Low-Radioactive Waste Repositories on the Mineral Composition of Surrounding Soils" (Article), https://www.mdpi.com/2071-1050/12/19/8259 (accessed on 20 July 2023).	LLW
K	"Calculation of Potential Radiation Doses Associated with Predisposal Management of Dismantled Steam Generators from Nuclear Power Plants" (Article), https://www.mdpi.com/2071-1050/12/12/5149 (accessed on 20 July 2023).	All
L	"Effect of Gamma Irradiation on Structural Features and Dissolution of Nuclear Waste Na–Al–P Glasses in Water" (Article), https://www.mdpi.com/2071-1050/12/10/4137 (accessed on 20 July 2023).	HLW
M	"On the Sustainable Utilization of Geopolymers for Safe Management of Radioactive Waste: A Review" (Review), https://www.mdpi.com/2071-1050/15/2/1117 (accessed on 20 July 2023).	LLW, ILW
N	"Recent Advances in Alternative Cementitious Materials for Nuclear Waste Immobilization: A Review" (Review), https://www.mdpi.com/2071-1050/15/1/689 (accessed on 20 July 2023).	LLW, ILW
O	"Toward Sustainable Cementitious Radioactive Waste Forms: Immobilization of Problematic Operational Wastes" (Review), https://www.mdpi.com/2071-1050/13/21/11992 (accessed on 20 July 2023).	LLW, ILW
P	"Glass Crystalline Materials as Advanced Nuclear Wasteforms" (Review), https://www.mdpi.com/2071-1050/13/8/4117 (accessed on 20 July 2023).	HLW, ILW

Cements are widely used within NWM starting from its inception stage and are used at an incomparable larger scale compared to other materials [7–9]. The cementation of both solid and liquid nuclear waste has become an important and developing part of the waste management system owing to its simplicity and versatility. Cements are inorganic materials that set and harden because of hydration reactions between its constituents and water, forming a composite material containing both crystalline (i.e., portlandite, ettringite, monosulfate, hydrogarnet) and amorphous (the tobermorite gel termed CSH phase) constituents. Geopolymers are used to a lesser extent in NWM. They are materials often termed alkali-activated cements and are made by mixing a reactive source of alumina and silica, such as fly ash, or by mixing metakaolin with an activator, typically in the form of concentrated aqueous solutions of NaOH or KOH, resulting in the formation of a solid material comprising a hydrous aluminosilicate gel, which binds much of the added alkali [7]. Cements and geopolymers allow for the microencapsulation of waste, although some physicochemical binding of waste cations can also occur. Modern cementitious wasteforms include Portland-based cements, calcium aluminate, calcium sulfoaluminate, phosphate, ceramicrete (hydrous potassium magnesium phosphate), magnesium silicate, and alkali-activated cement (also termed geopolymers). The overviews (N, O, P) provide data on the advances within the uses of cements and geopolymers including their application to so-called problematic waste. The continued development of the cementation technique is driven by the improvement and expansion of cementitious materials that are suitable and efficient for NWM, with advances that have significantly improved nuclear waste cementation technology and the quality of cementitious nuclear wasteforms.

Crystalline ceramics containing one or more crystalline phases are the most durable and reliable host materials for the retention of long-lived radionuclides [10–12]. Single-phase crystalline ceramics can be used to immobilize separated radionuclides such as weapon grade ^{239}Pu, minor actinides after nuclear fuel reprocessing, as well as more chemically complex waste streams including HLW from fuel reprocessing. The atomic structure of the ceramic phase must have multiple cations and sites that can accommodate the variety of radionuclides in the waste stream; therefore, multiphase crystalline ceramics are preferred for complex chemical compositions of waste. Chapter (I) investigates the use of ceramic materials obtained from natural bentonite clay as immobilizing matrices for radioactive waste in the form of a LiCl–KCl eutectic resulting from the pyrochemical reprocessing of used nuclear fuel.

Glasses are amorphous solid materials below the glass transition temperature T_g, which are typically produced via vitrification (liquid–glass transition) by quickly cooling them from a molten to a solid state to avoid crystallization [13]. Durable glasses of borosilicate and alumina phosphate families are the worldwide choice for HLW immobilization [3–6,10,14]. The vitrification of LLW and ILW is also deployed by including legacy and nuclear power plant operational waste [3,10,14]. Chapter (E) provides information on the HLW vitrification program in China, while Chapter (L) provides information on the radiation effects in glasses used in Russia. The results presented in Chapter (C) provide evidence on the enhanced retention properties of glasses in natural waters compared to the deionized water of testing protocols, whereas Chapter (P) supports the recent trend of using inhomogeneous glass crystalline materials (GCMs), which provide superior waste loading and durability compared to homogeneous vitreous materials, as illustrated by Figure 4.

Homogeneous vitreous materials are shown in the lower left corner of the triangle. If durable crystals such as spinel are allowed to form in the glass, then GCMs (left leg of the triangle) that have superior waste loading and durability relative to homogeneous glass can be produced. Fully crystalline materials (ceramics) are shown in the upper corner of the triangle and are the most durable wasteforms with high waste loadings. However, the incorporation of certain species (e.g., Mo, Cr, S, and Cl) into glass creates non-durable secondary phases (lower right corner or triangle) that may have unacceptable durability.

The multibarrier system preventing radionuclide migration into the biosphere includes natural barriers, which are key factors in the case of HLW disposal, accounting for the geological timescales needed to retain the long-lived radionuclides (see Figure 3). The disposal is always the end point of NWM, although it is its integral part [3,15,16]. The natural barrier materials were analyzed with a focus on the rock structure and migration of radioactive colloids in Chapter (G), the chemical and thermal impact of nuclear waste on the minerals in the repository near field in Chapter (J), and the effect of the HLW thermal load on the sealing characteristics of surrounding geological formations in Chapter (H).

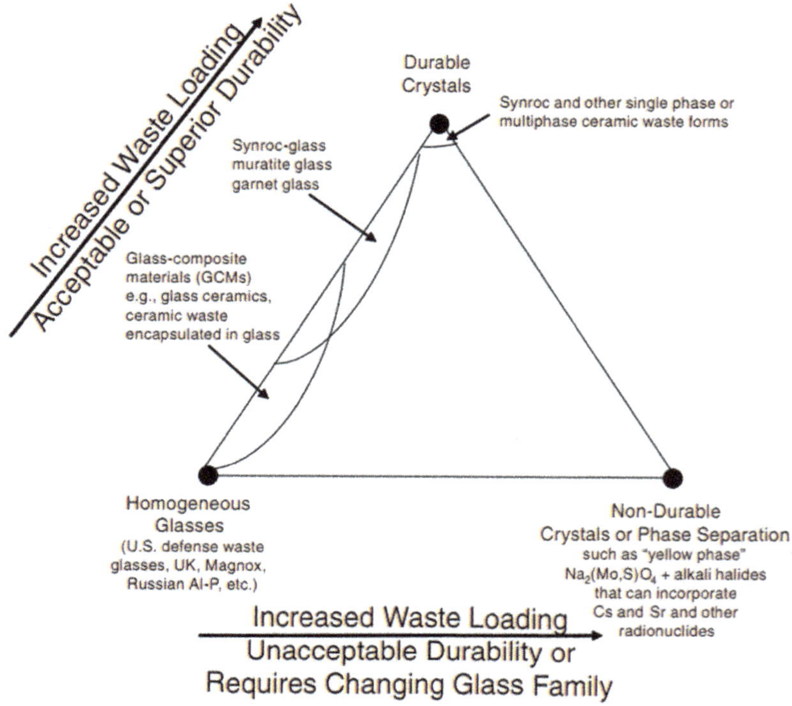

Figure 4. Schematic diagram illustrating the durability and waste loading of GCM wasteforms relative to homogeneous glass. Reproduced from [6], publicly available.

The technological aspect of NWM at both its predisposal and disposal stages [3] is analyzed within Chapter (K), focusing on the potential radiation doses associated with predisposal activities; Chapter (B) provides details on the removal of Cs-137 from alkaline HLW solutions; and Chapter (O) describes the decontamination of ILW in the form of spent ion-exchange resins.

Chapter (N) of this Special Issue introduces a novel discipline—nuclear industrial archaeology—which specializes in finding and characterizing abandoned nuclear sites.

Altogether, the papers published within this Special Issue (Table 1) present evidence on the progress achieved within NWM, aiming to support the sustainable and safe utilization of nuclear energy.

3. Conclusions

The *Sustainability* Special Issue "Nuclear Waste Management and Sustainability" provides a substantial addition to the literature on nuclear waste management, focusing on the material aspect and the technologies used.

Author Contributions: Conceptualization, V.A.P., M.I.O. and S.V.Y.; methodology, V.A.P., M.I.O. and S.V.Y.; validation, V.A.P., M.I.O. and S.V.Y.; formal analysis, V.A.P., M.I.O. and S.V.Y.; investigation, V.A.P., M.I.O. and S.V.Y.; resources, V.A.P.; data curation, V.A.P., M.I.O. and S.V.Y.; writing—original draft preparation, M.I.O.; writing—review and editing, V.A.P., M.I.O. and S.V.Y.; visualization, V.A.P., M.I.O. and S.V.Y.; supervision, V.A.P. and S.V.Y.; project administration, V.A.P. and S.V.Y.; funding acquisition, V.A.P. All authors have read and agreed to the published version of the manuscript.

Funding: This work was carried out within the framework of a state assignment for the Institute of Geology of Ore Deposits, Petrography, Mineralogy, and Geochemistry at the Russian Academy of Sciences.

Acknowledgments: The editors are grateful to all contributing authors.

Conflicts of Interest: The authors declare no conflict of interest.

References

1. IAEA. *Policies and Strategies for Radioactive Waste Management*; Nuclear Energy Series No. NW-G-1.1; IAEA: Vienna, Austria, 2009.
2. Corkhill, C.; Hyatt, N. *Nuclear Waste Management*; IOP Publishing: Bristol, UK, 2018; Available online: https://iopscience.iop.org/book/mono/978-0-7503-1638-5.pdf (accessed on 20 July 2023).
3. Ojovan, M.I.; Lee, W.E.; Kalmykov, S.N. *An Introduction to Nuclear Waste Immobilisation*, 3rd ed.; Elsevier: Amsterdam, The Netherlands, 2019; p. 497.
4. Jantzen, C.M.; Ojovan, M.I. On selection of matrix (wasteform) material for higher activity nuclear waste immobilisation (Review). *Russ. J. Inorg. Chem.* **2019**, *64*, 1611–1624. Available online: https://link.springer.com/article/10.1134/S0036023619130047 (accessed on 20 July 2023). [CrossRef]
5. Stefanovsky, S.V.; Yudintsev, S.V.; Vinokurov, S.E.; Myasoedov, B.F. Chemical-Technological and Mineralogical-Geochemical Aspects of the Radioactive Waste Management. *Geochem. Int.* **2016**, *54*, 1136–1156.
6. National Research Council. *Waste Forms Technology and Performance: Final Report*; National Academies Press: Washington, DC, USA, 2011; 308p. Available online: https://nap.nationalacademies.org/catalog/13100/waste-forms-technology-and-performance-final-report (accessed on 20 July 2023).
7. Glasser, F. Application of inorganic cements to the conditioning and immobilisation of radioactive wastes. In *Handbook of Advanced Radioactive Waste Conditioning Technologies*; Woodhead: Cambridge, UK, 2011; pp. 67–135.
8. Bart, F.; Cau-dit-Coumes, C.; Frizon, F.; Lorente, S. *Cement-Based Materials for Nuclear Waste Storage*; Springer: New York, NY, USA, 2013; 265p. [CrossRef]
9. Abdel Rahman, R.O.; Rahimov, R.Z.; Rahimova, N.R.; Ojovan, M.I. *Cementitious Materials for Nuclear Waste Immobilization*; Wiley: Chichester, UK, 2015; 232p.
10. Donald, I.W. *Waste Immobilisation in Glass and Ceramic Based Hosts*; Wiley: Chichester, UK, 2010; 507p.
11. Burakov, B.E.; Ojovan, M.I.; Lee, W.E. *Crystalline Materials for Actinide Immobilisation*; Imperial College Press: London, UK, 2010; 198p.
12. Orlova, A.I.; Ojovan, M.I. Ceramic Mineral Waste-Forms for Nuclear Waste Immobilization. *Materials* **2019**, *12*, 2638. [CrossRef] [PubMed]
13. Ojovan, M.I. Glass formation. In *Encyclopedia of Glass Science, Technology, History, and Culture*; Richet, P., Conradt, R., Takada, A., Dyon, J., Eds.; Wiley: Hoboken, NJ, USA, 2021; Volume I, Chapter 3.1; pp. 249–259. 156p, 11p, Available online: https://onlinelibrary.wiley.com/doi/book/10.1002/9781118801017 (accessed on 20 July 2023).
14. Pinet, O.; Vernaz, E.; Ladirat, C.; Gin, S. Nuclear waste vitrification. In *Encyclopedia of Glass Science, Technology, History, and Culture*; Richet, P., Conradt, R., Takada, A., Dyon, J., Eds.; Wiley: Hoboken, NJ, USA, 2021; Volume II, Chapter 9.11; pp. 1205–1218. Available online: https://onlinelibrary.wiley.com/doi/book/10.1002/9781118801017 (accessed on 20 July 2023).
15. Drace, Z.; Ojovan, M.I.; Samanta, S.K. Challenges in Planning of Integrated Nuclear Waste Management. *Sustainability* **2022**, *14*, 14204. [CrossRef]
16. Ojovan, M.I.; Steinmetz, H.J. Approaches to Disposal of Nuclear Waste. *Energies* **2022**, *15*, 7804. [CrossRef]

sustainability

MDPI

Article

Removal of Cs-137 from Liquid Alkaline High-Level Radwaste Simulated Solution by Sorbents of Various Classes

Vitaly Milyutin [1], Natalya Nekrasova [1], Pavel Kozlov [2], Arseni Slobodyuk [3], Darya Markova [2], Sergey Shaidullin [2], Kirill Feoktistov [2], Eduard Tokar [3], Mikhail Tutov [3] and Andrei Egorin [3,*]

[1] Frumkin Institute of Physical Chemistry and Electrochemistry, Russian Academy of Sciences, 31 Leninsky Prospect, Moscow 1119071, Russia; vmilyutin@mail.ru (V.M.); nnekrassova@gmail.com (N.N.)

[2] Mayak Production Assosiation, 31 Prospect Lenina, Ozersk 456780, Russia; kozlov_pavel@inbox.ru (P.K.); dvmarkova@mail.ru (D.M.); shaidullinsergey@gmail.com (S.S.); dendifak@gmail.com (K.F.)

[3] Institute of Chemistry, Far Eastern Branch, Russian Academy of Sciences, 159 Prosp. 100-letiya Vladivostoka, Vladivostok 690022, Russia; ampy@ich.dvo.ru (A.S.); d.edd@mail.ru (E.T.); thunderbird87@mail.ru (M.T.)

* Correspondence: andrey.egorin@gmail.com

Abstract: The present work describes the results of the removal of cesium by sorbents of various classes from highly mineralized alkaline solutions simulating the clarified phase of storage tanks with high-level radioactive waste (HLW) of the Mayak Production Association. Within the scope of the performed works, inorganic sorbents of the Clevasol® and Fersal brands, as well as resorcinol-formaldehyde ion-exchange resins (RFRs: RFR-i, RFR-Ca, and Axionit RCs), were used. The sorbents' characteristics under both static and dynamic conditions are presented. The Fersal sorbent has demonstrated the best sorption characteristics in the series of sorbents under study. The disadvantage of inorganic sorbents is the loss of mechanical strength upon cesium desorption, which complicates their repeated use. It has been demonstrated that RFRs, despite their lower selectivity towards cesium and adsorption capacity, can be used many times in repeated sorption-desorption cycles. The latter makes RFRs more technologically attractive in terms of the total volume of decontaminated HLW. However, RFRs tend to be oxidized during storage, which results in the formation of carboxyl groups and a decrease in sorption characteristics—this must be further taken into account in the real processes of liquid radioactive waste (LRW) management.

Keywords: selective sorbents; resorcinol–formaldehyde resin; adsorption; cesium; liquid radioactive waste

Citation: Milyutin, V.; Nekrasova, N.; Kozlov, P.; Slobodyuk, A.; Markova, D.; Shaidullin, S.; Feoktistov, K.; Tokar, E.; Tutov, M.; Egorin, A. Removal of Cs-137 from Liquid Alkaline High-Level Radwaste Simulated Solution by Sorbents of Various Classes. *Sustainability* **2023**, *15*, 8734. https://doi.org/10.3390/su15118734

Academic Editors: Michael I. Ojovan, Vladislav A. Petrov and Sergey V. Yudintsev

Received: 29 April 2023
Revised: 20 May 2023
Accepted: 25 May 2023
Published: 29 May 2023

1. Introduction

The necessity to process liquid alkaline high-level wastes (HLW) from storage tanks of the Mayak Production Association and, subsequently, transfer them to a stable matrix form is an important and urgent task. To process HLW, schemes were suggested involving the sorption removal of Cs-137 as the main dose-forming radionuclide of the waste from the supernatant. After removal of Cs-137, the decontaminated waste can be solidified by cementation, resulting in solid waste in the intermediate-level category (ILW). The ferrocyanide sorbent FS-10 was chosen for the selective removal of Cs-137 from HLW [1]. However, the disadvantage of using the FS-10 sorbent consists in the need for preliminary neutralization of HLW to pH 6–8 due to the fact that sorbents based on poorly soluble transition metal ferrocyanides undergo destruction in alkaline media [2]. For this reason, the search continues for sorption materials capable of selectively removing Cs-137 from alkaline HLW without prior neutralization.

As the above-mentioned sorption materials, one can consider resorcinol-formaldehyde resins (RFR), which at present are actively used to remove cesium from alkaline radioactive waste streams [3]. The selectivity of RFR towards some single-charge cations changes as follows: $Na^+ < K^+ < Rb^+ < Cs^+ < H^+$ [4]. This property makes it possible to reversibly remove cesium from alkaline LRW with subsequent desorption with suitable acid solutions [5–7].

The mechanism of cesium adsorption is related to the occurrence of ion exchange on deprotonated functional groups [8]. The selectivity of ion-exchange resins towards cesium can be explained by the predominance of the stage of transfer of alkaline metal cations from the liquid phase of the polymer to its solid phase over the stage of electrostatic binding with ion-exchange groups [9,10]. Another model explaining the nature of selectivity suggests that swelling in alkaline media results in an increase in the osmotic pressure in the resin and, consequently, a tension in the polymer network. Sorption of low-hydration cesium and rubidium is preferential in comparison with that of sodium and potassium, as it allows reducing the polymer network tension [11].

Sphere-granulated RFRs (sRFR) became widely used [12–14] due to their advantage of reducing hydrodynamic resistance as well as greater mechanical strength as compared to granular ion-exchangers [14–16]. To solve the problem of RFR oxidation in alkaline media, samples with improved chemical resistance and selectivity to cesium were prepared [7].

Inorganic sorbents can serve as a cheaper and more effective alternative for selective ion-exchange resins to solve the problem of treating LRW with complex chemical composition. The Clevasol® sorbent, commercially manufactured by the French company LEMER PAX Innovative, can be considered as such an alternative. The sorbent is marketed as an effective material for the removal of radionuclides from solutions in a broad pH range [17].

The Clevasol® sorbent comprises a polymer structure with high selectivity to a number of cations, especially cesium, silver, and thallium [18]. Due to its mechanical strength and large particle size, the sorbent can be used under dynamic conditions as a load for adsorption columns. It was shown that the Clevasol® sorbent effectively removed Cs^+ from nitric acid solutions at concentrations of 1–8 mol/dm^3 (K_d for Cs^+ greater than 10^4 cm^3/g) [19]. Composite sorbents were prepared by embedding the Clevasol® sorbent particles into the polyvinyl alcohol (PVA) cryogel [20]. According to the authors, this material can be used as part of anti-migration barriers due to its high distribution coefficients for Cs^+ and Sr^{2+}. The information on the sorbent composition and a detailed mechanism of the adsorption process is a commercial secret of the LEMER PAX Innovative company.

The Institute of Physical Chemistry and Electrochemistry RAS, in collaboration with Komfinservis LLC, has developed a novel inorganic sorbent of the Fersal brand based on modified nickel ferrocyanide selective to cesium ions. Insoluble ferrocyanides of transition metals and sorbents on their basis demonstrated good results in the processes of cesium ion removal. The mechanism of cesium adsorption on ferrocyanides of transition metals is known in detail and is related to the exchange of potassium and, partially, transition metals by cesium [21–23]. Despite the fact that the sorption-active phase of a sorbent is represented by nickel ferrocyanide, earlier performed tests demonstrated the possibility of cesium removal from alkaline solutions [24].

The above-mentioned materials can be used in the processes of decontamination of highly mineralized alkaline LRW from cesium. However, to choose the most effective sorbent in order to optimize the LRW treatment process, it is necessary to compare their sorption characteristics towards cesium under identical experimental conditions simulating the real ones to a maximal degree. The earlier published results on cesium removal from liquid media do not enable one to conduct this comparison since they were obtained under different experimental conditions using different criteria of efficiency assessment. At the same time, despite the encouraging prospects of the available sorption materials for HLW treatment, there is a necessity for a preliminary assessment of the sorption characteristics and features of the cesium extraction process on the mentioned sorption materials under practically real-life conditions. The present work was devoted to the study of the physical-chemical and sorption characteristics of RFRs: RFR-i, RFR-Ca, and Axionit RCs, as well as inorganic sorbents Clevasol® and Fersal, in order to determine the most promising of them in terms of the decontamination of alkaline HLW from cesium radionuclides.

2. Materials and Methods

2.1. Sorbents

The following resorcinol-formaldehyde ion exchange resins were used in the present work:

- Axionit RCs, the sample was synthesized and provided for experiments by JSC "Axion—Rare and Noble metals". Two batches of RFR, synthesized in 2017 (Axionit RCs 2017) and 2022 (Axionit RCs 2022), respectively, were used in the study.
- RFR-i is an experimental laboratory sample, synthesized and provided for testing by the Institute of Chemistry of the Far Eastern Branch of the Russian Academy of Sciences (Vladivostok, Russia).
- RFR-Ca is an experimental laboratory sample with a porous surface morphology, synthesized and provided for testing by the Institute of Chemistry of the Far Eastern branch of the Russian Academy of Sciences (Vladivostok, Russia).

The following inorganic sorbents were used in the present work:

- Sorbent of the Fersal brand, an inorganic composite sorbent based on modified nickel ferrocyanide. According to the X-ray fluorescence analysis data, the sorbent contains %: N—24.38; O—56.83; Si—9.46; K—0.20; Fe—1.87; Ni—2.86; Cs—3.65. It is supplied in the K^+—form. The cesium sorption proceeds through the exchange of K^+ ions in the sorbent phase with the Cs^+ ions in the solution. The sorbent is manufactured by the LEMER PAX Innovative company, France. A sample for tests was provided by the official distributor of this sorbent in Russia—Komfinservis LLC (Russia).
- Sorbent of the Clevasol® brand, a macroporous inorganic polymer. According to the X-ray fluorescence analysis data, the sorbent contains %: B—83.0; O—6.27; Na—0.46; S—6.86; Cl—3.44. It is supplied in the H^+—form. The cesium sorption proceeds through the exchange of H^+ ions in the sorbent phase by the Cs^+ ions in the solution. The sorbent is manufactured by the LEMER PAX Innovative company, France. A sample for testing was provided by the official distributor of this sorbent in Russia—Komfinservis LLC (Russia).

Figure 1 shows SEM images of the sorbents under study. Resorcinol-formaldehyde resins comprise monolithic, glasslike structures. The sorbent Axionit RCs 2022 comprises regular spherical granules with smooth surfaces. The sorbents RFR-i and RFR-Ca are represented by granules of an irregular shape with surface chips formed during grinding. A specific feature of the sorbent RFR-Ca is the presence of micropores formed as a result of the dissolution of a part of $CaCO_3$ preliminarily introduced into the oligomeric phase at the initial synthesis stage.

The sorbent of the Clevasol® brand is represented by granules of an irregular shape, containing large fissures and cracks. Granules of the Clevasol® sorbent are formed through the agglomeration of smaller particles of irregular and spherical shapes. Granules of the Fersal sorbent are monolithic and do not contain fissures and cracks. As can be seen on the insert with large magnification, the surface of the Fersal sorbent granules includes pores with an average diameter of 5–6.5 μm, which probably ensures high accessibility of adsorption sites.

The resorcinol-formaldehyde resins Axionit RCs, RFR-i, and RFR-Ca were converted to a working sodium form prior to the experiments and sequentially treated with the following under static conditions: HNO_3 solution with a concentration of 1 mol/dm^3, distilled water, and NaOH solution with a concentration of 1.0 mol/dm^3. Then the sorbents were washed with distilled water and dried in the air at a temperature of 60 °C to a constant weight. The Clevasol® and Fersal sorbents were preliminary sieved to obtain a fraction of 0.25–0.50 mm and then dried in air at a temperature of 60 °C to a constant weight.

Figure 1. SEM images of sorbents, (**a**) Clevsol, (**b**) Fersal, (**c**) RFR-i, (**d**) RFR-Ca, (**e**) Axionit RCs 2022.

The bulk weight (d_{bw}, g/cm^3) of the sorption materials was calculated by Formula (1):

$$d_{bw} = \frac{m_s}{V_{(dry)}}, \quad (1)$$

where m_s is the mass of the sorbent (g) and $V_{(dry)}$ is the volume of dry sorption material (cm^3). The arithmetic mean of the three parallel definitions was taken as the test result, and the relative discrepancy should not exceed 5%.

The specific volume value ($V_{specvol}$, cm^3/g) was calculated by Formula (2):

$$V_{(specvol)} = \frac{V_{(svel)}}{m_s}, \quad (2)$$

where $V_{(svel)}$ is the volume of the swollen sorption material (cm^3).

The arithmetic mean of three parallel definitions was taken as the test result, and the relative discrepancy must not exceed 5%

Table 1 shows the physical-chemical characteristics of the sorption materials; errors do not exceed 5%. A characteristic difference between ion-exchange resins and inorganic sorbents is their significant swelling, which increases during the transition from distilled water to an alkaline model solution and must be taken into account when using this class of sorption materials. Strong swelling may be accompanied by an increase in the hydrodynamic resistance of the stationary sorbent layer, as well as the rupture of the column; therefore, ion-exchange resins were soaked under a layer of distilled water before the experiment.

Table 1. Physical and chemical characteristics of the studied sorption material.

Indicator	The Value of the Indicator for the Sorbent				
	RFR-Ca	RFR-i	Axionit RCs	Fersal	Clevasol
Granule size (mm)	0.25–1.0	0.25–1.0	0.25–0.80	0.25–3.0	0.25–1.0
Bulk weight (g/cm^3)	0.61	0.68	0.82	0.53	0.52
d_{bw} (cm^3/g)	1.63	1.47	1.22	1.87	1.92
$V_{(specvol)}$ (cm^3/g)	2.73	2.08	3.29	1.87	1.92
$V_{(specvol)}$ in model solution * (cm^3/g)	2.87	2.16	3.40	1.87	1.92

* HLW simulated solution.

2.2. Compositions of Model Solutions

The sorption characteristics of Cs-137 were evaluated using model solutions of the following composition:

- $NaNO_3$ solutions of concentrations at 0.5, 1.0, 2.0, and 3.0 mol/dm^3, pH = 13. pH adjustment was carried out with NaOH solution;
- KNO_3 solutions of concentrations at 0.1, 0.25, 0.5, and 1.0 mol/dm^3, pH = 13. pH adjustment was carried out with NaOH solution;
- NaOH + $NaNO_3$ solutions with constant sums of concentrations of NaOH + $NaNO_3$ = 3.5 mol/dm^3, containing 0.1, 0.25, 0.50, 1.0, 1.5, and 2.0 mol/dm^3 of NaOH and 3.4, 3.25, 3.0, 2.5, 2.0, and 1.5 mol/dm^3 of $NaNO_3$;
- Model solution simulating the clarified phase of HLW storage tanks of the Mayak Production Association with the following composition, g/dm^3: NaOH—100, $NaNO_3$—128, $Al(NO_3)_3 \times 9H_2O$—82.5 (6.0 by Al), K_2CrO_4—7.72, $CsNO_3$—0.0733 (0.050 by Cs), density—1.190 g/cm^3 [1].

Model solutions were prepared by dissolving the corresponding salts in distilled water. Sodium hydroxide was introduced as a concentrated solution.

Prior to the sorption experiments, the model solutions were labeled with Cs radionuclide in an amount of about 10^5 Bq/dm^3 and then held for more than 3 days to achieve equilibrium between the active and inactive components of the solution. The specific activity of Cs-137 was determined by a direct radiometric method on the 661 keV gamma line using the SCS-50M spectrometric complex (Green Star Technologies, Russia). The geometry of the measuring sample was a Petri dish, the sample volume was 20 $\pm$ 2 cm^3, the measurement time was 300 s, and the relative error of measuring activity was no more than 15%.

The C-13 NMR spectra of the polymers in the solid phase were recorded using a Bruker Avance AV-300 spectrometer (Rheinstetten, Germany) with a proton resonance frequency of 300 MHz. To record the spectra, the method of rotating samples at a magic angle (MAS) was used. Tetramethylsilane was used as a standard for carbon nuclei. The chemical shift reference (CS) was set in a separate experiment. The determination error for chemical shift (CS) was 0.3 ppm, while for the intensity of the spectral line it was 5% of its area. The spectra were recorded at 300 K.

2.3. Evaluation of Sorption Characteristics under Static Conditions

Sorption characteristics under static conditions were evaluated as follows: A sample weight of air-dry sorption material of 0.10 $\pm$ 0.01 g was placed in a polyethylene container, into which 20 cm^3 of the model solution was poured (W/T ratio = 200 cm^3/mg), after which it was stirred for 24 h. After that, the solid phase was separated from the solution by filtering through a white ribbon paper filter, and the specific activity of Cs-137 was determined in the filtrate.

Based on the results of the analyses, the values of the distribution coefficient (K_d) Cs-137 were calculated; see Equation (3).

$$K_d = \frac{A_0 - A_{(\text{resact})}}{A_{(\text{resact})}} \times \frac{V_{(\text{solut})}}{m_s}, \tag{3}$$

where A_0 and $A_{(\text{resact})}$ are the initial specific activity of Cs-137 in the solution and the residual activity of Cs-137 in the solution after sorption, respectively (Bq/cm^3), and $V_{(\text{solut})}$ is the volume of the liquid phase (cm^3).

The value of static exchange capacity (SEC, mg/cm^3) in solutions with a macro concentration of stable cesium was calculated using Equation (4):

$$\text{SEC} = \frac{C_0 - C_{(\text{resid})} \times V_{(\text{solut})}}{m_s \times V_{(\text{specvol})}}, \tag{4}$$

where C_0 and $C_{(\text{resid})}$ are the initial and residual concentrations of cesium in solution, respectively (mg/cm^3).

The measurement error for the confidence probability 0.95 ($p < 0.05$) was estimated on the basis of three parallel measurements of the solution gamma-activity using the one-sample t-criterion.

2.4. Evaluation of Sorption Characteristics under Dynamic Conditions

Sorption characteristics under dynamic conditions were evaluated as follows: First, 3.0 cm^3 of the sorption material held under a layer of water for more than 12 h was transferred to a polypropylene sorption column with an inner diameter of 8.5 mm. Percolation of the HLW model solution was carried out by a peristaltic pump from Longerpump (China), model BQ-50J at a speed of 10.0 ± 0.5 cm^3/h. The rate of solution percolation was selected based on the fact that to establish equilibrium under dynamic conditions, the contact time must be at least $10t_{0,5}$, where $t_{0,5}$ is the time of half-exchange. Based on the data of [25], the time of half-exchange of cesium ions on selective absorbents is about 1–2 min, so the contact time must be at least 20 min, which corresponds to the specific filtration rate of 3 BV/h.

Filtrates after the column were collected by fractions, and the residual specific activity of Cs-137 was determined. Based on the results of the filtrate analysis, the output adsorption curves were constructed in coordinates f(DF) = BV, where BV is the bed volume calculated as the ratio of the sorbent volume in the column (cm^3) to the volume of the solution being passed (cm^3), and DF is the decontamination factor for dynamic conditions, which was calculated by the Formula (5):

$$\text{DF} = \frac{A_0}{A_{(\text{filtrate})}}, \tag{5}$$

where $A_{(\text{filtrate})}$ is the specific residual activity of Cs-137 in the filtrate (Bq/cm^3).

When using an HLW simulated solution containing stable cesium, the dynamic exchange capacity up to reaching a given slip (DEC, mg/cm^3) was calculated as an integral of the dependence of the amount of adsorbed cesium on the amount of the fed solution (6).

$$\text{DEC} = \int_0^n f\left(\frac{\left(C_0 - C_{(\text{filtrate})}\right) V_{(i)}}{V_{(\text{svel})}}\right) dV_{(t)}, \tag{6}$$

where $C_{(\text{filtrate})}$ is the residual concentration of cesium in the i-th fraction of the eluate (mg/cm^3), $V_{(i)}$ is the volume in the i-th fraction of the eluate (cm^3), m is the mass of the sorbent (g), an d$V_{(t)}$ is the volume of the fed solution (cm^3).

To determine the total dynamic exchange capacity (TDEC, mg/cm^3) for the output curves of cesium adsorption in coordinates $f(C_{(filtrate)}/C_0) = t$, the Thomas model (7)–(9) was used as follows:

$$\frac{C_{(filtrate)}}{C_0} = \frac{1}{1+e^{(a-b\times t)}}, \quad (7)$$

$$a = \frac{K_{Th} \times TDEC \times m}{Q}, \quad (8)$$

$$b = K_{Th} \times C_0, \quad (9)$$

where K_{Th} is the adsorption rate constant (cm^3/mg × h), m is the mass of the sorbent (g), Q is the volume rate of the solution (cm^3/h), and t is the duration of the experiment (h). The experimental data were processed using Veusz software (ver. 3.4) [26].

2.5. Desorption of Cs-137 and Regeneration of Sorbents

After the sorption of Cs-137 from the HLW simulated solution was completed, the sorbent in the column was washed with distilled water with a volume of 30 cm^3 at a rate of 6–9 cm^3/h to remove the residues of the solution. Then a desorbing solution was fed through the column with the washed sorbent using a peristaltic pump at a rate of 2.5–3.0 cm^3/h (0.8–1.0 BV/h). The eluates after the column were collected by fractions, and the residual specific activity of Cs-137 in them was determined. Based on the results of the filtrate analysis, the output desorption curve was constructed with the coordinates Des% = $f(BV)$. The degree of desorption of cesium (Des, %) under dynamic conditions was calculated by the Equation (10):

$$Des\% = \left(\frac{\sum_1^i V_{(el)} \times C_{(eluate)}}{C_{DEC}}\right) \times 100, \quad (10)$$

where $C_{(eluate)}$ is the concentration of the cesium in the i-th fraction of the eluate (mg/cm^3), $V_{(el)}$ is the volume of the eluate (cm^3), C_{DEC} *is* the amount of the adsorbed cesium (mg), and i is the ordinal number of the eluate fraction.

For the repeated sorption of cesium, the RFR-i and RFR-Ca ion exchange resins were washed with water to remove acid residues and treated with a 1.0 M NaOH solution to transform them into sodium form. The volume of a 1.0 M NaOH solution was 30 cm^3, and the transmission rate of the solution was 2.5 cm^3/h.

Under dynamic conditions, the error with the confidence probability 0.95 ($p < 0.05$) for a single measurement of the solution gamma-activity was determined using the Equation (11).

$$SD = \frac{\sqrt{n}}{t} \times 1.96 \quad (11)$$

where n is the quantity of registered impulses and t is the measurement time (min).

3. Results and Discussion

Figure 2 shows the results for the values of the Cs-137 distribution coefficients and the static cesium capacity for various sorbents during sorption from a model HLW solution containing a stable cesium isotope in macro concentrations. Based on the results obtained, the efficiency of the sorption materials in terms of cesium removal from the HLW simulated solution under static conditions can be put in the following order: Fersal > RFR-i > Axionit RCs 2022~RFR-Ca > Axionit RCs 2017 > Clevasol®. One should mention that the storage of Axionit RCs ion exchange resin for 5 years led to a decrease in SEC by 10% and K_d Cs-137 by 25% due to oxidation processes that negatively affected sorption characteristics [27]. In order to exclude the oxidation of RFR for a long time, the storage must be carried out in an inert nitrogen atmosphere at a pressure of 0.26 bar [28].

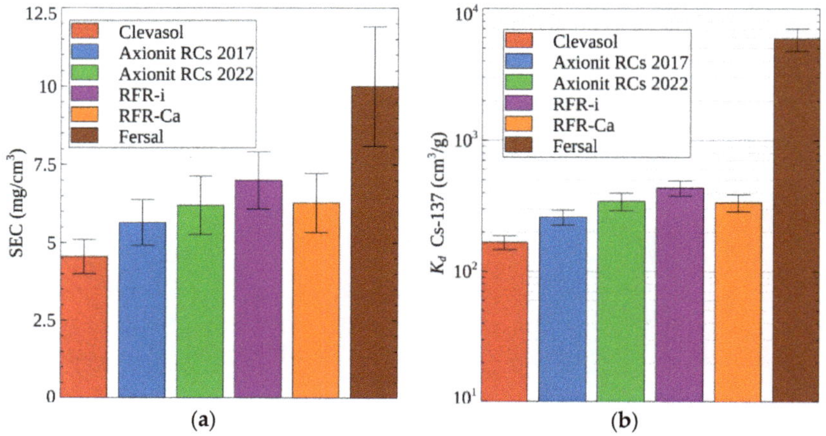

Figure 2. Sorption-selective characteristics of the sorption materials under study in the HLW simulated solution: (**a**) SEC, (**b**) value of K_d Cs-137.

The dependencies of the values of the Cs-137 distribution coefficients on the studied sorbents on the concentration of $NaNO_3$, KNO_3, and NaOH are shown in Figure 3. For the Axionit RCs 2022 ion exchange resin, curves are not presented since its sorption characteristics are similar to those of RFR-Ca. According to the presented results, when the concentration of $NaNO_3$ and KNO_3 increases, the values of K_d Cs-137 naturally decrease due to an increase in competitive adsorption. In $NaNO_3$ solutions, the highest values of K_d Cs-137 were obtained for the sorbents Fersal, RFR-i, and RFR-Ca. One can note that, in the presence of Na^+ ions, the values of K_d Cs-137 for different sorbents are similar except for Clevasol® and Axionit RCs 2017. K^+ ions manifest a greater competitive effect than Na^+, thus reducing the K_d Cs-137 values by 0.5 orders of magnitude on average. The exception is the Fersal sorbent, whose efficiency is equally high in both $NaNO_3$ and KNO_3 solutions due to its extremely high selectivity towards Cs^+ ions. With an increase in the concentration of NaOH, the K_d Cs-137 value for the Fersal sorbent gradually decreases, while for the ion-exchange resins Axionit RCs, RFR-I, and RFR-Ca, on the contrary, the efficiency of Cs-137 extraction increases. This is related to an increase in the degree of deprotonation of functional hydroxyl groups as the concentration of hydroxide ions increases [11].

Figure 4 shows the output curves of the dependence of the decontamination factor on the volume of the fed solution (bed volumes). The sorption characteristics of RFR-I and RFR-Ca ion exchange resins were additionally tested in three consecutive sorption cycles (Figure 3b). The dependence curves $f(\text{DF}) = BV$ were interpolated by the Akima spline, which was later used to determine the number of bed volumes fed before reaching a preset cesium breakthrough into the filtrate. For the sorbents Fersal, RFR-I, and RFR-Ca, 50% cesium breakthrough was calculated using the Thomas model (Equation (7)), since the obtained curves of the efficiency of Cs-137 extraction from the model HLW solution did not reach the set values.

To understand the processes of oxidation of RFRs during storage in air, NMR spectra of C-13 (Figure 5) were obtained, as well as those of RFR-I, which was stored in H-form in plastic containers for 6, 15, and 21 months.

Table 2 shows the values of chemical shifts and the corresponding peak areas. Relative integral intensities of the spectrum components were determined by fitting a calculated resonance line to the experimental one using the least squares method by means of independently developed software.

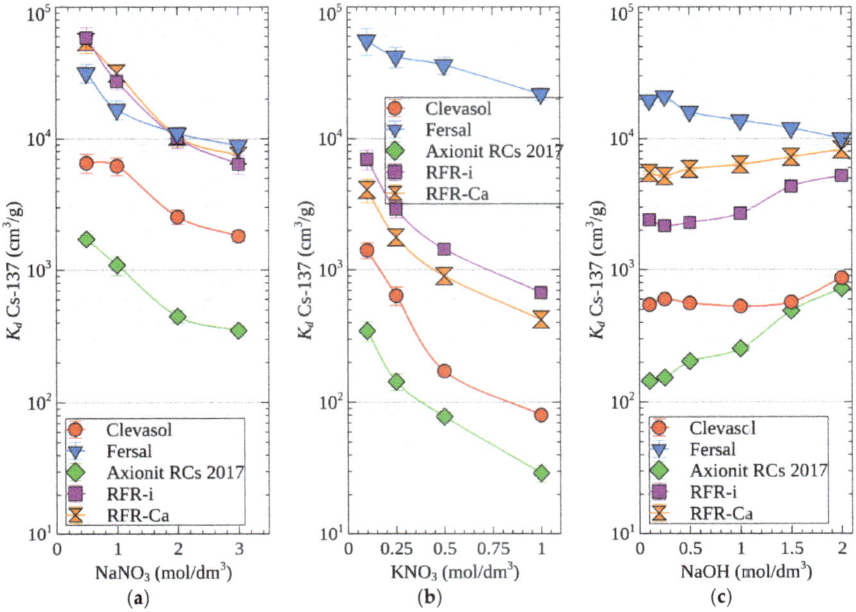

Figure 3. Dependence of the distribution coefficient of Cs-137 on the concentration of (**a**) $NaNO_3$, (**b**) KNO_3, (**c**) NaOH.

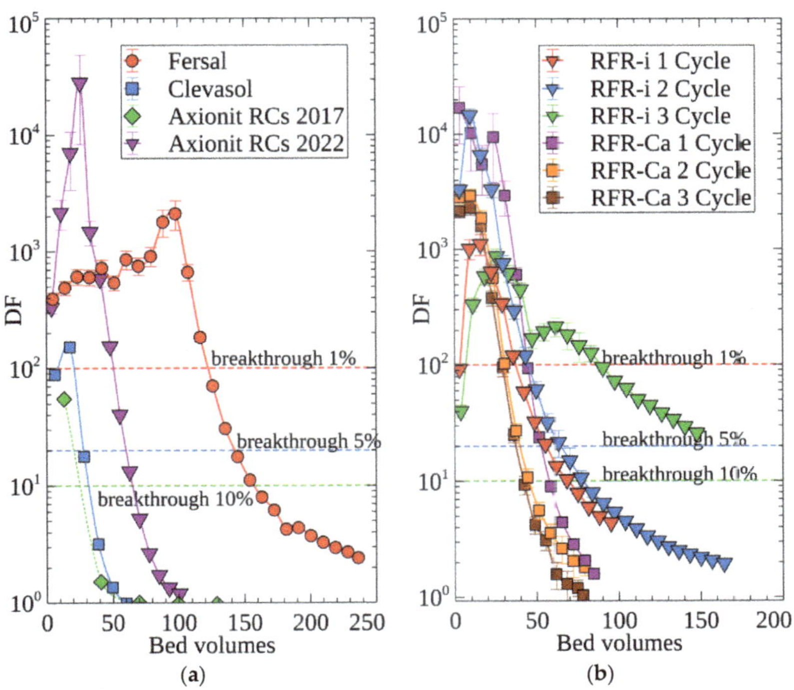

Figure 4. Curves of the dependence of the decontamination factor on the volume of the fed model solution: (**a**) sorbents Fersal, Clevasol®, Axionit RCs 2017, Axionit RCs 2022, (**b**) removal of Cs-137 in three repeated sorption cycles on RFR-I and RFR-Ca.

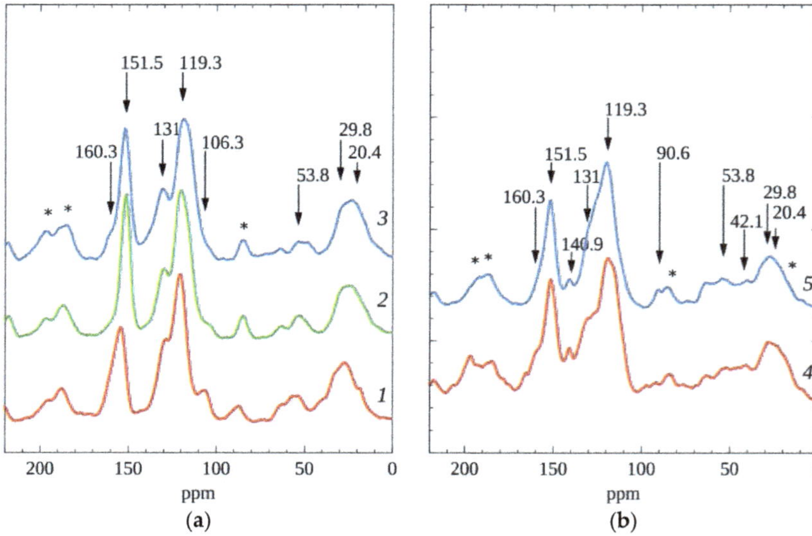

Figure 5. C-13 NMR spectra of RFRs; (**a**) RFR-i, (**b**) Axionit RCs, 1—RFR-i stored for 6 months, 2—RFR-i stored for 15 months, 3—RFR-i stored for 21 months, 4—Axionit RCs 2017, 5—Axionit RCs 2022. Asterisks mark the spinning sidebands.

Table 2. Values of chemical shifts and corresponding peak areas on the C-13 NMR spectra of the RFRs.

Organic Group	RFR-i Stored for 21 Months		RFR-I Stored for 15 Months		RFR-I Stored for 6 Months		Axionit RCs 2018		Axionit RCs 2022	
	Area (%) *	CS, ppm *	Area (%)	CS, ppm	Area (%)	CS, ppm	Area (%)	CS, ppm	Area (%)	CS, ppm
$-CH_2-(4,6')$	12.9	19.3	11.1	19.2	13.2	24.3	10.8	19.6	10.6	21
$-CH_2-(4,2')$	10.4	29.1	13.6	28.3	8.2	32.7	7.5	29.1	9.5	30.2
$-CH_2-C_6H_6$	-	-	-	-	-	-	6.7	40.2	7	42.1
$-CH_2OH$	8	50.6	6.5	52.2	10	56.2	5.9	51.7	6.8	53.8
hemiformals	-	-	-	-	-	-	-	-	2.1	91
C_2, C_6	2.7	105	3.9	104.8	6.7	106.6	-	-	-	-
C_4	32.3	117.9	22.1	120.8	26.9	120.1	31	117.6	29	119.2
C_5	12.2	131.4	15.6	129.9	14.8	129.5	12.9	131.4	10.4	130.7
$>C=C<$	-	-	-	-	-	-	3.7	141	4.6	140.7
C_1, C_3	17.5	151.8	23.2	151.2	15.4	154.2	16.4	151.2	16.6	151.7
R-COO-, >C=O	4.1	160.1	4.1	159.1	4.7	160.9	5.1	160.2	3.3	158.2

* Errors for the area fractions and CS values are ±2% and 0.3 ppm, respectively.

The C-13 NMR spectra of RFR-i and Axionit RCs revealed characteristic features indicating a change in the molecular structure of ion-exchangers during storage. Axionit RCs are characterized by the presence of an additional peak around 90.6 ppm related to hemiformals [29], which, however, disappears during storage, probably due to oxidation processes. In the spectrum of Axionit RCs, there are peaks around 42.1 and 140.9 ppm, which were not described in the works on RFRs and can be attributed to the benzyl-methylene radical [30] and alkenes [31], respectively. The presence of these peaks can be explained by the impurities in the form of organic compounds used as an emulsifier to produce spherical resin granules.

RFR-i is characterized by the presence of a peak around 106 ppm, which can be related to the carbon of the aromatic ring of resorcinol both at the C2 position not associated with the methylene bridge and at the C6 position [32]. The area of this peak gradually decreases during the storage of ion-exchange resin, which can be explained by the formation of para-quinone during oxidation with oxygen in the air [33].

A common feature of both ion exchange resin samples is a peak around 160 ppm, which appears after 21 months of storage. This peak is presumably determined by the formation of quinone and carboxylate groups [33,34]. The formation of carboxylates and quinones during oxidation is probably one of the main reasons for the decrease in the efficiency of Cs-137 extraction by RFRs.

Therefore, the negative effects of the RFR oxidation process are more pronounced under dynamic conditions, making it impossible to use resin as a sorption load in columns. Thus, during the decontamination of alkaline LRW, it is necessary to use fresh batches of RFR, and the long-term storage must be carried out in an inert atmosphere.

According to the volume of the fed model solution before attaining the 50% cesium breakthrough into the filtrate (DF 2, Table 3), the order of sorbents changes and looks as follows: Fersal > RFR-i > RFR-Ca~Axionit RCs 2022 > Clevasol®. This can be explained by the fact that, at high SEC values (Figure 2a), the efficiency of cesium removal by RFR-i resin at the initial stage is presumably limited by mass transfer in the grain volume.

Table 3. The number of bed volumes of the solution passed before reaching the preset cesium breakthrough into the filtrate.

Removal Efficiency		Axionit RCs 2022	Axionit RCs 2017	Clevasol®	Fersal	RFR-i			RFR-Ca		
						Adsorption Cycle			Adsorption Cycle		
%	DF					1	2	3	1	2	3
99	100	41 ± 2	0	22 ± 1	127 ± 6	30 ± 1	45 ± 2	89 ± 4	44 ± 2	31 ± 2	29 ± 1
95	20	56 ± 3	9 ± 0.5	27 ± 1	160 ± 8	50 ± 3	65 ± 3	157 ± 8	53 ± 3	40 ± 2	37 ± 2
90	10	63 ± 3	10 ± 0.5	32 ± 2	177 ± 9	63 ± 3	78 ± 4	195 ± 9	57 ± 3	45 ± 2	42 ± 2
50	2	83 ± 4	13 ± 0.5	45 ± 2	237 ± 12 *	117 ± 6 *	153 ± 8	298 ± 15 *	77 ± 4	73 ± 4	59 ± 3

* Values calculated using the Thomas model.

The disadvantage of the Fersal and Clevasol® sorbents is the loss of mechanical strength during the desorption of cesium, which makes it impossible to reuse them. Granules of the ion-exchange resins RFR-i and RFR-Ca after desorption remained in their original form and were used for the resorption of cesium.

Figure 6b shows the output sorption curves of Cs-137 on the ion exchange resins RFR-I and RFR-Ca during the first, second, and third sorption cycles. During three repeated sorption-desorption cycles using RFRs, changes in sorption characteristics are observed (Table 3). For RFR-Ca, there is a decrease in the volume of the solution fed before reaching a preset cesium breakthrough into the filtrate due to oxidation at low percolation rates of the model HLW solution. On the contrary, for RFR-I, an increase in the sorption resource was observed, which was associated with the gradual transition of the ion-exchange resin to the operating mode. This transition can be associated with a number of reasons, such as the gradual removal of potassium ions [35], non-reacted reaction products from the polymer matrix, and an increase in the mass transfer rate in the bulk of the resin grain.

To calculate the theoretical value of TDEC, we used the output curves of the dependence $f(C/C_0) = t$ (Figure 6), which were approximated by nonlinear regression using the Thomas equation (Equation (7)). Table 4 shows the calculated parameters of the Thomas model equation for the sorbents Fersal, RFR-I, and RFR-Ca. The coefficient of determination exceeds 0.95, which indicates a good agreement of the experimental values with the Thomas model.

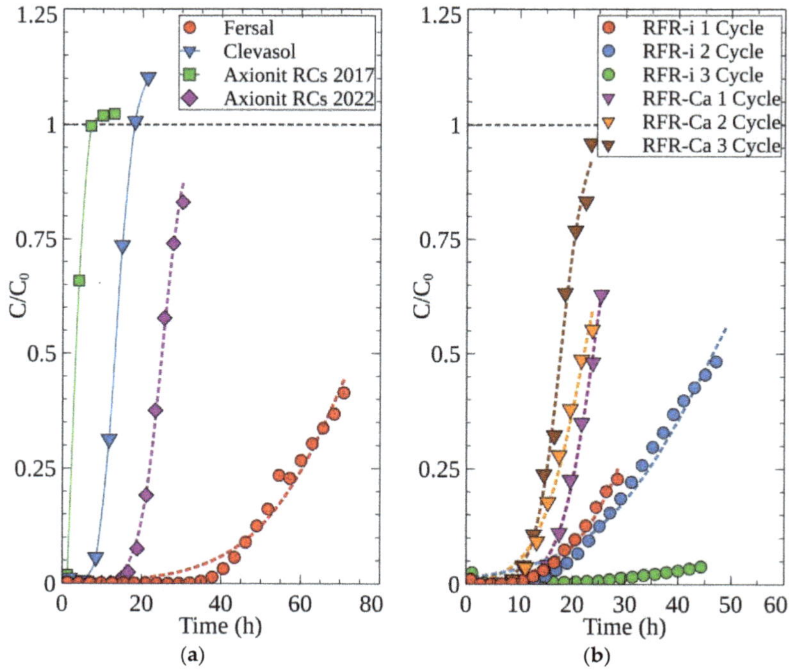

Figure 6. Dependence curves of C/C_0 on time: (**a**) sorbents Fersal, Clevasol®, Axionit RS 2017, Axionit RS 2022, (**b**) extraction of Cs-137 in three repeated sorption cycles on RFR-I and RFR-Ca; the dotted lines are curves obtained by the nonlinear regression equation of the Thomas model.

Table 4. Calculated parameters of the Thomas model equation.

Parameters of the Thomas Equation	RFR-Ca			RFR-i			Fersal
	Adsorption Cycle			Adsorption Cycle			
	1	2	3	1	2	3	
a	7.7 ± 0.5	5.3 ± 0.3	7.8 ± 0.3	5.4 ± 0.2	4.3 ± 0.2	6.3 ± 0.4	9.1 ± 0.3
b	0.43 ± 0.03	0.24 ± 0.01	0.33 ± 0.01	0.15 ± 0.01	0.09 ± 0.01	0.7 ± 0.1	0.13 ± 0.01
R^2	0.9961	0.9909	0.9972	0.9891	0.9886	0.9125	0.9943
K_{Th} ($cm^3/mg \times g$)	8.69	4.79	6.55	3.03	1.84	1.42	2.54

For the Clevasol® and Axionit RCs sorbents, as the volume of the fed solution increases, the value of the ratio C_0/C begins to exceed one (Figure 6a), which indicates the desorption process, presumably associated with the chemical destruction of the sorbents. For this reason, the Thomas model was not used for cesium output curves obtained with Clevasol® and Axionit RCs sorbents.

The presence of a macro concentration of a stable isotope of cesium in the LRW imposes high requirements for the sorption materials in regard to the value of the dynamic exchange capacity. The latter is caused by the fact that, despite the high selectivity of the sorption materials for cesium, along with a decrease in the value of the dynamic exchange capacity, the volume of liquid waste decontaminated from cesium will also decrease.

To calculate the total dynamic exchange capacity, the experimental values were extrapolated to the values of total cesium saturation using the Thomas equation with the parameters given in Table 3. Figure 7 shows the theoretical integral curves obtained for the saturation of sorbents with cesium under dynamic conditions.

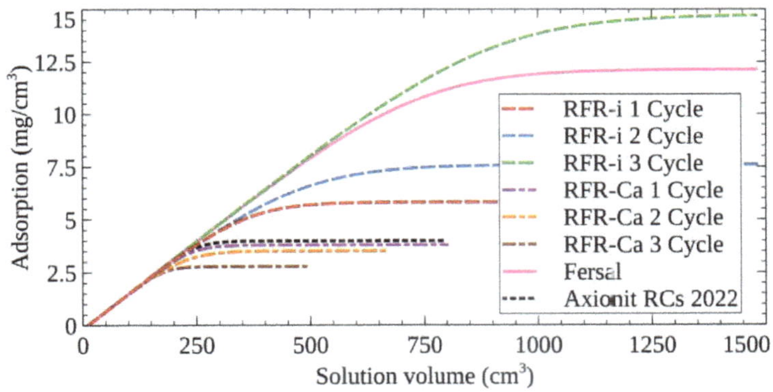

Figure 7. Integral curves of the adsorption of cesium from the HLW simulated solution.

Table 5 shows the calculated values of the dynamic exchange capacity before attaining a preset cesium breakthrough into the filtrate, as well as the theoretical values of the total dynamic capacity. According to the results obtained during the first cycle, the Fersal sorbent has the largest capacity for cesium among the other investigated sorption materials.

Table 5. Values of the dynamic exchange capacity (mg/cm^3) before reaching a preset cesium breakthrough into the filtrate, and values of the total dynamic exchange capacity calculated using the Thomas model.

Removal Efficiency		RFR-Ca			RFR-i			Axionit RCs 2022	Clevasol	Fersal
		Adsorption Cycle			Adsorption Cycle					
%	DF	1	2	3	1	2	3			
99	100	2.1 ± 0.1	1.5 ± 0.1	1.4 ± 0.1	1.5 ± 0.1	2.2 ± 0.1	2.2 ± 0.1	2.5 ± 0.1	1.0 ± 0.1	6.2 ± 0.3
95	20	2.6 ± 0.1	1.9 ± 0.1	1.8 ± 0.1	2.4 ± 0.1	3.1 ± 0.2	7.6 ± 0.4 [1]	2.9 ± 0.1	1.2 ± 0.1	7.8 ± 0.4
90	10	2.8 ± 0.1	2.2 ± 0.1	2.0 ± 0.1	3.0 ± 0.1	3.8 ± 0.2	9.3 ± 0.4 [1]	3.2 ± 0.2	1.5 ± 0.1	8.6 ± 0.5
50	2	3.4 ± 0.2	3.0 ± 0.2	2.5 ± 0.1	5.0 ± 0.3 [1]	6.31 ± 0.3	13.1 ± 0.8 [1]	3.8 ± 0.2	1.9 ± 0.1	10.8 ± 0.5 [1]
TDEC		3.8 ± 0.2 [1]	3.5 ± 0.2 [1]	2.8 ± 0.1 [1]	5.8 ± 0.3 [1]	7.61 ± 0.4	14.7 ± 0.8 [1]	4 ± 0.2 [1]	2 ± 0.1 [2]	11.7 ± 0.4 [1]

[1] The value calculated using the Thomas model. [2] Maximum value before desorption.

The RFR-Ca resin is characterized by a decrease in the theoretical value of TDEC, which is associated with chemical and physical degradation, whereas the accumulation of ionizing radiation dose can also intensify this process [36]. RFR-I, with the number of adsorption cycles increasing, is characterized by an increase in the theoretical value of TDEC, which, at the third adsorption cycle, is comparable to the value obtained for the Fersal sorbent. This indicates the high chemical stability of the RFR-I ion exchange resin and its gradual transition to operating mode. The increase in the total dynamic exchange capacity can also be explained by an increase in the number of exchange groups due to the oxidation of methyl groups to carboxyl groups [37].

Solutions of HNO_3 were used for the desorption of Cs-137. One should mention that, at room temperature, 2.0–3.0 M solutions of HNO_3 lead to changes in RFR after 24 h due to its nitration [38]. Nitration of RFRs with 0.5–1.0 M HNO_3 solutions occurs when heated to 55 °C or after prolonged exposure for 70 days [38,39]. Nitration of RFRs is accompanied by the destruction of the polymer matrix as well as a decrease in the sorption capacity [40]. Therefore, in terms of the volume of the desorbing solution consumed, the time of desorption, and the destruction of the polymer matrix, the optimal concentration of the HNO_3 solution is 1.0 mol/dm^3 [6]. The solution of HNO_3 at a concentration of 7.5 mol/dm^3 was used as a desorbing solution for the sorbents Fersal and Clevasol®. The rate of feeding of the desorbing solution was 2.3 cm^3/h (0.77 BV/h). The curves of desorption of Cs-137 from various sorbents are shown in Figure 8.

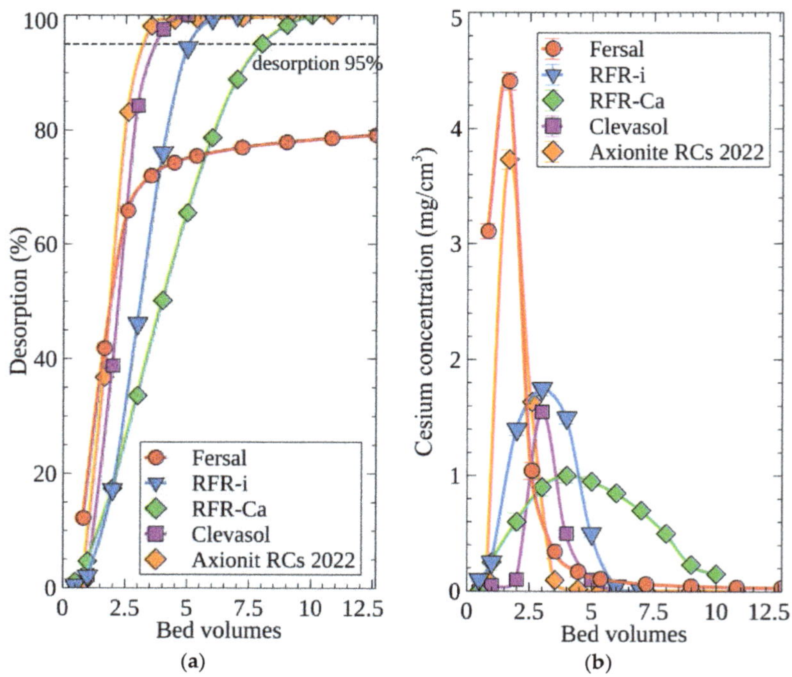

Figure 8. Cs-137 desorption: (**a**) integral desorption curves, (**b**) output desorption curves. Desorbing solutions: 7.5 mol/dm^3 HNO_3 used for Fersal and Clevasol® sorbents, 1.0 mol/dm^3 HNO_3 used for RFRs sorbents.

Ion-exchange resins have an advantage over inorganic sorbents in requiring solutions with lower concentrations of HNO_3, which is important for reducing the corrosion of equipment. The curves obtained for ion-exchange resins RFR-I and RFR-Ca indicate a wide desorption front. This feature could be associated with a low mass transfer rate inside the resin grain or a significant decrease of the resin volume due to a large difference between bulk weight and specific volume value, which leads to sorbent compaction in the column, thus preventing the normal flow of the desorbing solution. Despite some differences, the results presented in Figure 8 show that, with the exception of the Fersal sorbent, when 10 bed volumes of the HNO_3 solution are fed, the desorption efficiency exceeds 99%, which is consistent with the known published data [3].

4. Conclusions

The sorption characteristics of resorcinol-formaldehyde resins Axionit Cs, RFR-i, and RFR-Ca as well as inorganic sorbents of Clevasol® and Fersal brands towards cesium ions in model-simulated solutions of high-level alkaline waste streams from the Mayak Production Association have been evaluated. The studied sorbents can be arranged in the following order in terms of their sorption characteristics in the HLW simulated solution under static conditions: Fersal > RFR-i > Axionit RCs 2022~RFR-Ca Axionit RCs 2017 Clevasol. The Fersal sorbent is characterized by the highest selectivity towards cesium ions. Along with the increase in NaOH concentration in solution, the efficiency of the Fersal sorbent decreases, whereas that of resorcinol-formaldehyde resins increases. It has been demonstrated that, under dynamic conditions, the resource of the Fersal sorbent until the 1% cesium breakthrough is a minimum 2.5-fold larger in comparison with other sorption materials. The disadvantage of inorganic sorbents consists of a loss of mechanical strength upon cesium desorption by a 7.5 M solution of HNO_3. Using RFRs, cesium can be desorbed by a 1 V solution of HNO_3 without a loss of the granule's mechanical strength. Within three repeated sorption-desorption cycles, RFR-i has the best values of resource and full dynamic exchange capacity, as compared with the porous RFR-Ca. A common disadvantage of RFRs is worsening of sorption characteristics as a result of oxidation accompanied by the formation of quinone and carboxylate groups. Based on the obtained results, one can conclude that in the decontamination of high-level wastes in the storage tanks of the Mayak Production Association, the sorbent of the Fersal brand and the resorcinol-formaldehyde resin RFR-I can be used. However, taking into account repeated use in sorption-desorption cycles and the total amount of the decontaminated HLW, resorcinol-formaldehyde resins can turn out to be more preferential.

Author Contributions: Conceptualization, V.M., P.K. and A.E.; methodology, V.M., N.N. and P.K.; formal analysis, V.M., P.K. and A.E.; investigation, V.M., N.N., A.S., D.M., S.S., K.F., E.T. and M.T.; writing—original draft preparation, V.M. and A.E.; writing—review and editing, V.M., P.K., A.E. and E.T.; funding acquisition, P.K. and A.E. All authors have read and agreed to the published version of the manuscript.

Funding: Cs-137 adsorption studies were carried out within the framework of the agreement No. N.4d.241.20.22.1057 dated 04.04.2022, for the implementation of the state contract "Development and justification of options for processing highly active wastes of complex chemical composition, including experimental design work and experimental industrial testing of equipment". Synthesis of the samples of RFR-i and RFC-Ca ion-exchangers, obtaining NMR spectra of C-13 RFRs, as well as theoretical calculations were carried out within the framework of the State Order of the Institute of Chemistry of the Far Eastern Branch of the Russian Academy of Sciences, project No. FWFN(0205)-2022-0002.

Institutional Review Board Statement: Not applicable.

Informed Consent Statement: Not applicable.

Data Availability Statement: Not applicable.

Acknowledgments: Equipment of the CUC, "Far Eastern Center of Structural Investigations", was used in this work.

Conflicts of Interest: The authors declare no conflict of interest.

References

1. Kozlov, P.V.; Kazadaev, A.A.; Makarovskii, R.A.; Remizov, M.B.; Verbitskii, K.V.; Logunov, M.V. Development of a Process for Cesium Recovery from the Clarified Phase of High-Level Waste Storage Tanks of the Mayak Production Association with a Ferrocyanide Sorbent. *Radiochemistry* **2016**, *58*, 295–301. [CrossRef]
2. Milyutin, V.V.; Mikheev, S.V.; Gelis, V.M.; Kozlitin, E.A. Sorption of Cesium on Ferrocyanide Sorbents from Highly Saline Solutions. *Radiochemistry* **2009**, *51*, 298–300. [CrossRef]
3. Fiskum, S.K.; Pease, L.F.; Peterson, R.A. Review of Ion Exchange Technologies for Cesium Removal from Caustic Tank Waste. *Solvent Extr. Ion Exch.* **2020**, *38*, 573–611. [CrossRef]
4. Aleman, S.; Hamm, L.; Smith, F. *Ion Exchange Modeling Of Cesium Removal From Hanford Waste Using Spherical Resorcinol-Formaldehyde Resin*; Technical Report WSRC-STI-2007-00030; Washington Savannah River Site (SRS): Aiken, SC, USA, 2007.
5. Fiskum, S.K.; Blanchard, D.L.; Arm, S.T.; Peterson, R.A. Cesium Removal from Simulated and Actual Hanford Tank Waste Using Ion Exchange. *Sep. Sci. Technol.* **2005**, *40*, 51–67. [CrossRef]
6. Milyutin, V.V.; Zelenin, P.G.; Kozlov, P.V.; Remizov, M.B.; Kondrutskii, D.A. Sorption of Cesium from Alkaline Solutions onto Resorcinol-Formaldehyde Sorbents. *Radiochemistry* **2019**, *61*, 714–718. [CrossRef]
7. Tokar, E.; Tutov, M.; Kozlov, P.; Slobodyuk, A.; Egorin, A. Effect of the Resorcinol/Formaldehyde Ratio and the Temperature of the Resorcinol–Formaldehyde Gel Solidification on the Chemical Stability and Sorption Characteristics of Ion-Exchange Resins. *Gels* **2021**, *7*, 239. [CrossRef]
8. Miller, H.S.; Kline, G.E. Reactions of Cesium in Trace Amounts with Ion-Exchange Resins. *J. Am. Chem. Soc.* **1951**, *73*, 2741–2743. [CrossRef]
9. Kargov, S.I.; Shelkovnikova, L.A.; Ivanov, V.A. The Nature of Ion Exchange Selectivity of Phenol-Formaldehyde Sorbents with Respect to Cesium and Rubidium Ions. *Russ. J. Phys. Chem. A* **2012**, *86*, 860–866. [CrossRef]
10. Samanta, S.K.; Misra, B.M. Ion Exchange Selectivity of a Resorcinol-Formaldehyde Polycondensate Resin for Cesium in Relation to Other Alkali Metal Ions. *Solvent Extr. Ion Exch.* **1995**, *13*, 575–589. [CrossRef]

11. Hubler, T.L.; Franz, J.A.; Shaw, W.J.; Bryan, S.A.; Hallen, R.T.; Brown, G.N.; Bray, L.A.; Linehan, J.C. *Synthesis, Structural Characterization, and Performance Evaluation of Resorcinol-Formaldehyde (R-F) Ion-Exchange Resin*; Technical Report PNL-10744; Pacific Northwest Laboratory: Richland, WA, USA, 1995.
12. Duignan, M.R.; Nash, C.A. Removal of Cesium from Savannah River Site Waste with Spherical Resorcinol Formaldehyde Ion Exchange Resin: Experimental Tests. *Sep. Sci. Technol.* **2010**, *45*, 1828–1840. [CrossRef]
13. Adamson, D.J. Pilot-Scale Hydraulic Testing of Resorcinol Formaldehyde Ion Exchange Resin. In Proceedings of the Fluids Engineering Division Summer Meeting, Vail, CO, USA, 2–6 August 2009; Volume 43727, pp. 545–553. [CrossRef]
14. Duignan, M.R.; Nash, C.A.; Punch, T.M. High Aspect Ratio Ion Exchange Resin Bed—Hydraulic Results for Spherical Resin Beads. *Sep. Sci. Technol.* **2008**, *43*, 2943–2979. [CrossRef]
15. Brooks, K.P.; Augspurger, B.S.; Blanchard, D.L.; Cuta, J.M.; Fiskum, S.K.; Thorson, M.R. Hydraulic Testing of Ion Exchange Resins for Cesium Removal from Hanford Tank Waste. *Sep. Sci. Technol.* **2006**, *41*, 2391–2408. [CrossRef]
16. Fiskum, S.K.; Augspurger, B.S.; Brooks, K.P.; Buchmiller, W.C.; Russell, R.L.; Schweiger, M.J.; Snow, L.A.; Steele, M.J.; Thomas, K.K.; Wallace, D.E.; et al. *Comparison Testing of Multiple Resorcinol-Formaldehyde Resins for the River Protection Project—Waste Treatment Plant*; Technical Report PNWD-3387 WTP-RPT-103, Rev 0; Savannah River Technology Center, Department of Energy 1000 Independence Ave.: Washington, DC, USA, 2004; p. 2978.
17. Clevasol®—A Novel Cation Exchange Resin for Isotope Capture. Available online: https://www.lemerpax.com/en/products/clevasol-en/ (accessed on 25 April 2023).
18. Yakushev, A.; Turler, A.; Dvorakova, Z.; Bremen, K. CLEVASOL, a Novel Radiation Hard Cation Exchanger Suitable for Treatment of Liquid Radioactive Waste with High Salinity. In Proceedings of the APSORC'13—5th Asia-Pacific Symposium on Radiochemistry Kanazawa, Kanazawa, Japan, 22–27 September 2013; p. 133.
19. Lin, M.; Kajan, I.; Schumann, D.; Türler, A.; Fankhauser, A. Selective Cs-Removal from Highly Acidic Spent Nuclear Fuel Solutions. *Radiochim. Acta* **2020**, *108*, 615–626. [CrossRef]
20. Chaplin, J.D.; Berillo, D.; Purkis, J.M.; Byrne, M.L.; Tribolet, A.D.C.C.M.; Warwick, P.E.; Cundy, A.B. Effective $^{137}Cs^{+}$ and $^{90}Sr^{2+}$ Immobilisation from Groundwater by Inorganic Polymer Resin Clevasol® Embedded within a Macroporous Cryogel Host Matrix. *Mater. Today Sustain.* **2022**, *19*, 100190. [CrossRef]
21. Haas, P.A. A Review of Information on Ferrocyanide Solids for Removal of Cesium from Solutions. *Sep. Sci. Technol.* **1993**, *28*, 2479–2506. [CrossRef]
22. Loos-Neskovic, C.; Fedoroff, M. Fixation Mechanisms of Cesium on Nickel and Zinc Ferrocyanides. *Solvent Extr. Ion Exch.* **1989**, *7*, 131–158. [CrossRef]
23. Lee, E.F.T.; Streat, M. Sorption of Caesium by Complex Hexacyanoferrates. v. a Comparison of Some Cyanoferrates. *J. Chem. Tech. Biotechnol.* **1983**, *33*, 333–338. [CrossRef]
24. Milyutin, V.V.; Nekrasova, N.A.; Kozlov, P.V.; Markova, D.V.; Feoktistov, K.A. Extraction of cesium from model solutions of highly active alkaline wastes on an inorganic ferrocyanide sorbent of the brand Fersal. *Vopr. Radiatsionnoy Bezop.* **2022**, *3*, 22–27. (In Russian)
25. Milyutin, V.V.; Gelis, V.M.; Leonov, N.B. Study of the Sorption Kinetics of Cesium and Strontium Radionuclides by Sorbents of Different Classes. *Radiochemistry* **1998**, *40*, 418–420. (In Russian)
26. Sanders, J. Veusz—A Scientific Plotting Package. Available online: https://veusz.github.io/ (accessed on 25 April 2023).
27. Hubler, T.L.; Shaw, W.J.; Brown, G.N.; Linehan, J.C.; Franz, J.A.; Hart, T.R.; Hogan, M.O. *Chemical Derivation to Enhance the Chemical/Oxidative Stability of Resorcinol-Formaldehyde (R-F)*; Resin Technical Report No. PNNL–11327; Pacific Northwest National Laboratory: Richland, WA, USA, 1996.
28. Russell, R.L.; Fiskum, S.K.; Smoot, M.R.; Rinehart, D.E. *Cesium Isotherm Testing with Spherical Resorcinol-Formaldehyde Resin at High Sodium Concentrations*; Technical Report PNNL-25277; Pacific Northwest National Laboratory (PNNL): Richland, WA, USA, 2016.
29. Kim, M.G.; Wu, Y.; Amos, L.W. Polymer Structure of Cured Alkaline Phenol–Formaldehyde Resol Resins with Respect to Resin Synthesis Mole Ratio and Oxidative Side Reactions. *J. Polym. Sci. Part A Polym. Chem.* **1997**, *35*, 3275–3285. [CrossRef]
30. Christiansen, A.W. Resorcinol-Formaldehyde Reactions in Dilute Solution Observed by Carbon-13 NMR Spectroscopy. *J. Appl. Polym. Sci.* **2000**, *75*, 1760–1768. [CrossRef]
31. Pretsch, E.; Clerc, T.; Seibl, J.; Simon, W. *Tables of Spectral Data for Structure Determination of Organic Compounds*; Springer Science & Business Media: Berlin/Heidelberg, Germany, 2013; ISBN 978-3-662-10207-7.
32. Werstler, D.D. Quantitative ^{13}C n.m.r, Characterization of Aqueous Formaldehyde Resins: 2. Resorcinol- Formaldehyde Resins. *Polymer* **1986**, *27*, 757–764. [CrossRef]
33. Hubler, T.L.; Franz, J.A.; Shaw, W.J.; Hogan, M.O.; Hallen, R.T.; Brown, G.N.; Linehan, J.C. *Structure/Function Studies of Resorcinol-Formaldehyde (R-F) and Phenol-Formaldehyde (P-F) Copolymer Ion-Exchange Resins*; Technical Report PNNL-11347 Richland; Pacific Northwest National Laboratory: Washington, WA, USA, 1996.
34. Pratt, L.M.; Szostak, R.; Khan, I.M. Alkaline degradation of resorcinol- formaldehyde resins: Solid-State nmr, thermal adsorption and desorption analysis, and molecular modeling. *J. Macromol. Sci. Part A Pure Appl. Chem.* **1997**, *34*, 281–289. [CrossRef]
35. Arm, S.T.; Blanchard, D.L. *Pre-Conditioning and Regeneration Requirements of Ground Gel Resorcinol Formaldehyde Ion Exchange Resin*; Technical Report WTP-RPT-104, PNWD-3390; Battelle–Pacific Northwest Division, Richland: Washington, DC, USA, 2004.

36. Fiskum, S.K.; Colburn, H.A.; Rovira, A.M.; Allred, J.R.; Smoot, M.R.; Peterson, R.A.; Landon, M.R.; Colosi, K.A. Cesium Removal from AP-105 Hanford Tank Waste Using Spherical Resorcinol Formaldehyde Resin. *Sep. Sci. Technol.* **2019**, *54*, 1932–1941. [CrossRef]
37. Shelkovnikova, L.A.; Gavlina, O.T.; Ivanov, V.A. Stability of Phenol-Formaldehyde Ion-Exchange Sorbents in Aqueous Solutions. *Russ. J. Phys. Chem. A* **2011**, *85*, 1652–1659. [CrossRef]
38. King, W.D.; Fondeur, F.F.; Wilmarth, W.R.; Pettis, M.E.; McCollum, S.W. Reactivity of Resorcinol Formaldehyde Resin with Nitric Acid. *Sep. Sci. Technol.* **2006**, *41*, 2475–2486. [CrossRef]
39. Russell, R.L.; Rinehart, D.E.; Brown, G.N.; Schonewill, P.P.; Peterson, R.A. *Ion Exchange Kinetics Testing with SRF Resin*; Technical Report PNNL-21109; Pacific Northwest National Laboratory: Richland, WA, USA, 2012.
40. Arm, S.T.; Blanchard, D.L.; Fiskum, S.K.; Weier, D.R. *Chemical Degradation of SuperLig® 644 Ion Exchange Resin*; Technical Report PNWD-3315; Battelle-Pacific Northwest Division: Richland, WA, USA, 2003.

sustainability

MDPI

Article

Long-Term Chemical Alteration of ^{238}Pu-Doped Borosilicate Glass in a Simulated Geological Environment with Bentonite Buffer

Bella Zubekhina [1,*], Boris Burakov [2], Andrei Shiryaev [3], Xiaodong Liu [4] and Yury Petrov [5]

1 Japan Atomic Energy Agency, 2-4 Shirakata, Tokai-mura, Naka-gun, Ibaraki 319-1195, Japan
2 The Ioffe Institute, 26 Politekhnicheskaya St., St. Petersburg 194021, Russia
3 Frumkin Institute of Physical Chemistry and Electrochemistry, Russian Academy of Sciences, Leninsky Pr., 31 korp. 4, Moscow 119071, Russia
4 State Key Laboratory of Nuclear Resources and Environment, East China University of Technology, 418 Guanglan Ave, Nanchang 330013, China
5 Khlopin Radium Institute, 28 2-nd Murinskiy Ave., St. Petersburg 194021, Russia
* Correspondence: zubekhina.bella@jaea.go.jp

Abstract: Chemical degradation of borosilicate glass doped with ^{238}Pu was modelled in conditions of a failed underground radwaste repository in granite host rock with bentonite buffer material after penetration of aqueous solutions at temperature of 90 °C. The total duration of the experiment exceeded two years. It is shown that wet bentonite preserved its barrier function and prevents migration of plutonium to the solution. The total amount of plutonium adsorbed on bentonite clay during the experiment did not exceed 0.02% of the initial amount of plutonium in the glass sample. Estimated accumulated dose of self-irradiation of the glass sample after the experiment varies from 3.16×10^{15} to 3.39×10^{15} α-decays per gram, which is equivalent to more than 1000 years storage of ^{239}Pu doped sample with the same Pu content. Beishan granite remained intact, with no evidence of Pu penetration into the granite matrix along mineral grain boundaries.

Keywords: plutonium; bentonite; leaching; alteration; granite; waste

Citation: Zubekhina, B.; Burakov, B.; Shiryaev, A.; Liu, X.; Petrov, Y. Long-Term Chemical Alteration of ^{238}Pu-Doped Borosilicate Glass in a Simulated Geological Environment with Bentonite Buffer. *Sustainability* **2023**, *15*, 6306. https://doi.org/10.3390/su15076306

Academic Editor: Barry D. Solomon

Received: 14 February 2023
Revised: 17 March 2023
Accepted: 4 April 2023
Published: 6 April 2023

1. Introduction

Effective development of nuclear energy is strongly depending on implementation of safe and sustainable solutions on each stage of the nuclear fuel cycle, from uranium mining to waste treatment. Treatment of radioactive wastes containing long-lived actinides is an extremely important issue in terms of sustainability, since all the decisions made in this field will be affecting next generations for thousands of years. Final disposal of highly-radioactive waste is one of the most challenging issues in both open and closed nuclear fuel cycles. Two basic strategies are considered for high-level radioactive waste management in the closed nuclear fuel cycle. Nuclear waste streams after the fuel reprocessing can be treated by a solidification technology like vitrification or immobilization into crystalline ceramic matrix. Another approach suggests partitioning the waste to separate long-lived actinides and other relatively short-lived fission products followed by immobilization in more targeted ceramic or glass-ceramic matrix [1–3]. In both cases, deep geological disposal is considered for the final stage of the nuclear fuel cycle. A common well-known international approach to the immobilization of highly radioactive wastes (HLW) is based on their vitrification [1] followed by deep geological disposal. A suitable site for the long-term safe geological disposal has to meet set of requirements such as stable geological and hydrogeological settings (including their eventual changes in future), acceptable construction and engineering conditions, low human activities and land use, environmental protection, and good logistics. Among several types of geological formations suitable for the HLW repository, granite rock is one of the most promising candidates because of its good mechanical properties and low permeability. A repository in granite rocks is considered in many countries operating underground research facilities (Finland, Sweden, Czech Republic, Switzerland, Republic of Korea) or constructing them (China, Russia). A deep geological repository relies on multi-barrier concept comprising several geological and engineering barriers to minimize the migration of radionuclides during the intended period of disposal, lasting up to 1 million years. A swelling clay, for example, a bentonite buffer, is an important engineering barrier suggested for placement around metal containers filled with the vitrified HLW [4]. This barrier, so-called "buffer" zone, has very low permeability and can protect the waste canister during the rock movements. Bentonite has a high capacity for physical sorption of radionuclides and pronounced swelling upon contact with water helps suppressing migration of underground fluids coming from cracks in the host rock. However, the protective properties of bentonite can weaken because of physical and chemical factors affecting it under hydrothermal conditions of a geological repository. The effect of the bentonite degradation is determined by the presence of water and temperature conditions.

Despite all the components in the multibarrier concept being considered simultaneously, every single barrier material has to meet the safety requirements and acceptance criteria. For each candidate repository site, the safety case studies should include numerical modeling, long-term geological and geochemical investigation, small-scale laboratory experiments, and large-scale experiments in underground research facilities. It is critically important to take into account complex radiation effects affecting a HLW matrix behavior in contact with steel, bentonite, host rock, and underground water. These effects include chemical degradation of the glass matrix, thermal and radiation damages in bentonite, and local change of reducing and oxidizing conditions due to water radiolysis. Since experiments with real vitrified HLW are expensive and complicated, information about the long-term behavior of highly radioactive glass under self-irradiation and contact with water is very limited [5–10]. However, it was reported earlier that the chemical alteration of a real highly radioactive glass doped with ^{238}Pu in water is much more intense in comparison with a simulated non-radioactive glass of similar chemical composition; pronounced glass degradation is observed [9,10].

Radiation effects in bentonite and other clay materials had been studied mainly in terms of the stability of clay materials under external gamma-irradiation [9–11]. It was shown that bentonite has high radiation stability and preserves crystallinity with no significant changes in its structure at accumulated doses as high as 3×10^{10} rad at room temperature and at 3.5×10^9 rad at 300 °C [12,13]. However, the radiation stability of bentonite under external and internal alpha-irradiation is still unclear. It is expected, that under radiation damage from α-particles the lattice of montmorillonite—a basic mineral of a bentonite clay—can be fully amorphised. As a consequence of this process, the sorption capacity of newly formed products of the montmorillonite destruction can decrease dramatically in comparison with the fresh bentonite. Radiolysis of water may chemically destroy montmorillonite as well. In addition, water radiolysis decreases the pH level in the system which leads to bentonite degradation and changes in montmorillonite chemical and physical properties such as structure, composition, morphology, and sorption capacity [14–16].

Alteration of a highly radioactive glass surrounded by water-saturated bentonite is a complex process and can be properly addressed only in dedicated experiments. The presence of granite parent rock in the vitrified HLW repository adds additional uncertainty to the modeling of the long-term behavior of all components [17]. For simplification, we consider the interaction between granite, bentonite, metallic container, vitrified waste, and water as a static system. In this case, migration of radionuclides will be limited by ion diffusion along the grain boundaries in granite rock. However, in a long term (up to 100,000 years period), we should consider the formation of the cracks in the granite host rock and the convective transport of radionuclides along the cracks. In this case, the cracks will play a main role in the transport of the long-lived actinides in ionic or colloidal forms and their migration into the biosphere [18].

Experiments with trace amounts of radionuclides were performed in several underground research facilities (URF) [19–21], however, experiments with real vitrified radioactive waste were not yet performed in URF. Up to now, laboratory scale leaching and alteration tests is the only possibility to study complex interactions between the solidified waste, buffer, and host rock in static and dynamic conditions. Long-term experiments are very important for prediction of features of the Pu sorption behavior such as reaching the sorption equilibrium and process of Pu(V) reduction on montmorillonite clay [22].

The main objective of this work is to study the long-term behavior of a borosilicate glass doped with ^{238}Pu in the system "water—bentonite buffer—granite host rock". High specific α-activity of this isotope dramatically accelerates rate of radiation damage and, with certain caution, allows projection of the current results to relatively long periods. The results obtained may contribute to clarifying the model of radionuclide migration from corroded containers filled with vitrified HLW under conditions of a geological repository located in the granite massif.

2. Materials and Methods

The experiment is designed to simulate ageing of the Pu-bearing glass matrix in conditions of a failed container placed into a deep geological repository (DGR). In this experiment, we neglect the contribution of a metal container and model the situation when groundwater equilibrated with bentonite accessed the vitrified waste, e.g., via a crack or corrosion pit in the metal cask.

2.1. Pu-Doped Highly Radioactive Glass Synthesis

^{238}Pu-doped glass was used for the experiment as a simulated vitrified waste sample. ^{238}Pu isotope dramatically accelerates the radiation damage rate in comparison with ^{239}Pu because of its shorter half-life (88 years) and higher specific activity (6.32×10^{11} Bq/g for ^{238}Pu and 2.2×10^{9} Bq/g for ^{239}Pu). Thus, using ^{238}Pu, more pronounced aging of the glass can be achieved in a shorter period of time; more extensive radiation damage of the surrounding bentonite can be expected as well. The Pu-doped borosilicate glass has been synthesized in 2016 [9,10] by melting the oxide mixture with a suitable frit at temperature of 1400 °C for 2 h in air atmosphere (Figure 1). The glass was doped with 0.42–0.45 wt. % of ^{238}Pu (recalculated to Pu metal) to accelerate radiation damage in the glass matrix; Eu_2O_3 was added to simulate the presence of trivalent lanthanides. The chemical composition of the glass is presented in Table 1. After the synthesis, the crucible with the glass sample was broken and one solid fragment of the glass (1216 mg) was used for the experiment (Figure 1).

Table 1. The chemical composition of ^{238}Pu-doped borosilicate glass used for experiment.

Element/Oxide	Content, wt. %
SiO_2	47.86
Na_2O	14.60
B_2O_3	21.20
Al_2O_3	6.84
Eu_2O_3	3.02
CaO	5.87
PuO_2 (all isotopes)	0.58–0.65
$^{238}PuO_2$	0.47–0.53
^{238}Pu (as metal)	0.42–0.45

Figure 1. Sample of the borosilicate glass doped with ^{238}Pu prior to the experiment.

As the Pu-doped glass sample was synthesized two years before the experiment, it has already suffered radiation damage. For evaluation of number of alpha decays, we used Pu content of 0.42 wt. % as a minimum and 0.45 wt. % as a maximum content values. Then, for the initial glass sample (1216 mg) the Pu content varies from 5.11 to 5.47 mg of ^{238}Pu per whole sample. Using ^{238}Pu specific activity we can recalculate this content as (3.23–3.46) $\times$ 10^7 Bq per sample. For the entire storage period (22 months) the of self-irradiation dose can be estimated in a range from 1.51 $\times$ 10^{14} to 1.62 $\times$ 10^{14} α-decays per gram of the sample.

After two years of the experiment, the total accumulated dose of self-irradiation naturally increased and can be estimated in a range from 3.16 $\times$ 10^{15} to 3.39 $\times$ 10^{15} α-decays per gram of the sample. Mass loss caused by Pu leaching during the experiment was neglected in this estimation. The dose of self-irradiation calculated above is equivalent to the one accumulated by the sample doped with 0.45 wt. % of ^{239}Pu after more than 1000 years of storage.

2.2. Granite Rock

A sample of granite rock from the Beishan area of Gansu Province, China was used for the experiment. This sample represents the actual host rock of emerging Beishan HLW geological repository in China. Magmatic granites of the Beishan area mainly possess granitic and porphyritic structures. The rock body is characterized by batholiths, stocks, dykes, etc. According to surface geological mapping and boreholes investigations, the main rock types of the planning DGR are biotite monzonitic granite and biotite granodiorite [23]. The mineralogical and chemical compositions of Beishan granite are presented in Tables 2 and 3, respectively.

Table 2. Mineralogical composition of Beishan granite rock and Gaomiaozi (GMZ) Na-bentonite used for experiment, wt. %.

Mineral	Beishan Granite [24]	Bentonite GMZ [25]
Microcline	46.14	-
Albite	27.50	-
Biotite	15.65	-
Quartz	8.50	11.7
Cordierite	1.33	-
Amesite	0.89	-
Montmorillonite	-	75.4
Cristobalite	-	7.3
Feldspar	-	4.3
Kaolinite	-	0.8
Calcite	-	0.5

Table 3. Chemical composition of Beishan granite rock and Gaomiaozi (GMZ) Na-bentonite used for experiment, wt. %.

Oxide	Beishan Granite [24]	Bentonite GMZ [25]
SiO_2	57.78	67.43
Al_2O_3	15.42	14.20
Fe_2O_3	4.07	2.40
Na_2O	2.54	1.75
CaO	3.25	1.13
K_2O	6.15	0.73
FeO	-	0.29
TiO_2	0.79	0.12
MgO	2.12	0.10
MnO	-	0.02
P_2O_5	0.96	0.02
SrO	0.13	-
Loss of ignition	6.45	11.38

2.3. Bentonite Clay

A sodium bentonite powder from the Gaomiaozi (GMZ) deposit was used for the experiment. The GMZ bentonite is a Na-montmorillonite clay that was formed in the late Jurassic period. The formation of bentonite clay was caused by mineralization due to interaction with groundwater and weathering of newly formed volcanic deposits. The experimental samples of bentonite were excavated from a large clay deposit located in Inner Mongolia Chinese autonomous region, around 300 km northwest of Beijing. The deposit may contain up to 160 million tons of clay materials [24]. The high content of montmorillonite leads to a high cation exchange capacity (CEC = 77.30 meq/100 g), a large plasticity index (Ip = 275), and large specific surface area (S = 570 m^2/g). The major exchangeable cations are Na^+, Ca^{2+}, Mg^{2+} and K^+ [25,26].

A sample of the initial bentonite was received in form of a homogeneous powder, with a soft texture and waxy appearance. The mineralogical and chemical composition of GMZ bentonite is presented in Tables 2 and 3, respectively.

2.4. Leaching Solution

The distilled water was used as a leaching solution to allow a comparison of the results of the Pu-doped glass alteration in a water-saturated bentonite with previously published results on Pu-bearing glass leaching and alteration mostly performed in distilled or deionized water. Taking into account relatively long duration (2 years) and high temperature (90 °C) of the experiment, it was supposed that the initial distilled water will be saturated with chemical elements from granite and bentonite during the granite-bentonite-water interaction.

The experimental design is presented in Figure 2. The experiment was performed according to the following scheme:

- A small amount (15–20 g) of water saturated (wet) bentonite clay was placed into granite block 70 × 70 × 125 mm in size.
- A chip of the ^{238}Pu-doped glass (weight 1216 mg) was placed into the wet bentonite mass inside the container. The thickness of the bentonite layer around the glass sample was about 15–20 mm (Figure 2).
- The granite container was filled with distilled water up to the 10–20 mm above the bentonite surface, sealed with Teflon^{TM} lid, placed into a stainless steel clamp and tightened (Figure 3a).
- After that, the clamped container was immersed into a Teflon^{TM} case filled with water (Figure 3b) and sealed with a screw cap. The sealed container was placed into a thermostatic oven maintaining a temperature of 90 °C for 2 years.

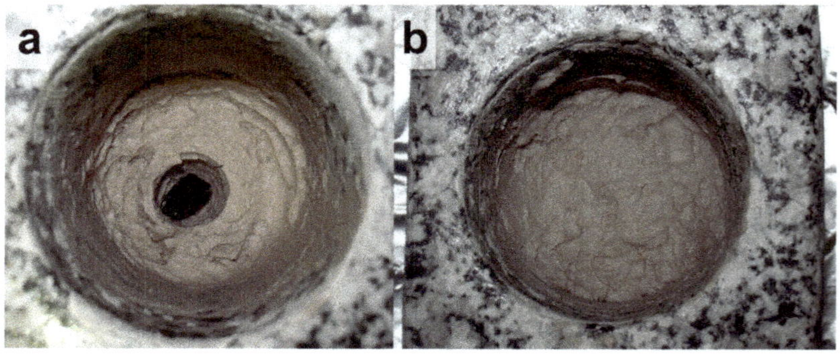

Figure 2. Preparation of the experiment: (**a**)—the Pu-doped borosilicate glass surrounded by wet bentonite inside the granite container; (**b**)—view after sealing filling with bentonite.

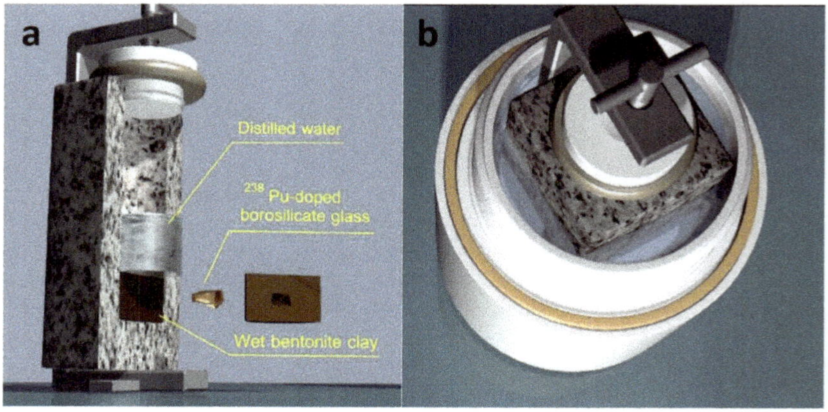

Figure 3. 3D model of the complete experimental setup. (**a**) virtual cross-section of the experimental assembly. (**b**) virtual top view of the clamped granite container in a protective Teflon^{TM} vessel.

Every 6 months sampling of ~5 mL of the leachate solution from the granite container was performed and an equivalent amount of distilled water was added after the sampling. The aqueous solution between the granite container surface and the inner wall of Teflon^{TM} cask was also sampled every 6 months to control eventual plutonium release through granite. The sampled solutions were centrifuged at 4500 rpm for 10 min and separated liquid phase was analysed by alpha- and gamma-spectrometry (Canberra-7401 and multichannel analyser DSA-1000 with Ge-detector, Canberra, respectively).

After completion of the experiment, the granite container was cut with a saw and a cross-section was put into direct contact with a Retina XBM film for 2 months. Wet bentonite was mechanically extracted from the container, intermixing of the bentonite layers was avoided as much as possible, see below. The extracted bentonite was examined with X-ray powder diffraction using Empyrean or Aeris diffractometers operated with Cu-Kα radiation in reflection (Bragg-Brentano) geometry. The sample was placed on a zero-background Si holder or, in case of abundant initial (reference) material, was backloaded. Note that the initial bentonite sample was analysed in fully dry state, whereas the bentonite from the container was measured both in partly wet state and after overnight drying at ambient conditions; deep drying was not performed to minimise amounts of radioactive dusting. Comparison of the same sample measured in wet and "dry" conditions show similarity of the diffraction patterns with exception of position of the reflection corresponding to the basal plane of montmorillonite, which shifts considerably. Of course, the presented results are only qualitative, since no attempts to separate clay minerals was attempted.

3. Results

The Pu content in the leachates is presented in Table 4. The lack of data after the first 6 months is due to the full water consumption in the granite container by the bentonite (Figure 4a).

Table 4. Results of Pu α-spectrometry measurements of water solution inside and outside the granite container during the experiment.

Sample	Pu Content, Bq/mL			
	6 Months	12 Months	18 Months	24 Months
Water sample from the inner space of granite container	No data	50–110	<50	<50
Water sample from outside area of granite container	<10			

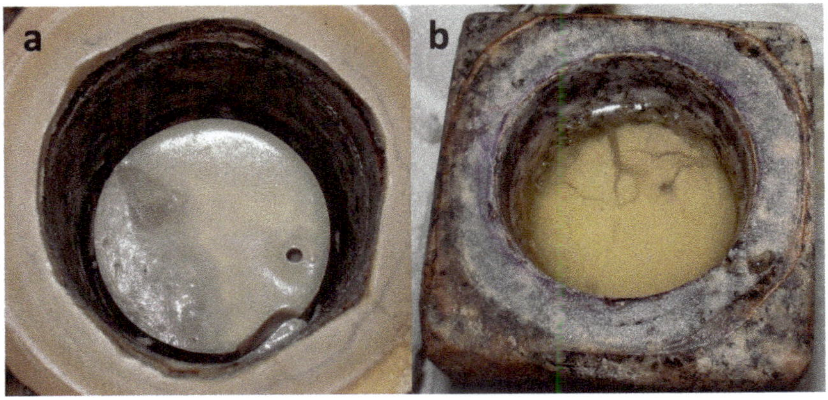

Figure 4. Photo of the wet bentonite inside granite container after 6 months (**a**) and 2 years (**b**) of the experiment at temperature 90 °C.

After two years of the experiment (Figure 4b) all water above the bentonite surface was removed and the bentonite was extracted layer by layer with a thickness of about 1 cm each (Figure 5). The glass sample was washed in distilled water and dried in air. Both inner and outer surfaces of the granite container were carefully washed, and the container was sawn across into several fragments using a low-speed saw. All surfaces of the granite samples used for further radiography tests were carefully ground and washed.

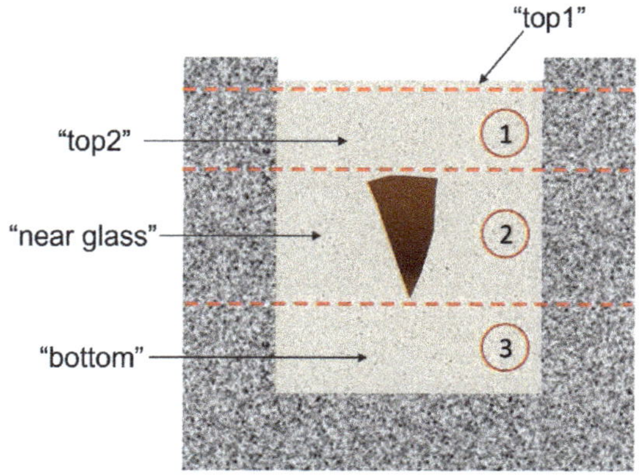

Figure 5. Scheme of the bentonite removal and sampling areas.

XRD patterns of initial dry bentonite ("start") as well as partly dried bentonite with unknown content of water (see Section 2 for details) from different locations inside the granite container are shown in Figure 6.

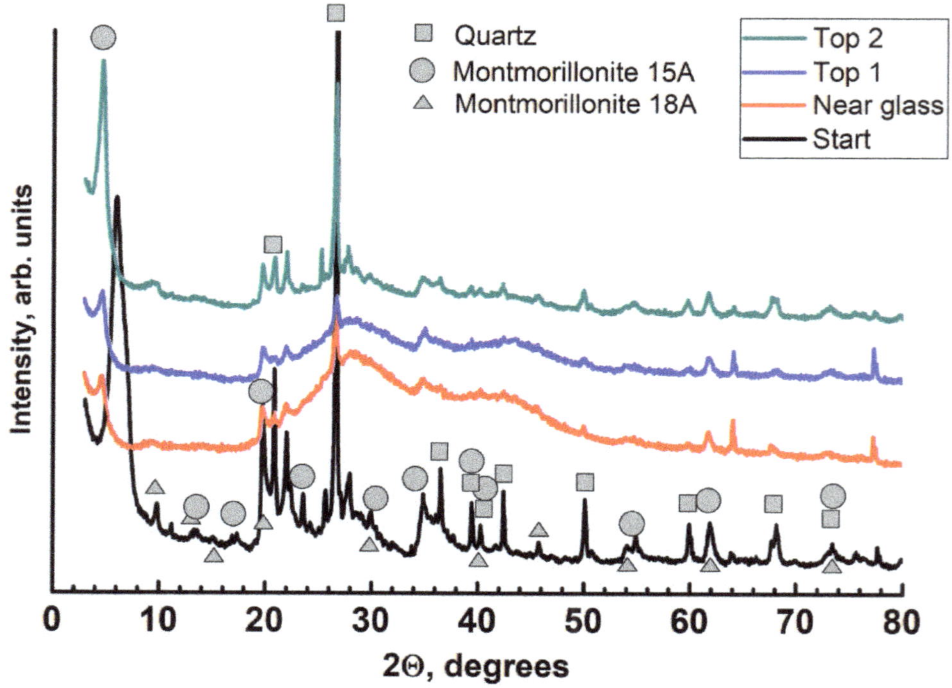

Figure 6. XRD patterns of initial dry bentonite ("start") and of moist bentonite from different locations inside the granite container ("top1", "top2", "near glass"). The 101 quartz reflection (2Θ = 26.627°) is off-scale. The curves are displaced vertically for clarity. Main reflections of principal mineral phases (ICDD cards): filled squares—quartz (01-079-1910), circle—montmorillonite 15A (00-013-0135); triangle—montmorillonite 18A (00-012-0219). Other peaks belong to feldspars and layered clay minerals.

For evaluation of the amount of ^{238}Pu adsorbed on bentonite clay, a desorption experiment was performed using a mixture of concentrated nitric acid and hydrofluoric acid.

Before the experiment, a sufficient amount of bentonite clay removed from different layers was placed into a quartz glass cup and dried in air at 40 °C until the constant weight of 14.3 g. The dry sample was brought in contact with the acids solution (10 mL of fluoric acid and 20 mL of nitric acid) for 3 weeks. The acid-resistant residue was less than 5 wt.%. α-spectroscopic measurements of the Pu-containing solution after the desorption showed that the estimated specific activity of ^{238}Pu is 0.41 MBq per gram of dry bentonite. The total amount of ^{238}Pu in the dry bentonite after the experiment makes 0.01–0.02% of initial amount of ^{238}Pu in the glass sample. As shown in Table 4, the Pu specific activity of the solution above the bentonite layer is insignificant in comparison with amount of Pu adsorbed by the clay.

4. Discussion

4.1. Alteration of the Glass Sample

During the first examination using optical microscope, formation of a secondary phase on the glass surface was clearly observed (Figure 7). However, the alteration of the glass observed in wet bentonite medium is much less pronounced than for borosilicate glass in contact with distilled water at temperature 90 °C [9,10]. The reasons for the discrepancy are not yet fully understood, but variations in pH and eventual partial radiation-induced destruction of Teflon™ container with release of fluorine compounds in experiments described in [9] may be responsible. Value of pH of natural bentonite suspension is close to neutral and varies from 5 to 7.2 with average of 6.2, which is comparable with those for the distilled water environment. However, further experiments are needed to reveal how does water radiolysis in a bentonite environment affect the degradation process of a highly radioactive glass. Also, for future research, using real or simulated groundwater seems to be reasonable to make test results more relevant to the actual URF or planned disposal site.

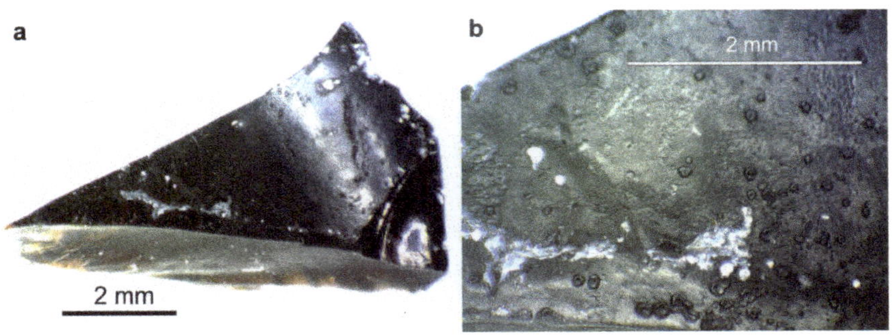

Figure 7. Optical microscopy of the surface of ^{238}Pu-doped borosilicate glass after 2 years of contact with wet bentonite. (**a**) general view. (**b**) higher magnification image.

4.2. Bentonite and Granite Alteration

The phase composition of the initial sample is dominated by quartz, different montmorillonite varieties (e.g., 15A, 18A) and feldspars (e.g., anorthoclase). In presence of a thermal gradient, leaching and reprecipitation of constituents from bentonite at different locations in the container may, in long term, induce formation of mineralogically distinct layers. In addition, intense radiolysis may contribute to phase changes. Examination of XRD patterns does not show clear difference between the bentonite samples from different locations in the container. The changes in the patterns are relatively minor and are mostly confined to reduced relative intensity of crystalline peaks, implying partial amorphisation and/or decrease of crystallite sizes. The largest relative changes are observed for the sample from the topmost layer, which demonstrates rather pronounced "amorphous" hump centred at ~28° (2Θ). This observation might be related to the least stable environment, for example, semi-annual contact with fresh solution and/or air above the clay.

Interestingly, we do not observe formation of abundant crystalline SiO_2 which may form a cement during illitisation of smectites in hydrothermal conditions (see, e.g., reviews [27,28]). Possibly, thermal gradients in our experiments were fairly small, precluding extensive dissolution-precipitation phenomena. According to the XRD patterns, the montmorillonite phase survived after two years of α-irradiation and related radiolysis at a relatively high temperature of 90 °C. It is known, that smectites may undergo structural changes in hydrothermal conditions. Whereas temperatures below 300 °C do not alter dry bentonite structure to noticeable extent [9,28,29], under hydrothermal conditions degrading of the bentonite structure can be observed at 130–150 °C [28] and even at 90 °C (in 3.5% $CaCl_2$ solution) [30]. That makes impossible to distinguish the effect of α-irradiation of wet bentonite at 90 °C and the effects caused by long-term bentonite-water interaction itself.

Even after 2 years of α-irradiation at 90 °C wet bentonite preserved its barrier function and prevents migration of the adsorbed Pu to the solution. Apparently, the influence of radiolysis and of α-irradiation on structure of comprising minerals is small. For some clay minerals, such as kaolin and attapulgite, significant structural changes were observed during alpha-irradiation by ^{253}Es but even after partial loss of crystallinity these materials were able to prevent Es release into the solution [31].

Autoradiography of the granite sample shows absence of noticeable penetration of Pu in granite rock matrix (Figure 8). It is possible to observe little black colour vein on the bottom of granite container, which is related to penetration of radioactive solution into a small crack caused by mechanical damage of the granite matrix during drilling of the cylindrical hole of the container inner space. Some amount of ^{238}Pu in a small crack can be also explained by contamination during the sawing, since no Pu penetration along grain boundaries was noted. This observation is in a good agreement with previously published data [32], showing that Pu transport in low-porosity granitic rocks from the Beishan site is mostly determined by migration along the fractures (especially in form of colloids [33]) but not by diffusion in granite matrix.

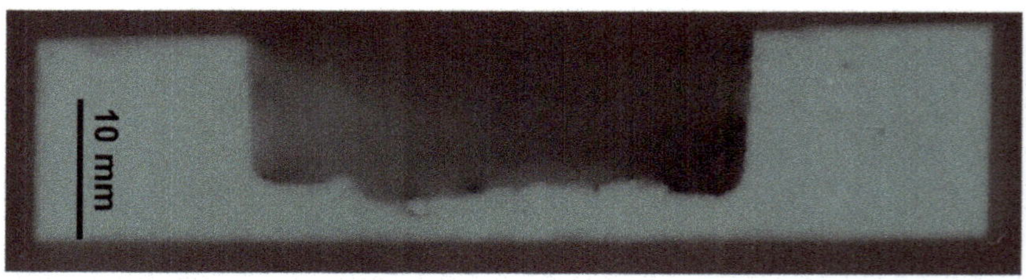

Figure 8. Autoradiography (2 months exposure) of the granite sample from the bottom of experimental granite container: cross section of lower part. Black rectangle appears due to α-particles reaching the film through air gap.

5. Conclusions

The results obtained contribute to laboratory scale safety-case studies of geological disposal for achieving a safe and sustainable way for actinide-bearing waste disposal. Based on experimental results, the following conclusions can be made:

(1) After two years of contact with ^{238}Pu-doped glass and water at temperature 90 °C Gaomiaozi Na-bentonite remains crystalline; no cementation by crystalline SiO_2 was observed.

(2) Chemical alteration of ^{238}Pu-doped borosilicate glass after long-term contact with wet bentonite at 90 °C is less extensive than for the identical sample in distilled water.

(3) The borosilicate glass is slightly altered during the experiment, but leached Pu was quantitatively adsorbed and retained by the surrounding bentonite. The total amount of plutonium adsorbed on bentonite clay for during two years of the experiment did not exceed 0.02% of the initial amount of plutonium in the glass sample.

(4) Beishan granite remained intact with no evidence of Pu penetration into the granite matrix along mineral grain boundaries.

Author Contributions: Conceptualization and methodology: B.B. and B.Z.; analysis: B.B., B.Z., Y.P. and A.S.; writing—original draft preparation: B.Z., A.S. and B.B.; writing—review and editing: B.Z., A.S., B.B. and X.L. All authors have read and agreed to the published version of the manuscript.

Funding: This research received no external funding.

Institutional Review Board Statement: Not applicable.

Informed Consent Statement: Not applicable.

Data Availability Statement: The data presented in this study are available on request from the corresponding author.

Acknowledgments: Authors are grateful to Vladimir Garbuzov for his valuable support with the 3D visualization. Authors highly appreciate the technical support of Nina Petrova, Oxana Azarchenko (V. G. Khlopin Radium Institute) and Ksenia Vaganova.

Conflicts of Interest: The authors declare no conflict of interest.

References

1. Lutze, W.; Ewing, R.C. *Radioactive Waste Forms for the Future*; North-Holland Physics Publishing: Amsterdam, The Netherlands, 1988; 778p.
2. Burakov, B.E.; Ojovan, M.I.; Lee, W.E. *Crystalline Materials for Actinide Immobilisation*; Imperial College Press: London, UK, 2011; 216p. [CrossRef]
3. Ojovan, M.I.; Petrov, V.A.; Yudintsev, S.V. Glass Crystalline Materials as Advanced Nuclear Wasteforms. *Sustainability* **2021**, *13*, 4117. [CrossRef]
4. International Atomic Energy Agency. *Characterization of Swelling Clays as Components of the Engineered Barrier System for Geological Repositories*; IAEA-TECDOC-1718; IAEA: Vienna, Austria, 2013.
5. Peuget, S.; Noël, P.-Y.; Loubet, J.-L.; Pavan, S.; Nivet, P.; Chenet, A. Effects of deposited nuclear and electronic energy on the hardness of R7T7-type containment glass. *Nucl. Instrum. Methods Phys. Res.* **2006**, *246*, 379–386. [CrossRef]
6. Peuget, S.; Broudic, V.; Jégou, C.; Frugier, P.; Roudil, D.; Deschanels, X.; Rabiller, H.; Noel, P. Effect of alpha radiation on the leaching behavior of nuclear glass. *J. Nuc. Mater.* **2007**, *362*, 474–479. [CrossRef]
7. Rolland, S.; Tribet, M.; Jégou, C.; Broudic, V.; Magnin, M.; Peuget, S.; Wiss, T.; Janssen, A.; Blondel, A.; Toulhoat, P. 99Tc- and 239Pu-doped glass leaching experiments: Residual alteration rate and radionuclide behavior. *Int. J. Appl. Glass Sci.* **2013**, *4*, 295–306. [CrossRef]
8. Peuget, S.; Delaye, J.-M.; Jégou, C. Specific outcomes of the research on the radiation stability of the French nuclear glass towards alpha decay accumulation. *J. Nuc. Mater.* **2014**, *444*, 76–91. [CrossRef]
9. Zubekhina, B.Y.; Shiryaev, A.; Burakov, B.E.; Vlasova, I.E.; Averin, A.A.; Yapaskurt, V.O.; Petrov, V.G. Chemical alteration of 238Pu-loaded borosilicate glass under saturated leaching conditions. *Radiochim. Acta* **2020**, *108*, 19–27. [CrossRef]
10. Zubekhina, B.Y.; Burakov, B.E.; Ojovan, M.I. Surface alteration of borosilicate and phosphate nuclear waste glasses by hydration and irradiation. *Challenges* **2020**, *11*, 14. [CrossRef]
11. Allen, C.C.; Wood, M.I. Bentonite in nuclear waste disposal: A review of research in support of the basalt waste isolation project. *Appl. Clay Sci.* **1988**, *3*, 11–30. [CrossRef]
12. Parab, H.; Mahadik, P.; Sengupta, P.; Vishwanadh, B.; Kumar, S.D. comparative study on native and gamma irradiated bentonite for cesium ion uptake. *Prog. Nucl. Energy* **2020**, *127*, 103419. [CrossRef]
13. Cheng, J.; Gu, R.; He, P.; Pan, Y.; Leng, Y.; Wang, Y.; Liu, Y.; Zhu, M.; Tuo, X. Effect of high-dose γ-ray irradiation on the structural stability and U(VI) adsorption ability of bentonite. *J. Radioanal. Nucl. Chem.* **2022**, *331*, 339–352. [CrossRef]
14. Nithya, K.; Gandhi, S.; Arnepalli, D.N. Effect of pH on Sorption Characteristics of Bentonite. In Proceedings of the 6th International Congress on Environmental Geotechnics (6ICEG), New Delhi, India, 7–11 November 2010. [CrossRef]
15. Choo, K.Y.; Bai, K. Effects of bentonite concentration and solution pH on the rheological properties and long-term stabilities of bentonite suspensions. *Appl. Clay Sci.* **2015**, *108*, 182–190. [CrossRef]
16. Cui, J.; Zhang, Z.; Han, F. Effects of pH on the gel properties of montmorillonite, palygorskite and montmorillonite-palygorskite composite clay. *Appl. Clay Sci.* **2020**, *190*, 105543. [CrossRef]
17. Jia, S.; Dai, Z.; Yang, Z.; Du, Z.; Zhang, X.; Ershadnia, R.; Soltanian, M.R. Uncertainty quantification of radionuclide migration in fractured granite. *J. Clean. Prod.* **2022**, *366*, 132944. [CrossRef]
18. Ménager, M.-T.; Petit, J.-C.; Brocandel, M. The migration of radionuclides in granite: A review based on natural analogues. *Appl. Geochem.* **1992**, *7*, 217–238. [CrossRef]

19. Chunli, L.; Zhiming, W.; Shushen, L.; Yuee, Y.; Hong, J.; Bing, L.; Ling, J.; Ling, W.; Dan, L.; Zhongde, D.; et al. The migration of radionuclides ^{237}Np, ^{238}Pu and ^{241}Am in a weak loess aquifer: A field column experiment. *Radiochim. Acta* **2001**, *89*, 519–522. [CrossRef]
20. Yamaguchi, T.; Nakayama, S.; Vandergraaf, T.T.; Drew, D.J.; Vilks, P. Radionuclide and Colloid Migration Experiments in Quarried Block of Granite under In-Situ Conditions at a Depth of 240 m. *J. Power Energy Syst.* **2008**, *2*, 186–197. [CrossRef]
21. Aertsens, M.; Maes, N.; Van Ravestyn, L.; Brassinnes, S. Overview of radionuclide migration experiments in the HADES Underground Research Facility at Mol (Belgium). *Clay Miner.* **2013**, *48*, 153–166. [CrossRef]
22. Begg, J.D.; Zavarin, M.; Tumey, S.J.; Kersting, A.B. Plutonium sorption and desorption behavior on bentonite. *J. Environ. Radioact.* **2015**, *141*, 106–114. [CrossRef]
23. Wang, J.; Chen, L.; Su, R.; Zhao, X. The Beishan underground research laboratory for geological disposal of high-level radioactive waste in China: Planning, site selection, site characterization and in situ tests. *J. Rock Mech. Geotech. Eng.* **2018**, *10*, 411–435. [CrossRef]
24. Li, C.; Zheng, Z.; Liu, X.Y.; Chen, T.; Tian, W.Y.; Wang, L.H.; Wang, C.L.; Liu, C.L. The Diffusion of Tc-99 in Beishan Granite—Temperature Effect. *J. Nucl. Sci. Technol.* **2013**, *3*, 33–39. [CrossRef]
25. Ye, W.M.; Cui, Y.J.; Qian, L.; Chen, B. An experimental study of the water transfer through confined compacted GMZ bentonite. *Eng. Geol.* **2009**, *108*, 169–176. [CrossRef]
26. Ye, W.M. Investigation on chemical effects on GMZ bentonite used as buffer materials. *E3S Web Conf.* **2016**, *9*, 02001. [CrossRef]
27. Meunier, A.; Velde, B.; Griffault, L. The reactivity of bentonites: A review. An application to clay barrier stability for nuclear waste storage. *Clay Miner.* **1998**, *33*, 187–196. [CrossRef]
28. Wersin, P.; Johnson, L.H.; McKinley, I.G. Performance of the bentonite barrier at temperatures beyond 100 °C: A critical review. *Phys. Chem. Earth* **2007**, *32*, 780–788. [CrossRef]
29. Laine, H.; Karttunen, P. *Long-Term Stability of Bentonite. A Literature Review*; POSIVA Working Report; Posiva WR: Eurajoki, Finland, 2010.
30. Liu, X.; Prikryl, R.; Pusch, R. THMC-testing of three expandable clays of potential use in HLW repositories. *Appl. Clay Sci.* **2011**, *52*, 419–427. [CrossRef]
31. Haire, R.G.; Beall, G.W. Consequences of radiation from sorbed transplutonium elements on clays selected for waste isolation. *ACS Symp. Ser.* **1979**, *100*, 291–295. [CrossRef]
32. Zhang, X.; Wang, Z.; Reimus, P.; Ma, F.; Soltanian, M.R.; Xing, B.; Zang, J.; Wang, Y.; Dai, Z. Plutonium reactive transport in fractured granite: Multi-species experiments and simulations. *Water Res.* **2022**, *224*, 119068. [CrossRef]
33. Malkovsky, V.I.; Petrov, V.A.; Yudintsev, S.V.; Ojovan, M.I.; Poluektov, V.V. Influence of Rock Structure on Migration of Radioactive Colloids from an Underground Repository of High-Level Radioactive Waste. *Sustainability* **2023**, *15*, 882. [CrossRef]

sustainability

Article

An Introduction to Nuclear Industrial Archaeology

Erin I. Holland [1,*], Yannick Verbelen [1], Dean T. Connor [1,2], Tomas Martin [1], Matthew Higginson [3] and Thomas B. Scott [1]

[1] Interface Analysis Centre, H.H. Wills Physics Laboratory, University of Bristol, Bristol BS8 1TL, UK
[2] National Nuclear Laboratory, Risley, Warrington WA3 6AE, UK
[3] Atomic Weapons Establishment, Aldermaston, Reading RG7 4PR, UK
* Correspondence: erin.holland@bristol.ac.uk

Abstract: The legacy of the early days of the Atomic Age consists of many problematic sites worldwide, including radioactive waste dumps, uranium mines, spent fuel reprocessing plants, and defunct processing and enrichment plants. Although nature quickly reclaims abandoned sites, any remaining radioisotopes can pose a threat for millennia to come, long after the benefits gained from nuclear technology have faded. The field of nuclear industrial archaeology specialises in finding and characterising these sites to support local communities and site owners. Where maps and building plans have been lost, nuclear archaeologists deploy state-of-the-art analysis techniques on the ground to unravel the current state of legacy sites, and quantify the remaining radioactive inventories to the standard required by the nation the site is located within. The objectives of nuclear industrial archaeology are varied and site dependent. Whether the objective is to puzzle the forgotten history of activity back together or safeguard and recover dangerous radioactive materials, nuclear industrial archaeology adapts radioanalytical laboratory and site-surveying techniques in order to understand the site and allow scientists to communicate this information to support remediation efforts. This paper discusses current methodologies alongside a case study.

Keywords: nuclear forensic analysis; legacy waste; nuclear forensics; XRF; isotopic fingerprinting; microscopy; sampling techniques; in-situ analysis; nuclear industrial archaeology; photogrammetry

Citation: Holland, E.I.; Verbelen, Y.; Connor, D.T; Martin, T.; Higginson, M.; Scott, T.B. An Introduction to Nuclear Industrial Archaeology. *Sustainability* **2023**, *15*, 6178. https://doi.org/10.3390/su15076178

Academic Editors: Michael I. Ojovan, Vladislav A. Petrov and Sergey V. Yudintsev

Received: 20 January 2023
Revised: 11 February 2023
Accepted: 9 March 2023
Published: 4 April 2023

1. Introduction

Nuclear forensics (NF) is the analysis of radioactive and other nuclear material in support of governmental priorities including national security [1,2]. Although NF work has been ongoing within the radioanalytical realm since World War II, modern nuclear forensics as a field of activity coalesced in earnest when several large finds of fissile materials were seized by law enforcement across Europe [3,4]. These discoveries created a subsequent need to formalise the assessment of the radiological hazard, intended use, and origin of the material, necessitating the development of a tailored suite of measurements, techniques and databases to store the details of each declared incident [5]. In the three decades since the first high profile seizure of a uranium fuel pellet outside of regulatory control, several countries have developed their forensic capabilities, dedicated facilities and specific programs for dealing with threats should they appear within or escape their borders.

The Incident and Trafficking Database (ITDB), maintained by the International Atomic Energy Agency (IAEA) since 1995 [6], serves as a nonpartisan record, increasing the ability of governments to track, find, and provenance nuclear materials that have been smuggled, lost, improperly disposed of, or stolen. It is a database that is voluntarily maintained by member governments, with over 3500 incidents recorded within it as of 2022. Approximately 10% of these have been confirmed as "acts related to trafficking or malicious use" [6]. Incidents reported to the ITDB involve radioactive materials such as uranium, plutonium, and thorium, as well as naturally occurring and artificially produced radioisotopes, and radioactively contaminated material.

In the last three decades, NF has expanded from subdisciplines of radiochemistry and geochemistry to a field in its own right, giving rise to numerous experimental techniques, methodologies, and microfields [7]. The methods that nuclear forensics encompasses are broad and varied, ranging from field-based instruments such as portable X-ray fluorescence spectroscopy (pXRF) and Geiger–Müller (GM) counters to specialised lab-based methodologies, including inductively coupled plasma mass spectrometry (ICP-MS) and radiochronometry [7]. These techniques do not only apply to forensic questions, however—they are also used in cases where no crime has been committed and where instead the investigatory focus is on other aspects, including process reconstruction, legacy waste analysis, medical/industrial radiation sources, releases from civilian nuclear reactors, accident investigation, or decommissioning and clean-up of legacy nuclear facilities.

Nuclear forensics is iterative, following the requirements of law enforcement, and accomplishes only the agreed-upon analytical plan to provide the necessary data aiming to answer a defined question [8]. All results are produced by using methods verified and approved within a legally recognised quality framework [9,10]. However, an investigation of historical radiological material is structured differently from a forensic investigation; scientists and prosecutors can have different requirements ranging from the chain of custody reporting requirements to the level of certainty required. "Nuclear forensics" is an incorrect designation unless there is a prosecutorial element. The phrase "forensic investigation" is more than a misnomer; it can be actively damaging to the overall aims of the scientists involved. Open and honest discussions about radiological holdings and clean-up efforts are hindered if the threat of prosecution is assumed [11]. Therefore, nuclear industrial archaeology (NIA) is a more accurate designation for the study of historical nuclear sites, free of many constraints placed on nuclear forensics [12]. Without an active crime scene and the need for findings to have confidence levels that would hold up in court, a more balanced view of the radiological material and the environment it is settled within can be taken. The more relaxed standards of accuracy and precision levels required can, in turn, inform future nuclear forensic investigations, allowing new techniques to enter the investigative process after extensive field and laboratory testing on low-risk, low time-pressured samples, although these results must still be accurate and defensible. Nuclear industrial archaeology, therefore, is also a developmental pipeline enabling the validation of novel scientific techniques before their adoption in NF.

2. Nuclear Industrial Archaeology

Definitions of nuclear forensics and nuclear archaeology have changed throughout history, with different authors adopting variations in scope and meaning depending on the context. Considerable overlap between definitions exists depending on the context. In the late 1970s and throughout the 1980s, the meaning of "nuclear archaeology" was almost identical to present-day "nuclear forensics". Nuclear forensics is defined by the IAEA as

> "... the analysis of intercepted illicit nuclear or radioactive material and any associated material to provide evidence for nuclear attribution. The goal of nuclear analysis is to identify forensic indicators in interdicted nuclear and radiological samples or the surrounding environment, e.g. the container or transport vehicle. These indicators arise from known relationships between material characteristics and process history. Thus, NF analysis includes the characterisation of the material and correlation with its production history."
>
> —*IAEA Nuclear Forensics Support Reference Manual, 2006* [13]. In this context, "analysis" includes radiometric and nonradiometric measurement techniques.

In common usage, nuclear forensics has been expanded to include any investigation of radiological material that uses techniques in common with a criminal investigation. This is seen in relevant paper titles and conference proceedings [14–16]. It is these authors' opinion that this is a misnomer and that in scientific endeavours which do not include an element of law enforcement (especially where the aim is to support safe decommissioning and characterisation of legacy sites) any reference to "forensics" is undesirable. The more appropriate term is, "nuclear industrial archaeology". This has parallels, interlinks, and joint interests with similar fields where investigations of legacy activities is relevant, such as industrial archaeology:

> "The systematic study of structures and artefacts as a means of enlarging our understanding of the industrial past."
>
> —*Industrial Archaeology Principles and Practice* [17] (p. 21)

The term "nuclear archaeology" itself has historically been associated with two different meanings. The earliest mention dates back to 1973 when radiocarbon dating became the main activity of the University of Missouri–Columbia (UMC) Laboratory for Nuclear Archaeology (LNA), and neutron activation analysis (NAA) gradually became a mainstream technique by which to analyse archaeological and palaeontological artefacts [18–21]. Afterward, "nuclear archaeology" was adopted in a deceiving nuclear forensics context, referring to the investigations related to dose rate estimations of atomic bomb survivors [22,23], and more frequently, efforts to estimate the amount of fissile materials produced or passed through reactors based on residual radiation signatures [24–26]. Neither meaning can be applied to the study of legacy installations and radioactive relics from the early nuclear era when the focus is on understanding and/or reconstructing processes in the nuclear industry regardless of law enforcement interests. Nuclear industrial archaeology (NIA) is thus defined as

> "the systematic study of material evidence from the nuclear industrial past, with the aim to understand and reconstruct nuclear processes, record nuclear history, preserve nuclear heritage and/or facilitate containment, decontamination and decommissioning of hazardous nuclear industrial legacies."

In this study, we use an investigation into a disused radiological processing site in Ukraine to demonstrate the key advantages of considering it as a nuclear industrial archaeology study, rather than a nuclear forensics case.

3. NIA Case Study: Pridnyprovsky Chemical Plant

Prydniprovsky Chemical Plant (PChP) is a large ex-Soviet materials and chemicals processing site of roughly 260 hectares in central Ukraine, on the outskirts of the city of Kamianske. Between 1948 and 1991 [27] (p. 217), the site processed uranium and thorium ore into ore concentrate, separated rare earth elements and produced other industrial chemicals such as fertiliser [28] and ion exchange resins [29]. Tens of millions of tonnes of radioactive waste were generated on the site during its operational lifetime. The site has fallen into disrepair but has no legal investigation into it. A NIA investigation therefore offers more potential to the site operators that a narrower NF investigation could not.

This industrial area comprises ∼100–200 buildings and structures, many of which are in a state of disrepair and dereliction. Approximately 45% of the southern part of the site has become radioactively contaminated as a direct consequence of the site's operation, compounded by the lack of historic management of the resulting radioactive byproducts. Approximately 20 buildings are heavily contaminated with radiation levels up to a maximum of 4 $mSv\,h^{-1}$ reported, which is equivalent to nearly twice the average annual dose received by members of the general public from natural sources in the UK, in one hour [30]. It is estimated that there are over 40 million tonnes of low-level radioactive waste stored in the site's tailings ponds, which is evidenced to be slowly leaching into the Dnieper river [31].

As of 2022, the European Commission is working with the Ukraine Ministry of Energy and Coal Use and the Nuclear Regulator to improve radiological safety and security. For remediation works to proceed, site management needs a detailed characterisation of the radioactive materials held there. As part of this activity, a team from the University of

Bristol surveyed several of the most heavily contaminated buildings, analysing a selection of in situ material samples.

The application of nondestructive characterisation methods field tested in and around the PChP buildings is explored below. It is the experience of these authors that in initial field surveys, less equipment and lower complexity of equipment allows more data to be collected. This is due to the reduced time spent carrying, testing, decontaminating, inventorying, and problem solving. There will always be a tradeoff in terms of the weight that operatives can carry and the amount of analytical equipment that can be carried. Equally, based on the limited carrying capacity, the toolkit for in situ investigations must be carefully planned and consider factors such as battery life, as well as mass of equipment and even the ease and cost of transportation in compliance with transport regulations.

4. Techniques

As in a traditional forensic investigation, preservation of life and, therefore, investigators' safety is paramount during nuclear industrial archaeology [32]. Assessment of ionising radiation on a site must include alpha, beta, and gamma rays as well as neutrons. Irrespective of radiological detection methods used, users must understand their personal, equipment and project limits and capabilities to avoid a site being inadvertently declared safe for lack of accurate assessment for radiation. For example, it cannot be assumed that all site workers will have access to an alarmed dosimeter calibrated to the isotopes expected to be present. Although these reduce the risk of accidental exposure, they can also be prohibitively expensive for legacy site management with constrained budgets.

Radioactive compounds may also be chemically toxic in addition to their radiotoxicity, and for isotopes with half lives in excess of 1 million years (Ma), chemical toxicity can be more dangerous than the emitted radiation. The chemical toxicity of uranium is similar to other heavy metals [33], and equivalent protection measures must be taken.

4.1. Nondestructive vs. Destructive Analysis

Nondestructive, in-situ analysis techniques are key components of complex site investigations and have several advantages over destructive techniques. Notably, they allow samples to be analysed quickly and in a safe manner, minimising operator exposure. The most important drawback of destructive analysis techniques is that analyses cannot be repeated because parts of the sample, and particularly features of interest, are destroyed during the analysis process. This is true for traditional chemical analysis methods such as chromatography, which typically require sample fragments to be dissolved, but also for stand-off methods such as laser-induced breakdown spectroscopy (LIBS). In the specific case of LIBS, a high-power laser pulse is used to vaporise a sample and generate a light-emitting plasma, of which the spectral lines are subsequently analysed with a spectrometer. The U-associated emissions, for example, are then easily identified at 385.464 nm, 385.957 nm, and 386.592 nm respectively [34]. Because the amount of vaporised material and the resulting plasma wavefront is slightly different every time a LIBS analysis is conducted, the obtained data tend to drift. Firing the laser at the same location multiple times leads to cavitation of the sample, affecting plasma formation. If the sample is heterogeneous (for example, if it has a surface composition that is slightly different due to oxidation effects), then subsequent LIBS measurements will show varying results. This is because the first plasma will vaporise surface oxides whereas the following measurements reflect the composition of the now exposed interior of the sample. Analysis techniques destroying samples pose a challenge for scientific falsification. This scientific research principle dictates that peer reviewers must be able to replicate research results to independently verify the research data. It can also have consequences for forensic investigations if samples later turn out to have been criminal evidence [35] (pp. 455–456).

A third drawback of using destructive analysis methods is that, by definition, a part of the sample transforms into waste products. When analysing radioactive samples, this leads to radioactive waste products that must be disposed of, and such disposal process is regulatorily complex if the composition of the sample and its resulting waste is unknown. Radioactive waste produced during the destructive analysis of radioactive samples may also contaminate analysis instrumentation, increasing its background radiation signature, and further jeopardising repeatability of the analysis. In the case of wet chemical analysis, they are in the form of a radioactive effluent. LIBS, for example, can cause aerosolised radioactive particles found in the plasma to be dispersed over a wider area. This leads to possible contamination of the equipment and scientists performing in situ measurements, with radiologically contaminated PPE requiring additional remediation as a result. Nonetheless, techniques such as LIBS have been used for in situ nuclear industrial archaeology [36], but when the subject is related to radioactive contamination, such techniques can thus contribute to spreading contamination or causing cross-contamination.

4.2. *Geiger–Müller Counting*

Geiger–Müller (GM) counters are among the oldest radiation detectors available and remain useful in field-deployed nuclear industrial archaeology as the first line of intervention on site. GM counters can be fitted with probes sensitive to alpha [37], beta [38], gamma [39–41], positron [42], or a combination of those ionising radiation types [43]. GM counters have several advantages in comparison to more sophisticated scintillators or semiconductor detectors that can perform gamma energy spectrometry: they are low-cost [44], [45] (p. 103), can be tuned to different radiation types easily by fitting filters, and produce digital pulses that are processed electronically in a straightforward manner [46,47].

GM tubes are often wired straight to a buzzer, which produces characteristic ticks when ionising radiation strikes the tube. The ticking rate (per minute or second) is proportional to the number of detected events in a given period. It can also be calibrated over a known distance to allow for initial triaging of samples and areas. The instrument's ease of use, simplicity, and resulting reliability make it a standard instrument in the nuclear industrial archaeology toolkit.

What makes a GM counter particularly useful in the field is the instant response of the tick rate, as opposed to analogue or digital displays, which make use of moving averages over a set period to calculate and display a count rate in counts per second (cps) or counts per minute (cpm). Because GM counters with a mica window are sensitive to α, β, and γ radiation and have such a fast response, they are the best method to track down the location of radioactive contamination of unknown origin. This is typically done by sweeping a pancake-type GM tube with a large window over a surface and listening to the tick rate. Using this method, a trained operator can quickly locate the source of ionising radiation. If the source is physically small, such as a single droplet of a solution containing a radioactive salt, a GM counter is often the fastest and most reliable way to track it down. An additional advantage is that operators do not need to look at a display or gauge to interpret measurements, which improves safety in the field in nuclear industrial archaeology environments such as disused industrial facilities that present numerous trip hazards.

GM counters also have two drawbacks. The first limitation is the inability of a tube to differentiate between α, β, or γ radiation. Any of these radiation types will trigger the electron cascade (the Geiger–Müller effect), and the resulting current pulse contains no information on the type of radiation or its energy. This can be partially mitigated by fitting different filters, exposing the tube either unfiltered ($\alpha + \beta + \gamma$), through a thin Al window ($\beta + \gamma$), or through a thick Al window (γ only) [48]. Some models have built-in filters, whereas others require different probes to be fitted. A second limitation of GM tubes is the low density of their fill gas, which makes them more sensitive to α and β radiation than to γ radiation, of which most passes through the tube without significant attenuation, especially at energies above 100 keV. However, this is rarely an issue in practice because pure γ-emitters are of synthetic origin and usually sufficiently concentrated to produce detectable changes in tick rates even at low concentrations and short distances.

Because GM counters are low-cost, compact, and lightweight, they should be among the first instruments deployed in an unknown field environment to assess the radiological situation. First and foremost, they are used to establish whether there is a radiological anomaly present—that is, whether the area of interest shows count rates that are significantly above natural background levels. Secondly, if ionising radiation is detected, it should be used to qualitatively assess the radiation levels in preparation for further field nuclear industrial archaeology analysis. This takes advantage of the fast impulse response and relative directionality of the sensor to measure whether radiation levels are low enough to accommodate safe further inspections or too high (thereby putting operators at risk). For most instruments, the saturation limit—the point at which subsequent pulses overlap—is between 2×10^3–10×10^3 cps. As a general rule of thumb, if the GM counter saturates during the survey, radiation levels are potentially hazardous, and the survey should be aborted as a precaution. In such cases, a robot may need to be deployed instead.

The efficiency and speed with which GM counters can help locate radiation sources in the field means that further analysis methods can be deployed more quickly and effectively (at specific points of interest rather than needlessly sampling large areas). The resulting reduction in time required to perform analyses also reduces the radiation dose of workers.

A consequence of the inability of GM counters to differentiate between ionising radiation types or their respective energies is that personal electronic dosimeters (PED), which are often based on internal GM tubes, are intrinsically unable to measure actual dose rates. This is because the radiation dose is a function of radiation type and energy, neither of which the GM tube can measure [49]. PEDs are typically calibrated under the assumption of specific isotopic composition (such as ^{137}Cs or ^{90}Sr in a nuclear power station environment), which leads to drastically different results when surveying a processing facility for naturally occurring radioactive materials (NORM), for example. If accurate dosimetry is desired, a gamma spectrometer must be used instead.

The final field utilisation of GM counters is during postdeployment decontamination. Regardless of how careful nuclear industrial archaeology field analysts are during their work, there is always a risk of contamination of protective equipment (PPE) or instrumentation. For overshoes, gloves, or hard hats, such contamination is even expected. GM counters are the ideal instrument for contamination assessments on personnel and equipment, facilitating targeted decontamination if needed.

4.3. Gamma Spectroscopy

Gamma spectroscopy, the identification and quantitative study of the energy spectra of radioisotopes that emit γ-rays, is a rapid and nondestructive standoff measurement technique that has been used and developed since Rutherford and Andrade's work on emission lines in 1914 [50].

Portable gamma spectrometers are comparatively less accurate and precise than laboratory detectors such as high purity germanium (HPGe), but as they are small and portable, they are beneficial for in-field analysis. Even where sample geometries are unknown, reliable conclusions can be drawn about which gamma-emitting isotopes are present, which can inform further decommissioning work and allow a more detailed assessment of radiological risk for site workers than that offered by GM counters. This is especially useful for uncategorised loose radioactive debris and sealed tanks, both of which prevent significant issues for analysis. Micro gamma spectrometers can easily be mounted onto a probe or telescopic boom, allowing analysts to maintain a safe distance from unknown samples, or deployed on a mobile robotic platform [51]. Gamma spectrometers require careful calibration before field applications to ensure that isotopic peak assignments are reliably made. Examples of gamma spectrometers deployed in the field are shown in Figure 1.

Gamma spectrometry is an important tool in the initial phases of an investigation where identifying the nuclides present is of paramount importance. Some radionuclides present a more substantial toxicological risk than radiological; hence, a device that can discern nuclides is of key importance.

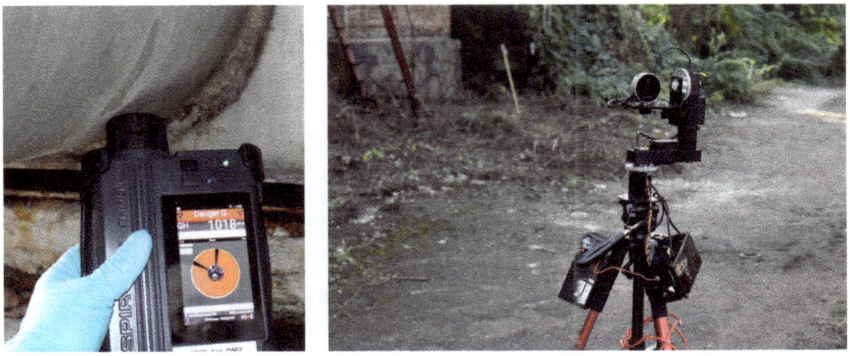

Figure 1. Gamma spectroscopy in field-deployed nuclear industrial archaeology. (**Left**) a portable NaI(Tl) gamma spectrometer of type Mirion SPIR-Ace [52] used to measure gamma radiation dose rates in excess of 1.0 mSv h^{-1} on the outside of a pipe connecting a mixer–settler tank in a legacy uranium ore processing facility. (**Right**) a collimated gamma spectrometer [53] mounted on a pan-tilt unit and suspended on a tripod to map contamination in an industrial environment [51].

4.4. X-ray Fluorescence

X-rays are well suited to probe the composition of matter because they can penetrate the surface of a sample to a depth inversely proportional to the density of that sample. The most popular analysis methods using X-rays are X-ray diffraction (XRD) and X-ray fluorescence (XRF). Both are available in a portable form for in-situ (handheld) analysis: portable XRF analysers are in use to determine alloy compositions in metal recycling industry, and portable XRD analysers are in use for metal alloy phase identification in junkyards. Combined XRD/XRF instruments have been developed for space missions, such as the CheMin instrument onboard the *Curiosity* Mars rover [54,55]. XRF gives information on the sample's elemental composition and is the most simple of the X-ray analysis techniques.

For analysis of samples in nuclear industrial archaeology applications, XRF as an analysis technique works well for samples containing heavy elements because these elements have electrons in high-energy orbits, which thus produce high-energy fluorescence. Light elements (LE; H to Na) have loosely bound electrons and give rise to high Compton scattering rather than fluorescence. The fluorescence effect is maximal when the incident X-rays are only marginally higher in energy than the electron's binding energy. If the photons have an energy that is too high, there is a higher chance of scattering than fluorescence. Likewise, if the photons have an energy that is under the binding energy, electrons cannot be ejected, and no fluorescence can occur. Because binding energy is a function of the electron shell, and these, in turn, are a function of the number of electrons the atom has, the ideal X-ray excitation energy depends on the sample's elemental composition. This is an essential consideration in the design of an XRF instrument. Even when fluorescence occurs in LE, the emitted secondary photons are of such low energy (in the order of a few eV) that they are either absorbed by the sample, the air, or the window of the spectrometer itself. The lightest identifiable element using XRF can vary from Be to Ti depending on spectrometer sensitivity. All heavier elements can be identified up to Am [56,57] in concentrations ranging from ppm to percentage levels, although noble gases are rarely calibrated for.

The capability of a pXRF to identify elemental presence from ppm to per cent concentrations makes it an attractive third nuclear industrial archaeology analysis tool in the field—after gamma spectroscopy and GM-counting. Depending on configuration and energy range, the preliminary elemental composition can be obtained after a few seconds, and quantitative analysis results after 1–3 min. Because operators can potentially accumulate a substantial and potentially harmful radiation dose in 1–3 min if the radiation flux is high enough, a preliminary sweep with other dosimetry, such as a GM counter, is a necessary first step to ensure dose rates are sufficiently low to allow further investigation. Handheld instruments such as the Olympus Vanta XRF or ThermoScientific Niton XL3T XRF are sealed instruments with an IP rating of 54 or higher, making them resistant to dust ingress and splashing liquids from all directions. This enables their use within contaminated environments with high dust or humidity levels. The Olympus Vanta XRF used in this study weighs 1.70 kg with battery installed and can be operated with one hand by an operator, as shown in Figure 2. Miniature X-ray tubes mean these instruments no longer contain radioisotopes such as ^{55}Fe, ^{109}Cd, or ^{241}Am like their predecessors [58] (p. 94), [59]. Therefore, the produced excitation beam is an order of magnitude higher, reducing the time needed for analysis by an order of magnitude, but reducing the residual dose to 0 when the instrument is turned off. During operation, the dose is quite substantial, at ca. 18.5 mGy h^{-1} at 10 cm from the aperture, as opposed to only 267 µGy h^{-1} for the TN 9000 XRF analyzer from 1995 [58] (p. 94). The relatively high dose rate is accompanied by numerous safety features, such as a password-protected user interface, X-ray ON lights, proximity sensors, quick battery release, and dead man switch [58] (p. 95). The X-ray tube itself is capped in power to 4 W [60] (p. 2).

Figure 2. A portable XRF analyser (type: Olympus Vanta) used for in situ nuclear industrial archaeology of samples in a legacy uranium ore processing facility. (**Left**) Nuclear industrial archaeology of contamination on construction material. (**Right**) Nuclear industrial archaeology of a spill of ionic exchange resins.

As in-situ measurements typically do not require sample preparation, there is no setup time needed before NIA measurements can commence. There are, however, several limitations. First, the fluorescence effect peaks at excitation energies slightly higher than electron binding energy levels. If the X-ray energy is either too high or too low, Compton or Rayleigh scattering are the dominant interaction modes for the X-rays [61]. In the case of Compton scattering, a lower energy X-ray will be scattered into the detector, and in the case of Rayleigh scattering, the scattered X-ray will have identical energy. These produce additional counts in the detector, swamping weak signals and lowering measurement results. If the density of the sample is unknown, or the sample contains both high- and low-density fractions, it is necessary to sweep the X-ray source between ca. 8 keV and 50 keV to enhance the measurement accuracy over a range of material densities.

Secondly, as the penetration power of the source X-rays is proportional to their energy and the sample's composition, acquiring accurate data from in situ measurements requires samples of sufficient thickness to either absorb, fluoresce, or scatter the source X-rays completely. The minimum sample thickness to produce homogeneous results ranges from 55 µm for solid Pb to 0.9 cm for sand [61] (p. 16, Table 1). When samples are insufficiently thick or heterogeneous in depth in comparison to the materials with which the instrument has been calibrated, the elemental composition reported by the instrument will be inaccurate. When measuring samples of insufficient thickness on a substrate, the elemental composition of the substrate will also be included in the measurement results, which is a drawback to the technique. This is inevitable for in situ measurements of crusts or oxidation on pipes or tanks, or thin layers of spilt substances in solid or liquid form. There are two ways around this issue that use postprocessing of data. One option is to measure the sample on its substrate or matrix, then measure the "clean" substrate or matrix separately and subtract them from each other. Doing so invalidates concentration data but allows for quick identification of foreign elements on common industrial surfaces such as 304L or 316L stainless steel [62]. The drawback is those sample elements present (including a variety of application-specific trace elements) in stainless steel are also removed. Another option is to measure samples on different substrates or matrix materials and use subtraction to find the elements both datasets have in common. The drawbacks of this method are identical.

As EDXRF instruments cannot detect LE, a way around the substrate problem is by measuring samples on a substrate that exclusively contains LE. An example of these is disposable cardboard plates or scoops, which can retain a small amount of material for analysis. As neither cardboard nor air induces X-ray fluorescence, measurements made this way will only show the sample's elemental composition. Alternatively, Kapton tape [57], or mylar pouches and carry-straps can also be used [60]. The practical feasibility of each approach depends on the type of material and its specific activity. Tools such as files, paint knives, or scalpels may be needed to scrape off a small amount of material onto a cardboard holder for analysis.

Thirdly, a consequence of analysing radioactive materials that undergo alpha decay is that X-rays produced due to the alpha decay process can interfere with the EDXRF measurements. For isotopes undergoing gamma decay, this is usually not a problem because the gamma energies are beyond those of X-rays (over 100 keV). However, for isotopes undergoing alpha decay, this may lead to a more complex spectrum and therefore complicate the interpretation of the measurement results [57]. For example, ^{239}Pu exhibits a prominent uranium L X-ray emission peak series as a result of alpha decay into ^{235}U. The ^{235}U daughter nucleus has a probability of being left in an unstable state, transferring excess energy to an electron through internal conversion. This could lead to the ejection of the electron, just like it would have been ejected by a higher-energy X-ray originating from the XRF analyser source [63]. The secondary X-rays emitted by higher shell electrons filling the vacancy are therefore indistinguishable, and the XRF detector would interpret the additional counts as higher concentrations of the corresponding elements in the sample. This is a problem for radioactive samples containing Pu and Am, such as nuclear industrial archaeology of radioisotope thermoelectric generators, for which XRF analysis cannot accurately determine the elemental composition in a sample. High-resolution gamma-ray spectroscopy (HRGS) is the preferred technique for studying actinides [64]. This also allows the use of coincidence methods to enhance the signal-to-noise ratio. This is discussed in further detail in the Gamma Spectroscopy techniques section above.

Finally, as portable XRF measurements for in situ nuclear industrial archaeology are made in the air rather than in vacuum or helium, absorption of low-energy fluorescence is inevitable, and elements with exclusively low electron binding energy can therefore not be measured. The XRF analyser instrument should ideally be brought in physical contact with the sample surface to eliminate as much air absorption as possible and, at the same time, maximise the detection of fluorescent X-rays. When measuring samples with both light and heavy elements, measurement results can be misleading. For example, an organic ionic exchange resin for separation of Th from REEs (shown in Figure 2) will typically be composed of over 95% resin, and only a few per cent heavy ions. However, because the XRF does not obtain a signal from the LE in the resin, it may falsely indicate disproportionally large concentrations of Th or REEs in such a sample. Some portable XRF analysers such as the Olympus Vanta mitigate this effect by comparing received X-ray intensity with emitted X-ray intensity, and assuming the difference is due to LE it cannot detect. In nuclear environments, it is not always possible or desirable to establish physical contact with a sample, as this could result in radiological contamination of the instrument itself. If prior GM-counter sweeps have established that a surface is contaminated with radioactive compounds or otherwise emitting ionising radiation, it is left to the operator's judgement to decide whether or not physical contact with the sample is justified. Instruments such as the Olympus Vanta XRF have Kapton windows that can be replaced if they become contaminated, if necessary. The instrument is equipped with a proximity sensor that cuts off the X-ray beam if the aperture is further than 15 mm away from the sample. Standoff measurements take considerable practice from the operator to keep the instrument's aperture suspended and steady for the measurement duration, which can be as long as 3 min.

Although it is possible to operate a portable XRF instrument as a single operator, it is instead advisable to opt for an operator pair instead for a nuclear deployment scenario. In such a team, one member can focus on identifying sample locations for in situ measurements or collecting samples on cardboard plates, whereas the other member can focus on performing the XRF measurements with the portable XRF analyser. To avoid contaminating the instrument, pouches are available to carry it hands-free. If a large number of samples are analysed, mounts are also available [65] (p. 4, Figure 1). It is possible to make up to 15–20 measurements per hour in handheld mode. A hard hat-mounted camera is recommended to facilitate documentation of the measurement conditions. Figure 2 is an example of a photograph made automatically with a GoPro Hero 8 camera mounted on a hard hat. The GoPro was configured in "photo" mode and set to take a photograph every 0.5 s.

Decontamination of the portable XRF analyser instrument is necessary between measurements in contact with samples. As XRF is a sensitive nuclear industrial archaeology technique, minor quantities of material in the order of a few grains of sand could result in the instrument falsely reporting elemental composition in the next sample due to contamination. Avoiding cross-contamination is therefore of extreme importance. The best way to avoid cross-contamination is to avoid physical contact with samples completely; if this is not possible or accidental contact was made with any surface, the instrument must be decontaminated meticulously. This can be done inside an active environment by wiping the instrument's front with industrial wet wipes. Surfactants such as Decon-90 [66] may be used to remove more persistent contamination. After the survey, the portable XRF analyser must be inspected with a handheld GM counter for any residual contamination and decontaminated as deemed necessary.

4.5. Digital Optical Microscope

Although optical microscopy is one of the oldest experimental techniques, it is easily forgotten during investigations that have access to higher magnification and more advanced imaging instrumentation, such as scanning electron microscopes (SEMs). Even if it were possible to take a scanning electron microscope (SEM) into the field and get it to an appropriate level of vacuum, it would not be worth the effort to decontaminate it. However, digital microscopes for scientific applications are cheap, portable, and highly effective for field use. At a low cost, 200× magnification, 5+ megapixel digital microscopes designed for scientific use can be bought from commercial and scientific suppliers, with an example of a result captured in the field shown in Figure 3. Some come with a liquid crystal display (LCD), negating the need for laptops or external screens to be brought into the field. At this low price range, any microscopes that become contaminated or otherwise damaged during their duties can be disposed of with minimal loss.

Like photogrammetry, optical microscopy is not a resource-heavy technique. It requires minimal training, and micrometre calibration stages give images taken in the field precise and accurate measurements. Digital microscopes produce true-colour images, allowing a wide array of data to be taken in the field. These images can be quantitatively analysed in the lab postfieldwork, giving them broad scientific value beyond illustrative and metrological purposes [67].

Figure 3. Optical microscopy of ion exchange resin spill, observed in situ. To the naked eye, this was indistinguishable from finely ground ore sand, but its real nature became clear under the microscope (see Figure 2) which also enabled a particle size distribution and morphology to be characterised.

Limitations of Optical Microscopy

Light conditions inside facilities (some of which have no electricity and therefore no built-in lighting, easily contaminated parts (especially lenses), the use of calibrated scale bars that can be bulky, requiring an LED screen or laptop into the area, are all limitations to the use of optical microscopes in the field.

4.6. Photogrammetric Survey

Photogrammetry is the trigonometric stitching together of photographs taken from one or multiple sources to recreate a 2D or 3D environment in digital form (coordinates and derived geometric elements) or graphical form (images, drawings and maps) [68]. Multiple research fields are involved, including optics, projective geometry and cloud computing. One of the many benefits of photogrammetric models is that they allow 2D photo stills to be compiled into a 3D rendering, sometimes referred to as a digital twin, which enables researchers, workers, and other interested parties to easily visualise the layout and conditions within a contaminated building as a virtual walkthrough and without requiring PPE or exposure to a high radiation dose. It also allows precise measurements to be made of structures that would otherwise be difficult to reach or access. Anything that can be photographed can be rendered into a photogrammetric representation, from sand grains to landscapes.

Photogrammetry straddles the line between qualitative and quantitative data. Although its primary use is in recreating environments for users to explore safely, it can also determine some length measurements such as the height and width of infrastructure features such as tanks, furniture, and walkways. These are not as accurate as using a properly calibrated measurement tool but can be very helpful in obtaining approximate dimensions of internal and external structures. These digital twins, and the photographic records from which they are constructed, can be used to compare changes such as structural degradation, damage, and destabilisation over time in a quantifiable manner. They also enable the study of details which may have been missed in the field due to time constraints, poor lighting, or investigative priority decisions.

This has broad implications for inspection protocol and informed decision-making within hazardous facilities.

4.6.1. Analysis Principle

Image coordinates from *n* views, *n* exterior orientations, 3D coordinates, inner orientations of the camera, camera models, and other further observations such as measurements and pattern recognition all combine to produce a photogrammetric model.

Image coordinates define the locations of the object points' images on the camera. The exterior orientation of a camera defines both its location in space and its view direction. The 3D coordinates define the locations of object points within the 3D space.

The inner orientation defines the geometric parameters of the imaging process. This is primarily the focal length and lens distortions.

These are especially seen in wide-angle lenses such as those used by consumer-grade "action cameras", used commonly in the field due to their rugged build and ease of use, but every camera lens will have flaws.

These sets of points are then processed within histogram software and analysed for similarities and combined into a point cloud showing the likely position of all photographs. This is then further analysed to produce a wireframe mesh, as seen in Figure 4. This mesh is then turned into a solid model, as seen in Figure 5. The texture is then mapped onto this mesh to create an image that a viewer can interpret, as seen in Figure 6.

Further additional observations play an essential role in adding value to photogrammetric models. For example, scale bars, annotated hazards, overlain radiation maps and marked sample locations vastly increase the amount of information shown in a model.

Photogrammetry is an excellent technique for use in the field because it requires no specialist equipment. Any modern smartphone with a camera can capture photos suitable for photogrammetry, as can any basic point-and-shoot/compact digital camera, bridge camera, digital single-lens reflex (DSLR) camera or image-capable robot or drone. It is easy for researchers to carry or pilot at least one of these devices into all but the most secure or sensitive environments, and minimal training is required to take photos in an efficient grid. Multiple heights, orientations, and positions add complexity and depth to a survey. Still, valuable results can be produced even when taken from a singular position at multiple heights and angles, such as on a gantry or accessway.

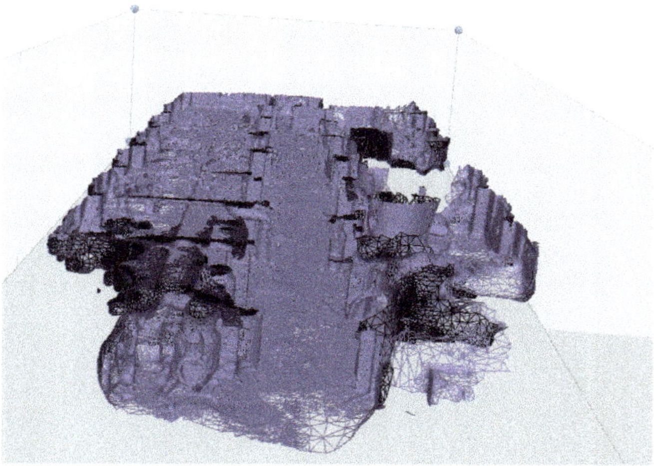

Figure 4. A cross-section of the wireframe photogrammetry survey of floor two of one of the legacy ore processing buildings, PChP.

Figure 5. A cross-section of the solid model photogrammetry survey of floor two of one of the legacy ore processing buildings, PChP.

Figure 6. A cross-section of the textured photogrammetry survey of floor two of one of the legacy ore processing buildings, PChP.

Photogrammetric models can be used to create animated "flythroughs" of hazardous sites, used to train new personnel on the layout and dangers ahead of them before undertaking an entry, thereby improving safety and reducing the amount of time personnel need to spend inside and therefore the dose they are exposed to.

Minor specialist training is required to establish a sampling grid method that increases the utility of photographs taken. Once trained, any member of a field-deployed nuclear industrial archaeology team wearing sufficient PPE and personal dosimetry can be entrusted with creating a photogrammetric record. The decision to carry it out should only be taken when dose rates are relatively well understood. It is a form of data collection that can be undertaken between other resource-heavy investigative methods—for example, while waiting for batteries to charge between robot deployments. Operator fatigue is minimal, although can become an issue during time-consuming investigations that involve large, multistorey buildings with complex internal features.

Photogrammetry can also yield evidence that was missed in the field due to poor lighting, time constraints, and other confounding factors. For example, Figure 7 shows a poster that was spotted during an examination of the 3D reconstruction of the site. This allowed a photo to be found retrospectively, which had sufficient clarity for some translation.

Figure 7. A poster which was spotted after the fact, during an examination of the 3D reconstruction of the site. The image has been enhanced for clarity.

4.6.2. Limitations of Photogrammetric Surveys

Barriers to accurate photogrammetric processing include inconsistencies between frames. Within well-lit scenes, this can come from reflective surfaces and mobile surfaces (such as vegetation moving in the wind). Surfaces with limited texture cause reconstruction problems unless there are other observable features in the photographs. Within poorly lit scenes, low light, moving shadows, and variable lighting produced by head torches and camera flashes all have an impact. Camera blur, image exposure, and other internal inconsistencies also prevent appropriate matching on the ability of photogrammetry software to match images and place them accurately within the 3D model. The algorithms used can average out occasional photographs containing team members. Ideally, this will be eliminated by using clear communication and taking surveys while other members of the team are working elsewhere.

To create as static an environment as possible, floors were photographed by two analysts "back-to-back" around noon whenever sunlight can reach the scene. This ensured that lighting was as full and consistent as possible, facilitating the use of two different camera models. This method also enabled analysts to stay out of photographs.

The point cloud in photogrammetry derives from extra processing done after collection. In contrast, a LiDAR point cloud is a direct product, available in seconds when the position of the sensor is understood.

4.7. LiDAR Mapping

Using 3D scanning light detection and ranging (LiDAR) is increasingly commonplace in providing rapid measurements of complex 3D spaces such as building sites or civil engineering projects. In a nuclear industrial archaeology context, LiDAR provides an enhanced capability over photogrammetry in terms of accuracy (mm versus cm) and the ability to operate in very low (or nonexistent) external light conditions.

Photogrammetry can generate high-quality 3D maps, but one limiting factor is the computing power requirements for stitching together the images. This can be a time-consuming process and is difficult to perform accurately in real time; it is normally done in post-processing. A LiDAR unit uses the reflection of lasers to find the distance to the sensor's surroundings and can use this data to perform simultaneous localisation and mapping (SLAM), where successive scans are compared to work out the location of the unit in space and build up a map. In this way, the position of radiation measurements can be ascertained in real time. A LiDAR unit can be carried by a human operator, mounted on an unmanned aerial vehicle (UAV) [69] or unmanned ground vehicle (UGV) [70], mounted on a robotic manipulator, or mounted to a human operator, and information from other coincident or adjoined sensors, e.g., gamma spectrometers, can be position stamped by the LiDAR [70]. An example of an UGV is shown in Figure 8.

If such a sensor is not used, the location of measurements would need to be manually recorded, increasing burdens on the operator in a hazardous environment and potential error. As the LiDAR mapping requires no input from the operator, they are free to perform other tasks if the LiDAR unit can be mounted appropriately on an operator or UAV/UGV, with the position of each task being accurately recorded.

Two-dimensional LiDARs are more cost effective, but a 3D LiDAR, which consists of multiple sets of lasers scanning at different angles, can generate a 3D point cloud of a space. This point cloud can be used to take accurate measurements, and sensor measurements can be expressed within it. As this is a real-time approach, it can be used to influence strategy and decision making immediately in the field, as compared to photogrammetry, which requires some postprocessing.

A LiDAR mapping unit can also be uploaded with existing maps, which the unit can then use to attempt to localise itself within its current environment. This means that measurements taken at multiple different times can share the same reference frame, and thus be directly compared to each other. Photogrammetry models will not have a consistent reference frame, meaning any comparisons must be done manually.

Figure 8. Deploying a remote-controlled wheeled ground vehicle equipped with gamma spectrometer and neutron detector in an environment with radioactive contamination.

4.8. Unoccupied Aerial Vehicles and Unoccupied Ground Vehicles

Consumer drones are relatively cheap compared to some scientific instruments and can be modified for use in the field. Unoccupied vehicles, whether they are aerial or ground based, provide the possibility of delivering sensors into areas that would be unfavourable or hazardous to send a human operator. Depending on the variant of unoccupied vehicle chosen and its associated payload capacity, multiple sensor packages can be included to collect multiple data types simultaneously, allowing for real-time data fusion and near-immediate interpretation of the results. This can include sensors that are unsafe for human operators. Remote surveys and inspections carried out by aerial or ground-based robotic systems have become significantly more prevalent since 2005, partly due to the significant reduction in the cost of the robotic platforms themselves [71].

UAVs provide a terrain-independent method of collecting standoff measurements, allowing operators to access areas that would otherwise be inaccessible. Three primary types of UAV exist, consisting of multirotor, single-rotor, and fixed-wing, although a further subdivision can be made for vertical takeoff and landing (VTOL) vehicles that takeoff and land like a multirotor, but operate as a fixed wing in flight [72]. Most UAVs operated within a commercial environment are of the multirotor type, which is generally the easiest to pilot and offers the most stable platform for data collection, and so the following section will primarily focus on these systems [72].

There exists a large variation in total size and weight within the commercial offerings of UAVs on the market. These can range from as little as 250 g to more than 20 kg. However, the operation of vehicles with a takeoff weight (including batteries and payloads) above 20–25 kg are often more heavily restricted within aviation regulations throughout the world, making their operation more heavily regulated in many environments. Commercial UAVs with an overall mass at the lower end of this scale are generally equipped with a single, integrated camera as a payload, allowing for the collection of visual information through still images or video. As described above, these can be used for simple visual inspection or for being processed into photorealistic 3D models through photogrammetry [69]. Generally, as the UAV increases in size or mass, the weight of the payload that it can also carry increases. Larger UAVs, therefore, provide an increased capacity to collect data from multiple payloads simultaneously [73]. Common combinations within radiological and wider nuclear UAV surveys include the combination of a visual camera, some form of gamma spectrometer system and, often, a LiDAR scanning system [73–76].

UGVs, while more limited than UAVs in the areas they can cover, can detect low-level or diffuse sources that would have been invisible to a UAV due to the standoff distance and the inverse square law [70].

Although a large focus in nuclear industrial archaeology is placed on identifying the chemical or radiological nature of the material within a site, acquiring a contextual visual representation of the site and ensuring the continuing safety of operators should be one of the first objectives for operating workforces. UAVs and UGVs are crucial in this regard, as they can more easily reach and operate in locations that are physically problematic or dangerous for humans, and at the same time, offer greater repeatability and measurement reliability. When coincident radiometric and 3D data is recorded by using adequate methodologies, there is also an opportunity to postprocess the recorded data by using inversion techniques, which often utilise Kaczmarz methods [77] to provide a localisation calculation of possible "hot" emitters in the survey zone. Although this capability is currently performed in postprocessing, it is expected that near real-time localisations may be achieved in the near future in concert with advances in the computing power of microprocessors used in handheld field instrumentation.

Passing through airport security and customs is more straightforward when there is a basic understanding of the transported equipment, which is nowhere more acutely felt than in the recent rise of consumer-grade drones. Ten years ago, transporting a drone battery involved a discussion with security personnel regarding drones and why they need to be transported; these days, drone batteries and parts in hand luggage do not provoke much interest.

4.9. Discussion

In this work, we have delineated a toolkit of portable analysis technologies for NIA applications. As with any toolkit, tools can be added and taken out depending on the application scenario. The expectation is that the technologies we have outlined will, as a function of time, evolve to increase their effectiveness and useability, and decrease their size, energy consumption, and mass.

We expect that more portable analytical tools will become available within the next decade, for example, Raman, LIBS, and standoff alpha imagers.

In this work, we have put a series of NIA tools into a case study context. The buildings assessed in this work presented a series of challenges and hazards, ranging from the obvious (holes in the floor) to the subtle (doses of up to 8.2 mSv h^{-1}). One of the reasons that this investigation is important and relevant is that many buildings on the PChP site are still in use, with complex ownership status leading to previously unknown issues being uncovered in recent years. For example, the designated outdoor smoking area for one of the factories was for many years situated next to a highly active, abandoned uranium leachate storage tank from the building next door. As a direct result of the NIA investigation, this situation has been resolved by management and the smoking area moved.

For this work at PChP, data from individual instruments was recorded and then later combined in postmortem analysis. This meant that there was a delay in the fusion of the data, and therefore a delay in fully understanding the disposition of materials and hazards within the facility. It is expected that within this decade, technological advancements will enable a combination of data streams in near real time and with advances in data transmission technology (5G), the ability to relay large sets of information to a central repository as it is collected. This will ensure that all data recovered is protected and archived. With advances in computing power, techniques such as photogrammetry will be compiled and outputted onsite in minutes, allowing more advanced visualisations of 3D datasets in real time. Advancements in real-time data visualisation will enhance the relay and explanation of important arising information to site stakeholders, decision makers, and workers operating in these environments. Importantly, it will enhance communal understanding of the whereabouts and significance of any residual radiological risks in a visually intuitive manner, which is independent of written or spoken language.

Over the next decade, with the continued sensorisation of the industrial world, it is expected that the field of NIA stands to continually benefit from advances that will be driven by technology development within other industrial fields of application. Equally, NIA methodologies could be beneficially applied to other relevant industrial areas, for example verification of NORM contamination of oil and gas/mining infrastructure.

Technological advances validated in the field of NIA will have downstream benefits for nuclear forensics applications of the same tools. Accordingly, as NIA technology advances, so does NF.

At present, there are many nuclear facilities worldwide reaching the end of their operational lifetimes, and many countries are planning substantial new nuclear energy infrastructure to contribute toward net zero carbon emissions targets by the middle of the century. This means that the field of NIA will apply to an increasing number of sites. These sites are known to be diverse and contain unique challenges and materials, as no two sites are the same. It is our assertion that by understanding and cataloguing the nuclear challenges of the past, it is possible to avoid the repetition of these issues in the development of this planned nuclear infrastructure. At the same time, NIA enables objective assessment of the historic value of nuclear industrial heritage, and NIA research provides the scientific foundations to initiate often much-needed protection and conservation work.

5. Conclusions

In this paper, we have established that NIA is a distinct and separate branch of science from nuclear forensics, although the two share a common history and set of techniques. This distinction has clear benefits for practitioners and clients. A toolkit of in situ analysis techniques has been presented for NIA applications, with the selection based on prior experience from field trips and industrial investigations.

A thorough technical plan is key to any scientific investigation, especially those under time pressure and weight limits. To that end, field NIA aims to provide a systematic, semistandardised approach to data collection within a complex radiological environment.

This paper defines a methodology for combined application of this toolkit of analysis techniques, which if applied to other sites in a standardised manner, will allow comparisons to be made between investigations.

Author Contributions: Conceptualisation, E.I.H. and Y.V.; methodology, E.I.H., Y.V. and T.B.S.; validation, E.I.H., Y.V. and T.B.S.; investigation; E.I.H., Y.V., D.T.C. and T.B.S.; data curation, E.I.H.; writing, E.I.H., Y.V. D.T.C. and T.B.S.; visualisation, E.I.H.; illustrations: E.I.H. and Y.V.; draft review: E.I.H., Y.V., D.T.C., T.M., M.H. and T.B.S.; project administration: T.B.S.; funding acquisition: T.M., M.H. and T.B.S. All authors have read and agreed to the published version of the manuscript.

Funding: D.T.C. acknowledges funding from UK Research and Innovation (UKRI) through the National Center for Nuclear Robotics (NCNR) (EP/R02572X/1) and from National Nuclear Laboratory (NNL) through the R&D Decontamination Science Core Science Theme and NNL Post-doctorate Programme. Y.V. acknowledges funding from UKRI though the NCNR and Robotics and Artificial Intelligence for Nuclear Applications (RAIN) research programs.

Data Availability Statement: The data presented in this study are available on request from the corresponding author. The data are not publicly available due to the sensitive nature of nuclear facilities where data was collected.

Acknowledgments: The authors wish to thank the Ukrainian Government and the management and staff of the Prydniprovksy Chemical Plant in Kamianske for their hospitality, and generously provided access to nuclear facilities on-site. Additionally, the authors thank Alex for the logistic support during field deployments in difficult circumstances. The authors thank Matthew Ryan Tucker of the University of Bristol Interface Analysis Centre for his input into the discussion on robot mounted LiDAR mapping. The authors also extend their gratitude to Sofia Leadbetter of the National Nuclear User Facility for Hot Robotics (NNUF-HR) for the administrative support and access to research tools such as the Vanta pXRF used in this study.

Conflicts of Interest: The authors declare no conflict of interest. The funders had no role in the design of the study; in the collection, analyses, or interpretation of data; in the writing of the manuscript; or in the decision to publish the results.

Abbreviations

The following abbreviations are used in this manuscript:

2D	2-dimensional
3D	3-dimensional
DSLR	Digital single lens reflex camera
EDXRF	Energy dispersive X-ray fluorescence spectroscopy
GM	Geiger-Müller
HPGe	High-purity germanium detector
HRGS	High-resolution gamma-ray spectroscopy
IAEA	International Atomic Energy Agency
ICP-MS	Inductively coupled plasma mass spectroscopy
ITDB	Incident and Trafficking Database
LCD	Liquid crystal display
LE	Light elements
LIBS	Laser induced breakdown spectroscopy
LiDAR	Light detection and ranging
NAA	Neutron activation analysis
NF	Nuclear forensics
NIA	Nuclear industrial archaeology
NORM	Naturally occurring radioactive materials
PChP	Pryniprovsky Chemical Plant
PED	Personal electronic dosimeter
PPE	Personal protection equipment
pXRF	Portable X-ray fluorescence spectroscopy
REE	Rare earth element
SEM	Scanning electron microscope
SLAM	Simultaneous localisation and mapping

UAV	Unoccupied aerial vehicle
UGV	Unoccupied ground vehicle
VTOL	Vertical take-off and landing
XRD	X-ray diffraction
XRF	X-ray fluorescence spectroscopy

References

1. Mayer, K.; Wallenius, M.; Ray, I. Nuclear forensics–a methodology providing clues on the origin of illicitly trafficked nuclear materials. *Analyst* **2005**, *130*, 433–441. [CrossRef] [PubMed]
2. Keegan, E.; Kristo, M.J.; Toole, K.; Kips, R.; Young, E. Nuclear Forensics: Scientific Analysis Supporting Law Enforcement and Nuclear Security Investigations. *Anal. Chem.* **2016**, *88*, 1496–1505. [CrossRef] [PubMed]
3. Schmid, A.P.; Spencer-Smith, C. Illicit Radiological and Nuclear Trafficking, Smuggling and Security Incidents in the Black Sea Region since the Fall of the Iron Curtain—An Open Source Inventory. *Perspect. Terror.* **2012**, *6*, 117–157.
4. Potter, W.C. Nuclear Smuggling from the Former Soviet Union 1. In *Nuclear Energy and Security in the Former Soviet Union*; Routledge: Oxfordshire, UK , 2018; pp. 139–165.
5. Niemeyer, S.; Koch, L. The historical evolution of nuclear forensics: A technical viewpoint. In Proceedings of the International Conference on Advances in Nuclear Forensics, Vienna, Austria, 7 July 2014.
6. International Atomic Energy Agency. Incident and Trafficking Database. 1995. Available online: https://www.iaea.org/sites/default/files/22/01/itdb-factsheet.pdf (accessed on 19 January 2023).
7. Kristo, M.J. Chapter 13—Nuclear forensics. In *Handbook of Radioactivity Analysis: Volume 2*, 4th ed.; L'Annunziata, M.F., Ed.; Academic Press: Cambridge, MA, USA, 2020; pp. 921–951. [CrossRef]
8. Grant, P.; Moody, K.; Hutcheon, I.; Phinney, D.; Whipple, R.; Haas, J.; Alcaraz, A.; Andrews, J.; Klunder, G.; Russo, R.; et al. Nuclear forensics in law enforcement applications. *J. Radioanal. Nucl. Chem.* **1998**, *235*, 129–132. [CrossRef]
9. Leggitt, J.; Inn, K.; Goldberg, S.; Essex, R.; LaMont, S.; Chase, S. Nuclear forensics—metrological basis for legal defensibility. *J. Radioanal. Nucl. Chem.* **2009**, *282*, 997–1001. [CrossRef]
10. Wallenius, M.; Mayer, K. Age determination of plutonium material in nuclear forensics by thermal ionisation mass spectrometry. *Fresenius' J. Anal. Chem.* **2000**, *366*, 234–238. [CrossRef]
11. Fedchenko, V. *The New Nuclear Forensics: Analysis of Nuclear Materials for Security Purposes*; Oxford University Press: Oxford, UK, 2015.
12. Nitrean, A.I.; Stebelkov, V.A. Main Problems and Development Trends in Nuclear Forensics. *Russ. J. Leg. Stud.* **2022**, *9*, 97–106. [CrossRef]
13. International Atomic Energy Agency. *Nuclear Forensics Support, Reference Manual*; IAEA: Vienna, Austria, 2006; p. 67.
14. Choi, S.U.; Han, S.C.; Yun, J.I. In situ detection of neodymium isotopes using tunable diode laser absorption spectroscopy for nuclear forensic analysis. *J. Anal. At. Spectrom.* **2023**, *38*, 166–173. [CrossRef]
15. Weller, A. Environmental Impact of Noble Metal Radionuclides and their use in Nuclear Forensics. Ph.D. Thesis, Leibniz Universität Hannover, Hannover, Germany, 2021. [CrossRef]
16. Vesterlund, A.; Canaday, J.; Chamberlain, D.B.; Curry, M.R.; Sandström, B.; Schnaars, D.D.; Ramebäck, H. National nuclear forensics libraries: A case study on benefits and possibilities for identification of sealed radioactive sources. *J. Radioanal. Nucl. Chem.* **2022**, *331*, 639–643. [CrossRef]
17. Palmer, M.; Neaverson, P. *Industrial Archaeology Principles and Practice*; Routledge: Oxfordshire, UK, 1998.
18. Shaw, R.J. Activation analysis at the UMC Laboratory for nuclear archaeology. *Newsl. Lithic Technol.* **1974**, *3*, 61–64.
19. Price, T.D.; Chappell, S.; Ives, D.J. Thermal alteration in Mesolithic assemblages. In *Proceedings of the Prehistoric Society*; Cambridge University Press: Cambridge, MA, USA, 1982; Volume 48, pp. 467–485.
20. Neff, H. The present state of nuclear archaeology. *Trans. Am. Nucl. Soc.* **1994**, *71*. Available online: https://www.osti.gov/biblio/88951 (accessed on 19 January 2023).
21. Neff, H.; Glascock, M.D. The state of nuclear archaeology in North America. *J. Radioanal. Nucl. Chem.* **1995**, *196*, 275–286. [CrossRef]
22. Auxier, J.A. A. Physical Dose Estimates for A-bomb Survivors—Studies at Oak Ridge, USA. *J. Radiat. Res.* **1975**, *16*, 1–11. [CrossRef]
23. Loewe, W. *Initial Radiation Dosimetry at Hiroshima and Nagasaki*; Technical Report; Lawrence Livermore National Lab: Livermore, CA, USA, 1983.
24. Fetter, S. Nuclear archaeology: Verifying declarations of fissile-material production. *Sci. Glob. Secur.* **1993**, *3*, 237–259. [CrossRef]
25. Gaines, L.L. *Start II: Thinking One Move Ahead*; Technical Report, Arms Control and Tready Verification Program; Argonne National Lab: Lemont, IL, USA, 1991.
26. Kalinowski, M.B.; Liebert, W.; Scheffran, J. Beyond technical verification: Transparency, verification, and preventive control for the Nuclear Weapons Convention. In Proceedings of the NPT PrepCom Geneva, New York, NY, USA, 27 April–8 May 1998; p. 4.
27. Dudar, T. Uranium Mining and Milling Facilities Legacy Sites: Ukraine Case Study. *Environ. Probl.* **2019**, *4*, 212–218. [CrossRef]
28. Ivanchenko, A.; Yelatontsev, D.; Savenkov, A.; Nazarenko, O.; Kundirenko, V. Obtaining of complex mineral fertilizer by phosphogypsum conversion with ammonium nitrate. *Chem. J. Mold.* **2021**, *1*, 18–23. [CrossRef]

29. Mukhachev, A.; Nefedov, V.; Yelatontsev, D. Physicochemical foundations of the pioneering technology for the carbonaceous uranium ore processing. *IOP Conf. Ser. Earth Environ. Sci.* **2022**, *970*, 012027. [CrossRef]
30. Hughes, J.; Watson, S.; Jones, A.; Oatway, W. Review of the radiation exposure of the UK population. *J. Radiol. Prot.* **2005**, *25*, 493. [CrossRef]
31. Voitsekhovych, O.; Lavrova, T. Remediation Strategy for Former PChP Uranium Production Site in Ukraine - 18217. In Proceedings of the Waste Management Symposia, Phoenix, AZ, USA, 18–22 March 2018.
32. Drielak, S.C. *Hot Zone Forensics: Chemical, Biological, and Radiological Evidence Collection*; Charles C Thomas Publisher: St. Springfield, IL, USA, 2004.
33. McDiarmid, M.A.; Gaitens, J.M.; Squibb, K.S. Uranium and Thorium. In *Patty's Toxicology*; John Wiley & Sons: Hoboken, NJ, USA, 2012; pp. 769–816. [CrossRef]
34. Bhatt, B.; Angeyo, K.H.; Dehayem-Kamadjeu, A. LIBS development methodology for forensic nuclear materials analysis. *Anal. Methods* **2018**, *10*, 791–798. [CrossRef]
35. Moody, K.J.; Grant, P.M.; Hutcheon, I.D. *Nuclear Forensic Analysis*, 2nd ed.; CRC Press: Boca Raton, FL, USA, 2015.
36. Manard, B.T.; Wylie, E.M.; Willson, S.P. Analysis of Rare Earth Elements in Uranium Using Handheld Laser-Induced Breakdown Spectroscopy (HH LIBS). *Appl. Spectrosc.* **2018**, *72*, 1653–1660. [CrossRef]
37. Shampo, M.A.; Kyle, R.A.; Steensma, D.P. Hans Geiger—German Physicist and the Geiger Counter. *Mayo Clin. Proc.* **2011**, *86*, e54. [CrossRef] [PubMed]
38. Papp, Z.; Daróczy, S. Measurement of radon decay products and thoron decay products in air by beta counting using end-window Geiger-Muller counter. *Health Phys.* **1997**, *72*, 601–610. [CrossRef] [PubMed]
39. Ivasheva, A.Y.; Borisov, V.S.; Semenikhin, P.V. Adjusting of the detection efficiency of Geiger-Muller counters in the energy range of gamma radiation from 50 to 3000 keV. *J. Physics Conf. Ser.* **2018**, *1135*, 012042. [CrossRef]
40. Arvela, H.; Markkanen, M.; Lemmelä, H. Mobile Survey of Environmental Gamma Radiation and Fall-Out Levels in Finland After the Chernobyl Accident. *Radiat. Prot. Dosim.* **1990**, *32*, 177–184. [CrossRef]
41. Habrman, P. Directional Geiger-Müller Detector with Improved Response to Gamma Radiation. *J. Instrum.* **2019**, *14*, P09018. [CrossRef]
42. McDicken, W.N. A position sensitive Geiger counter. *Nucl. Instruments Methods* **1967**, *54*, 157–158. [CrossRef]
43. Maulany, G.J.; Manggau, F.X.; Jayadi, J.; Waremra, R.S.; Fenanlampir, C.A. Radiation Detection of Alfa, Beta, and Gamma Rays With Geiger Muller Detector. *Int. J. Mech. Eng. Technol.* **2018**, *9*, 21–27.
44. Taamté, J.M.; Saïdou; Bodo, B.; Noubé, M.K.; Signing, V.R.F.; Younui, S.Y.N.; Yerima, A.H.; Iamsumang, C.; Massey, C. Low-cost wireless smart device design for radiation protection and nuclear security. *J. Instrum.* **2021**, *16*, T06011. [CrossRef]
45. Becker, E.M.; Farsoni, A.T. Wireless, low-cost, FPGA-based miniature gamma ray spectrometer. *Nucl. Instrum. Methods Phys. Res. Sect. A Accel. Spectrometers Detect. Assoc. Equip.* **2014**, *761*, 99–104. [CrossRef]
46. Neher, H.V.; Pickering, W.H. Modified High Speed Geiger Counter Circuit. *Phys. Rev.* **1938**, *53*, 316. [CrossRef]
47. Eichholz, G.G. *Principles of Nuclear Radiation Detection*; CRC Press: Boca Raton, FL, USA, 2018.
48. Silva, A.; Dinis, M.L. Assessment of the Indoor Gamma Dose Rates in 15 Portuguese Thermal Spas. In *Occupational and Environmental Safety and Health*; Springer: Berlin/Heidelberg, Germany, 2019; pp. 221–230. [CrossRef]
49. Martin, A.; Harbison, S.A.; Beach, K.; Cole, P. *An Introduction to Radiation Protection*; CRC Press: Hoboken, NJ, USA, 2018.
50. Rutherford, E.; Andrade, E.d. The wave-length of the soft γ rays from Radium B. *Lond. Edinb. Dublin Philos. Mag. J. Sci.* **1914**, *27*, 854–868. [CrossRef]
51. Verbelen, Y.; Megson-Smith, D.; Russell-Pavier, F.; Martin, P.; Connor, D.; Tucker, M.R.; Scott, T.B. A Flexible Power Delivery System for Remote Nuclear Inspection Instruments. In Proceedings of the 2022 8th International Conference on Automation, Robotics and Applications (ICARA), Prague, Czech Republic, 18–20 February 2022; pp. 170–175. [CrossRef]
52. Mirion Technologies. SPIR-Ace: Radio Isotope Identification Device (RIID). 2020. Available online: https://www.southernscientific.co.uk/products-by-manufacturer/mirion-technologies/spir-ace (accessed on 19 January 2023).
53. Verbelen, Y.; Martin, P.G.; Ahmad, K.; Kaluvan, S.; Scott, T.B. Miniaturised Low-Cost Gamma Scanning Platform for Contamination Identification, Localisation and Characterisation: A New Instrument in the Decommissioning Toolkit. *Sensors* **2021**, *21*, 2884. [CrossRef]
54. Blake, D.; Vaniman, D.; Achilles, C.; Anderson, R.; Bish, D.; Bristow, T.; Chen, C.; Chipera, S.; Crisp, J.; Des Marais, D.; et al. Characterization and calibration of the CheMin mineralogical instrument on Mars Science Laboratory. *Space Sci. Rev.* **2012**, *170*, 341–399. [CrossRef]
55. Rampe, E.; Blake, D.; Sarrazin, P.; Bristow, T.; Gailhanou, M.; Lafuente, B.; Tu, V.; Zacny, K.; Downs, R. CheMinX: A Next Generation XRD/XRF for Quantitative Mineralogy and Geochemistry on Mars. 2020. Available online: https://ntrs.nasa.gov/api/citations/20205005011/downloads/Rampe_etal_CheMinX_DS_white_paper_2020.pdf (accessed on 19 January 2023).
56. Izumoto, Y.; Fukutsu, K.; Takamura, K.; Sakai, Y.; Oguri, Y.; Yoshii, H. Rapid detection of plutonium contamination with and without uranium contamination in wounds by x-ray fluorescence. *J. Radiol. Prot.* **2020**, *40*, 692. [CrossRef]
57. Worley, C.G. Analysis of nuclear materials by energy dispersive X-ray fluorescence and spectral effects of alpha decay. *J. Radioanal. Nucl. Chem.* **2009**, *282*, 539–542. [CrossRef]
58. Brand, C. History of portable XRF and radiation safety issues. Are we keeping up with the times? In Proceedings of the Australasian Radiation Protection Society, Canberra, Australia, 6–9 October 2015; p. 102.

59. Thomsen, V.; Schatzlein, D. Advances in Field-Portable XRF. *Spectroscopy* **2002**, *17*, 14–21.
60. Fry, K.L.; Wheeler, C.A.; Gillings, M.M.; Flegal, A.R.; Taylor, M.P. Anthropogenic contamination of residential environments from smelter As, Cu and Pb emissions: Implications for human health. *Environ. Pollut.* **2020**, *262*, 114235. [CrossRef]
61. Brouwer, P. *Theory of XRF*, 3rd ed.; PANalytical BV: Almelo, The Netherlands, 2010.
62. Onoyama, M.; Nakata, M.; Hirose, Y.; Nakagawa, Y. Development of a high performance type 304L stainless steel for nuclear fuel reprocessing facilities. In Proceedings of the Third International Conference on Nuclear Fuel Reprocessing and Waste Management, RECOD'91, Sendai, Japan, 14–18 April 1991; pp. 1066–1071.
63. Reilly, D.; Ensslin, N.; Smith, H.J.; Kreiner, S. *Passive Nondestructive Assay of Nuclear Materials*; Los Alamos National Laboratory: Los Alamos, NM, USA, 1991.
64. Higginson, M.; Dawkins, B.; Shaw, T.; Taylor, F.; Ingman, L.; Dunn, S.; Thompson, P.; Kaye, P. Development of assay, isotopic and trace actinide measurements on solid plutonium and uranium samples by high resolution gamma spectrometry. *J. Radioanal. Nucl. Chem.* **2021**, *330*, 901–911. [CrossRef]
65. McDowell, D.; Thurston, A. Screening of ceramic and leaded contaminants in glass recycling streams via hardheld X-ray fluorescence (HHXRF) analyzers. In Proceedings of the 15th Asia Pacific Conference for Non-Destructive Testing (APCNDT2017), Singapore, 13–17 November 2017; p. 11.
66. Fisher Scientific. Decon 90. 2018. Available online: https://decon.co.uk/wp-content/uploads/2018/11/Decon-90.pdf (accessed on 19 January 2023).
67. Wolf, D.E.; Samarasekera, C.; Swedlow, J.R. Chapter 14—Quantitative Analysis of Digital Microscope Images. In *Digital Microscopy*; Methods in Cell Biology; Sluder, G., Wolf, D.E., Eds.; Academic Press: Cambridge, MA, USA, 2013; Volume 114, pp. 337–367. [CrossRef]
68. Luhmann, T.; Robson, S.; Kyle, S. *Close Range Photogrammetry: Principles, Techniques and Applications*; Whittles Publishing: Dunbeath, UK, 2006; pp. 2–8.
69. Martin, P.; Scott, T. Glimpsing Chernobyl's hidden hotspots. *Phys. World* **2019**, *32*, 29. [CrossRef]
70. Banos, A.; Verbelen, Y.; Kaluvan, S.; Hutson, C.; Tucker, M.R.; Scott, T.B. Hexapod robotic system for indoor neutron and gamma radiation mapping and inspection. In Proceedings of the 2022 25th International Symposium on Design and Diagnostics of Electronic Circuits and Systems (DDECS), Prague, Czech Republic, 6–8 April 2022; pp. 48–53. [CrossRef]
71. Tsitsimpelis, I.; Taylor, C.J.; Lennox, B.; Joyce, M.J. A review of ground-based robotic systems for the characterization of nuclear environments. *Prog. Nucl. Energy* **2019**, *111*, 109–124. [CrossRef]
72. Connor, D.; Martin, P.G.; Scott, T.B. Airborne radiation mapping: overview and application of current and future aerial systems. *Int. J. Remote. Sens.* **2016**, *37*, 5953–5987. [CrossRef]
73. Connor, D.T.; Wood, K.; Martin, P.G.; Goren, S.; Megson-Smith, D.; Verbelen, Y.; Chyzhevskyi, I.; Kirieiev, S.; Smith, N.T.; Richardson, T.; et al. Radiological mapping of post-disaster nuclear environments using fixed-wing unmanned aerial systems: A study from Chornobyl. *Front. Robot.* **2020**, *6*, 149. [CrossRef] [PubMed]
74. Hellfeld, D.; Barton, P.; Gunter, D.; Haefner, A.; Mihailescu, L.; Vetter, K. Real-time free-moving active coded mask 3D gamma-ray imaging. *IEEE Trans. Nucl. Sci.* **2019**, *66*, 2252–2260. [CrossRef]
75. Vetter, K.; Barnowksi, R.; Haefner, A.; Joshi, T.H.; Pavlovsky, R.; Quiter, B.J. Gamma-Ray imaging for nuclear security and safety: Towards 3D gamma-ray vision. *Nucl. Instrum. Methods Phys. Res. Sect. A Accel. Spectrometers Detect. Assoc. Equip.* **2018**, *878*, 159–168. [CrossRef]
76. Haefner, A.; Barnowski, R.; Luke, P.; Amman, M.; Vetter, K. Handheld real-time volumetric 3D gamma-ray imaging. *Nucl. Instrum. Methods Phys. Res. Sect. A Accel. Spectrometers Detect. Assoc. Equip.* **2017**, *857*, 42–49. [CrossRef]
77. White, S.R.; Wood, K.T.; Martin, P.G.; Connor, D.T.; Scott, T.B.; Megson-Smith, D.A. Radioactive Source Localisation via Projective Linear Reconstruction. *Sensors* **2021**, *21*, 807. [CrossRef]

sustainability

MDPI

Article

Influence of Radioactive Sludge Content on Vitrification of High-Level Liquid Waste

Shengheng Tan [1,*], Jiong Chang [1], Xiao Liu [1], Shikuan Sun [2], Liang Xian [1] and Shengdong Zhang [1,*]

1 Department of Radiochemistry, China Institute of Atomic Energy, Beijing 102413, China
2 School of Materials Science and Energy Engineering, Foshan University, Foshan 528000, China
* Correspondence: tanshengheng@ciae.ac.cn (S.T.); zhangsd@ciae.ac.cn (S.Z.)

Abstract: The radioactive sludges formed at the bottom of high-level liquid waste (HLW) storage tanks pose challenges when the HLWs are vitrified. This study aims to determine the influence of the sludge content (enriched in Na_2O, Al_2O_3, NiO, Fe_2O_3, and $BaSO_4$) on the structure and properties of waste glasses in order to find the optimal ratio of sludges to HLW during vitrification. In the experiments, the simulated sludge and simulated HLW were mixed at different ratios from 0:8 to 4:4, with an overall waste content of 16 wt %, in a borosilicate glass wasteform. It is found that the glass density, molar volume, sulfur retention, and glass transition temperature changed little when increasing the sludge content of the glasses, while the viscosity, chemical durability, and crystallization features of the glasses varied notably. The crystals formed in the glasses during the thermal treatment were exclusively Fe-substituted diopside (Ca, Mg, Fe)$_2$Si$_2$O$_6$. An increase in the Al_2O_3 and NiO content of the glasses may have been responsible for the increased crystallinity at high temperatures. The leaching rate of Si, B, Na, and Cs from the glasses declined with the increasing addition of sludge to the glasses. Although all the glasses fulfilled the requirements for vitrification processing and glass-product performance, it is recommended that the sludge content of the whole waste should not exceed 25 wt %. This study guides further research on the immobilization of high-level sludges.

Keywords: high-level liquid waste; radioactive sludge; vitrification; glass crystallization

Citation: Tan, S.; Chang, J.; Liu, X.; Sun, S.; Xian, L.; Zhang, S. Influence of Radioactive Sludge Content on Vitrification of High-Level Liquid Waste. *Sustainability* **2023**, *15*, 4937. https://doi.org/10.3390/su15064937

Academic Editors: Michael I. Ojovan, Vladislav A. Petrov and Sergey V. Yudintsev

Received: 30 January 2023
Revised: 7 March 2023
Accepted: 8 March 2023
Published: 10 March 2023

1. Introduction

High-level liquid waste (HLW) from the reprocessing of spent nuclear fuels is highly radioactive, toxic, and mobile; thus, it is extremely detrimental to human beings and the environment. Therefore, to achieve the sustainable development of nuclear energy, HLWs must be appropriately immobilized into reliable hosts before geological disposal. Vitrification is recognized worldwide as the primary choice for immobilizing high-level nuclear waste [1]. Among the vitrification technologies developed so far, Joule-heated ceramic melter (JHCM) technology is the most widely applied and borosilicate glasses are the most commonly used immobilization hosts [2,3]. China operated a vitrification plant, i.e., the Vitrification Plant of China (VPC), since 2021 to produce borosilicate nuclear-waste glasses based on JHCM technology for the immobilization of legacy HLWs generated at early ages.

These legacy HLWs are usually rich in Na, Fe, Al, and S, with some traces of Ln and U [4,5]. During their long-term storage in tanks, various insoluble substances can form due to hydrolysis and salt precipitation, and the precipitates gradually sink into the bottom to form sludge layers [6]. Because the sludges are categorized as high-level nuclear waste due to their significant content of radioactive cesium and actinides, they must be immobilized into reliable matrices for deep geological disposal. Ideally, these sludges should be vitrified with the liquid HLW during the operation; however, because of the significant differences between the compositions of sludges and liquid HLWs, blending them may have significant effects on the vitrification process and the properties of borosilicate waste glasses. Furthermore, vitrifying radioactive sludges alone is also considered a possible method for the treatment of these sludges; however, the waste loading is limited by the formation of sulfates and spinels in glass melts, according to our preliminary results. Similar sludges are also abundant at the Hanford sites in the USA, and both glass and ceramic waste forms have been developed to immobilize sludges alone [7–9]. In addition, many studies have focused on the thermal treatment of other types of radioactive sludge, such as Magnox sludge [10,11] and Fukushima sludge [12], to form glass or ceramic wasteforms.

Unlike HLWs, sludges are prone enrichment in some specific elements. In this study, the targeted sludge was composed of Na_2O, Al_2O_3, Fe_2O_3, NiO, La_2O_3, BaO, and SO_3, with Al_2O_3, NiO, and BaO contents much higher than those of legacy HLWs. Furthermore, Al_2O_3 is a common component in nuclear waste glasses. Functioning as a network former at low contents, it usually gives rise to enhanced glass-network connectivity and chemical durability, but with an increased crystallization tendency and viscosity [13–16]. Nickel(II) oxide is not commonly found in borosilicate glasses; it is reported to be present as nanoparticles within glass matrices, and the doping of NiO could lead to increased glass crystallization [17,18]. Furthermore, BaO has been introduced to nuclear-waste glass to improve sulfate solubility, although it may result in deteriorated glass-chemical durability [5,19]. When these differences are combined, the influences might be complex and should be further investigated.

This study is part of a research project whose aim is to develop glass formulations for the vitrification of radioactive HLW sludges. The study investigates the influence of sludge contents on the structure and properties of borosilicate-nuclear-waste glasses by mixing simulated sludges with HLW simulants at different ratios to produce a series of glasses containing varying sludge contents. It then investigates the changes in the glass properties using these variations.

2. Materials and Methods

2.1. Raw Materials

2.1.1. Simulated HLW and Sludge

This study used inactive waste simulants instead of natural waste. The chemical compositions of the HLW simulants in Table 1 originated from our previous paper [20], representing a natural HLW stream to be vitrified in China. During this study, a few minor components were omitted to simplify the experiments. The compositions of the simplified sludge simulants are also listed in Table 1. In order to evaluate the influence of sludge additions to HLW, the sludge simulant was mixed with HLW simulant at a ratio of 0:8, 1:7, 2:6, 3:5, and 4:4 (on an oxide basis), with the overall waste loaded constantly at 16.0 wt % in the glass.

Table 1. The chemical composition of the simulated HLW and sludge.

Simulated HLW		Simulated Sludge	
Oxide	**Content (wt %)**	**Oxide**	**Content (wt %)**
CeO_2	0.35		
Cr_2O_3	2.02		
Cs_2O	0.09		
K_2O	0.61		
La_2O_3	11.63	La_2O_3	5.0
Fe_2O_3	21.46	Fe_2O_3	20.0
Al_2O_3	8.77	Al_2O_3	25.0
BaO	0.13	BaO	5.0
Na_2O	45.41	Na_2O	20.0
MoO_3	0.78		
Nd_2O_3	0.69		
NiO	2.56	NiO	20.0
P_2O_5	0.44		
SO_3	4.62	SO_3	5.0
SrO	0.05		
TiO_2	0.32		
Y_2O_3	0.09		
Total	100.00	Total	100.00

2.1.2. Basic Glass

The basic glass was composed of 53.44 wt % SiO_2, 14.60 wt % B_2O_3, 4.40 wt % Al_2O_3, 2.60 wt % Li_2O, 5.21 wt % Na_2O, 5.20 wt % MgO, 8.00 wt % CaO, 4.16 wt % BaO, 1.79 wt % V_2O_5, and 0.60 wt % Sb_2O_3, which was identical to the proportion currently used in the VPC project [20]. The basic glass contributed to 84.0 wt % of the final glass composition.

2.2. Glass Preparation

The glasses were labeled NJn, namely NJ0, NJ1, NJ2, NJ3, and NJ4, where n denotes the ratio of the sludge to the sum of the sludge and HLW. All the HLW simulants, sludge simulants, and primary glasses were made from the corresponding oxides, carbonates, and sulfates of analytic-grade purity purchased from Beijing Sinopharm Co. Ltd., Beijing, China.

Appropriate amounts of chemicals were weighed to make 300 g glass products, which were then mixed and mechanically homogenized for 20 min. The batches were placed in corundum crucibles and heated in an electric furnace subjected to the following heating profile: 1 h from room temperature to 200 °C, 4 h from 200 to 1150 °C, and 3 h retention at 1150 °C. Subsequently, the glass melts were poured into a preheated graphite mold and annealed in an annealing furnace at 500 °C for 1 h before the furnace was turned off.

2.3. Sample Characterization

Part of the as-cast glasses was crushed into pieces for glass-viscosity measurement, which was carried out with a Brookfield DV2T Viscometer equipped with an electric furnace for heating. The measurement started at 1150 °C and ended at 950 °C, with an interval of 50 °C and a period of about 10 min for temperature stability. The rotor speed was adjusted to keep the torque range constant at 40–50%. The viscometer was calibrated with a borosilicate-glass standard prior to use.

Part of the as-cast glasses was crushed and ground into fine powders ($\leq$75 μm) for X-ray diffraction (XRD), X-ray fluorescence (XRF), and differential scanning calorimetry (DSC) analyses. The XRD analysis was performed on a Bruker D8 Advance X-ray diffractometer operating at 40 kV and 40 mA (CuKα), with a 2θ angle of 10°–70°, a step size of 0.02°, and a dwell time of 0.4 s. The XRF was employed to analyze glass compositions with an ARL Advant XP+ X-ray fluorescence spectrometer (Thermo Scientific, Waltham, MA, USA). The DSC curves were recorded with a Netzsch STA 449 F3 Thermal Analyzer, from room temperature to 1000 °C, at a heating rate of 10 °C·min^{-1}, under a nitrogen flow. Glass powders between 75 and 150 μm were collected for chemical-durability analysis (the performance-consistency test, PCT-B). The experiments followed the procedures described in ASTM 1285-14: the glass powders were immersed in deionized water in poly(tetrafluoroethylene) (PTFE) vessels for 7 days at 90 °C with a ratio of the surface area to the leachant volume of 1200 $\pm$ 50 m^{-1}. The concentrations of Si, B, Na, Cs, and other elements in the leachant after leaching were measured using inductively coupled plasma–atomic emission spectroscopy (ICP–AES) or inductively coupled plasma–mass spectrometry (ICP–MS).

The crystallization behaviors and the liquidus temperatures of the glasses were analyzed by the isothermal-heat-treatment method following ASTM C1720-17. Fine glass powders (~5 g) were placed in a Pt95-Rh5 disk, covered by a Pt95-Rh5 cap, and heated in an annealing furnace for 24 h in a temperature range of 850–1025 °C using intervals of 25 °C. The heating period of 24 h was selected to allow sufficient time for crystal growth within glasses. However, it should be noted that the crystallization process did not necessarily reach equilibrium during this period. After the heat treatment, the samples were taken out of the furnace and cooled to room temperature in the air. The samples were cut diagonally into two halves, one for the XRD analysis and the other for the scanning-electron microscopy (SEM) observation. The XRD analysis was identical to that mentioned above; the fraction of the crystals in the glass was calculated based on the Rietveld refinement method using CaF_2 as the internal standard and the GSAS EXPGUI interface. The SEM observations were performed on polished glass pieces mounted into epoxy resin and polished with 1 μm of diamond suspension using an FEI Inspect Quanta650 scanning-electron microscope equipped with an energy-dispersive X-ray spectrometer (EDXS, Oxford Instruments, Abingdon, UK). The polished samples were coated with gold prior to analysis to allow electric conductivity.

3. Results

3.1. Appearance and Homogeneity

The prepared glasses (NJ0–NJ4) were visibly homogeneous. While the bulk of all the glasses appeared dark green, the color of the powdered-glass samples gradually changed from light yellow to yellow-brownish with the addition of the sludge content. The color changes in the glass products are attributable to the high content of transition metals, such as Fe (Fe_2O_3) and Ni (NiO), in the sludge. Meanwhile, according to Figure 1, all the glasses were X-ray-amorphous, and the various samples differed negligibly.

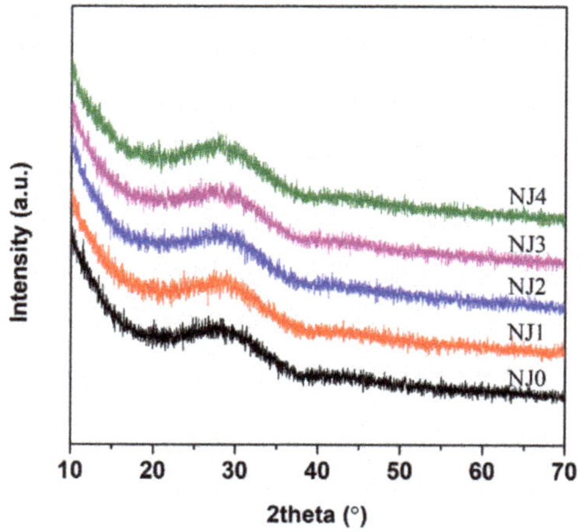

Figure 1. The XRD patterns of prepared glasses NJ0–NJ4.

3.2. Glass Composition and Density

The prepared glasses were compositionally close to the nominal values, as listed in Table 2. The higher-than-expected Al_2O_3 content (~1 wt %) is likely to have arisen through the corrosion of corundum crucibles during the glass melting, while the lower-than-expected Na_2O content might have been due to the evaporation of the glass batches and glass melts at high temperatures. Furthermore, the gaps between the nominal and measured Al_2O_3 contents were similar among all the glasses, suggesting that the corrosivity of glass melts is not affected by the addition of sludge content.

Table 2. The nominal and XRF-measured normalized glass compositions (wt %).

Oxide	NJ0		NJ1		NJ2		NJ3		NJ4	
	Nominal	Measured	Nominal	Measured	Nominal	Measured	Nominal	Measured	Nominal	Measured
SiO_2	44.89	45.20	44.89	44.70	44.89	45.25	44.89	44.76	44.89	44.60
B_2O_3 *	12.26	12.26	12.26	12.26	12.26	12.26	12.26	12.26	12.26	12.26
Al_2O_3	5.10	6.15	5.43	6.44	5.75	6.74	6.08	6.99	6.40	7.23
Li_2O *	2.18	2.18	2.18	2.18	2.18	2.18	2.18	2.18	2.18	2.18
Na_2O	11.65	11.10	11.14	10.67	10.63	10.11	10.12	9.78	9.61	9.25
MgO	4.37	4.24	4.37	4.08	4.37	4.28	4.37	4.35	4.37	4.45
CaO	6.72	6.31	6.72	6.68	6.72	6.53	6.72	6.56	6.72	6.87
BaO	3.51	3.39	3.61	3.37	3.71	3.53	3.80	3.55	3.90	3.75
Fe_2O_3	3.43	3.58	3.40	3.47	3.38	3.35	3.35	3.43	3.32	3.28
La_2O_3	1.86	2.01	1.73	2.06	1.60	1.73	1.46	1.58	1.33	1.21
NiO	0.41	0.32	0.76	0.82	1.11	1.05	1.46	1.35	1.80	1.84
V_2O_5	1.50	1.54	1.50	1.43	1.50	1.38	1.50	1.53	1.50	1.55
SO_3	0.74	0.54	0.75	0.62	0.75	0.57	0.76	0.61	0.77	0.53
Sb_2O_3	0.50	0.56	0.50	0.59	0.50	0.53	0.50	0.56	0.50	0.48
Others **	0.88	0.62	0.76	0.63	0.65	0.51	0.55	0.58	0.45	0.52
Total	100.00	100.00	100.00	100.00	100.00	100.00	100.00	100.00	100.00	100.00

Note: * The content of B_2O_3 and Li_2O is not measured; ** Others include Cr_2O_3, K_2O, ZrO_2, P_2O_5, MoO_3, Cs_2O, SrO, CeO_2, and Y_2O_3.

Sulfur retention is always a concern in vitrification because of the formation of detrimental yellow phases [1,3,21]. In this study, the sulfur content retained in the glasses fluctuated from 0.53 to 0.62 wt %, with a retention rate ranging from 68.8% to 82.7%, as plotted in Figure 2. As plotted in Figure 3, the glass density and the molar volume remained constant in the range of 2.72–2.74 g·cm^{-3} and 23.8–24.0 cm^3·mol^{-1}, respectively, as the sludge content of the glass increased.

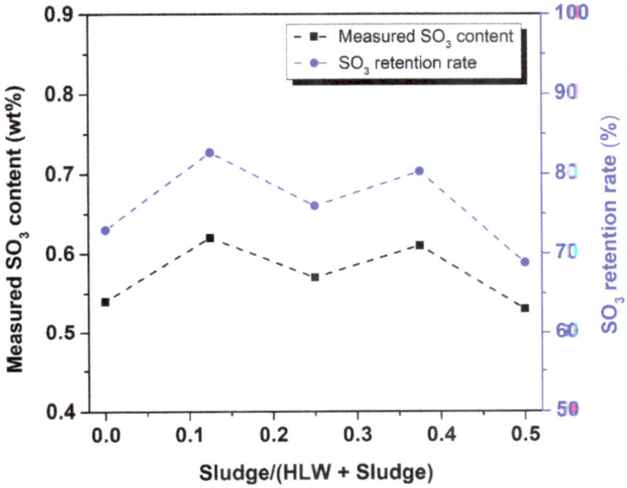

Figure 2. The measured SO_3 retention in glasses NJ0–NJ4.

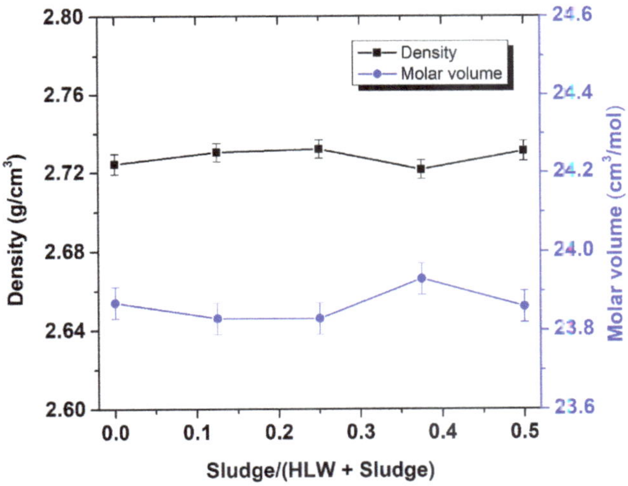

Figure 3. The density and molar volume of glasses NJ0–NJ4.

3.3. DSC Curves

Figure 4a illustrates the DSC curves of the prepared glasses. The glasses remained thermally stable up to the glass-transition temperature (T_g), estimated from the onset of the first endothermic peak. Moreover, while the T_g of glasses NJ0–NJ2 was constant around 500 °C, it increased to 506 and 510 °C for glasses NJ3 and NJ4, respectively, as plotted in Figure 4b. The exothermic peak, indicating the glass crystallization temperature (T_c), was at 800–900 °C. In addition, the intensity of the T_c peak gradually increased from glass NJ0 to glass NJ4, implying an increased tendency towards glass crystallization caused by the addition of the sludge.

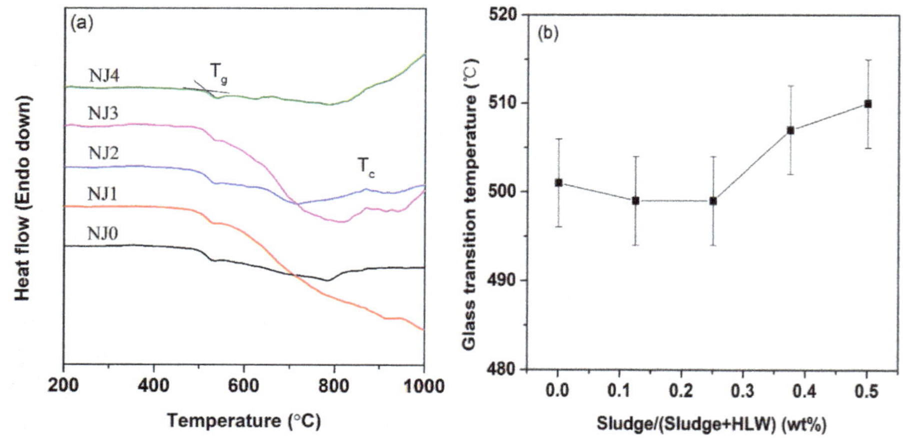

Figure 4. (**a**) The DSC curves of glasses NJ0–NJ4 and (**b**) the estimated T_g values.

3.4. Viscosity

Figure 5 presents the viscosity of the prepared glasses at 950–1150 °C. The viscosity of all five glasses steadily decreased exponentially with the increasing temperature, which agreed well with the widely used Arrhenius-type formula for the viscosity of Newtonian glass melts [1,22,23], as expressed below:

$$\ln(\eta) = A + B/T \tag{1}$$

where η represents the viscosity of the glass melts, A and B are the composition-related constants, and T indicates the temperature. The introduction of sludge to glass does not affect the Newtonian nature of glass melts at high temperatures.

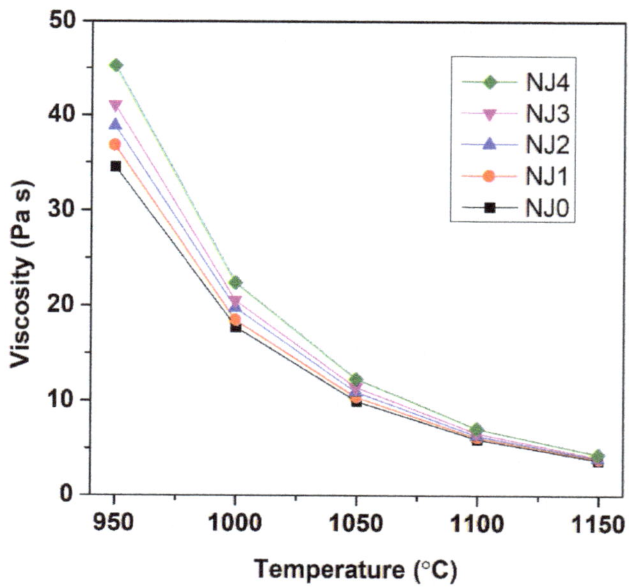

Figure 5. The high-temperature viscosity of glasses NJ0–NJ4.

Meanwhile, the viscosity of the glasses almost linearly increased with the increases in the sludge content: The viscosity increased from 3.74 Pa·s for glass NJ0 to 4.37 Pa·s for glass NJ4 at 1150 °C; at 950 °C, it increased from 34.52 Pa·s for glass NJ0 to 45.24 Pa·s for glass NJ4.

3.5. Glass-Crystallization Features

Figure 6 illustrates the XRD patterns of glasses NJ0–NJ4 after 24 h of heat treatment at different temperatures. Significant crystalline peaks can be observed for all the glasses thermally treated at 850–950 °C, and these peaks (at a 2θ of 19.8°, 27.6°, 29.9°, 30.3°, 30.9°, 35.0°, 35.5°, 35.8°, 39.2°, 40.7°, 42.0°, 42.5°, 43.0°, 44.3°, 44.9°, 52.4°, 56.7°, and 65.8°) agree well with the patterns of the diopside ($CaMgSi_2O_6$, PDF#72-1497), which was also the dominant crystalline phase in VPC glass [24,25]. No other crystalline phases were observed.

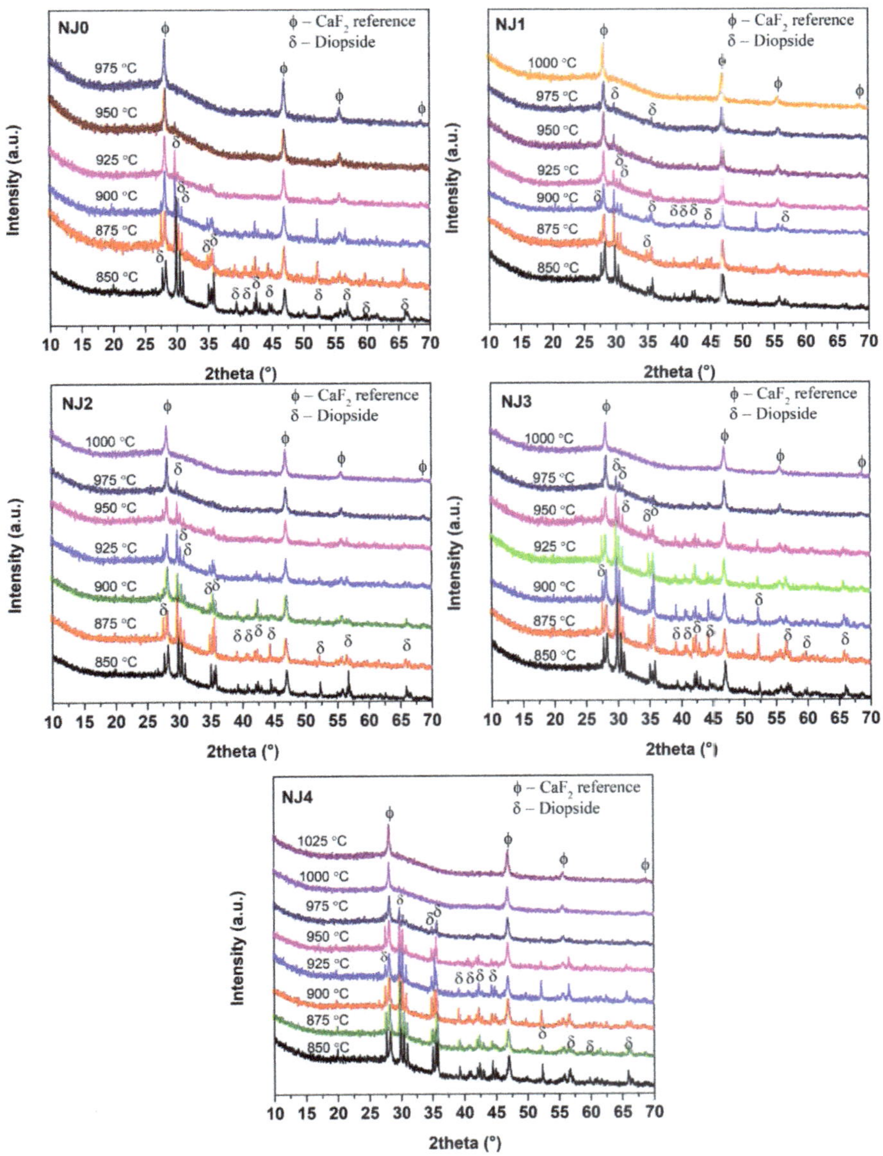

Figure 6. The XRD patterns of glasses NJ0–NJ4 when thermally treated for 24 h at different temperatures.

Figure 7 plots the results of the quantitative analysis of the crystallinity in the glasses. The content of diopside crystals in glass NJ0 linearly decreased from 12.3 wt % at 850 °C to 2.5 wt % at 950 °C, and, no crystals were formed during the heat treatment at 975 °C. The crystal content of glass NJ1 slightly increased compared to that of glass NJ0 at all the treatment temperatures, and the temperature at which the crystalline peaks disappeared changed to 1000 °C. The increasing tendency towards diopside crystallization continued with the increase in the simulated sludge contents of the glasses, and glass NJ4 had diopside contents of 26.0 wt % at 850 °C and 11.3 wt % at 950 °C, coupled with a glass liquidus temperature (T_L) higher than 1000 °C.

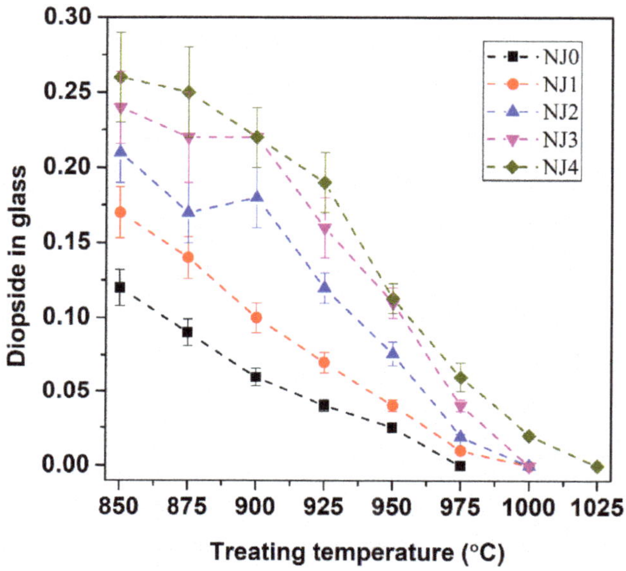

Figure 7. The weight percentages of diopside in different crystallized glasses, as determined by the quantitative XRD analysis.

Figure 8 illustrates the backscattered electron (BSE) images of glasses NJ0–NJ4 after the heat treatment at 900 °C for 24 h. The formed crystals of all the glasses were tabular or plate-like and were randomly distributed within the glass matrices. These particles were generally greater than or equal to 100 μm in length and tens of micrometers in width, with a tendency to agglomerate and form larger crystals. Figure 8f is a typical EDXS spectrum taken from the particles in glass NJ0 treated at 900 °C, demonstrating that these particles primarily comprised Si, Ca, Mg, Fe, Na, and O. Table 3 lists the normalized compositions of the particles formed in the different glasses. It should be pointed out that only the crystals in the glasses treated at 900 °C were analyzed, and our previous unpublished work showed that the compositions of diopside crystals formed in one glass treated at different temperatures are essentially the same. All the crystals had a stoichiometry close to that of the diopside $CaMgSi_2O_6$, with a significant substitution of Fe with Ca and Mg. In addition, the crystals contained small amounts of Al, Cr, Ni, and La.

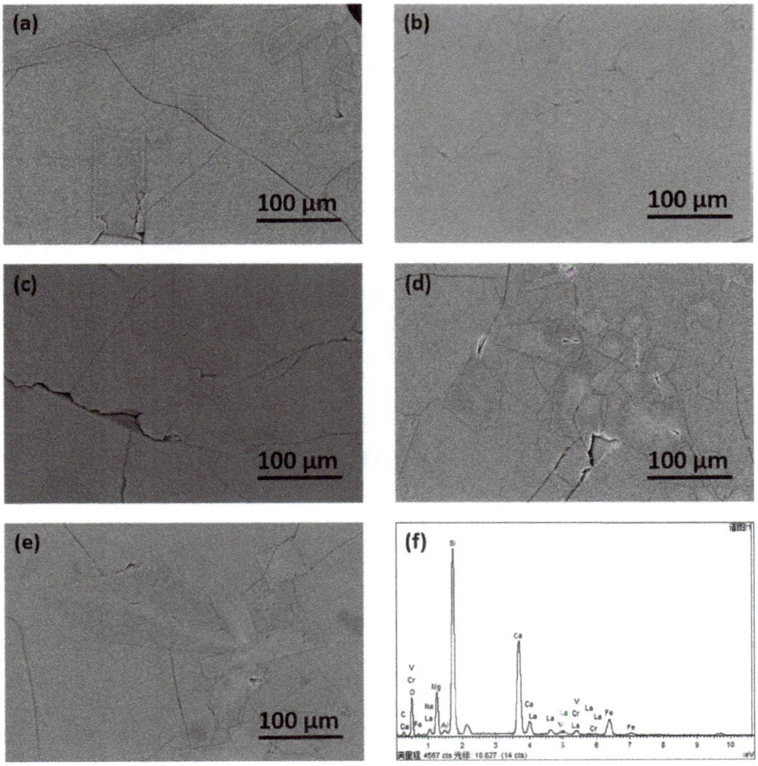

Figure 8. The BSE images of the crystallized regions in glasses (**a**) NJ0, (**b**) NJ1, (**c**) NJ2, (**d**) NJ3, and (**e**) NJ4; (**f**) an EDXS spectrum taken from diopside crystals in the crystallized regions in glass NJ0. The Chinese word on the top-right means "Spectrum 1". At the bottom the words mean "Full scale 4567 cts Cursor 10.627 (14 cts)".

Table 3. The chemical compositions of the crystals in the glasses analyzed by EDXS.

Sample	Atomic Content (Normalized to O = 6; Fe Assumed to Be Present as Fe^{2+})								
	Si	Ca	Mg	Fe	Al	Cr	Ni	La	O
NJ0	2.05	0.77	0.80	0.20	0.02	0.02	0.01	0.03	6.00
NJ1	2.07	0.78	0.77	0.17	0.04	0.01	0.02	0.03	6.00
NJ2	2.07	0.76	0.78	0.19	0.03	0.02	0.02	0.02	6.00
NJ3	2.07	0.78	0.74	0.18	0.03	0.02	0.04	0.03	6.00
NJ4	2.06	0.75	0.77	0.18	0.04	0.01	0.04	0.04	6.00

3.6. PCT-B Test

Figure 9 plots the changes in the normalized leaching rates (NR_i) of the Si, B, Na, and Cs from the different glasses after the 7-day PCT-B leaching test. As was typically found, B and Na were the major leaching elements, and the Si and Cs leached less. Further, increasing the sludge content of the glass steadily reduced the normalized leaching rates of these elements: The NR_{Na} was 0.26 $g{\cdot}m^{-2}{\cdot}d^{-1}$ for glass NJ0 and rapidly dropped to 0.12 $g{\cdot}m^{-2}{\cdot}d^{-1}$ for glass NJ4; however, NR_B was 0.23 $g{\cdot}m^{-2}{\cdot}d^{-1}$ for glass NJ0 and gradually declined to 0.15 $g{\cdot}m^{-2}{\cdot}d^{-1}$ for glass NJ4. The NR_{Si} also showed a decreasing tendency similar to that of the NR_B, but to a much lesser extent, declining from 0.062 to 0.026 $g{\cdot}m^{-2}{\cdot}d^{-1}$. The NR_{Cs} slightly decreased from 0.12 $g{\cdot}m^{-2}{\cdot}d^{-1}$ for glass NJ0 to 0.08 $g{\cdot}m^{-2}{\cdot}d^{-1}$ for glass NJ3, but it increased to 0.09 $g{\cdot}m^{-2}{\cdot}d^{-1}$ for glass NJ4.

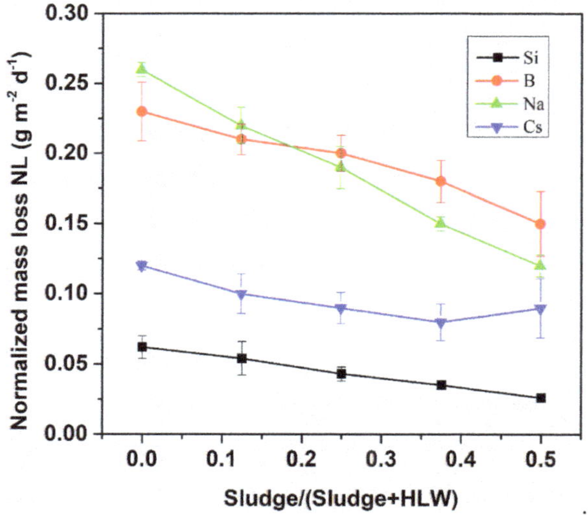

Figure 9. The normalized leaching-mass loss of Si, B, Na, and Cs of glasses NJ0–NJ4, according to the 7−day PCT−B leaching test.

4. Discussion

4.1. Influence of Sludge Content on Glass Products' Properties

All the prepared glasses were homogeneous and X-ray-amorphous, indicating that basic glass can tolerate the addition of at least 50 wt % of sludge to HLW at a given overall waste-oxide loading of 16 wt % (the level that is currently applied in VPC). Therefore, the capacity of glass to accommodate sludge content is not the factor that restricts the addition of sludge to HLW during vitrification.

Introducing sludge into glass leads to minor changes in glass density and molar volume. Compared with the HLW simulant, the sludge simulant contained more Al_2O_3 and BaO and less Na_2O and La_2O_3. Both BaO and La_2O_3 are prone to increasing glass densities due to the fact that they have greater mass than other components, and the influence of the addition of sludge to HLW on glass density is balanced by the increase in BaO content and the decrease in La_2O_3 content.

It appears that the sulfur retention was not correlated with the increasing sludge content, although it is widely reported that sulfur solubility and retention behavior are determined by glass composition [26,27]. One possible explanation for this is that the fractions of BaO, CaO, MgO, and V_2O_5 remained stable among the glasses, and they functioned as the principal oxides for determining the dissolution of the sulfur in the glass melts. The compositional variation herein, primarily Al_2O_3, Na_2O, and NiO, was not influential. In addition, the sulfur-retention rates (0.5–0.6 wt %) in this study were close to the values reported in [5,28,29], in which sulfate-bearing wastes were vitrified in similar conditions. It is also worth noting that, although the sulfur retention rate was not particularly high, the "cold cap" and the bubbling stirrer applied in JHCMs can improve the further dissolution of sulfur into glass melts during vitrification.

The T_gs of all the glasses ranged between 500 and 510 °C, which suggests that the influence of sludge content on the thermal stability of glass is limited. However, the addition of sludge content does affect the crystallization behavior of glass. The crystals formed in the glass during the heat treatment were compositionally similar among the glasses; this was because the major compositional variations among the glasses were Al_2O_3, NiO, Na_2O, and La_2O_3, none of which play a vital role in diopside crystallization. The crystallization tendency linearly increased with the increasing addition of sludge content, which is likely to have been due to the increased Al_2O_3 and NiO contents of the glass, which affected the glass-network stability. Borosilicate glasses containing high Al_2O_3 contents tend to crystallize when heated at specific temperatures [14,16,30], resulting in phase separations within the glass matrices, regardless of whether the phases contain Al. A small amount of NiO was found in the formed Fe-diopside crystals, according to the EDXS; thus, increasing NiO contents may enhance the overall crystallization of glasses. In addition, NiO is reported to degrade the thermal stability of glass networks and, therefore, support the crystallization of alkaline-earth borosilicate in CaO-SrO-Al_2O_3-B_2O_3-SiO_2 glasses [18,31], although it does not join the crystals.

The chemical durability of glass is improved by the addition of sludge content. This can be attributed to the increase in Al_2O_3 content and the decrease in Na_2O content in glass, which gives rise to enhanced glass-network connectivity [13]. Although BaO is reported to have a reduced impact on the chemical durability of glass [18], it may be overridden by the notable changes in Al_2O_3 and Na_2O contents. In addition, since the least durable glass (NJ0) had already been verified for chemical durability [20], all the glasses in this study were deemed to be sufficiently durable as high-level wasteforms.

In summary, basic glass can tolerate the addition of up to 50% sludge to HLW while maintaining similar or better glass properties.

4.2. Influence of Sludge Content on Glass Melting

The increase in the viscosity of the glasses (~17% at 1150 °C and ~31% at 950 °C, from NJ0 to NJ4, respectively) with the addition of sludge can be attributed to the much higher content of Al_2O_3 and lower content of Na_2O in the simulated sludge, as reported elsewhere [23,32,33]. However, despite the apparent increase, the viscosity of all the glasses still fulfilled the suggested processing requirements (3–6 Pa·s at 1150 °C and 20–65 Pa·s at 950 °C). Nevertheless, this is likely to have caused the decreased fluidity of the glass melt in the melter and a decreased pouring rate for glass production, resulting in a longer duration requirement for glass residence and a long duration of glass pouring. Therefore, although the current processing parameters may still apply to the addition of sludge, the processing methods need to be modified along, with their operation. If a variation in viscosity of less than 10% is needed, NJ1 and NJ2 glasses are more desirable.

The liquidus temperature T_L of the glass should be sufficiently low (e.g., ≤950 °C) for no or few crystals to be present in glass melts at the temperatures set for pouring, so as to prevent the emergence of residual crystals in glass melts from the clogging of the discharge valve. In this study, the addition of sludge to the glass led to a notable increase in T_L (from 950–975 °C for glass NJ0 to 975–1000 °C for glasses NJ1–NJ3 and to 1000–1025 °C for glass NJ4). This increase was probably due to the increase in the Al_2O_3 and NiO contents in the glass; NiO is reported to strongly increase T_L at ~80 °C per wt % in nuclear-waste glasses, and Al_2O_3 increases T_L to a lesser extent [34,35]. Increases in T_L increases the risk of clogging glass melters, especially when glass melts are readily crystallized. Therefore, glasses NJ0–NJ2 are considered more acceptable, from the point of view of the liquidus temperature, for maintaining a low crystal content in glass melts at 950–975 °C.

As a result, the use of NJ0–NJ2 glasses, in which the ratio of sludge to HLW is from 0:8 to 2:6, is recommended to facilitate the glass melting process.

5. Conclusions

Borosilicate glasses containing different amounts of simulated HLW and sludge were successfully produced, with all the glasses being amorphous and homogeneous. From the above results and discussion, the following conclusions can be drawn:

- The substitution of sludge for HLW was conducive to minor changes in the glass density, molar volume, glass-transition temperature, and sulfur retention.
- The substitution of sludge for HLW gave rise to a notable increase in the viscosity and chemical durability of the glasses. Upon heat treatment, the crystallinity increased with the increasing sludge content in the glass, although the formed crystalline phases in the glasses remained the same.

- The glasses NJ0–NJ2 were proposed to vitrify the HLW blended with the sludge from the perspective of melter processing; that is, the sludge content in the overall waste should not exceed 25 wt %.

However, considering the varying compositions of the sludges in and among HLW tanks, the influence of individual elements in sludge on glass properties should also be determined. An acceptable range of compositional fluctuation is needed for application. In addition, optimization of basic glass formulations for mixed sludge and HLW vitrification was also studied in our recent work.

Author Contributions: Conceptualization, S.T.; methodology, J.C. and S.T.; data curation, X.L., J.C., S.T. and S.S.; writing—original draft preparation, S.T. and J.C.; writing—review and editing, S.T., S.S. and S.Z.; supervision, S.Z.; funding acquisition, L.X. All authors have read and agreed to the published version of the manuscript.

Funding: This research was funded by the China National Nuclear Corporation (CNNC) Youth Innovation Team Project, grant number FC192012000403, and SHIKUAN SUN appreciates the financial support from the National Natural Science Foundation of China under grant number 52172064.

Institutional Review Board Statement: Not applicable.

Informed Consent Statement: Not applicable.

Data Availability Statement: The data that support the findings of this study are available from the corresponding authors upon reasonable request.

Acknowledgments: We thank Tao Yu and Dongsheng Qie for the use of the laboratories for this study.

Conflicts of Interest: The authors declare no conflict of interest. The funders had no role in the design of the study; in the collection, analyses, or interpretation of data; in the writing of the manuscript; or in the decision to publish the results.

References

1. Ojovan, M.I. Immobilisation of Radioactive Wastes in Glass. In *An Introduction to Nuclear Waste Immobilisation*, 3rd ed.; Ojovan, M.I., Lee, W.E., Kalmykov, S.N., Eds.; Elsevier: Amsterdam, The Netherlands, 2019; pp. 319–368.
2. Aloy, A.S. Calcination and vitrification processes for conditioning of radioactive wastes. In *Handbook of Advanced Radioactive Waste Conditioning Technologies*, 1st ed.; Ojovan, M.I., Ed.; Woodhead Publishing: Oxford, UK, 2011; pp. 136–158.
3. Tan, S. Glass-based stabilization/solidification of radioactive waste. In *Low Carbon Stabilization and Solidification of Hazardous Wastes*; Tsang, D., Wang, L., Eds.; Elsevier: Amsterdam, The Netherlands, 2021; pp. 433–447.
4. Wang, X.; Tuo, X.; Zhou, H.; Zhang, W.; Li, Z.; Chen, X.; Luo, H. Glass formulation development on high sodium and high sulfur bearing high level liquid waste for vitrification process. *J. Nucl. Radiochem.* **2013**, *35*, 180–192. (In Chinese)
5. Wu, L.; Xiao, J.; Wang, X.; Teng, Y.; Li, Y.; Liao, Q. Crystalline phase, microstructure, and aqueous stability of zirconolite–barium borosilicate glass-ceramics for immobilization of simulated sulfate bearing high-level liquid waste. *J. Nucl. Mater.* **2018**, *498*, 241–248. [CrossRef]
6. Liang, J.; Song, C.; Pan, C.; Ding, G.; Yang, W. The Preliminery Study of High-Level Radioactive Sludge I. Preparation and Property Study of Simulated High Level Radioactive Sludge. *J. Nucl. Radiochem.* **2000**, *22*, 37–44. (In Chinese)
7. Fox, K.M.; Edwards, T.B.; Peeler, D.K. *High Level Waste (HLW) Sludge Batch (SB4)-Selecting Glasses for a Variability Study*; Savannah River National Laboratory: Aiken, SC, USA, 2006; p. 29.
8. Sayenko, S.Y.; Shkuropatenko, V.A.; Pylypenko, O.V.; Karsim, S.O.; Zykova, A.V.; Kutnii, D.V.; Wagh, A.S. Radioactive waste immobilization of Hanford sludge in magnesium potassium phosphate ceramic forms. *Prog. Nucl. Energy* **2022**, *152*, 104315. [CrossRef]
9. Donald, I.W. *Waste Immobilisation in Glass and Ceramic Based Hosts: Radioactive, Toxic and Hazardous Wastes*; Wiley-Blackwell: Chichester, UK, 2010; p. 527.
10. Heath, P.G.; Stewart, M.W.A.; Moricca, S.; Hyatt, N.C. Hot-isostatically pressed wasteforms for Magnox sludge immobilisation. *J. Nucl. Mater.* **2018**, *499*, 233–241. [CrossRef]
11. Tan, S.; Kirk, N.; Marshall, M.; McGann, O.; Hand, R.J. Vitrification of an intermediate level Magnox sludge waste. *J. Nucl. Mater.* **2019**, *515*, 392–400. [CrossRef]
12. Amamoto, I.; Kobayashi, H.; Kitamura, N.; Takebe, H.; Mitamura, N.; Tsuzuki, T.; Fukayama, D.; Nagano, Y.; Jantzen, T.; Hack, K. Research on vitrification technology to immobilize radioactive sludge generated from Fukushima Daiichi power plant: Enhanced glass medium. *J. Nucl. Sci. Technol.* **2016**, *53*, 1467–1475. [CrossRef]
13. Piovesan, V.; Bardez-Giboire, I.; Fournier, M.; Frugier, P.; Jollivet, P.; Montouillout, V.; Pellerin, N.; Gin, S. Chemical durability of peraluminous glasses for nuclear waste conditioning. *NPJ Mater. Degrad.* **2018**, *2*, 1–10. [CrossRef]

14. Mohd Fadzil, S.; Hrma, P.; Schweiger, M.J.; Riley, B.J. Component effects on crystallization of RE-containing aluminoborosilicate glass. *J. Nucl. Mater.* **2016**, *478*, 261–267. [CrossRef]
15. Mysen, B.; Richet, P. *Silicate Glasses and Melts*, 2nd ed.; Elsevier: Amsterdam, The Netherlands, 2018; p. 720.
16. Fox, K.M.; Peeler, D.K.; Edwards, T.B.; Best, D.R.; Reamer, I.A.; Workman, R.J.; Marra, J.C.; Riley, B.J.; Vienna, J.D.; Crum, J.V.; et al. *International Study of Aluminum Impacts on Crystallization in U.S. High Level Waste Glass*; Savannah River National Laboratory: Aiken, SC, USA, 2008; p. 172.
17. Cizman, A.; Idczak, K.; Krupinski, M.; Girsova, M.; Zarzycki, A.; Rysiakiewicz-Pasek, E.; Zielony, E.; Staniorowski, P.; Wrzesinska, P.; Perlikowski, I.; et al. Comprehensive studies of activity of Ni in inorganic sodium borosilicate glasses doped with nickel oxide. *Appl. Surf. Sci.* **2021**, *558*, 149891. [CrossRef]
18. Ren, M.; Yang, P.; Xu, J.; Zhang, Q.; Chen, K.; Tang, D.; Zhuang, H.; Sa, B.; Zhang, T. Effect of nickel doping on structure and suppressing boron volatility of borosilicate glass sealants in solid oxide fuel cells. *J. Eur. Ceram. Soc.* **2019**, *39*, 2179–2185. [CrossRef]
19. Mishra, R.K.; Sudarsan, K.V.; Sengupta, P.; Vatsa, R.K.; Tyagi, A.K.; Kaushik, C.P.; Das, D.; Raj, K. Role of Sulfate in Structural Modifications of Sodium Barium Borosilicate Glasses Developed for Nuclear Waste Immobilization. *J. Am. Ceram Soc.* **2008**, *91*, 3903–3907. [CrossRef]
20. Liu, L.; Qie, D.; Zhou, H.; Li, B.; Li, Y.; Xu, J.; Zhang, H. Verification experiment research of glass formualtion for high sodium and high sulfur bearing HLLW. *J. Nucl. Radiochem.* **2014**, *36*, 163–168. (In Chinese)
21. Goel, A.; McCloy, J.S.; Pokorny, R.; Kruger, A.A. Challenges with vitrification of Hanford High-Level Waste (HLW) to borosilicate glass–An overview. *J. Non-Cryst. Solids X* **2019**, *4*, 100033. [CrossRef]
22. Ojovan, M.I. Viscosity and Glass Transition in Amorphous Oxides. *Adv. Condens. Matter Phys.* **2008**, *2008*, 1–23. [CrossRef]
23. Hrma, P.; Kruger, A.A. High-temperature viscosity of many-component glass melts. *J. Non-Cryst. Solids* **2016**, *437*, 17–25. [CrossRef]
24. Meng, B.; Zhu, Y.; Yang, D.; Cui, Z.; Jiao, Y.; Liu, H.; Dai, C. Effect of alkaline earth metal oxide content on crystallization behavior of simulated high level liquid waste glass. *Bull. Chin. Ceram. Soc.* **2022**, *41*, 4081–4088.
25. Yang, Y.-G.; Zhang, Z.-T.; Hua, X.-H.; Wang, L. Crystallization of the simulated HLW glass. *J. Nucl. Radiochem.* **2014**, *36*, 317–320. (In Chinese)
26. Manara, D.; Grandjean, A.; Pinet, O.; Dussossoy, J.L.; Neuville, D.R. Sulfur behavior in silicate glasses and melts: Implications for sulfate incorporation in nuclear waste glasses as a function of alkali cation and V_2O_5 content. *J. Non-Cryst. Solids* **2007**, *353*, 12–23. [CrossRef]
27. Lenoir, M.; Grandjean, A.; Poissonnet, S.; Neuville, D.R. Quantitation of sulfate solubility in borosilicate glasses using Raman spectroscopy. *J. Non-Cryst. Solids* **2009**, *355*, 1468–1473. [CrossRef]
28. Billings, A.L.; Fox, K.M. Retention of Sulfate in Savannah River Site High-Level Radioactive Waste Glass. *Int. J. Appl. Glass Sci.* **2010**, *1*, 388–400. [CrossRef]
29. Lorier, T.H.; Reamer, I.A.; Workman, R.J. *Initial Sulfate Solubility Study for Sludge Batch 4 (Sb4)*; Savannah River National Laboratory: Aiken, SC, USA, 2005; p. 22.
30. McClane, D.L.; Amoroso, J.W.; Fox, K.M.; Kruger, A.A. Nepheline crystallization behavior in simulated high-level waste glasses. *J. Non-Cryst. Solids* **2019**, *505*, 215–224. [CrossRef]
31. Mahapatra, M.K.; Lu, K. Effects of nickel on network structure and thermal properties of a new solid oxide cell seal glass. *J. Power Sources* **2008**, *185*, 993–1000. [CrossRef]
32. Hrma, P.; Han, S.-S. Effect of glass composition on activation energy of viscosity in glass-melting-temperature range. *J. Non-Cryst. Solids* **2012**, *358*, 1818–1829. [CrossRef]
33. Kroll, J.O.; Nelson, Z.J.; Skidmore, C.H.; Dixon, D.R.; Vienna, J.D. Formulation of high-Al2O3 waste glasses from projected Hanford waste compositions. *J. Non-Cryst. Solids* **2019**, *517*, 17–25. [CrossRef]
34. Hrma, P.; Vienna, J.; Crum, J.; Piepel, G.; Mika, M. Liquidus Temperature of High-Level Waste Borosilicate Glasses With Spinel Primary Phase. *MRS Online Proc. Libr.* **1999**, *608*, 671. [CrossRef]
35. Jantzen, C.M.; Brown, K.G. Predicting the Spinel?Nepheline Liquidus for Application to Nuclear Waste Glass Processing. Part I: Primary Phase Analysis, Liquidus Measurment, and Quasicrystalline Approach. *J. Am. Ceram. Soc.* **2007**, *90*, 1866–1879. [CrossRef]

sustainability

MDPI

Article

Toward Deep Decontamination of Intermediate-Level-Activity Spent Ion-Exchange Resins Containing Poorly Soluble Inorganic Deposits

Marina Palamarchuk [1], Maxim Chervonetskiy [1], Natalya Polkanova [2] and Svetlana Bratskaya [1,*]

1 Institute of Chemistry, Far Eastern Branch, Russian Academy of Sciences, 159, Prosp. 100-letya Vladivostoka, 690022 Vladivostok, Russia
2 Limited Liability Company "Spetsatomservice", 75 Novouglichskoe sh., Office 305, Sergiev Posad, 141301 Moscow, Russia
* Correspondence: s.bratskaya@gmail.com

Abstract: Spent ion-exchange resins (SIERs) generated yearly in large volumes in nuclear power plants (NPPs) require particular predisposal handling and treatment with the primary objectives of waste volume reduction and lowering the disposal class. Deep decontamination of the SIERs using solution chemistry is a promising approach to reduce the amount of intermediate-level radioactive waste (ILW) and, thus, SIER disposal costs. However, the entrapment of nonexchangeable radionuclides in poorly soluble inorganic deposits on SIERs significantly complicates the implementation of this approach. In this work, the elemental and radiochemical compositions of inorganic deposits in an intermediate-level-activity SIER sample with an activity of 310 kBq/g have been analyzed, and a feasibility study of SIER decontamination using solution chemistry has been conducted. The suggested approach included the magnetic separation of crud, removal of cesium radionuclides using alkaline solutions in the presence of magnetic resorcinol-formaldehyde resin, removal of cobalt radionuclides using acidic EDTA-containing solutions, and hydrothermal oxidation of EDTA-containing liquid wastes with immobilization of radionuclides in poorly soluble oxides. The decontamination factors for ^{137}Cs, ^{60}Co, and ^{94}Nb radionuclides were 3.9×10^3, 7.6×10^2, and 1.3×10^2, respectively, whereas the activity of the decontaminated SIER was 17 Bq/g, which allows us to classify it as very low-level waste.

Keywords: spent ion-exchange resins; resorcinol-formaldehyde resins; radionuclides; activated corrosion products; decontaminating solutions; EDTA; hydrothermal oxidation

Citation: Palamarchuk, M.; Chervonetskiy, M.; Polkanova, N.; Bratskaya, S. Toward Deep Decontamination of Intermediate-Level-Activity Spent Ion-Exchange Resins Containing Poorly Soluble Inorganic Deposits. *Sustainability* **2023**, *15*, 3990. https://doi.org/10.3390/su15053990

Academic Editors: Michael I. Ojovan, Vladislav A. Petrov and Sergey V. Yudintsev

Received: 15 December 2022
Revised: 14 February 2023
Accepted: 20 February 2023
Published: 22 February 2023

1. Introduction

Radioactively contaminated spent ion-exchange resins (SIERs) are a specific type of radioactive waste generated at various stages of the nuclear power cycle. In nuclear power plants (NPPs), SIERs are accumulated as a result of the operation of a cleanup system for the primary coolant, decontamination of water from cooling ponds, deactivation solutions, trap water, and water streams from special water decontamination systems. The total volume of SIERs generated in NPPs is estimated as 0.02–0.025 m^3/MW of nominal power for two-circuit reactors of WWER-type NPPs and 0.08–0.025 m^3/MW of nominal power for single-circuit reactors of RBMK-type NPPs [1].

SIERs, which accumulate radionuclides and chemical impurities and cannot be regenerated and reused, are present in the form of low-level radioactive waste (LLW) and intermediate-level radioactive waste (ILW) and have to be conditioned to meet the waste acceptance criteria for a disposal site. The primary objectives of SIER management are based on the IAEA Technical Document for Management of Spent Ion-Exchange Resins from Nuclear Power Plants [2] and include volume reduction, immobilization, storage, disposal, and economic issues. Currently, applied SIER conditioning methods include direct solidification using inorganic and organic compounds (cement, bitumen, and plastics) [3–6] and the destruction of the organic matrix via combustion, pyrolysis, liquid phase oxidation, supercritical water oxidation, or plasma combustion [7–10]. While destructive methods allow a significant reduction in the volume of solid radioactive wastes to be sent for disposal, this approach is complicated by the high-temperature release of gaseous destruction products and the demand for expensive corrosion-resistant equipment. The direct solidification of SIERs is the first-choice option in many cases, but it leads, as a rule, to an increase in waste volume. Considering the high difference in disposal costs between LLW and ILW, blending or concentration averaging intermediate- and low-activity SIERs was shown to be an efficient way to reduce the amount of waste stored, lower the disposal class, and minimize disposal costs [11,12].

From our point of view, a very promising approach to safe and economically sound SIER management is based on the deep decontamination of resins using specially selected solutions. This will enable one to lower SIERs' disposal class or, in the ideal case, to reach a level meeting criteria for industrial (nonradioactive) solid waste. Spent deactivation solutions containing radionuclides can be processed with the common methods used for conditioning liquid radioactive wastes (LRWs) to assure safe radionuclide immobilization. The deep decontamination approach has been previously tested on low-level-activity SIERs at the Balakovo and Kalinin NPPs (Russia). The decontamination scheme included the treatment of SIERs with highly acidic solutions with subsequent immobilization of cesium radionuclides on selective ferrocyanide sorbents. The specific activity of SIERs was reduced to the level of nonradioactive wastes [13], which demonstrated the applicability of the proposed strategy to the reduction in radioactive waste volume without the destruction of the SIER matrix. However, despite the obvious advantages of this approach, it has been scarcely investigated and reported in the literature so far.

One of the problems in SIER decontamination using a deactivation solution is the presence of poorly soluble inorganic deposits. Such deposits can be formed during the purification of the coolant and condensate (crud removed from the nuclear reactor core) and the processing of SIERs from the intermediate storage together with inorganic filter materials (for example, perlite) [14,15]. Moreover, the deposition of the dissolved impurities on resins used in water purification systems is also possible during operation. In this case, different types of deposits can be formed depending on the pH and water composition, which is rather typical for mixed-bed ion-exchange filters. For example, the hydrolysis of dissolved iron ions leads to the precipitation of iron hydroxides [16,17], which is accompanied by the coprecipitation of corrosion radionuclides (e.g., ^{60}Co) and a subsequent formation of the mixed oxide phase. Since many of such oxides are insoluble in nitric acid, which is used for the regeneration of ion-exchange resins, they are accumulated in resins from cycle to cycle. The radionuclides occluded by oxide particles are also accumulated. The presence of Si and Al in the water streams, as a rule, leads to the formation of polysilicic and aluminum-polysilicic acids and their subsequent transformation to poorly soluble aluminosilicate deposits both on the surface and inside of ion-exchanger beads [18–20]. A characteristic feature of such deposits consists of the ability to capture cesium ions at the stage of deposit formation and after via adsorption. Despite a strict restriction on the Si content in NPP waters, it continuously enters the circuit and other water streams from the following potential sources: corrosion of stainless steels (containing up to 1% Si); suction of technical water from coolers through leaks in pipe fittings; perlite from precoat filters; defoamers used in evaporation system (bentonite clay and organosilicon compounds); and technical water from special laundry and sanitary facilities [14,15,21,22].

The binding of radionuclides by the insoluble deposits significantly reduces the efficiency of SIER decontamination by acid–base regeneration. Thus, high decontamination factors for SIERs can be achieved only via the removal of non-ion-exchangeable radionuclides.

We have shown earlier that cesium radionuclides can be removed from resins contaminated with silicate deposits using alkaline deactivation solutions (DSs). This approach was applied for the decontamination of a low-level-activity SIER sample from the research reactor in a scheme including a transfer of the released cesium radionuclides to resorcinol-formaldehyde resins and then to a ferrocyanide sorbent [23].

In the present work, further development of the earlier proposed scheme and its application to the decontamination of an intermediate-level-activity SIER sample from the storage facility of Units 1 and 2 of the Kursk NPP with ^{137}Cs activity of 10^7 Bq/kg are reported. This task was additionally complicated by the presence of iron oxides and corrosion radionuclides with ^{60}Co activity above 10^6 Bq/kg in the SIER sample.

2. Materials and Methods

2.1. Materials

2.1.1. Reagents

All the reagents used in the present work (sodium nitrate $NaNO_3$, sodium hydroxide NaOH, disodium ethylenediaminetetraacetate dihydrate $Na_2EDTA{\cdot}2H_2O$, zinc nitrate hexahydrate $ZnNO_3{\cdot}6H_2O$, resorcinol $C_6H_4(OH)_2$, formaldehyde CH_2O, iron(III) chloride $FeCl_3{\cdot}6H_2O$, iron(II) sulfate $FeSO_4{\cdot}7H_2O$, nitric acid HNO_3, ammonium hydroxide NH_4OH, and hydrogen peroxide H_2O_2) of a chemically pure grade were purchased from NevaReaktiv LLC (St. Petersburg, Russia) and used without further purification. Siloxane-acrylate emulsion (KE 13–36), with a particle size of 160 nm and solid phase content of 50%, was produced by the scientific production association "Astrokhim" (Elektrostal', Russia).

2.1.2. Spent Ion-Exchange Resin (SIER)

SIER was sampled from the temporary storage tanks of Units 1 and 2 of Kursk NPP with an RBMK-1000 reactor (high-power channel reactor). Samples were taken using a cylindrical sampler at 1/2 of the depth from the surface of the tank and placed into clean plastic containers for characterization. The SIER sample was a polydispersed mixture consisting of brown spherical resin beads (cationite KU-2 and anionite AV-17) and fine dark-brown powder. The radionuclide composition of the sample was determined as described in Section 2.7.

2.1.3. Ferrocyanide Sorbents

Termoksid-35, a spherically granulated sorbent based on nickel-potassium ferrocyanide (32–36 wt %) immobilized in the pores of highly dispersed amorphous zirconium hydroxide, produced under Technical Conditions 6200-305-12342266–98, was purchased from Termoksid Ltd., Zarechny, Sverdlovsk Region, Russia, and used as a 0.25–0.5 mm fraction with a bulk density of 1.3 g/cm^3.

FNK-50, nickel-potassium ferrocyanide (50 wt %) coprecipitated with silicic acid gel, was purchased from JSK "Alliance Gamma", Russia, and used as a 0.2–0.25 mm fraction with a bulk density of 0.6 g/cm^3.

2.1.4. Synthesis of Resorcinol-Formaldehyde Resin (RFR) and Iron Oxides

Magnetic composite sorbent RFR-M was synthesized via the polycondensation of resorcinol and formalin in alkaline medium in the presence of synthetic magnetite (10 wt %) with subsequent solidification at 210 °C;. The coercive force of RFR-M at 300 K was 0.8 kA/m. The synthesis and characterization of the composite sorbent RFR-M are described in [23]. The 0.2–0.25 mm fraction of a bulk density of 0.59 g/cm^3 was used.

A model iron oxide labeled with ^{57}Co was synthesized as follows: 50 mL of 0.5 M $FeSO_4$ solution spiked with $^{57}CoCl_2$ was added to 100 mL of 0.5 M $FeCl_3$ solution, and then 3% ammonia hydroxide was added dropwise to reach pH 9. After stirring for 20 min, the colloidal solution was kept for 12 h, and then the precipitate was separated and washed with distilled water using a Buechner funnel. The precipitate was dried at 50 °C until constant weight, crushed in a mortar, and sieved to obtain a 0.1–0.05 mm fraction. The activity of the sample was 42.7×10^3 Bq/g. Phases of magnetite (Fe_3O_4) and goethite ($FeO(OH)$) were found in the iron oxide synthesized under the same conditions without a ^{57}Co tracer.

Porous hematite to be used as a catalyst for hydrothermal oxidation (HTO) was synthesized by the sol-gel method using siloxane-acrylate latex as a template. The synthesis was performed as follows: 40 mL of 0.5 M $FeCl_3$ solution was added to 200 mL of KE 13–36 emulsion with a latex content of 5%, and then 60 mL of 2 M NaOH solution was added dropwise until $Fe(OH)_3$ formation. After 30 min of mixing followed by the precipitate sedimentation, $Fe(OH)_3$ gel was separated using a "blue ribbon" filter paper, washed with distilled water, and dried at 90 °C; for 2 h until xerogel formation. The organic template was removed via xerogel annealing at 900 °C; for 1 h in an air atmosphere. The material obtained was crushed in a mortar, washed with distilled water to separate the fine fraction by decantation, and dried at 90 °C;. The fraction of a size of 0.25–0.5 mm and a bulk density of 0.9 g/cm^3 was used as an HTO catalyst.

2.2. SIER Decontamination Using Alkaline Deactivation Solution (DS)

2.2.1. SIER Decontamination Using Alkaline DS under Static Conditions

The decontamination of SIER with alkaline DS (2.25 mol/L $NaNO_3$, 0.75 mol/L NaOH) under static conditions was performed in hermetically sealed 250 mL polypropylene vials as follows: 50 mL of the DS was added to 5 mL of SIER placed in vial no. 1; after mixing dispersion at 50 rpm for 20 h, the DS was removed with the pipette and transferred to vial no. 2, and then the activities of the SIER and DS were measured. In the next cycle, a fresh portion of the DS (50 mL) was added to SIER in vial no. 1. The full cycle of decontamination and activity measurements was repeated 5 times, and the total volume of DS used was 250 mL.

2.2.2. RFR-M-Assisted SIER Decontamination Using Alkaline DS under Static Conditions

The decontamination of SIER in the presence of RFR-M was performed in two cycles using 250 mL vials and an alkaline DS as described in Section 2.2.1. In the first cycle, 50 mL of the DS was added to four no. 1 vials containing 1, 2, 5, and 5 mL of SIER and 0.1, 0.1, 0.25, and 0.1 g of RFR-M, respectively. After shaking dispersions at 50 rpm for 20 h, RFR-M was localized in the bottom near the vial wall using a flat permanent magnet, and then pulp (the SIER and DS mixture) was accurately removed and transferred to no. 2 vials. Thereafter, DSs were removed from no. 2 vials with a pipette and transferred to no. 3 vials. After measuring activity in vials no. 1–3, DSs were returned to the no. 2 vials (with partially decontaminated SIER), and fresh portions of RFR-M of the same weight as in the first cycle were added. Mixing, separation, and activity measurements were performed as described for the first cycle.

2.2.3. SIER Decontamination Using Alkaline DS under Dynamic Conditions

A total of 150 mL of a DS of the same composition as given in Section 2.2.1 was used for SIER decontamination under dynamic conditions as follows: 5 mL of SIER was placed in a sorption column with an inner diameter of 10 mm, then DS was fed at a flow rate of 1 mL/min, and eluate aliquots were taken every 25 mL for activity measurements.

2.3. SIER Decontamination Using Alkaline DS (Scheme 1)

2.3.1. RFR-Assisted SIER Decontamination

To separate the magnetic fraction, 100 mL of an alkaline DS (2.25 mol/L $NaNO_3$, 0.75 mol/L NaOH, and 0.02 mol/L EDTA) was added to 10 mL of SIER in a polypropylene vial with a flat magnet mounted to the wall. After stirring the dispersion for 5 h, magnetic particles, SIER, and the DS were sequentially separated for activity measurement as described in Section 2.2.2. Then, the DS was returned to the vial with SIER, and three cycles of decontamination were performed in the presence of RFR-M as described in Section 2.2.2. Fresh portions (0.4 g) of RFR-M were used in the 1st and 2nd cycles. In the 3rd cycle, RFR-M, regenerated with nitric acid as described in Section 2.3.2 after the 1st cycle, was used. Thus, the final ratio (mL:mL:g) of SIER:DS:RFR-M after three cycles of decontamination was 10:100:0.8. After the 3rd cycle, the DS was separated and additionally decontaminated using RFR-M collected in the 2nd cycle and regenerated with nitric acid. SIER was further decontaminated using an acidic DS as described in Section 2.4.

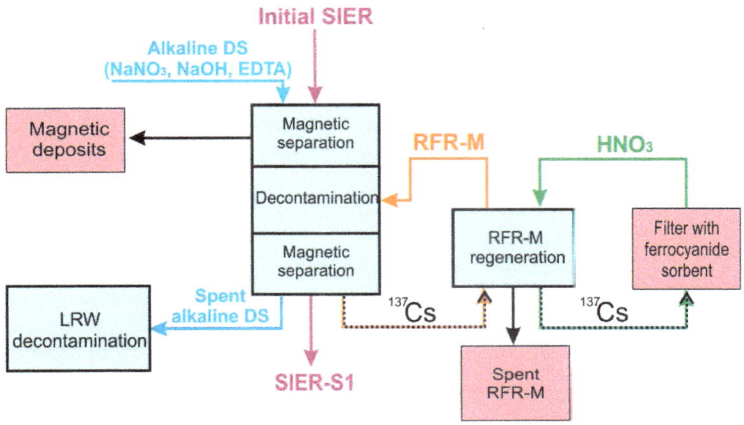

Scheme 1. RFR-assisted SIER decontamination using alkaline DS.

2.3.2. RFR-M Regeneration

The regeneration of RFR-M was performed under static conditions as follows: 5 mL of 0.5 M HNO_3 solution was added to 0.4 g of RFR-M and kept for 30 min under stirring. The procedure was repeated 4 times with fresh portions of the eluent. Nitric acid containing ^{137}Cs radionuclides was decontaminated in a column with an inner diameter of 6 mm filled with 1 mL of FNK-50 sorbent, and the flow rate was varied in the range of 3.5–13 bed volumes (BV/h) to optimize sorption conditions.

2.4. SIER Decontamination Using Acidic DS (Scheme 2)

A total of 25 mL of an acidic DS containing 0.05 mol/L $ZnNO_3$, 0.05 mol/L Na_2H_2EDTA, and 2 mol/L $NaNO_3$ (pH 1.2) was added to 10 mL of SIER decontaminated as described in Section 2.3.1. After 20 min of contact under constant stirring, the DS was separated, and the activities of the SIER and DS were measured. Then, a fresh portion of the DS was added to a vial with SIER. Six cycles of decontamination were conducted with a total DS volume of 150 mL.

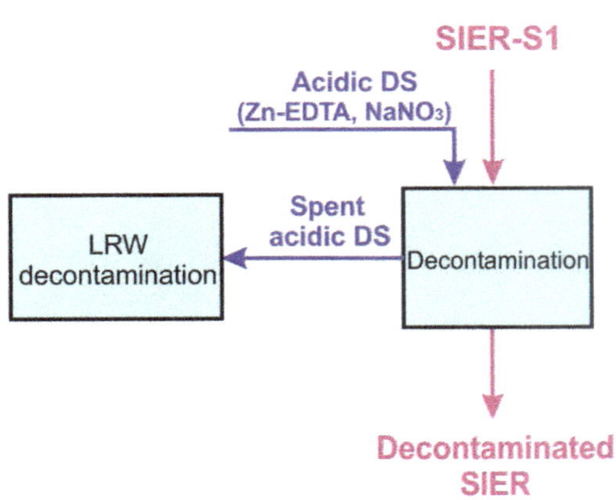

Scheme 2. SIER decontamination using acidic DS.

2.5. ^{57}Co Leaching from Model Iron Oxide Using Acidic DS

^{57}Co leaching from model iron oxide was investigated as follows: 0.05 g of iron oxide was brought in contact with 25 mL of solutions 0.05 M EDTA, 0.05 M Zn-EDTA, and 0.01 M Zn-EDTA under constant stirring. Zn-EDTA solutions were prepared by mixing Na_2H_2EDTA and $Zn(NO_3)_2$ solutions at an equimolar ratio. In 5 days, the activities of the solution and iron oxide were measured after the phase separation by centrifugation.

2.6. Decontamination of Liquid Radioactive Wastes (LRWs) Generated during SIER Decontamination

LRWs generated after SIER decontamination, i.e., spent acidic DS (Scheme 2) or a mixture of spent alkaline DS (Scheme 1) and spent acidic DS (Scheme 2), were decontaminated from ^{60}Co radionuclides in a flow-type hydrothermal installation with a stainless steel reactor with an inner diameter of 8 mm (Figure 1). A total of 1 mL of the catalyst, obtained as described in Section 2.1.4, was placed in the maximal heat zone (the catalyst beads were fixed using the stainless cage with a rod thickness of 0.1 mm and mesh size of 0.05 mm). LRWs and an oxidizer (3% H_2O_2) were fed into the reactor separately using high-pressure Shimadzu LC-20AT chromatography pumps at flow rates of 0.25–0.75 mL/min and 0.25 mL/min, respectively. Hydrothermal oxidation was performed at a pressure of 10 MPa and a temperature of 190–240 °C; to determine the optimal temperature and flow rate. The outlet solution was collected, the γ-activity of 10 mL aliquots of inlet and outlet solutions were measured, and the values were recalculated taking into account dilution in the reactor.

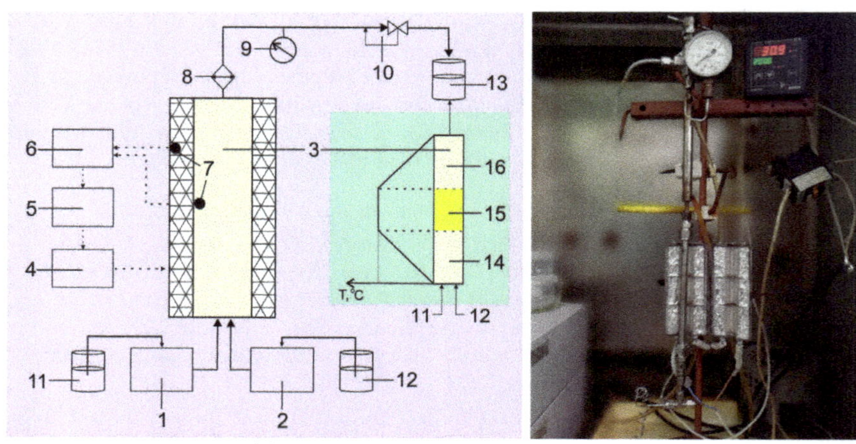

(a) (b)

Figure 1. (**a**) Scheme of the laboratory installation for hydrothermal oxidation: 1 and 2—high-pressure pumps, 3—hydrothermal reactor, 4—single-phase power transformer, 5—automatic voltage regulator, 6—heat regulator, 7—thermocouples, 8—flow-type water cooler, 9—pressure gauge, 10—high-pressure relief valve, 11—vessel with contaminated solution, 12—vessel with hydrogen peroxide, and PV—receiver vessel. Insert: 14—heating zone, 15—zone of the catalyst location, and 16—cooling zone. (**b**) Photo of the laboratory installation for hydrothermal oxidation.

Then, LRW was decontaminated from ^{137}Cs radionuclides in a column with an inner diameter of 6 mm filled with 1 mL of Termoksid-35 sorbent at a flow rate of 15 BV/h.

2.7. Methods of Analysis

Radionuclide composition and specific γ-activities were determined using an AT 1315 Gamma–Beta spectrometer with a 63 × 63 mm NaI (Tl) detector (ATOMTEX, Minsk, Belarus) and multichannel gamma-spectrometer CANBERRA with a high-purity germanium semiconductor detector of the SEGe series and Genie-2000 software (Canberra Industries, Inc., Meriden, CT, USA).

Gamma (γ) spectra were recorded using an AT 1315 Gamma–Beta spectrometer with a 63 $\times$ 63 mm NaI (Tl) detector (ATOMTEX, Minsk, Belarus).

The specific activity of ^{57}Co (photopeak energy: 122 keV) leached from the model iron oxide was determined by the direct radiometric method using an RKG-AT1320 gamma-radiometer equipped with a NaI(Tl) detector measuring 63 $\times$ 63 mm (ATOMTEX, Minsk, Belarus).

Micrographs of the SIER surface were taken, and elemental analysis was performed using a Hitachi TM-3000 scanning electron microscope (Hitachi, Japan) with a Bruker energy dispersive X-ray spectrometer detector (Bruker Nano GmbH, Germany).

The elemental composition of deposits was determined by energy-dispersive X-ray fluorescence spectroscopy using a Shimadzu EDX-800-HS spectrometer equipped with an X-ray tube with Rh-anode.

X-ray diffraction spectra were recorded using a D8 Advance diffractometer (Bruker, Germany).

UV-vis spectra were recorded using a Shimadzu 1650PC scanning UV-vis spectrophotometer (Shimadzu, Japan).

3. Results and Discussion

3.1. SIER Characterization

An SEM investigation has revealed an inorganic nature of the fine fraction, which was mechanically separated from SIER beads or deposited at the beads' surface (Figure 2). Due to the inhomogeneity of inorganic particle distribution, the specific activity of different SIER samples varied, but the average specific activity was about 3 $\times$ 10^4 Bq/g, with the main contributions from ^{137}Cs-137 (3 $\times$ 10^4 Bq/g), ^{60}Co (3 $\times$ 10^3 Bq/g), and ^{94}Nb (7 $\times$ 10^2 Bq/g) radionuclides.

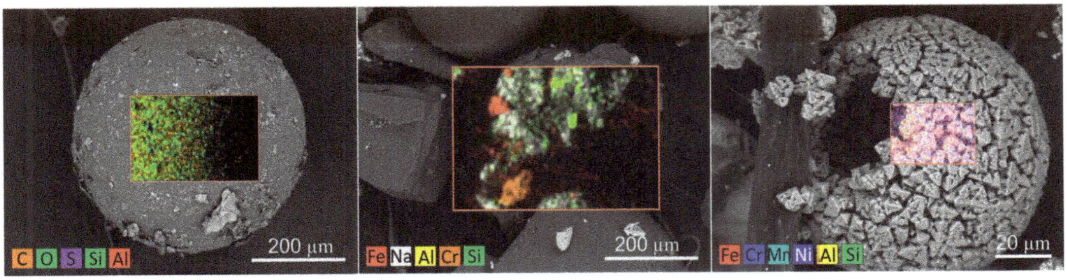

Figure 2. SEM images and elemental mapping of SIER beads and particles of inorganic deposits.

To investigate the nature of the inorganic deposits, the magnetic fraction was separated from a dry SIER sample using a flat magnet and analyzed by X-ray fluorescence spectroscopy: the following elemental composition was found (%): Fe—76, Si—10, Al—3.8, S—2.1, Zr—2.1, Na—1.8, K—0.96, Cr—0.82, Cu—0.81, Ca—0.48, Mn—0.26, Ni—0.25, Ti—0.17, Zn—0.07, Nb—0.03, and Sr—0.02. These data suggest that SIER was contaminated with silicate and corrosive deposits (crud from the active zone) as well as fragments of fuel element claddings (Zr and Nb). The specific activity of deposits was ^{137}Cs—27.5 $\times$ 10^3 Bq/g, ^{60}Co—64.3 $\times$ 10^3 Bq/g, and ^{94}Nb—5 $\times$ 10^3 Bq/g, which indicates the relatively homogeneous distribution of ^{137}Cs activity between organic and inorganic phases, while ^{60}Co and ^{94}Nb radionuclides were concentrated in the deposits. Thus, efficient mechanical separation or solubilization of the deposits from the SIER matrix is required to reach high decontamination factors and lower the class of waste.

3.2. RFR-Assisted SIER Decontamination from ^{137}Cs Radionuclides Using Alkaline Solutions

We have previously shown that alkaline $NaNO_3$ deactivation solutions (DSs) were efficient to release cesium radionuclides from silicate deposits in low-level-activity SIER samples [23,24]. Here, the feasibility and efficiency of this strategy for the decontamination of intermediate-level-activity SIER from an NPP storage tank were evaluated. Preliminary experiments (Table 1) have shown that washing SIER with an alkaline DS provides an efficiency of ^{137}Cs removal of up to 96.3%.

Table 1. Efficiency (%) of ^{137}Cs removal from SIER with dependence on decontamination method and DS: SIER ratio (mL:mL). DS composition—2.25 mol/L $NaNO_3$ and 0.75 mol/L NaOH.

Decontamination Method *	DS:SIER Ratio				
	10	20	30	40	50
Static conditions (2.2.1)	61	75	80	82	82
Dynamic conditions (2.2.3)	89	95.5	96.3	-	-
Static conditions + 0.5 g RFR-M (2.2.2)	99.8	-	-	-	-

* Corresponding section of the experimental part with detailed description of the decontamination method is given in parentheses.

The addition of sorbents selective to cesium ions to the alkaline DS at this stage is beneficial to shift the equilibrium of cesium distribution between SIER and the deactivation solution, i.e., to prevent readsorption of the released cesium ions on SIER, which increases the decontamination factor [23]. Since the application of highly selective ferrocyanide-based sorbents is limited by their low chemical stability at pH > 12 [25], resorcinol-formaldehyde resins (RFRs) with high chemical stability [26] and high selectivity to cesium ions [27] in highly mineralized alkaline media can serve as an alternative. Indeed, in the presence of RFR, which concentrates ^{137}Cs radionuclides released from SIER, the efficiency of ^{137}Cs removal increased up to 99.8% at a significantly lower volume of the alkaline DS (Table 1).

To solve the problem of SIER and RFR separation, magnetite-containing composite resin (RFR-M) was synthesized and used. Since the idea of RFR-assisted SIER decontamination is based on the shift of the cesium distribution equilibrium, the efficiency of ^{137}Cs transfer from SIER to RFR-M will be determined by the competition of different processes characterized by different distribution coefficients (K_d), i.e., $K_{d(silicate/DS)}$, $K_{d(SIER/DS)}$, and $K_{d(RFR\text{-}M/DS)}$, and, thus, by the SIER:DS:RFR-M ratio [23]. The effect of the selected SIER:DS:RFR-M ratios on the residual activity of ^{137}Cs in SIER has shown (Table 2) that an increase in RFR-M:DS and RFR-M:SIER ratios facilitates ^{137}Cs transfer from SIER to RFR-M. A significant reduction in the ^{137}Cs residual activity was attained already after the first cycle of SIER washing with an alkaline DS, and this result was improved in the 2nd cycle with a maximal efficiency of cesium removal of up to 99.96%.

Table 2. Residual ^{137}Cs activity (%) on SIER after RFR-M-assisted decontamination at different SIER:DS:RFR-M (mL:mL:g) ratios (see Section 2.2.2 for details).

SIER:DS:RFR-M *	1:50:0.2	2:50:0.2	5:50:0.5	5:50:0.2
1st cycle	0.4	0.98	1.2	3.9
2nd cycle	0.04	0.08	0.2	1.2

* Total weight of RFR-M after two cycles of decontamination in alkaline DS.

To minimize the volume of the secondary wastes, RFR-M can be regenerated with acidic solutions and reused. It was earlier shown that selective sorption characteristics of RFR toward ^{137}Cs radionuclide in highly mineralized alkaline solutions insignificantly decreased in several sorption/regeneration cycles [23,28,29].

The general scheme used in the feasibility study of intermediate-level-activity SIER decontamination from ^{137}Cs radionuclides is shown below (Scheme 1). Although a significant part of SIER magnetic deposits containing ^{60}Co and ^{94}Nb can be virtually removed at the stage of SIER/RFR-M magnetic separation, it is not recommended, since the accumulation in RFR-M of insoluble in HNO_3 impurities can negatively affect the RFR-M performance in consecutive sorption/regeneration cycles. Thus, we suggest performing magnetic separation of the deposits after the swelling of SIER in an alkaline DS but before RFR-M addition.

Scheme 1 was applied for the decontamination of the SIER sample (volume of 10 mL, weight of 8.7 g, and specific activities: ^{137}Cs 2.7 × 10^4 Bq/g, ^{60}Co 3.1 × 10^3 Bq/g, and ^{94}Nb 7.6 × 10^2 Bq/g), and the γ-spectrum shown in Figure 3 (spectrum 1). Decontamination was performed as follows: 100 mL of an alkaline DS (DS:SIER ratio 10) of the above mentioned composition with the addition of 0.02 mol/L EDTA (EDTA was added to eliminate hydroxide precipitation) was prepared, and then magnetic particles were separated with a flat magnet. In this case, aside from the dissolution of silicate deposits, SIER swelling in alkaline medium facilitates the mechanical detachment of magnetic particles attached to the beads' surface, which, in comparison with "dry" magnetic separation, decreases the probability of capturing SIER beads with the shell of magnetic deposits. At the stage of "wet" separation, 30% of ^{60}Co, 76% of ^{94}Nb, and about 1% of ^{137}Cs were removed with magnetic particles (Figure 3, spectrum 2). About 20% of ^{137}Cs, 15% of ^{60}Co, and 2% ^{94}Nb relative to the initial activity were transferred from SIER to the alkaline DS. Separated magnetic particles varied in morphology and composition (Figure 4).

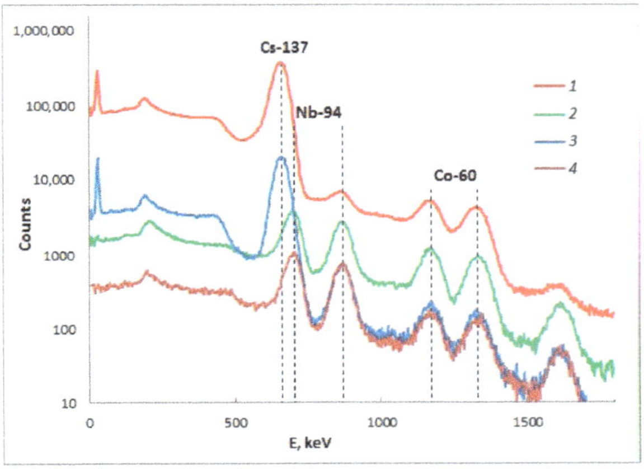

Figure 3. γ-spectra of initial SIER sample (1), magnetically separated deposits (2), and RFR-M after the 1st cycle of SIER decontamination before (3) and after (4) regeneration with 0.5 M HNO_3.

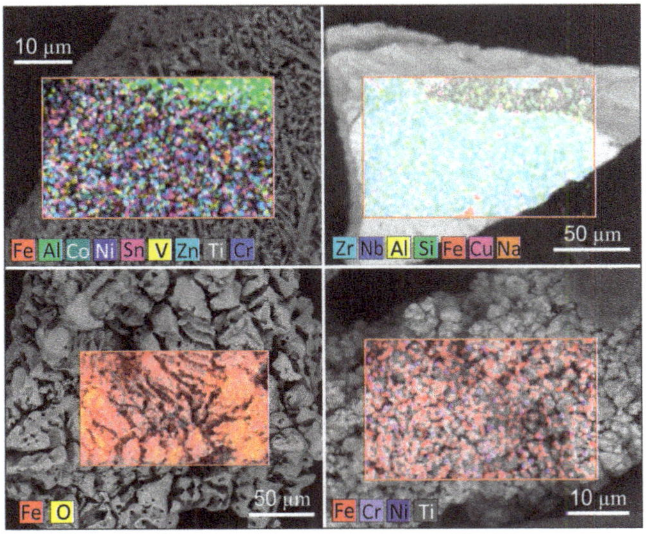

Figure 4. SEM images and elemental mapping of magnetic deposits on SIER.

After magnetic separation, RFR-M was added, and three cycles of RFR-M-assisted SIER decontamination using an alkaline DS were performed. The residual activity of the SIER was ^{137}Cs—127 Bq/g, ^{60}Co—1520 Bq/g, and ^{94}Nb—95 Bq/g, which corresponds to 0.46, 49, and 12.5% relative to the initial activity of SIER, respectively. The efficiency of cesium radionuclides elution from RFR-M with 0.5 M nitric acid was very high, whereas contamination with corrosion radionuclides was insignificant due to the preliminary magnetic separation of magnetic deposits (Figure 3, spectra 3 and 4). Nitric acid with a specific ^{137}Cs activity of 3.8×10^6 Bq/L was decontaminated using selective sorption on a ferrocyanide sorbent, FNK-50, under dynamic conditions at a flow rate of 3.5 BV/h. The activity of the decontaminated HNO_3 solution was 860 Bq/L, which corresponds to a decontamination factor of 4.4×10^3. This confirmed the possibility of cyclic HNO_3 usage in Scheme 1; however, periodical adjustment of its concentration may be required due to the H^+ consumption for RFR-M regeneration.

3.3. SIER Decontamination from Radionuclides Entrapped in Metal Oxide Deposits

Despite the high efficiency of Scheme 1 for cesium removal from SIER, ~49% of ^{60}Co activity remained in the resin. ^{60}Co removal was achieved mostly via the preliminary magnetic separation stage and the cocapture of ^{60}Co-containing crud during SIER/RFR-M separation (Figure 3, spectra 2 and 4). The remaining ^{60}Co activity in SIER was probably located in deposits, which cannot be separated magnetically. For instance, iron-containing particles were still found in SIER after decontamination with the alkaline DS (Figure 5). Thus, modification of Scheme 1 was required to increase the efficiency of intermediate-level-activity SIER decontamination.

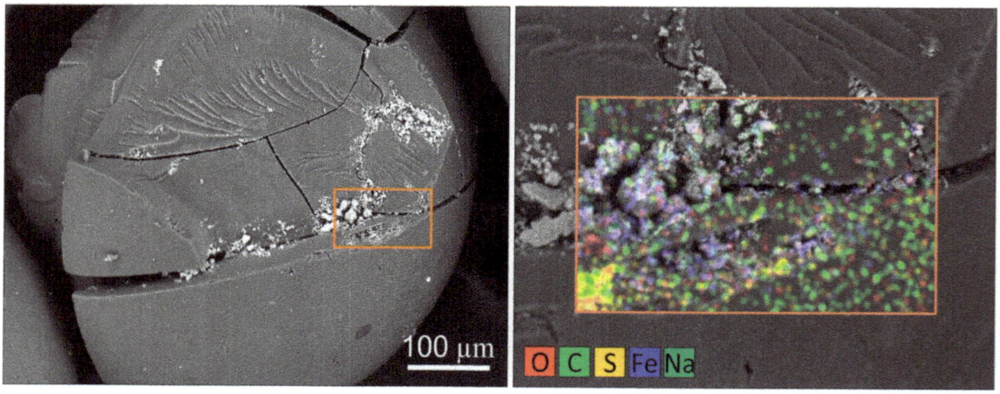

Figure 5. SEM images and elemental mapping of deposits on SIER after decontamination with alkaline DS.

To release ^{60}Co from SIER, organic acids with good chelating properties, such as oxalic, citric, and ascorbic acids as well as aminopolycarboxylic acids, are more appropriate alternatives to mineral acids, which are characterized by higher corrosion activity and lower potential to leach cobalt radionuclides from the iron oxide phase [13,30]. Among organic acids, EDTA is the best choice to meet the criteria of stability in nitrate solutions (nitrates are added to deactivation solutions to prevent readsorption of the anionic complexes on anion-exchangers during the decontamination of SIERs from mixed-bed filters [31]) and the ability to chelate transition metal ions, including corrosion radionuclides. Logarithms of stability constants of EDTA complexes are 16.4 and 41.4 with Co(II) and Co(III) ions, respectively, and 14.3 and 25 with Fe(II) and Fe(III) ions, respectively [32]. EDTA and its salt solutions, mostly Na_2EDTA, are used in nuclear power facilities to remove corrosion products from construction materials and cation-exchangers.

The dissolution of iron (oxy)hydroxide is known to be promoted at lower pH. However, the addition of inorganic acids to EDTA salt solutions results in the precipitation of poorly soluble EDTA on the SIER surface (the minimum of EDTA solubility falls into a pH range of 1.6–2) [33,34]. Avoiding EDTA precipitation at low pH is possible when deactivation solutions contain EDTA complexes with metal ions. In this case, upon complex formation, the pH decreases as a result of H^+ release from EDTA. For instance, the usage of Zn-EDTA solution enhances the release of cobalt radionuclide from synthetic iron oxide labeled with ^{57}Co (Table 3). One can see that the dissolution process is accompanied by the consumption of H^+ and proceeds more efficiently in Zn-EDTA than in Na_2EDTA solution.

Table 3. Leaching ^{57}Co from the model iron oxide (see Section 2.1.4 for the synthesis) at solid:liquid ratio of 1:500.

Solution	pH_0	^{57}Co in Solution, %	pH_{eqv}
Na_2EDTA 0.05 M	4.4	19.7	5.9
Zn-EDTA 0.05 M	1.8	80.2	2.4
Zn-EDTA 0.01 M	2.2	20.8	3.9

Thus, the SIER sample decontaminated using Scheme 1 (labeled SIER-S1) can be further treated with an acidic solution containing the complex of Zn(II) with EDTA (Scheme 2). The advantage of this complex is determined by the amphoteric nature of Zn^{2-} ions, which prevents the formation of insoluble hydroxides at a high pH.

The decontamination of the SIER-S1 sample with an acidic DS, containing 0.05 mol/L Zn-EDTA and 2 mol/L $NaNO_3$, was performed in six steps, using, in each step, a fresh DS aliquot at a DS:SIER ratio of 2.5:1 (v/v) to elucidate decontamination kinetics in acidic medium. The release of radionuclides to the DS occurred stepwise (Table 4): ^{137}Cs radionuclides were mainly detected in the first aliquot (pH was 8.9 as a result of the neutralization of the acidic DS with the alkaline DS from the swollen SIER), and a further decrease in pH resulted in the release of ^{60}Co; moreover, an intensive blue color of the second aliquot typical for Cu(II) complexes indicated possible desorption of Cu(II) ions from SIER-S1 (Figure 6, Table 4). The γ-spectrum of the solution obtained by mixing aliquots 1–6 is shown in Figure 7, spectrum 1. Subsequent treatment with 2.25 M $NaNO_3$ solution or an aliquot of a fresh alkaline DS followed by an aliquot of the acidic DS had no effect on the SIER residual activity. The specific residual activity of decontaminated SIER was 16.9 Bq/g, with a 6.9 Bq/g of ^{137}Cs, 4.1 Bq/g of ^{60}Co, and 5.9 Bq/g of ^{94}Nb, and its γ-spectrum is shown in Figure 7, spectrum 2. Thus, SIER decontamination factors after treatment using Schemes 1 and 2 were >10^3 for ^{137}Cs and >10^2 for ^{60}Co and ^{94}Nb.

Table 4. Radionuclide distribution in acidic DS aliquots with DS:SIER-S1 ratio of 2.5:1 (mL/mL).

Aliquot No.	Radionuclide Content, %		pH
	^{137}Cs	^{60}Co	
1	41.9	2.1	8.9
2	18.7	40.5	2.76
3	17.3	31.7	1.87
4	13	12.4	1.38
5	6.1	10.3	1.36
6	3.0	3.0	1.34

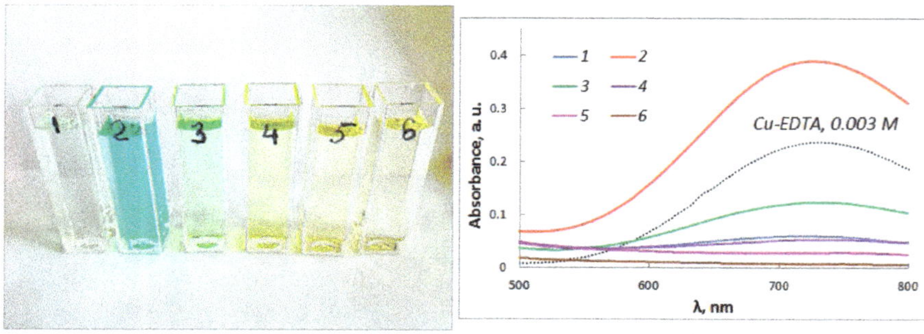

(**a**) (**b**)

Figure 6. Photo (**a**) and UV-vis spectra (**b**) of acidic DS after stepwise SIER-S1 decontamination.

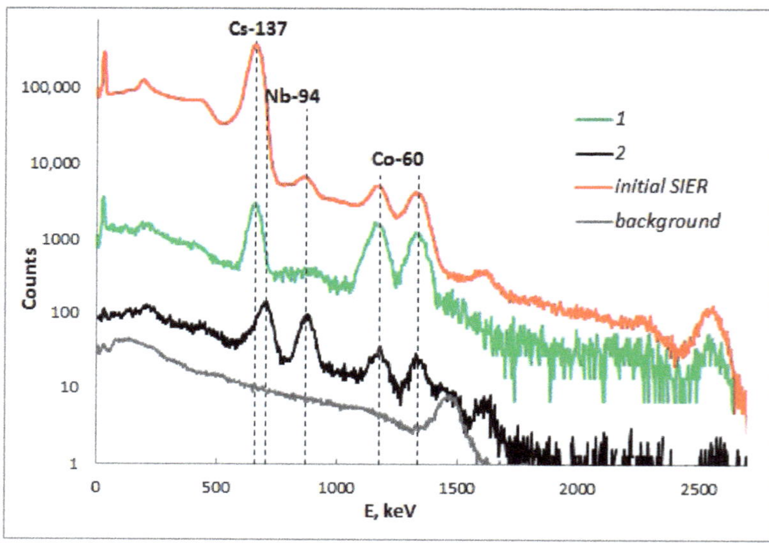

Figure 7. γ-spectra of acidic DS after SIER-S1 decontamination (1) and of decontaminated SIER (2). Spectra of initial SIER sample and background are given for comparison.

3.4. Decontamination of Liquid Radioactive Wastes (LRWs) Generated during SIER Decontamination

LRW generated in the process of SIER decontamination, including spent alkaline DS (Scheme 1) and spent acidic DS (Scheme 2), must be processed to safely immobilize radionuclides in solid matrices.

Spent acidic DS was contaminated with ^{60}Co-EDTA complexes, which are very stable under normal conditions. Since the removal of radionuclides from this type of LRW is complicated, usually, methods known as advanced oxidation and targeted to the destruction of the organic part of the complex are used in combination with subsequent ^{60}Co sorption or precipitation [35–37]. A promising method of such wastes processing consists of hydrothermal oxidation (HTO) using hydrogen peroxide as an oxidizer in the flow-type installation. After the destruction of the complexes, cobalt radionuclides remained in the reactor of HTO installation in the form of oxide Co_3O_4. The mechanism of Co-EDTA destruction under hydrothermal conditions was discussed earlier [38]: the chelate structure degrades as a result of the electron transfer from the ligand to Co(III), while the hydrogen peroxide is responsible for the generation of Co(III) from Co(II) and oxidation of organics not bound to complexes. The application of transition metal (iron, manganese, and nickel) oxides as catalysts increases the Co-EDTA decomposition rate and decreases the temperature of the HTO process [39]: ^{60}Co radionuclides in this case are immobilized in a newly formed oxide phase of the catalyst. Thanks to the very low cobalt leaching rate from the formed oxides (<10^{-6} g/cm^2·day), the HTO reactor with a spent catalyst can be considered a solid radioactive waste (SRW) ready for disposal [40–42].

Hydrothermal oxidation of spent acidic DS was performed with 1 M H_2O_2 solution. Figure 8 shows the γ-spectra of the initial solution (spectrum 1) and the solution after HTO at 220 °C; at a volumetric ratio of LRW:H_2O_2 flow rates of 0.5:0.25 (spectrum 2). Further decontamination of the outlet solution using the ferrocyanide sorbent Termoksid-35 provides efficient removal of ^{137}Cs radionuclides (Figure 8, spectrum 3).

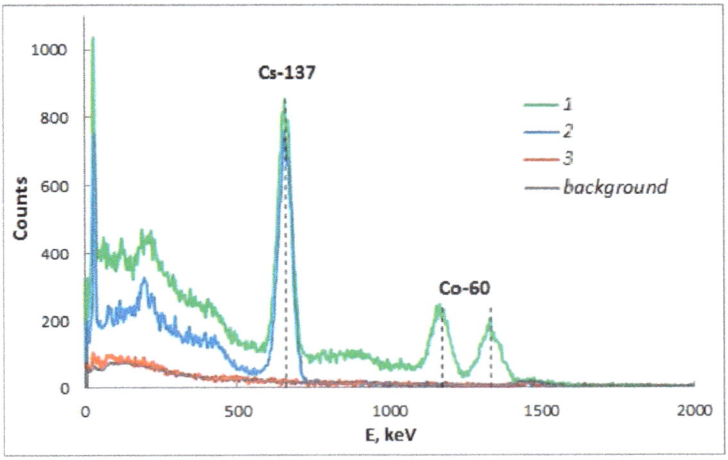

Figure 8. γ-spectra of spent acidic DS (pH 2.12) before purification (1), after HTO at 220 °C; at volumetric flow rates of 1 M H_2O_2 and LRW of 0.25 mL/min and 0.5 mL/min, respectively (2), and after post-treatment using Termoksid-35 sorbent (3).

Spent alkaline DS was contaminated with ^{137}Cs radionuclides, since RFR-M did not provide complete removal of ^{137}Cs under static sorption conditions. Since EDTA was added to the solution to prevent hydroxide precipitation in Scheme 1, spent alkaline DS also contained ^{60}Co-EDTA complexes. Moreover, alkaline solutions after long-term contact with RFR contain products of RFR oxidative destruction, which is enhanced from one application cycle to another [43]. Therefore, an appropriate purification scheme for such solutions is the same as for the spent acidic DS, i.e., HTO and post-treatment with ferrocyanide sorbent. This approach can be applied to spent alkaline DS and their mixtures with spent acidic DS.

Figure 9 shows the γ-spectrum of an aliquot of the LWR obtained by mixing spent alkaline and acidic DSs with a final pH 9.5 (spectrum 1) and the result of its decontamination by HTO at 220 °C; at a volumetric ratio of LRW:H_2O_2 flow rates of 0.5:0.25 (spectrum 2). When the volumetric ratio was increased to 0.75:0.25, complete removal of ^{60}Co was not attained (spectrum 3). The decrease in the decontamination efficiency is related to both the decrease in the oxidizer concentration and the increase in the cumulative flow rate of the mixture in the reactor. The advantage of the flow-type HTO installation consists of the possibility to tune the flow rates of LWR and the oxidizer, as well as the temperature, to attain the required decontamination factor. Post-treatment of the solution decontaminated by HTO using a column with Termoksid-35 sorbent provided activity reduction down to the background level (Figure 9, spectrum 4).

As was shown above, both types of LRW contained complexes of EDTA and cesium radionuclides. It is worth mentioning that the decontamination of such LRW from ^{137}Cs on selective inorganic sorbents should be performed after HTO, since the presence of EDTA in solutions can lead to the decomposition of ferrocyanides due to the binding of metal ions from the sorbent phase by EDTA with the formation of stable complexes. This is evident from the emergence of absorption bands typical for a Ni-EDTA complex in the electronic spectra of LRW passed through Termoksid-35 (Ni-K ferrocyanide) without HTO pretreatment (Figure 10).

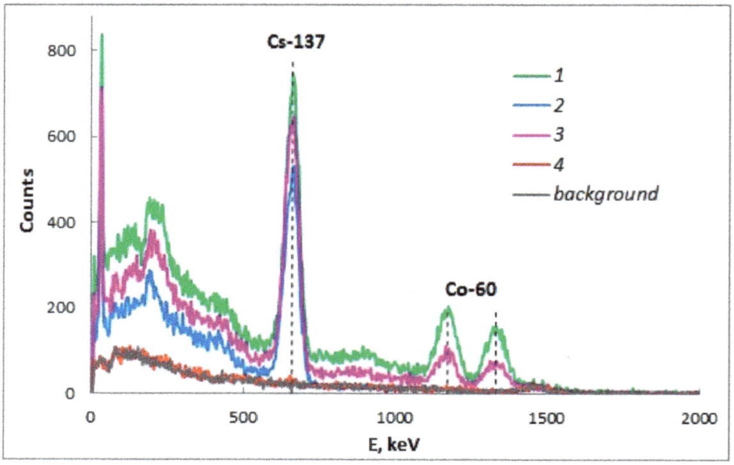

Figure 9. γ-spectra of LRW with pH 9.5 before purification (1), after HTO at 220 °C; at volumetric flow rates of 1 M H_2O_2 of 0.25 mL/min and of LRW of 0.5 (2) and 0.75 (3) mL/min, and after post-treatment using Termoksid-35 sorbent (4).

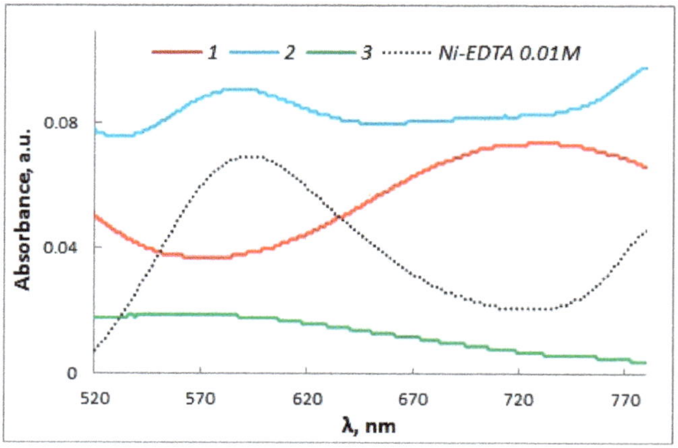

Figure 10. Electronic spectra of LRW obtained by mixing spent alkaline and acidic DSs before (1) and after filtration through Termoksid-35 without (2) and with (3) HTO pretreatment. Electronic spectra of 0.01 M Ni-EDTA solution is given for comparison.

To sum up, it can be concluded that LRW formed in the SIER decontamination process can be decontaminated by a combination of HTO and selective sorption to the level at which we can consider them nonradioactive wastes.

3.5. The Optimized Scheme of SIER Decontamination

Based on the results obtained, an optimized SIER decontamination scheme was suggested (Scheme 3). This scheme was obtained by the supplementation of Scheme 1 (RFR-assisted alkaline decontamination) with acidic decontamination (Scheme 2) and with an LRW treatment unit combining HTO and selective sorption in a column with a ferrocyanide sorbent. The contributions of each stage to the overall SIER decontamination efficiency are summarized in Table 5. The decontamination factors for ^{137}Cs, ^{60}Co, and ^{94}Nb radionuclides were 3.9×10^3, 7.6×10^2, and 1.3×10^2, respectively, and the activity of the decontaminated SIER was 17 Bq/g. According to the Classification of Radioactive Wastes, this residual activity is higher than that of exempt waste (EW), which is a few Bq/g, but meets criteria of very low-level waste (VLLW) [44], which do not need a high level of containment and isolation and, therefore, are suitable for disposal in near-surface landfill-type facilities with limited regulatory control [45].

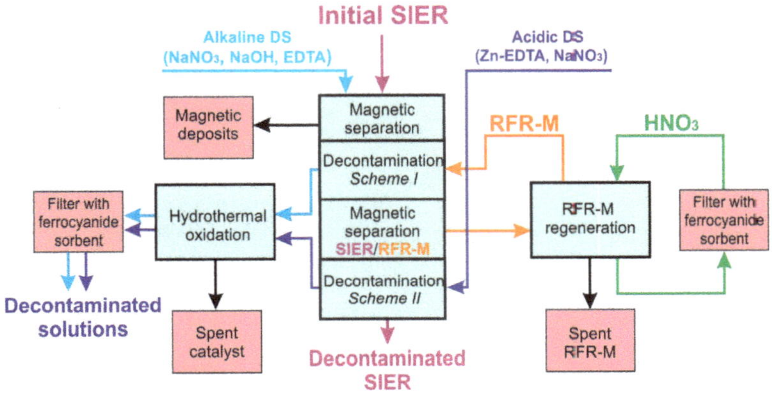

Scheme 3. The optimized scheme of SIER decontamination.

Table 5. Specific activities (Bq/g) of the air-dry SIER decontaminated using Scheme 3.

	^{137}Cs	^{60}Co	^{94}Nb	Σ
Initial SIER	2.7×10^4	3.1×10^3	7.6×10^2	3.1×10^4
SIER after wet magnetic separation	2.1×10^4	1.7×10^3	1.7×10^2	2.3×10^4
SIER after RFR-assisted decontamination using alkaline DS	127	1.5×10^3	95	1.7×10^3
SIER after decontamination using alkaline and acidic DS	6.9	4.1	5.9	1.7×10^1

3.6. Solid Radioactive Wastes Generated after SIER Decontamination

Solid radioactive wastes (SRWs) formed as a result of SIER decontamination using Scheme 3 include filters with ferrocyanide sorbents, a spent HTO catalyst, and powdered deposits magnetically separated from SIER. Spent RFR-M after cesium elution can be dissolved, and the solution formed can be further processed using HTO and selective sorption [43]; the insoluble inorganic fraction from RFR-M (synthetic magnetite and magnetic deposits) can be disposed of as SWR. As the volume of the decontaminated solutions significantly exceeds the volume of the sorbents and the catalyst used in this process, the reduction in the radioactive waste amount to be sent for the final disposal will be determined by the volumetric LRW/SRW ratio.

The radionuclide distribution between SRW types generated from the application of Scheme 3 to the decontamination of 10 mL of SIER is shown in Figure 11: ^{137}Cs is immobilized in ferrocyanide sorbents, ^{60}Co is immobilized in an HTO catalyst, and magnetic deposits contain ^{60}Co and ^{94}Nb.

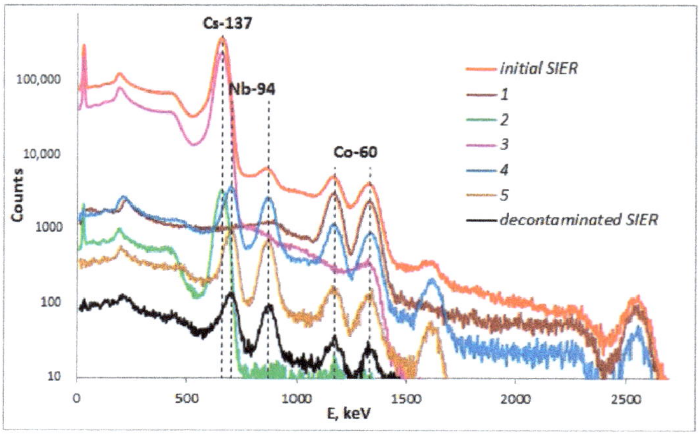

Figure 11. γ-spectra of solid wastes generated after decontamination of 10 mL of SIER: 1 mL of HTO catalyst (1), 1 mL of Termoksid-35 (2), 1 mL of FNK-50 sorbent (3), 0.1 g of magnetic deposits (4), and 0.8 g of spent RFR-M after regeneration with nitric acid (5). Spectra of initial and decontaminated SIER samples are given for comparison.

4. Conclusions

A feasibility study of decontamination of intermediate-level-activity spent ion-exchange resins (SIERs) from the Kursk NPP has been conducted with the primary objective to transfer SIER to the category of exempt waste (EW) or very low-level waste (VLLW) to reduce disposal costs. It has been shown that nonexchangeable radionuclides, which cannot be removed during resin regeneration, are concentrated in poorly soluble inorganic deposits represented by aluminosilicates, activated corrosion products, and fragments of the fuel element cladding. The suggested conceptual scheme of SIER decontamination included "wet" magnetic separation and sequential washing with alkaline and acidic deactivation solutions containing EDTA. Magnetic separation provided an efficiency of ^{94}Nb and ^{60}Co radionuclide removal of ~77% and ~45%, respectively. The decontamination of SIER using an alkaline deactivation solution (2.25 mol/L $NaNO_3$, 0.75 mol/L NaOH, and 0.02 mol/L EDTA) in the presence of magnetic resorcinol-formaldehyde resin (RFR-M) allowed the removal of >99% of ^{137}Cs radionuclides and a significant reduction in ^{94}Nb activity. ^{137}Cs radionuclides concentrated in RFR-M could be further eluted with nitric acid and immobilized in a ferrocyanide sorbent for disposal. SIER decontamination from ^{60}Co radionuclides, which are both ion-exchangeable and captured in poorly soluble deposits and were not separated magnetically, was conducted using acidic solutions containing Zn(II)-EDTA complexes, and the efficiency exceeded 99%. The activity removed from SIER was transferred to the inorganic solid matrices using selective sorption and hydrothermal oxidation (HTO), which provided an efficient reduction in the solid waste volume, virtually avoiding the generation of additional liquid radioactive wastes.

To sum up, the application of the suggested strategy allowed a decrease in the specific activity of the SIER sample from 3.1×10^4 (2.7×10^4 Bq/g of ^{137}Cs, 3.1×10^3 Bq/g of ^{60}Co, and 7.6×10^2 Bq/g of ^{94}Nb) to 16.9 Bq/g (6.9 Bq/g of ^{137}Cs, 4.1 Bq/g of ^{60}Co, and 5.9 Bq/g of ^{94}Nb), with the final decontamination factors above 10^3 and 10^2 for ^{137}Cs and ^{60}Co and ^{94}Nb, respectively. Although the decontaminated SIER did not meet criteria for EW (a few Bq/g), it can be considered VLLW (tens of Bq/g) [44] and, according to IAEA recommendations [45], is suitable for disposal in near-surface landfill-type facilities with limited regulatory control.

Author Contributions: Conceptualization, M.P.; methodology, M.P.; validation, N.P.; investigation, M.P., M.C. and N.P.; writing—original draft preparation, M.P.; writing—review and editing, S.B., M.P.; visualization, M.P. All authors have read and agreed to the published version of the manuscript.

Funding: The studies of SIER decontamination were funded by the Russian Science Foundation, Project 18-73-10066. The research on the purification of liquid radioactive wastes generated in SIER decontamination was funded by the Ministry of Higher Education and Science of the Russian Federation, State Order Project No. FWFN (0205)-2022-0002 of the Institute of Chemistry FEBRAS.

Institutional Review Board Statement: Not applicable.

Informed Consent Statement: Not applicable.

Data Availability Statement: Not applicable.

Acknowledgments: Elemental compositions and XRD patterns were determined using equipment of the Far East Center of Structural Studies (Institute of Chemistry, FEB RAS, Vladivostok, Russia). The authors are thankful to Elena Kaplun (Institute of Chemistry, FEB RAS, Vladivostok, Russia) for the HTO catalyst synthesis.

Conflicts of Interest: The authors declare no conflict of interest.

References

1. International Atomic Energy Agency (IAEA). *New Development and Improvements in Processing of Problematic Radioactive Waste*; International Atomic Energy Agency (IAEA): Vienna, Austria, 2007.
2. International Atomic Energy Agency (IAEA). *Management of Spent Ion-Exchange Resins from Nuclear Power Plants*; International Atomic Energy Agency (IAEA): Vienna, Austria, 1981.
3. Wang, J.; Wan, Z. Treatment and Disposal of Spent Radioactive Ion-Exchange Resins Produced in the Nuclear Industry. *Prog. Nucl. Energy* **2015**, *78*, 47–55. [CrossRef]
4. Osmanlioglu, A.E. Progress in Cementation of Reactor Resins. *Prog. Nucl. Energy* **2007**, *49*, 20–26. [CrossRef]
5. Abdel Rahman, R.O.; Zaki, A.A. Comparative Analysis of Nuclear Waste Solidification Performance Models: Spent Ion Exchanger-Cement Based Wasteforms. *Process Saf. Environ. Prot.* **2020**, *136*, 115–125. [CrossRef]
6. Abdel Rahman, R.O.; Ojovan, M.I. Toward Sustainable Cementitious Radioactive Waste Forms: Immobilization of Problematic Operational Wastes. *Sustainability* **2021**, *13*, 11992. [CrossRef]
7. Wan, Z.; Xu, L.; Wang, J. Treatment of Spent Radioactive Anionic Exchange Resins Using Fenton-like Oxidation Process. *Chem. Eng. J.* **2016**, *284*, 733–740. [CrossRef]
8. Gunale, T.L.; Mahajani, V.V.; Wattal, P.K.; Srinivas, C. Studies in Liquid Phase Mineralization of Cation Exchange Resin by a Hybrid Process of Fenton Dissolution Followed by Wet Oxidation. *Chem. Eng. J.* **2009**, *148*, 371–377. [CrossRef]
9. Leybros, A.; Roubaud, A.; Guichardon, P.; Boutin, O. Supercritical Water Oxidation of Ion Exchange Resins: Degradation Mechanisms. *Process Saf. Environ. Prot.* **2010**, *88*, 213–222. [CrossRef]
10. Nezu, A.; Morishima, T.; Watanabe, T. Thermal Plasma Treatment of Waste Ion-Exchange Resins Doped with Metals. *Thin Solid Film.* **2003**, *435*, 335–339. [CrossRef]
11. Kamaruzaman, N.S.; Kessel, D.S.; Kim, C.-L. Management of Spent Ion-Exchange Resins From Nuclear Power Plant by Blending Method. *J. Nucl. Fuel Cycle Waste Technol.* **2018**, *16*, 65–82. [CrossRef]
12. United States National Reguratory Commission. *Final Comparative Environmental Evaluation of Alternatives for Handling Low-Level Radioactive Waste Spent Ion Exchange Resins from Commercial Nuclear Power Plants*; Office of the Federal Register, National Archives and Records Administration: College Park, MD, USA, 2013; Volume 78.
13. Korchagin, Y.P.; Aref'Ev, E.K.; Korchagin, E.Y. Improvement of Technology for Treatment of Spent Radioactive Ion-Exchange Resins at Nuclear Power Stations. *Therm. Eng.* **2010**, *57*, 593–597. [CrossRef]
14. Yeon, J.W.; Choi, I.K.; Park, K.K.; Kwon, H.M.; Song, K. Chemical Analysis of Fuel Crud Obtained from Korean Nuclear Power Plants. *J. Nucl. Mater.* **2010**, *404*, 160–164. [CrossRef]
15. Vaidotas, A. Waste Inventory for Near Surface Repository (NSR)—13482. In Proceedings of the WM2013—Waste Management Conference: International Collaboration and Continuous Improvemen, Phoenix, AZ, USA, 24–28 February 2013.
16. Chaturvedi, S.; Dave, P.N. Removal of Iron for Safe Drinking Water. *Desalination* **2012**, *303*, 1–11. [CrossRef]
17. Becker, F.L.; Rodríguez, D.; Schwab, M. Magnetic Removal of Cobalt from Waste Water by Ferrite Co-Precipitation. *Procedia Mater. Sci.* **2012**, *1*, 644–650. [CrossRef]
18. Hingston, F.J.; Raupach, M. The Reaction between Monosilicic Acid and Aluminium Hydroxide: I. Kinetics of Adsorption of Silicic Acid by Aluminium Hydroxide. *Aust. J. Soil Res.* **1967**, *5*, 295–309. [CrossRef]
19. Ritcey, G.M. Silica Fouling in Ion Exchange, Carbon-in-Pulp and Solvent Extraction Circuits. *Can. Metall. Q.* **1986**, *25*, 31–43. [CrossRef]
20. Peter, B.; Resintech, M. Behavior of Silica in Ion Exchange and Other Systems. In Proceedings of the International Water Conference, 60th Annual Meeting, Spokane, WA, USA, 4–8 August 1999; pp. 1–10.
21. Kikuchi, M.; Sugimoto, Y.; Yusa, H.; Ebara, K.; Takeshima, M. Development of a Laundry Waste Treatment System. *Nucl. Eng. Des.* **1977**, *44*, 413–420. [CrossRef]
22. Fujii, S.; Murakami, R. Smart Particles as Foam and Liquid Marble Stabilizers. *KONA Powder Part. J.* **2008**, *26*, 153–166. [CrossRef]

23. Palamarchuk, M.; Egorin, A.; Tokar, E.; Tutov, M.; Marinin, D.; Avramenko, V. Decontamination of Spent Ion-Exchangers Contaminated with Cesium Radionuclides Using Resorcinol-Formaldehyde Resins. *J. Hazard. Mater.* **2017**, *321*, 326–334. [CrossRef]
24. Palamarchuk, M.S.; Tokar, E.A.; Tutov, M.V.; Yegorin, A.M. Deactivation of Spent Ion-Exchange Resins Contaminated by Cesium and Cobalt Radionuclides. *Ecol. Ind. Russ.* **2019**, *23*, 20–24. [CrossRef]
25. Wang, J.; Zhuang, S. Removal of Cesium Ions from Aqueous Solutions Using Various Separation Technologies. *Rev. Environ. Sci. Biotechnol.* **2019**, *18*, 231–269. [CrossRef]
26. Hassan, N.M.; Adu-Wusu, K.; Marra, J.C. Resorcinol—Formaldehyde Adsorption of Cesium from Hanford Waste Solutions: Part I. Batch Equilibrium Study. *J. Radioanal. Nucl. Chem.* **2004**, *262*, 579–586. [CrossRef]
27. Dumont, N.; Favre-Réguillon, A.; Dunjic, B.; Lemaire, M. Extraction of Cesium from an Alkaline Leaching Solution of Spent Catalysts Using an Ion-Exchange Column. *Sep. Sci. Technol.* **1996**, *31*, 1001–1010. [CrossRef]
28. Burgeson, I.E.; Deschane, J.R.; Cook, B.J.; Blanchard, D.L.; Weier, D.L. Evaluation of Elution Parameters for Cesium Ion Exchange Resins. *Sep. Sci. Technol.* **2006**, *41*, 2373–2390. [CrossRef]
29. Fiskum, S.K.; Colburn, H.A.; Rovira, A.M.; Allred, J.R.; Smoot, M.R.; Peterson, R.A.; Landon, M.R.; Colosi, K.A. Cesium Removal from AP-105 Hanford Tank Waste Using Spherical Resorcinol Formaldehyde Resin. *Sep. Sci. Technol.* **2019**, *54*, 1932–1941. [CrossRef]
30. Tokar, E.A.; Matskevich, A.I.; Palamarchuk, M.S.; Parotkina, Y.A.; Egorin, A.M. Decontamination of Spent Ion Exchange Resins Contaminated with Iron-Oxide Deposits Using Mineral Acid Solutions. *Nucl. Eng. Technol.* **2021**, *53*, 2918–2925. [CrossRef]
31. Palamarchuk, M.S.; Shelestyuk, E.A.; Tokar', E.A.; Egorin, A.M. Sorption of Complex Co(II)-EDTA by Anion Exchange Resin AB-17-8 from Model Deactivating Solutions Containing Fe(III)-EDTA. *Ecol. Ind. Russ.* **2020**, *24*, 20–23. [CrossRef]
32. Smith, R.M.; Martell, A.E. Critical Stability Constants, Enthalpies and Entropies for the Formation of Metal Complexes of Aminopolycarboxylic Acids and Carboxylic Acids. *Sci. Total Environ.* **1987**, *64*, 125–147. [CrossRef]
33. Karhu, J.; Harju, L.; Ivaska, A. Determination of the Solubility Products of Nitrilotriacetic Acid, Ethylenediaminetetraacetic Acid and Diethylenetriaminepentaacetic Acid. *Anal. Chim. Acta* **1999**, *380*, 105–111. [CrossRef]
34. Anderegg, G. Komplexone, X. Die Protonierungskonstanten Einiger Komplexone in Verschiedenen Wässerigen Salzmedien ($NaClO_4$, $(CH_3)_4NCl$, KNO_3). *Helv. Chim. Acta* **1967**, *50*, 2333–2340. [CrossRef]
35. Du, J.; Zhang, B.; Li, J.; Lai, B. Decontamination of Heavy Metal Complexes by Advanced Oxidation Processes: A Review. *Chin. Chem. Lett.* **2020**, *31*, 2575–2582. [CrossRef]
36. Walling, S.A.; Um, W.; Corkhill, C.L.; Hyatt, N.C. Fenton and Fenton-like Wet Oxidation for Degradation and Destruction of Organic Radioactive Wastes. *NPJ Mater. Degrad.* **2021**, *5*, 1–20. [CrossRef]
37. Rekab, K.; Lepeytre, C.; Goettmann, F.; Dunand, M.; Guillard, C.; Herrmann, J.M. Degradation of a Cobalt(II)–EDTA Complex by Photocatalysis and H2O2/UV-C. Application to Nuclear Wastes Containing60Co. *J. Radioanal. Nucl. Chem.* **2015**, *303*, 131–137. [CrossRef]
38. Palamarchuk, M.; Voit, A.; Papynov, E.; Marinin, D.; Bratskaya, S.; Avramenko, V. Quantum Chemistry and Experimental Studies of Hydrothermal Destruction of Co-EDTA Complexes. *J. Hazard. Mater.* **2019**, *363*, 233–241. [CrossRef]
39. Avramenko, V.; Mayorov, V.; Marinin, D.; Mironenko, A.; Palamarchuk, M.; Sergienko, V. Macroporous Catalysts for Hydrothermal Oxidation of MetalIorganic Complexes at Liquid Radioactive Waste Treatment. In Proceedings of the ASME 2010 13th International Conference on Environmental Remediation and Radioactive Waste Management, Tsukuba, Japan, 3–7 October 2010; Volume 1, pp. 183–187.
40. Avramenko, V.A.; Voit, A.V.; Dmitrieva, E.E.; Dobrzhanskii, V.G.; Maiorov, V.S.; Sergienko, V.I.; Shmatko, S.I. Hydrothermal Oxidation of Co-EDTA Complexes. *Dokl. Chem.* **2008**, *418*, 19–21. [CrossRef]
41. Avramenko, V.A.; Bratskaya, S.Y.; Voit, A.V.; Dobrzhanskiy, V.G.; Egorin, A.M.; Zadorozhniy, P.A.; Mayorov, V.Y.; Sergienko, V.I. Implementation of the Continuous-Flow Hydrothermal Technology of the Treatment of Concentrated Liquid Radioactive Wastesat Nuclear Power Plants. *Theor. Found. Chem. Eng.* **2010**, *44*, 592–599. [CrossRef]
42. Avramenko, V.A.; Bratskaya, S.Y.; Karpov, P.A.; Mayorov, V.Y.; Mironenko, A.Y.; Palamarchuk, M.S.; Sergienko, V.I. Macroporous Catalysts for Liquid-Phase Oxidation on the Basis of Manganese Oxides Containing Gold Nanoparticles. *Dokl. Phys. Chem.* **2010**, *435*, 193–197. [CrossRef]
43. Palamarchuk, M.; Egorin, A.; Golikov, A.; Trukhin, I.; Bratskaya, S. Hydrothermal Oxidation of Pre-Dissolved Resorcinol-Formaldehyde Resins as a New Approach to Safe Processing of Spent Cesium-Selective Organic Ion-Exchangers. *J. Hazard. Mater.* **2021**, *416*, 125880. [CrossRef]
44. Crossland, I. Disposal of Radioactive Waste. In *Nuclear Fuel Cycle Science and Engineering*; Elsevier: Amsterdam, The Netherlands, 2012; pp. 531–557. ISBN 9780857090737.
45. International Atomic Energy Agency (IAEA). *Classification of Radioactive Waste*; No. GSG-1; International Atomic Energy Agency (IAEA): Vienna, Austria, 2009; p. 48.

sustainability

Article

Influence of Rock Structure on Migration of Radioactive Colloids from an Underground Repository of High-Level Radioactive Waste

Victor I. Malkovsky, Vladislav A. Petrov, Sergey V. Yudintsev, Michael I. Ojovan * and Valeri V. Poluektov

Institute of Geology of Ore Deposits, Petrography, Mineralogy, and Geochemistry Russian Academy of Sciences, Staromonetny Lane, 35, 119017 Moscow, Russia
* Correspondence: m.i.ojovan@gmail.com

Abstract: Studies of leaching of vitrified simulated high-level radioactive waste (HLW) evidence that most of actinides or their simulators enter leaching water in a colloidal form. In this paper, we consider a mechanism of colloid-facilitated migration of radionuclides from an underground repository of HLW located at a depth of a few hundreds of meters in fractured crystalline rocks. The comparison between data of field and laboratory measurements showed that the bulk permeability of the rock massif in field tests is much greater than the permeability of rock samples in laboratory experiments due to an influence of a network of fractures in the rock massif. Our theoretical analysis presents evidence that this difference can take place even in a case when the network is not continuous, and the fractures are isolated with each other through a porous low-permeable matrix of the rock. Results of modelling revealed a possibility of mechanical retention of radionuclide-bearing colloid particles in the frame of rock during their underground migration.

Keywords: radioactive waste; underground repository; radiocolloid; migration; ground water; fractures; mechanical retention

Citation: Malkovsky, V.I.; Petrov, V.A.; Yudintsev, S.V.; Ojovan, M.I.; Poluektov, V.V. Influence of Rock Structure on Migration of Radioactive Colloids from an Underground Repository of High-Level Radioactive Waste. *Sustainability* **2023**, *15*, 882. https://doi.org/10.3390/su15010882

Academic Editor: Xiaojun Feng

Received: 29 November 2022
Revised: 29 December 2022
Accepted: 30 December 2022
Published: 3 January 2023

1. Introduction

Observed climate changes and the necessity to reduce emission of greenhouse gases increases the significance of nuclear power engineering. Sustainable development of the atomic industry calls for a safe solution of high-level radioactive waste (HLW) management. The most effective and reliable approach to solving the problem at present is disposal of conditioned (solidified) HLW in underground repositories at a depth of several hundred meters [1–3]. Safety of such geological repositories is based on the multibarrier concept, which implies that both engineered and natural barriers provide for a reliable isolation of HLW in the repository from the biosphere. The engineered barriers include waste forms, canisters, and containers for solid HLW, backfill for holes in the repository where the canisters are disposed of, and repository construction units. The natural barrier is a rock massif between the loaded part of the repository and the biosphere [2,4]. The engineered barriers can degrade during the time period that is comparable with the half-life of many radionuclides from the HLW's composition [5]. As a result, after an initial period, the rocks can be considered as the main isolation barrier protecting the biosphere from HLW. The main hazard of radionuclides ingress from HLW to the biosphere is caused by their transport by groundwater, which flows in rocks through systems of connected pore and fracture voids [6–8]. These systems are called flow channels, the scales of which are considered in [9]. Reliability of HLW isolation depends on the time of radionuclides migration from the loaded part of the repository to the biosphere. Concentration of radionuclides in groundwater decreases during their migration to the biosphere due to radioactive decay. If the concentration in the groundwater at its leakage to the earth surface (or to a water reservoir or a river net) is less than a maximum allowable level, the natural barrier is reliable, and the repository is safe [1]. Therefore, duration of radionuclides migration through the rocks and then the migration velocity is of paramount importance for the repository safety.

This paper analyses the process of radionuclide migration from underground repositories to the biosphere via colloid-facilitated transport of radionuclides by flowing groundwater accounting for the mechanical retention of colloidal radionuclide-bearing particles.

Continuum-scale modeling of groundwater flow in fractured rock with an explicit treatment of rock fracturing is used in many safety assessments of hazardous underground radiation objects. One of the first such models was proposed by Barenblatt et al. [10]. Development of this approach as applied to contaminant transport in fractured rocks was presented by J. Bear [6], I. Neretnieks [7], and C.-F. Tsang [11]. Reviews of methods for continuum-scale modeling flow in fractured rocks are provided in [12,13]. Most attention is paid to the influence of fractures that form a connected network. However, there is an option that the fractures do not form a linked structure even in a case when elevated permeability of the rock indicates directly a significant influence of rock fracturing. The fracture network channelizes the groundwater flow in this case, but the fractures can be disconnected, and parts of the groundwater streamlines between the fractures run through the rock matrix.

A substantial ecological hazard can be caused by a release of radionuclides from the underground repository of HLW and their transport from the repository to the biosphere. The groundwater can carry radionuclides as ions and colloid particles (radiocolloids). Since the radiocolloids can be much more mobile in the underground medium than the radionuclide ions they can represent the most hazardous form of radionuclides migration [8]. The intervals of the rock matrix between the disconnected fractures along the same streamlines of the groundwater can serve as filters for the particles of radionuclide-bearing colloids. Analysis of this option is the objective of this study.

We show that (i) even a system of disconnected fractures can cause a difference by a few orders of magnitude between bulk permeability of rock massif and permeability of the nonfractured rock matrix; (ii) analysis of permeability measurement at the site of potential federal repository of HLW indicates that fractures in the rock are, at least, partially disconnected; (iii) intervals between the disconnected fractures along the same streamlines of the groundwater can be effective filters retaining radionuclide-bearing colloids. Analysis of the influence of disconnected fractures on the bulk permeability of rocks is carried out by methods of computer modelling.

2. Colloid-Facilitated Transport of Radionuclides by Groundwater

As a result of sorption on walls of the flow channels, cations of many radionuclides should move at a velocity that is much less than velocity of the groundwater [14,15]. Let us denote velocities of any contaminant and the groundwater as V_c and V_f, respectively. As a first approximation, one can assume that sorption is reversible and satisfies the linear equation

$$C_r = \rho K_d C, \tag{1}$$

where C_r and C are mass fractions of the contaminant in the rock and in the groundwater, ρ is groundwater density, K_d is a coefficient which characterizes sorption properties of the rock to the contaminant.

Then ratio between V_c and V_f satisfies the expression

$$V_c/V_f = 1/(1 + \rho_r K_d/\varphi), \tag{2}$$

where ρ_r is rock density and φ is rock porosity.

The higher sorption properties of the rocks to the contaminant (K_d), the less is the ratio V_c/V_f. If sorption of radionuclides on the rocks is absent, the ratio V_c/V_f tends to 1, which is the maximum value of this ratio except of particular cases that are considered in [16–20]. However, results of radiation monitoring at the sites of significant radioactive pollution showed that values of V_c/V_f were much higher than it was predicted on the basis of sorption properties of the rocks to the radionuclides [16–23]. Similar result was obtained in laboratory experiments [24]. Elevated values of the radionuclide's migration velocity as compared to its predicted values were explained by the assumption that groundwater carries radionuclides not only as a solute, but also in bound form as colloid particles (by definition, particles are called colloidal if their size ranges from 1 to 1000 nm), which are sorbed by the rocks to a lower extent than the radionuclide ions [25]. Colloid particles that carry radionuclides in groundwater are called radiocolloids. The colloids are subdivided into three main groups according to their origin: intrinsic colloids, primary colloids and pseudocolloids [26,27]. Intrinsic colloids consist of particles of colloid size that are composed to a significant extent of radioactive isotopes and their oxi- and hydroxides. Primary colloids represent colloid particles composed of leaching products of HLW vitreous form at its contact with groundwater. Pseudocolloids consist of colloid particles existing in the groundwater before contact with radioactive materials. At contact of the particles with polluted water radionuclides will be sorbed on these particles.

At present, immobilization of HLW on industrial scale is carried out using Al-P- or B-Si-glass [3,28,29]. Experiments on leaching of aged Na-Al-P-glass by water show that more than 95% of actinide simulators in the leaching products are attached to particles with diameter between 450 and 100 nm [30–32]. Thus, primary colloids represent more than 95% of actinide simulators' ingress into the groundwater. Former studies have demonstrated that formation of two kinds of radioactive colloids: (1) primary—at hydrothermal alteration of vitreous waste forms and (2) pseudocolloids—at expense of initial colloids of waters or derived from eroded bentonite buffer are typical processes in the environment of underground HLW repository [17–25,33–35]. Therefore, one can expect that practically all actinides resulting from nuclear waste forms corrosion can be carried by the groundwater in highly mobile colloidal form, which will potentially decrease the safety of the repository [32]. However, the probability exists that diameters of the primary colloid particles can be larger than apertures of the filtration channels, and the coarsest fraction of primary colloids will be mechanically retained. Analysis of this possibility is the main subject of this study.

3. Influence of Rock Fracturing

It is known that the bulk permeability of crystalline rocks is usually higher by 1–3 orders of magnitude than the permeability of rock matrix, which is obtained in laboratory measurements on small ($\cong 10^{-2}$ m in size) samples [36,37]. This is caused by the influence of fractures, the lengths of which are much larger than dimensions of interstitial voids and pores of the rock matrix. The latter focuses on the mainstream of the groundwater, which flows through the rock massif. Since they are much larger in size than rock samples used for laboratory measurement of permeability, laboratory measurements do not take into account the influence of fractures, and the permeability of small samples (i.e., permeability of the rock matrix) is much less than the bulk permeability of the massif. One can suppose in this case that mechanical retention of radiocolloid should be absent because the significant difference between bulk and sample permeabilities suggests that the fracture network and relatively large apertures of fractures are quite sufficient for free movement of radiocolloid particles along the fractures with the mainstream of the groundwater.

However, the elevated bulk permeability compared with the rock matrix permeability do only indicate on presence of fractures, although separate fractures can be disconnected and do not form a linked network through the rock massif.

Let us examine the influence of disconnected fractures on bulk permeability of rocks. We first consider a 2D cross-section of a fractured rock with porous matrix and chessboard ordering of the disconnected fractures directed along a mean flow of groundwater in the rock massif (Figure 1a) noting that the cross-section can be inclined in a general case.

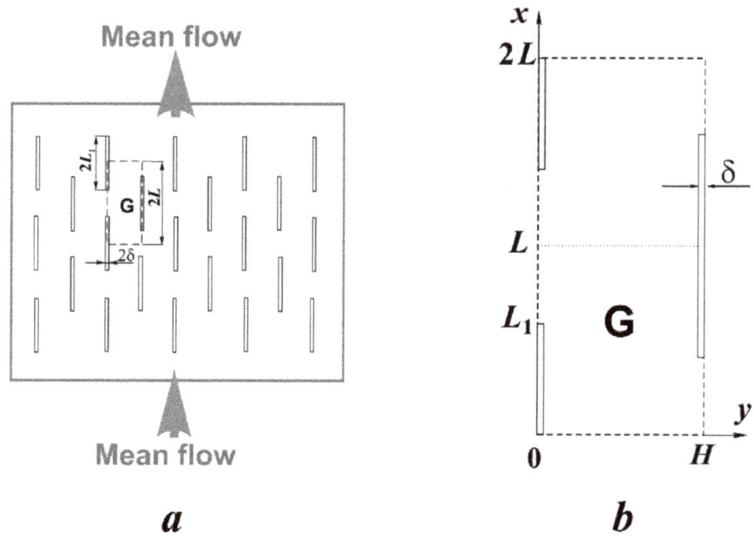

Figure 1. Diagram of porous–fractured rocks: (**a**) chessboard ordering of fractures in a porous permeable rock matrix; (**b**) recurrent cell of the porous–fractured medium at chessboard ordering of the fractures in the rock matrix.

Let us consider the groundwater flow in the recurrent cell G and introduce in it Cartesian coordinates as shown in Figure 1b.

The water flow in the porous matrix (outside the fractures) is governed by Darcy's law [15]

$$v_x = -\frac{k_m}{\mu}\frac{\partial p}{\partial x},\ v_y = -\frac{k_m}{\mu}\frac{\partial p}{\partial y}, \tag{3}$$

where $v_x,\ v_y$ are components of the Darcy's velocity, k_m is permeability of the porous matrix, μ is dynamic viscosity of the groundwater, $p = P + \rho g z(x,y)$, P is pressure, ρ is groundwater density, g is acceleration due to gravity, $z(x,y)$ is altitude of the point with coordinates x and y above any fixed horizontal plane. Since the cross-section is a plane, $z(x,y)$ is a linear function. If the cross-section is horizontal, then $p = P + const$.

Since the groundwater is practically incompressible, components of the velocity satisfy the continuity equation in the form

$$\frac{\partial v_x}{\partial x} + \frac{\partial v_y}{\partial y} = 0. \tag{4}$$

Hence, p satisfies the Laplace equation

$$\frac{\partial^2 p}{\partial x^2} + \frac{\partial^2 p}{\partial y^2} = 0. \tag{5}$$

We assume that the average velocity in cross-sections of fractures is governed by 1D Darcy's equation [6]. Then, the flow in the fractures is governed by mass balance equations in the form

$$\begin{aligned} -\frac{\delta k_f}{k_m}\frac{\partial^2 p}{\partial x^2}\bigg|_{y=0} &= \frac{\partial p}{\partial y}\bigg|_{y=\delta+0},\ 0 < x < L_1 \text{ or } 2L - L_1 < x < 2L; \\ \frac{\delta k_f}{k_m}\frac{\partial^2 p}{\partial x^2}\bigg|_{y=H} &= \frac{\partial p}{\partial y}\bigg|_{y=H-\delta-0},\ L - L_1 < x < L + L_1 \end{aligned}, \tag{6}$$

where 2δ is fracture aperture, $2L_1$ is fracture length, $2L$ is length of the recurrent cell G.

As a result of mirror symmetry of the streamlines,

$$\begin{array}{c} \frac{\partial p}{\partial y} = 0,\ y = 0,\ L_1 < x < 2L - L_1, \\ \frac{\partial p}{\partial y} = 0,\ y = H,\ \{x < L - L_1 \text{ or } x > L + L_1\}.' \end{array} \quad (7)$$

Let us denote pressures at $x = 0,\ 2L$ as p_0 and p_1, respectively. It follows from the mirror symmetry of streamlines that p_0 and p_1 do not depend on y. Hence,

$$x = 0,\ p = p_0;\ x = 2L,\ p = p_1. \quad (8)$$

Equalities (6)–(8) are boundary conditions for the Equation (5). The boundary problem (5)–(8) was solved numerically by finite differences method of successive over relaxation. Since the considered porous–fractured medium consists of recurrent cells G, we can express the bulk permeability of the medium on the basis of the obtained numerical solution as

$$k_{bulk} = \frac{\mu}{H}\frac{2L}{p_0 - p_1}\int_0^H v_x dy. \quad (9)$$

Results of k_{bulk} calculations can be approximated by the dimensionless expression

$$\frac{k_{bulk}}{k_m} = F(h, l, M) = \frac{h^2 F_1(h, l, M) + 0.328 F_2(h, l, M)}{h^2 + 0.328}, \quad (10)$$

where

$$\begin{array}{c} l = \frac{L}{L_1},\ h = \frac{H}{L_1},\ M = \frac{\delta k_f}{L_1 k_m}, \\ F_1 = 1 - \frac{4l}{\pi h} F_M \left\{ \cos\left(\frac{\pi}{2l}\right) + \exp\left[-\frac{\pi(M+2.62)}{2l}\right] \right\}, \\ F_2 = 1 + \left[\frac{h}{2M} + \frac{0.439h^2}{h^{1.46}/(l-1)^{0.138} + 0.439 F_M F_l} \right], \\ F_M = M/(M+2.62),\ F_l = \exp\left\{ -(l-1)^2 \left[1 + (l-1)^{3.57} \right] \right\} \end{array} \quad (11)$$

Comparison of k_{bulk}/k_m obtained numerically and calculated by approximating Formula (10) is shown in Figure 2.

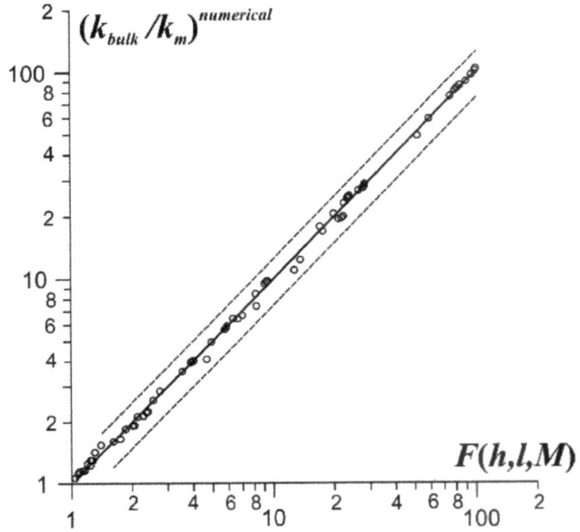

Figure 2. Comparison of bulk permeability values calculated numerically with approximating Formula (10). Dashed lines correspond to errors of ±25%.

The chessboard ordering of fractures is of particular importance for estimation of the influence of disconnected fracturing on bulk permeability of rock massifs. Let us consider disconnected fractures of equal length and aperture, which are disposed with equal spacing in rows, and distances between neighboring rows are equal. An example of this type of fracturing is the considered case of chessboard ordering of fractures. Another example is a rectangle-cluster ordering of the fractures (Figure 3).

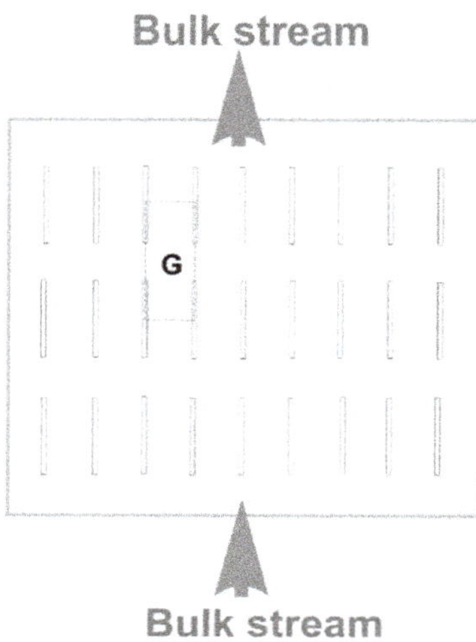

Figure 3. Rectangle-cluster ordering of fractures in porous rock matrix.

We show that chessboard ordering corresponds to an extremum in influence of such fracturing on bulk permeability of rocks. Let us enumerate fracture rows. In the case of chessboard ordering, middles of fractures in odd rows correspond to middles of intervals between fractures in even rows (Figure 4a).

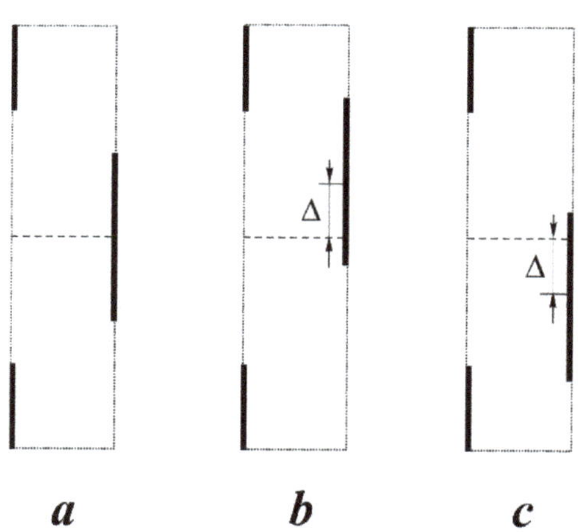

Figure 4. Modifications of chessboard ordering of the fractures. (**a**) chessboard ordering of fractures; (**b**) chessboard ordering shifted by Δ along the row direction; (**c**) chessboard ordering shifted by Δ in the opposite direction.

Let us shift odd rows by a distance Δ along the row directions (Figure 4b). We denote the bulk permeability of obtained medium as $k_{bulk}(\Delta)$. Then we consider shift of the odd rows from their initial position by a distance Δ in opposite direction, i.e., by $-\Delta$ (Figure 4c).

Permeability of the medium in this case is $k_{bulk}(-\Delta)$. Boundary problem (5)–(8) is linear. This implies in particular that the value of bulk permeability does not change if direction of the mean flow becomes opposite. Therefore, $k_{bulk}(-\Delta) = k_{bulk}(\Delta)$. Hence, $k_{bulk}(\Delta)$ is an even function, and

$$\frac{dk_{bulk}}{d\Delta}(0) = 0. \tag{12}$$

Therefore, $\Delta = 0$ (which corresponds to chessboard ordering of the fractures) is an extremum point of k_{bulk}.

In absolutely the same manner, we can show that rectangle-cluster ordering of the fractures is also an extremum point of k_{bulk}. Since k_{bulk} at chess-board ordering is much higher than in the case of rectangle-cluster ordering at the same values of $h,\ l,\ M$, it is reasonable to assume that chess-board and rectangle-cluster ordering correspond respectively to maximum and minimum influence of fracturing on k_{bulk} at the specified $h,\ l,\ M$.

One can see from Figure 3 that the bulk permeability can exceed the rock matrix permeability by more than two orders of magnitude even in the case of disconnected fractures. This implies that a significant difference between bulk permeability of the rocks and matrix permeability (which is measured in laboratory tests [30,31]) is not an argument in favor of an existence of a network of hydraulically connected fractures, which extends throughout the rock massif.

4. Characteristics of Fracturing of Nizhnekansky Massif Rocks (Eniseisky Site, Krasnoyarsk Region, Russia)

Granitoid massif Nizhnekansky is located in the Krasnoyarsk region (Russia). It is considered a potential territory for the development of a federal underground repository of high-level and intermediate-level nuclear waste with long-lived radionuclides [3,38]. Studies on the rock matrix were carried out on six samples from different sites and different depths of the granitoid massif. The core samples were 38–52 mm in diameter and about 150 mm in length. Petrographic and mineral–chemical studies have shown that the rocks are liable to metamorphic (amphibolite facies with quartz, feldspars, biotite, amphiboles) and low-temperature hydrothermal–metasomatic alterations (chloritization, sericitization and argillization), which are correspondingly accompanied by ductile (gneiss texture with characteristic foliation) and brittle (cataclastic, brecciated textures and microcracks filled by carbonate, chlorite, sericite and clay minerals) deformations [39,40]. A brief description of sample compositions and data on their permeability are presented in the Table 1, from which one can see that permeability of the samples (i.e., permeability of the rock matrix) does not exceed 3×10^{-18} m^2.

Table 1. Composition and permeability of rock samples [1].

Nos	Sample Index	Composition	Permeability, m^2
1	K 560.8	Granodiorite	1.488×10^{-18}
2	K 613.1	Porphyric adamellite	2.307×10^{-18}
3	I 142.6	Gneissic granite with metasomatic alterations	3.712×10^{-20}
4	I 491.7	Gneissic granite	8.201×10^{-19}
5	I 357.2	Quartz diorite	3.092×10^{-19}
6	I 504.6	Quartz diorite	9.595×10^{-19}

[1] Note: the sample number corresponds to depth of selection at the Kamenny (K) or Itatsky (I) sites.

Data of bulk permeability of rocks of the Yeniseisky site which is considered at present as the most promising place for development of the vitrified HLW repository are provided in [41]. Measurements were carried out in exploratory boreholes by pumping tests at depths up to 700 m. Results of the measurements are shown in Figure 5 (a and b where permeability k is expressed though hydraulic conductivity f as $k = f\mu/(\rho g)$, where ρ is groundwater density, g is acceleration due to gravity; $k \cong 1.16 \cdot 10^{-13} f$ in the considered case if permeability unit is m^2, and water conductivity unit is m/day, as in Figure 5b).

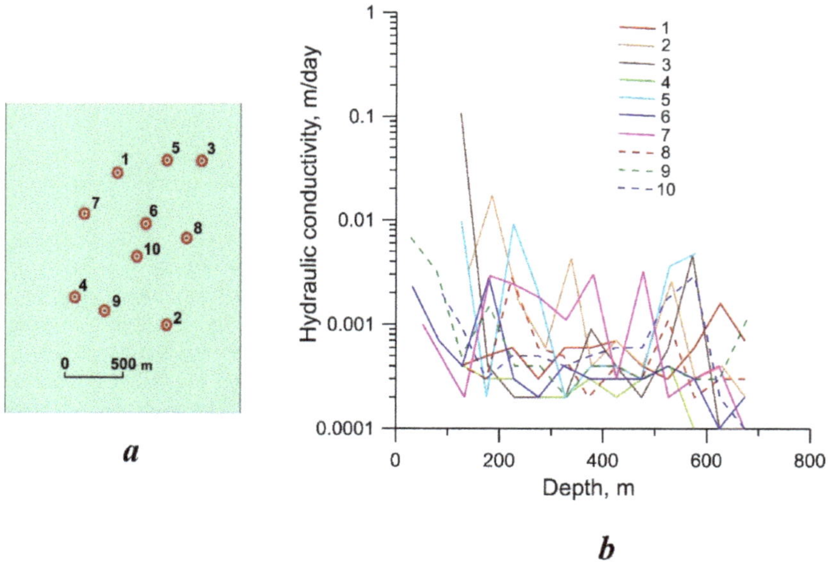

Figure 5. Data of pumping test at the site of the granitoid massif Nizhnekansky, which is selected for the development of the first federal underground repository of high-level radioactive waste in Russia. (**a**) Positions of the boreholes; (**b**) hydraulic conductivity of rocks.

5. Discussion

Recalculation of the data in Figure 5b shows that permeability of rocks does not exceed 10^{-15} m^2 beneath the depth of 300 m. This is higher by almost three orders of magnitude than the rock matrix permeability. Hence, one can expect that the bulk permeability is caused mostly by fractures. However, comparison of different curves in Figure 5b shows that peaks on one curve are absent at the same depth on curves obtained by tests in neighboring boreholes though the distance between them does not usually exceed 500 m. This evidences that fractures do not form hydraulically connected clusters (network) that extend throughout the whole massif [41]. Hence, a part of the groundwater flow path is within the rock matrix. Therefore, mechanical retention of radiocolloid is quite probable in this part of the Nizhnekansky granitoid massif, where research is being conducted to create the first federal underground repository of high-level nuclear waste in Russia.

6. Conclusions

The main mechanism of radioactive pollution propagation from an underground repository is caused by the transport of radionuclides by groundwaters to the biosphere. The groundwater flows in the earth's crust through connected systems of pore and fracture voids. These systems are called flow channels. Radionuclides can be carried by the groundwater as a dissolved component or in the form of colloid particles with attached radionuclides. Colloidal form can be more mobile in geological media than radionuclide ions. However, the rocks can mechanically retain radioactive colloid particles if dimensions of cross-sections of the filtration channels are less than dimensions of colloid particles. Hence, the possibility of mechanical retention of radioactive colloid remains even in the case of fracturing increasing bulk permeability of the rocks by a few orders of magnitude. This statement agrees with data from laboratory examination of rocks from the Nizhnekansky granitoid massif, which has been selected as a site for the construction of the first federal underground repository of vitreous high-level radioactive waste in Russia.

Author Contributions: Conceptualization, V.I.M., V.A.P. and S.V.Y.; methodology, M.I.O.; formal analysis, V.I.M.; software, V.I.M.; experimental studies, V.A.P. and S.V.Y.; resources, V.V.P.; data curation, V.A.P., S.V.Y. and V.V.P.; validation, S.V.Y.; writing—original draft preparation, V.I.M., V.A.P., S.V.Y. and M.I.O.; writing—review and editing, M.I.O. and S.V.Y.; supervision, V.A.P.; funding acquisition, V.A.P.; project administration, V.A.P.; All authors have read and agreed to the published version of the manuscript.

Funding: This research was funded by the Government of the Russian Federation, state assignment "Geoecological problems of nuclear power engineering" (01 January 2021–30 December 2023).

Institutional Review Board Statement: Not applicable.

Informed Consent Statement: Not applicable.

Data Availability Statement: Data supporting reported results can be found in referenced publications.

Conflicts of Interest: The authors declare no conflict of interest.

References

1. Krauskopf, K.B. Geology of high-level nuclear waste disposal. *Ann. Rev. Earth Planet. Sci.* **1988**, *16*, 173–200. [CrossRef]
2. National Research Council. *End Points for Spent Nuclear Fuel and High-Level Radioactive Waste in Russia and the United States*; Committee on End Points for Spent Nuclear Fuel and High-Level Radioactive Waste in Russia and the United States; National Academies Press: Washington, DC, USA, 2003; p. 137. [CrossRef]
3. Laverov, N.P.; Yudintsev, S.V.; Kochkin, B.T.; Malkovsky, V.I. The Russian Strategy of using Crystalline Rock as a Repository for Nuclear Waste. *Elements* **2016**, *12*, 253–256. [CrossRef]
4. Savage, D. *The Scientific and Regulatory Basis for the Geological Disposal of Radioactive Waste*; John Wiley&Sons: Chichester, UK, 1995; p. 437.
5. Niemeyer, M.J.; Hugi, M.; Smith, P.; Zuidema, P. Kristallin-I performance assessment: First results from sensitivity studies. In *Geological Disposal of Spent Fuel, High Level and Alpha Bearing Wastes*; International Atomic Energy Agency: Vienna, Austria, 1993; pp. 297–308.
6. Bear, J. Modeling flow and contaminant transport in fractured rocks. In *Flow and Contaminant Transport in Fractured Rocks*; Bear, J., Tsang, C.-F., De Marsily, G., Eds.; Academic Press, Inc.: San Diego, CA, USA, 1993; pp. 1–38.
7. Neretnieks, I. Solute transport in fractured rock–applications to radionuclide waste repositories. In *Flow and Contaminant Transport in Fractured Rocks*; Bear, J., Tsang, C.-F., De Marsily, G., Eds.; Academic Press, Inc.: San Diego, CA, USA, 1993; pp. 39–128.
8. Ewing, R.C.; Whittleston, R.A.; Yardley, B.W.D. Geological disposal of nuclear waste: A primer. *Elements* **2016**, *12*, 233–237. [CrossRef]
9. Petrov, V.A.; Lespinasse, M.; Poluektov, V.V.; Ustinov, S.A.; Minaev, V.A. Scale effect in a fluid-conducting fault network. *Geol. Ore Depos.* **2019**, *61*, 293–305. [CrossRef]
10. Barenblatt, G.; Zheltov, I.; Kochina, I. Basic concepts in the theory of seepage of homogeneous liquids in fissured rocks [strata]. *J. Appl. Math. Mech.* **1960**, *24*, 1286–1303. [CrossRef]
11. Tsang, C.-F. Tracer transport in fracture systems. In *Flow and Contaminant Transport in Fractured Rocks*; Bear, J., Tsang, C.-F., De Marsily, G., Eds.; Academic Press, Inc.: San Diego, CA, USA, 1993; pp. 237–266.
12. Berre, I.; Doster, F.; Keilegavlen, E. Flow in fractured porous media: A review of conceptual models and discretization approaches. *Transp. Porous Media* **2019**, *130*, 215–236. [CrossRef]
13. Wong, D.L.; Doster, F.; Geiger, S.; Francot, E.; Gouth, F. Fluid flow characterization framework for naturally fractured reservoirs using small-scale fully explicit models. *Transp. Porous Media* **2020**, *134*, 399–434. [CrossRef]
14. De Marsily, G. *Quantitative Hydrogeology*; Academic Press: Orlando, FL, USA, 1986; p. 440.
15. McKinley, I.G.; Alexander, W.R. Assessment of radionuclide retardation: Uses and abuses of natural analogue studies. *J. Contam. Hydrol.* **1993**, *13*, 249–259. [CrossRef]
16. Malkovsky, V.I.; Pek, A.A. Effect of elevated velocity of particles in groundwater flow and its role in colloid-facilitated transport of radionuclides in underground medium. *Transp. Porous Media* **2009**, *78*, 277–294. [CrossRef]
17. Penrose, W.R.; Polzer, W.L.; Essington, E.H.; Nelson, D.M.; Orlandini, K.A. Mobility of plutonium and americium through a shallow aquifer in a semiarid region. *Environ. Sci. Technol.* **1990**, *24*, 228–234. [CrossRef]
18. McCarthy, J.F.; Czerwinski, K.R.; Sanford, W.E.; Jardine, P.M.; Marsh, J.D. Mobilization of transuranic radionuclides from disposal trenches by natural organic matter. *J. Contam. Hydrol.* **1998**, *30*, 49–77. [CrossRef]
19. McCarthy, J.F.; Sanford, W.E.; Stafford, P.L. Lanthanide field tracers demonstrate enhanced transport of transuranic radionuclides by natural organic matter. *Environ. Sci. Technol.* **1998**, *32*, 3901–3906. [CrossRef]
20. Kersting, A.B.; Efurd, D.W.; Finnegan, D.L.; Rokop, D.J.; Smith, D.K.; Thompson, J.L. Migration of plutonium in ground water at the Nevada Test Site. *Nature* **1999**, *397*, 56–59. [CrossRef] [PubMed]
21. Nyhan, J.W.; Drennon, B.J.; Abeele, W.V.; Wheeler, M.L.; Purtymun, W.D.; Trujillo, G.; Herrera, W.J.; Booth, J.W. Distribution of plutonium and americium beneath a 33-yr-old liquid waste disposal site. *J. Environ. Qual.* **1985**, *14*, 501–509. [CrossRef]

22. Zachara, J.M.; Smith, S.C.; Liu Ch McKinley, J.P.; Serne, R.J.; Gassman, P.L. Sorption of Cs+ to micaceous subsurface sediments from the Hanford site, U.S.A. *Geochim. Cosmochim. Acta* **2002**, *66*, 193–211. [CrossRef]
23. Smith, D.K.; Finnegan, D.L.; Bowen, S.M. An inventory of long-lived radionuclides residual from underground nuclear testing at the Nevada test site, 1951–1992. *J. Environ. Radioact.* **2003**, *67*, 35–51. [CrossRef]
24. Malkovsky, V.I.; Yudintsev, S.V.; Aleksandrova, E.V. Leaching of radioactive waste surrogates from a glassy matrix and migration of the leaching products in gneisses. *Radiochemistry* **2018**, *60*, 648–656. [CrossRef]
25. Honeyman, B.D. Colloidal culprits in contamination. *Nature* **1999**, *397*, 23–24. [CrossRef] [PubMed]
26. Inagaki, Y.; Sakata, H.; Furuya, H.; Idemitsu, K.; Arima, T.; Banba, T.; Maeda, T.; Matsumoto, S.; Tamura, Y.; Kikkawa, S. Effects of water redox conditions and presence of magnetite on leaching of Pu and Np from HLW glass. *Mater. Res. Soc. Proc.* **1998**, *506*, 177–184. [CrossRef]
27. Buck, E.C.; Bates, J.K. Microanalysis of colloids and suspended particles from nuclear waste glass alteration. *Appl. Geochem.* **1999**, *14*, 635–659. [CrossRef]
28. Ojovan, M.; Lee, W.E. Glassy wasteforms for nuclear waste immobilization. *Metall. Mater. Trans. A* **2011**, *42*, 837–851. [CrossRef]
29. Ojovan, M.I.; Petrov, V.A.; Yudintsev, S.V. Glass crystalline materials as advanced nuclear wasteforms. *Sustainability* **2021**, *13*, 4117. [CrossRef]
30. Yudintsev, S.V.; Pervukhina, A.M.; Mokhov, A.V.; Malkovsky, V.I.; Stefanovsky, S.V. Influence of phosphate glass recrystallization on the stability of a waste matrix to leaching. *Dokl. Earth Sci.* **2017**, *473*, 427–432. [CrossRef]
31. Mal'kovskii, V.I.; Yudintsev, S.V.; Pervukhina, A.M. Leaching of degraded preservative matrices, based on sodium aluminophosphate glasses, for high-level wastes. *At. Energy* **2018**, *123*, 177–182. [CrossRef]
32. Malkovsky, V.I.; Yudintsev, S.V.; Aleksandrova, E.V. Influence of Na-Al-Fe-P glass alteration in hot non-saturated vapor on leaching of vitrified radioactive wastes in water. *J. Nucl. Mater.* **2018**, *508*, 212–218. [CrossRef]
33. Degueldre, C.; Benedicto, A. Colloid generation during water flow transients. *Appl. Geochem.* **2012**, *27*, 1220–1225. [CrossRef]
34. Missana, T.; Alonso, U.; Fernández, A.M.; García-Gutiérrez, M. Analysis of the stability behaviour of colloids obtained from different smectite clays. *Appl. Geochem.* **2018**, *92*, 180–187. [CrossRef]
35. Malkovsky, V.I.; Yudintsev, S.V.; Ojovan, M.I.; Petrov, V.A. The influence of radiation on confinement properties of nuclear waste glasses. *Sci. Technol. Nucl. Install.* **2020**, *2020*, 8875723. [CrossRef]
36. Bredehoeft, J.D.; Norton, D.L. Mass and energy transport in a deforming Earth's crust. In *The Role of Fluids in Crustal Processes*; Geophysics Study Committee, Comission on Geosciences, Environment and Resources, National Research Council; National Academy Press: Washington, DC, USA, 1990; pp. 27–41.
37. Shmonov, V.M.; Mal'kovskii, V.I.; Zharikov, A.V. A Technique for measuring permeability of samples of anisotropic rocks for water and gas. *Instrum. Exp. Tech.* **2011**, *54*, 722–728. [CrossRef]
38. Lobanov, N.F.; Beigul, V.P.; Lopatin, P.V.; Ozersky, A.Y. Selection and validation of area for underground research laboratory in Nizhnekansky Massif. *Min. J.* **2015**, *10*, 59–63. (In Russian) [CrossRef]
39. Petrov, V.A.; Poluektov, V.V.; Hammer, J.; Zulauf, G. Analysis of mineralogical and deformation-induced transformations of Nizhnekansky Massif rocks to estimate their retention capacity in geological disposal and isolation of radioactive waste. *Gorn. Zhurnal—Min. J.* **2015**, *10*, 67–74. (In Russian) [CrossRef]
40. Petrov, V.A.; Siitati-Kauppi, M.; Tikkanen, O.; Sardini, P.; Poluektov, V.V. Preliminary results of C14-PMMA impregnation of Krasnoyarsk rock samples for evaluating the porosity related properties. In Proceedings of the 21th International Conference "Physico-Chemical and Petrophysical Researches in Earth Sciences", IGEM RAS, Moscow, Russia, 21–23 September 2020; pp. 310–315.
41. Malkovsky, V.I.; Ozerskiy, A.Y. Stochastic model of filtration properties distribution for enclosing rocks of an underground repository of radioactive waste on the basis of pumping tests. In Proceedings of the 15th International Conference "Physico-Chemical and Petrophysical Researches in Earth Sciences", IGEM RAS, Moscow, Russia, 29 September–1 October 2014; pp. 159–162.

sustainability

MDPI

Article

Evaluation of a Long-Term Thermal Load on the Sealing Characteristics of Potential Sediments for a Deep Radioactive Waste Disposal

Norbert Clauer [1,*], Miroslav Honty [2], Lander Frederickx [2] and Christophe Nussbaum [3]

1 Institut des Sciences de la Terre et de l'Environnement (UdS/CNRS), Université de Strasbourg, 67084 Strasbourg, France
2 Belgian Nuclear Research Centre (SCK-CEN), B-2400 Mol, Belgium
3 Swiss Geological Survey, Federal Office of Topography, 3084 Wabern, Switzerland
* Correspondence: nclauer@unistra.fr

Abstract: An in situ and a batch heating experiment were applied on the fine-grained sediments of the Opalinus Clay from Mont Terri (Switzerland) and the Boom Clay of Mol (Belgium), both being currently studied as potential host formations for deep nuclear waste disposal. The purpose was here to test the impact of a 100 °C temperature rise that is expected to be produced by nuclear waste in deep repositories. The experiment on the Opalinus Clay mimicked real conditions with 8-months operating heating devices stored in core drillings into the rock. The comparison of the major, trace, rare-earth elemental contents and of the whole-rock K-Ar data before and after heating shows only a few variations beyond analytical uncertainty. However, the necessary drillings for collecting control samples after the experiment added an unexpected uncertainty to the analyses due to the natural heterogeneity of the rock formation, even if very limited. To overcome this aspect, Boom Clay ground material was subjected to a batch experiment in sealed containers during several years. The drawback being here the fact that controls were limited with, however, similar reproducible results that also suggest limited elemental transfers from rock size into that of the <2 µm material, unless the whole rocks lost more elements than the fine fractions. The analyses generated by the two experiments point to identical conclusions: a visible degassing and dewatering of the minerals that did not induce a visible alteration/degradation of the host-rock safety characteristics after the short-term temperature increase.

Keywords: long-term heat experiments; fine-grained sediments; deep disposal of nuclear waste; major; trace and rare-earth elements; K-Ar tracing; Opalinus Clay; Boom Clay formations

Citation: Clauer, N.; Honty, M.; Frederickx, L.; Nussbaum, C. Evaluation of a Long-Term Thermal Load on the Sealing Characteristics of Potential Sediments for a Deep Radioactive Waste Disposal. *Sustainability* **2022**, *14*, 14004. https://doi.org/10.3390/su142114004

Academic Editors: Michael I. Ojovan, Vladislav A. Petrov and Sergey V. Yudintsev

Received: 13 September 2022
Accepted: 21 October 2022
Published: 27 October 2022

1. Introduction

Petrophysical, mineralogical and geochemical studies of deep-seated geological sediments that might be selected as host formations for nuclear waste disposal require careful controls of their safety characteristics. In order to check these conditions, specific experiments were designed and performed in dedicated experimental underground laboratories. The data generated during such experiments need to be as representative as possible of the intrinsic characteristics of the target rocks before any human interference. The challenge is then the completion of reference analyses and observations as representative as possible of the original intact rocks.

In this context, appropriate drilling techniques were used in and around the Mont-Terri rock-laboratory to minimize the exchange reactions between the compositional minerals of the drilled rocks and the incoming atmosphere. However, despite all the care taken during such drillings and the conditioning under vacuum of the core pieces, the sampled material underwent various alteration reactions [1,2]. Oxidation by contact with the atmosphere resulted from almost instantaneous supply of O_2, CO_2 and water to rocks that were confined in reduced conditions with very limited amounts of fluids and gases for millions of years. Local disturbances can also occur by an instantaneous dehydration due to temperature increase, for instance during coring, possibly followed by some rehydration depending on how the recovered samples were handled, or on the conditions applied to the excavated sites. It has already been shown, in this context, that deeply buried sediments excavated for shafts and galleries yield modifications on their walls that were called "excavated damaged zones" (labeled EDZ hereafter) and "excavated disturbed zones" (EdZ) [3–9]. Horseman [10] suggested even that this recent evolution in the stress path also leads to a reduction in the permeability of the sediments.

The present study was designed in this context to evaluate analytically the impact of a long-lasting heating experiment on the geochemical and mineralogical signatures of host formations and more precisely on the sealing potential of rocks immediately around containers that are expected to release spontaneously heat due to the nuclear waste itself. The purpose was not to construct a model based on theoretical hypotheses, but to identify and quantify the changes on the basis of factual analyses. Indeed, mathematical models elaborated specifically for deep-disposal of nuclear waste are already available based on the fact that the storage will induce a moderate but extremely long-lasting temperature increase on the host rocks (e.g., [11,12]). As the spontaneous heat produced by the radioactive waste on its immediate environment has already been modeled, analytical controls of such an impact on the immediate environmental host rocks are now needed to check and consolidate these models. The present study is based on two experiments to evaluate such an impact of a long-term artificial heating: (1) on the Aalenian-Toarcian (183–170 Ma) Opalinus Clay Formation (called OPA hereafter) at the Mont Terri in Switzerland where an in situ experiment called HE-D was completed, and (2) on the Rupelian (34–28 Ma) Boom Clay Formation at Mol in Belgium, which was subjected to a batch experiment during 7 years at an ambient temperature of 80 °C. The reason for combining two different heating experiments on two different sedimentary formations is because the analytical controls of the OPA rocks after the experiment needed a supplementary coring to obtain the heated samples. In turn, as even very close samples can be slightly different in their mineralogy and chemistry, some of the possible analytical differences may be due to the sample mineralogy and not to the heating experiment. This aspect was not of concern in the case of the batch experiment as the same samples were analyzed along a "closed" experiment. However, the number of controls had to remain limited because each needed the opening of a container, which was lost for the continuation of the experiment.

2. Description of the Experiments, the Sampling and the Analytical Methods

2.1. The In Situ Experiment in the Opalinus Clay of the Mont-Terri Rock Laboratory

The in situ heat experiment was based on the study of the 8-m long BHE-D0 borehole drilled into the OPA from HE niche of the so-called 'new gallery' in the rock laboratory (Figure 1). The concept was heating the clay formation at a steady temperature in an undisturbed zone with no protecting backfill material around two heating devices inserted into the borehole. The mineralogical and geochemical characteristics of the drilled sequence were examined as references before heating on five samples collected along a second borehole BHE-D5 drilled perpendicularly to the BHE-D0 borehole from nearby MI niche through the EDZ and the unsaturated zone of the niche, both zones being considered to be about 1 m and 1.5 m thick, respectively [13,14]. During the initial 3 months, the two heaters in BHE-D0 borehole were set to diffuse a progressive heat flux from 15 °C to 43 °C and, then, to propagate a temperature of 100 °C during 8 months. After these 8 months of heating, 9 more months were allowed for the cooling of the rocks before collection, by drilling, of the 'heated' samples, close to and far from heaters. This sampling was completed along the BHE-D5 borehole and along a second BHE-D26 borehole cored parallel to the previous one (Figure 1). The samples from beyond the identified EDZ and EdZ zones were stored 8 more months under vacuum, before analysis. Those of these EDZ and EdZ volumes were analyzed previously in a study dedicated to the 'aging' effect after a regular excavation of galleries and niches [2].

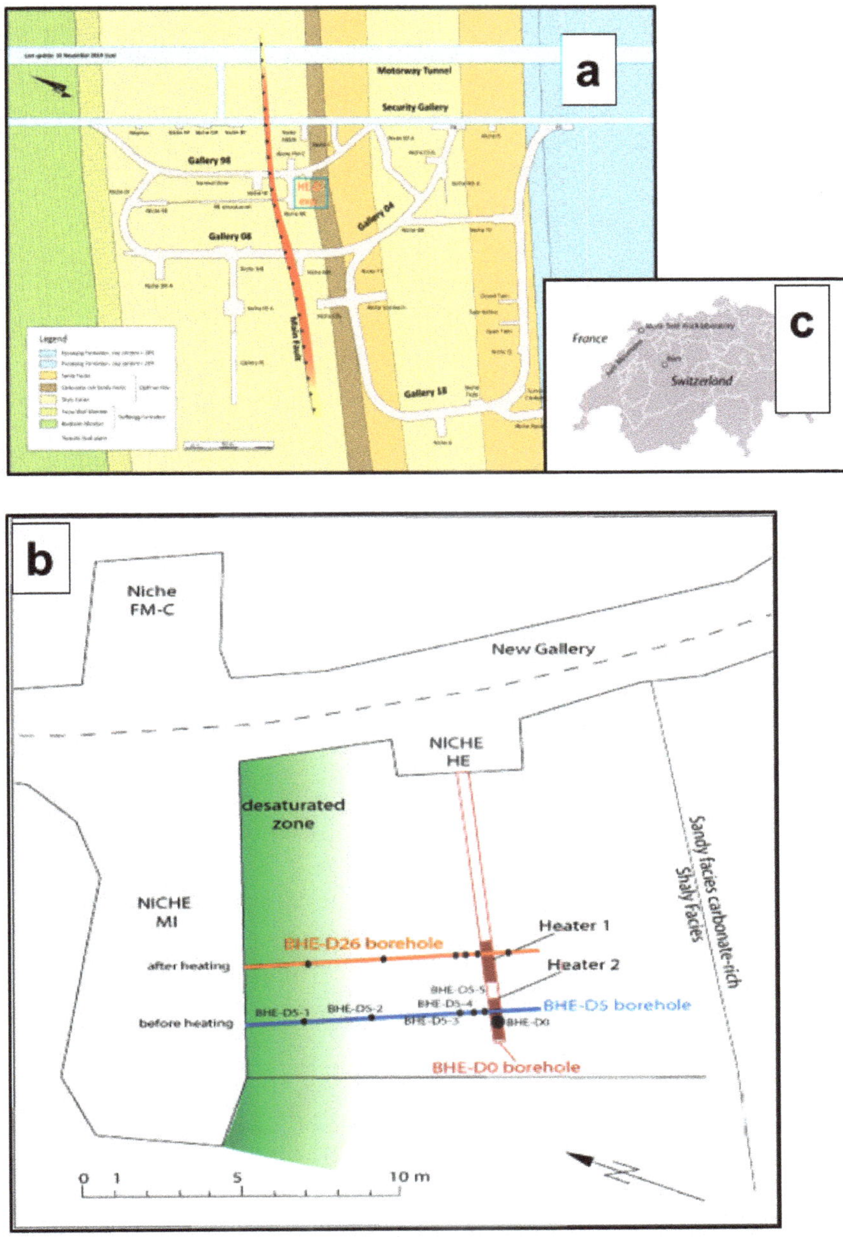

Figure 1. (**a**) Location of the HE-D experiment in the Opalinus Clay of the Mont Terri rock laboratory; (**b**) A closer view of the experiment in the MI niche with the different drill holes; (**c**) The geographic location of the rock laboratory in Switzerland.

The 14 to 21 cm long samples collected before the heating experiment along the BHE-D5 borehole between 2.24 and 8.00 m from heater and at the tip of the BHE-D0 borehole were also stored under vacuum after recovery. After 11 months of heating and 9 months of cooling, six samples were collected perpendicularly to borehole BHE-D0 at an increasing distance from heaters, as well as along the BHE-D26 borehole drilled after heating in the same rock volume. To keep the collected volume as small as possible and to allow the after-heat sampling to be as close as possible to that before heating for most reliable comparisons, micro-cores of 25 mm in diameter were drilled into each of the selected samples for the planned mineralogical, chemical and isotopic analyses.

2.2. The Batch Experiment of the Boom Clay

The main objective of the longer batch experiment completed on ground Boom Clay rocks under closed conditions at 80 °C during several years (Figure 2) was a precise analysis of the released CO_2 gas yields during heating of potential host material in the conditions of a deep waste disposal (e.g., [15–17]). Analyses were carried out on whole rocks and <2 µm-sized residues after dismantling of a batch container at each controlling step. The initial gas pressure was set at 20 bars and was controlled by pressure sensors connected to the pressure vessels with Ar as the cushion gas. The complete experiment lasted 2529 days with an opening for sample collections of two containers after 830 and 1799 days. The initial rock sample and its <2 µm fraction, and those from the two intermediate steps were analyzed, before and after an additional leaching with dilute (1M) hydrochloric acid, for the major, trace and rare-earth elements (REEs) and for the K-Ar isotopic systematics to which a special attention was given about the desorption of the atmospheric Ar used during the experiment as the cushion material.

Figure 2. View of the long-term batch experiment on the Boom Clay rock ships at 80 °C.

2.3. The Analytical Methods

All samples collected for the two experiments were observed by scanning electron microscope (SEM) and then were crushed gently, a powder aliquot of each being taken for their mineralogical, chemical and isotopic analyses. For the Boom material, the samples were subjected to quantitative X-Ray diffraction (XRD), were mixed with 10 wt% of ZnO as an internal standard, and milled in methanol for 5 min in a McCrone Micronizing Mill. Then, the powder was scanned on a Bruker D8 instrument equipped with a CuKα radiation and a Bruker LynxEye detector (Bruker, Billerica, MA, USA). The obtained spectra were analyzed by the Profex 4.3.2 release [18] and the BRGM mineral database. The XRD analyses of the OPA material were completed on a Bruker D5 instrument with the same radiation, but no LynxEye detector and no special quantification program. Still for the Boom Clay, the recovered rock powder was settled in distilled water for a classical separation of the <2 μm size fraction by sedimentation following the classical Stokes' law. The smeared <2 μm specimens were XRD analyzed two times in an air-dried and ethylene-glycol state on a Philips PW1830 diffractometer with a CuKα radiation and a PW3011/00 proportional detector. In turn, the obtained data of the clay minerals were similar to those described by Zeelmaekers et al. [19] and Frederickx et al. [20]. The leachates of the rocks and size fractions were also analyzed for their major, trace and rare-earth elements by Inductively Coupled Plasma-Atomic Emission Spectrometry (ICP-AES) and Inductively Coupled Plasma-Mass Spectrometry (ICP-MS), following the procedure of Samuel et al. [21]. The reproducibility and accuracy tests of the BE-N and GL-O international standards gave recommended values for the contents with standard deviations better, on the basis of 9 independent runs during the time of the analyses, than the given average of 2.5% for the major elements, 4.5% for the trace elements and 10% for the rare-earth elements (REEs).

To constrain best the chemical and mineral variations due to heating and, therefore, the amounts of the mobilized elements, some of the collected samples were also leached with dilute (1M) hydrochloric acid during 15 min at room temperature. After leaching, the mixtures were centrifuged and the leachates evaporated and analyzed for their chemical compositions by ICP-AES and/or by ICP-MS. The K-Ar procedure applied to the fresh and heated samples and to some <2 μm fractions is similar to that presented by Bonhomme et al. [22]. The Ar extractions were completed in a glass line coupled directly with the gas mass-spectrometer; a preliminary degassing at 80 °C during 24 h being systematically applied to remove any adsorbed atmospheric during the sample preparation. Potassium was measured by atomic absorption with an accuracy of ±1.5%, which was controlled by periodic analysis of the same B-EN and GL-O international standards. The quality of the Ar extractions was controlled by a weekly analysis of the standards, as well as of the atmospheric $^{40}Ar/^{36}Ar$ ratio and the blank of the extraction line coupled to the mass spectrometer. There was no need for data corrections as the average content of the radiogenic ^{40}Ar from GL-O standard and the atmospheric $^{40}Ar/^{36}Ar$ ratio were close to the theoretical values at 24.57 ± 0.60 (2σ) cm^3/g [23] and 298.6 ± 0.4 (2σ) [24], respectively. The blanks of the gas line and the gas spectrometer were also far below the measured contents of the samples. The usual decay constants were used for the age calculations [25] with an overall error of the K-Ar age determinations better than 2%.

3. Results

3.1. The Characteristics of the Reference Opalinus Clay Material

Previous XRD data gave a classical shaly composition for the OPA sediments [26,27]. Here, the samples appeared quite homogeneous along the 8 m long profile at the SEM scale. However, observations of pyrite that represents a potential mineral sensitive to heating and external air supply, showed some discrete changes, even before the heating experiment. In most samples from cores drilled for the emplacement of the heaters, pyrite occurs either as visually 'clean' framboïdal aggregates or large individual minerals, as for instance in sample BHE-D5-1 collected 7 m away from heater (Figure 3A). About 2.1 m away, and still before the heating experiment, some of the framboïds appear slightly erased and 'polished' (Figure 3C). After the 8 months heating experiment, the pyrite crystals at 5.6 m from heater were covered by a veil that looks as it has been torn somehow (Figure 3B), while the surficial crystals of the aggregates appeared altered and even peeled off with a pristine central core at 2.0 m from heater (Figure 3D). These modifications suggest some alteration by oxidation of the aggregated or the individual pyrite crystals, even before the heating experiment. Janots et al. [28] related similar veils to the activity of sulfate reducing bacteria, which action could have started right after drilling of the core due to an atmospheric/bacterial contamination that could have been amplified by the heating experiment.

Figure 3. SEM pictures of pyrite crystals and agglomerates from OPA sediments. The sample identifications are in the photographs, as well as the locations, the timing (t = 0 means before heating and t = 1 means after heating) and the scale (modified from Techer et al. [2]). (**A**) Picture of large pyrite and aggregated minuscule pyrite crystals before heating; (**B**) the same small crystals wrapped in a veil after heating; (**C**) an enlargement of the pyrite crystal balls before heating; (**D**) the same kind of crystal ball after heating.

Before heating, the major elements from reference rocks along the BHE-D5 drilling vary only slightly, except an increase in the loss on ignition (LOI) when sampling approaches the MI niche. This change corresponds probably to the above-identified EDZ/EdZ that was exposed to atmosphere during the excavations of the gallery and the niche. While most major elemental contents appear quite constant towards the heater before the experiment, there is a clear decrease in the CaO content towards the EDZ/EdZ zone of the niche, together with a decrease of the SiO_2 content (Table 1, Figure 4). This decrease in Ca changes into ups and downs after heating, while the contents of the other elements remain quite stable. The changing Ca of the rocks along the collection trend suggests discrete natural mineral heterogeneities due to how the samples were collected and stored after heating. They were taken as close as possible to the initial samples, as it could not be done differently, which does not guarantee a strict mineral and, therefore, a strict chemical identity. This weakness had to be kept in mind, while comparing the chemical data of the samples collected before and after the heating experiment. In this context, the analytical aspect of very uniform total contents of the major elements around 100% for the sums of the analyzed major elements is of importance for overall mineral and chemical uniformities, confirming also an analytical reproducibility better than the routinely assumed ±2.5% uncertainty. In sum, it is plausible that some of the changing elemental contents result from a discrete heterogeneity in the mineral compositions due to the additional drillings rather than to the long-term heating. In the case of the trace elements, the Sr contents increase significantly towards the niche and therefore away from heater, while those in Zr, Zn and Cu decrease, but to a lesser extent. For all other elements, the contents remain quite homogeneous, again beyond the EDZ and EdZ zones (Figure 5). In fact, the slight changes along the profile from heater to the MI niche wall decompose into a fair steady content between

2 to 7-m away from the niche wall, except for Si and Ca, and Sr, Ba and Y. The scatter becomes larger beyond 7 m away from the wall, which could be due to already reported lithological variations in the shaly OPA facies [26,27]. Significant elemental variations were also detected in the analyses of the soluble mineral phases of the samples recovered by a gentle leaching with dilute hydrochloric (1M) acid during 15 min at room temperature (Table 2). Indeed, the leachates become clearly enriched in Si, Al and K, while Mg, Fe, Na and P remain quite stable, and Ca is significantly depleted close to the MI niche before heating in the EDZ/EdZ zone.

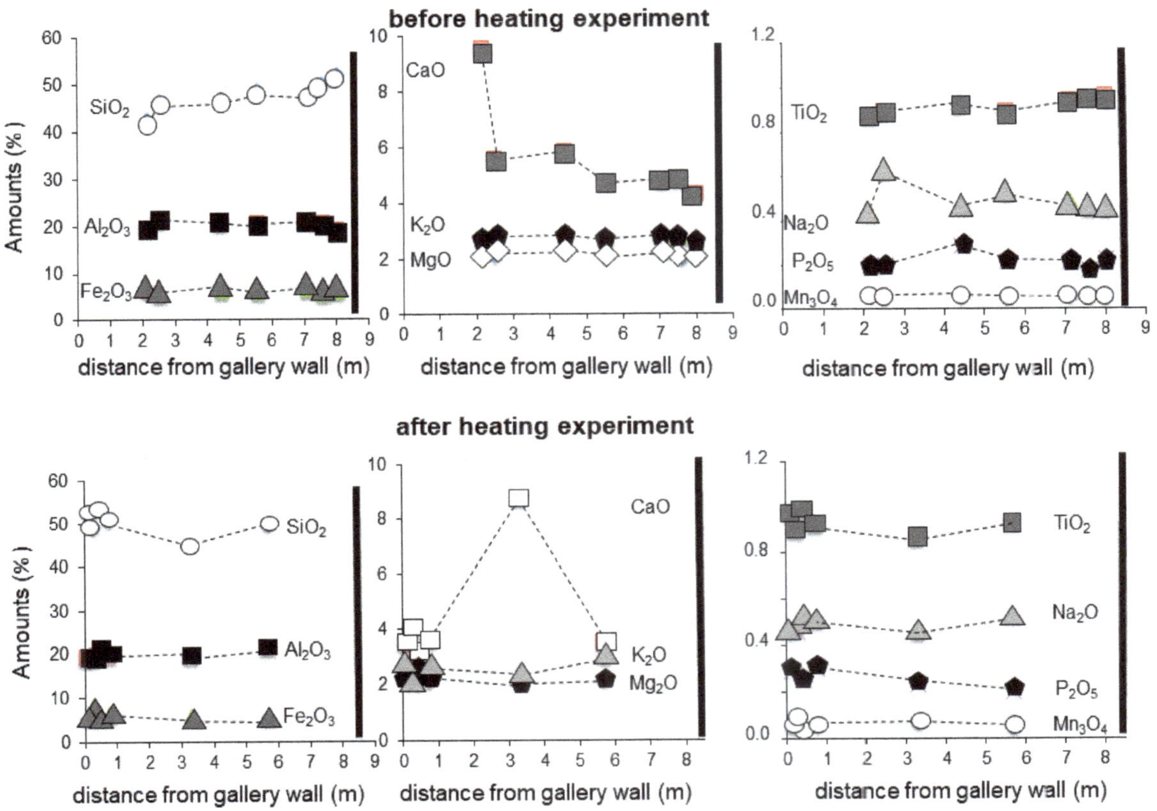

Figure 4. Variations of the major elements from OPA whole rocks during the heating experiment. The thick bar to the right represents the heater.

Table 1. The major and trace elemental contents of the OPA whole-rocks before and after the heating experiment.

Sample IDs	Place/MI Niche (m)	SiO_2 (%)	Al_2O_3 (%)	MgO (%)	CaO (%)	Fe_2O_3 (%)	Mn_3O_4 (%)	TiO_2 (%)	Na_2O (%)	K_2O (%)	P_2O_5 (%)	LOI (%)	Total (%)	Sr (mg/g)	Ba (mg/g)	Rb (mg/g)	Pb (mg/g)	Th (mg/g)	U (mg/g)
initial																			
BHE-D5 0	8.00	50.7	20.6	2.27	3.47	6.54	0.052	1.01	0.79	3.06	0.24	11.32	100.1	201	302	133	23.9	13.9	3.00
BHE-D5 5	7.96–8.15	52.0	18.8	2.10	4.30	5.94	0.054	0.96	0.46	2.63	0.23	11.56	99.12	216	291	65.4	14.6	8.46	1.85
BHE-D5 4	7.50–7.66	49.9	20.4	2.05	4.84	5.56	0.054	0.95	0.46	2.81	0.18	12.08	99.28	217	282	68.8	14.5	8.50	1.94
BHE-D5 3	7.00–7.21	47.4	21.0	2.27	4.79	6.63	0.060	0.94	0.49	2.87	0.22	12.79	99.52	239	306	74.8	13.8	8.37	1.87
BHE-D5 2	4.36–4.50	46.8	20.5	2.24	5.83	6.31	0.059	0.92	0.47	2.92	0.29	12.91	99.31	282	281	76.2	13.6	8.32	1.83
BHE-D5 1	2.10–2.24	42.5	19.5	2.11	9.52	6.59	0.068	0.87	0.44	2.76	0.20	14.72	99.24	282	267	70.4	12.1	7.30	1.64
BHE-D5 8	5.50–5.60	48.7	20.5	2.13	4.72	6.00	0.056	0.89	0.53	2.83	0.22	12.17	98.79	209	367	129	28.3	12.8	2.82
BHE-D5 6	2.45–2.60	46.4	21.2	2.15	5.56	5.43	0.055	0.89	0.63	2.88	0.20	13.38	98.78	258	349	133	36.6	13.2	2.87
after heating																			
BHE-D26 2	5.70	49.7	21.2	2.15	3.50	5.42	0.046	0.93	0.53	3.04	0.20	11.48	98.22	222	287	153	46.4	15.6	3.47
BHE-D26 3	3.30	44.9	19.2	2.13	8.70	5.82	0.069	0.86	0.46	2.30	0.23	13.92	98.62	258	271	123	34.1	12.4	2.70
BHE-D26 8	0.75	50.7	19.7	2.15	3.57	6.40	0.056	0.92	0.51	2.62	0.30	11.54	98.41	198	279	136	43.2	16.1	3.29
BHE-D26 14	0.40	53.3	19.5	2.10	3.20	5.69	0.041	0.98	0.53	2.95	0.25	10.80	99.32	198	290	124	48.0	14.2	2.93
BHE-D26 10	0.27	49.0	19.0	2.42	4.00	8.20	0.070	0.89	0.49	2.14	0.46	12.21	98.95	205	261	108	38.8	12.5	2.64
BHE-D26 12	0.10	52.7	19.0	2.20	3.50	5.88	0.050	0.97	0.47	2.63	0.29	11.26	98.94	193	284	117	37.8	13.5	2.75

IDs stand for identities.

Table 2. The major and trace elemental contents of the OPA whole-rock leachates by dilute acid before and after the heating experiment.

Sample IDs	Place/MI Niche (m)	Si (µg/g)	Al (µg/g)	Mg (µg/g)	Ca (µg/g)	Fe (µg/g)	Na (µg/g)	K (µg/g)	P (µg/g)	Sr (µg/g)	Ba (µg/g)	Rb (µg/g)	Th (µg/g)	U (µg/g)
initial														
BHE D5 5B	7.96–8.15	7.27	15.1	9.05	374	6.45	29.6	15.5	0.29	1.17	0.04	13.8	8.12	1.43
BHE D5 4B	7.50–7.66	8.25	16.4	9.62	429	7.10	31.5	16.8	0.10	1.28	0.05	15.0	8.41	1.48
BHE D5 3B	7.00–7.21	7.79	13.4	9.29	403	6.15	33.3	16.7	0.31	1.41	0.05	15.9	9.93	1.97
BHE D5 2B	4.36–4.50	7.69	13.8	11.2	557	7.94	37.9	18.2	0.38	2.02	0.06	16.1	11.9	2.43
BHE D5 1B	2.10–2.24	1.26	2.54	9.32	691	7.44	34.9	6.25	0.25	1.67	0.02	6.23	11.1	2.09
after heating														
BHE D5-8	5.50–5.60	1.29	2.61	7.69	282	3.11	25.7	4.40	0.29	0.88	0.02	3.91	6.52	1.00
BHE D5-6	2.45–2.60	0.97	2.24	7.25	282	3.09	21.8	3.85	0.16	0.96	0.01	4.12	5.00	0.94
BHE D5-7	1.90–2.00	0.46	1.24	5.54	310	3.22	15.7	2.44	0.05	0.66	0.01	2.57	6.60	1.11

IDs stand for identities and m for meters.

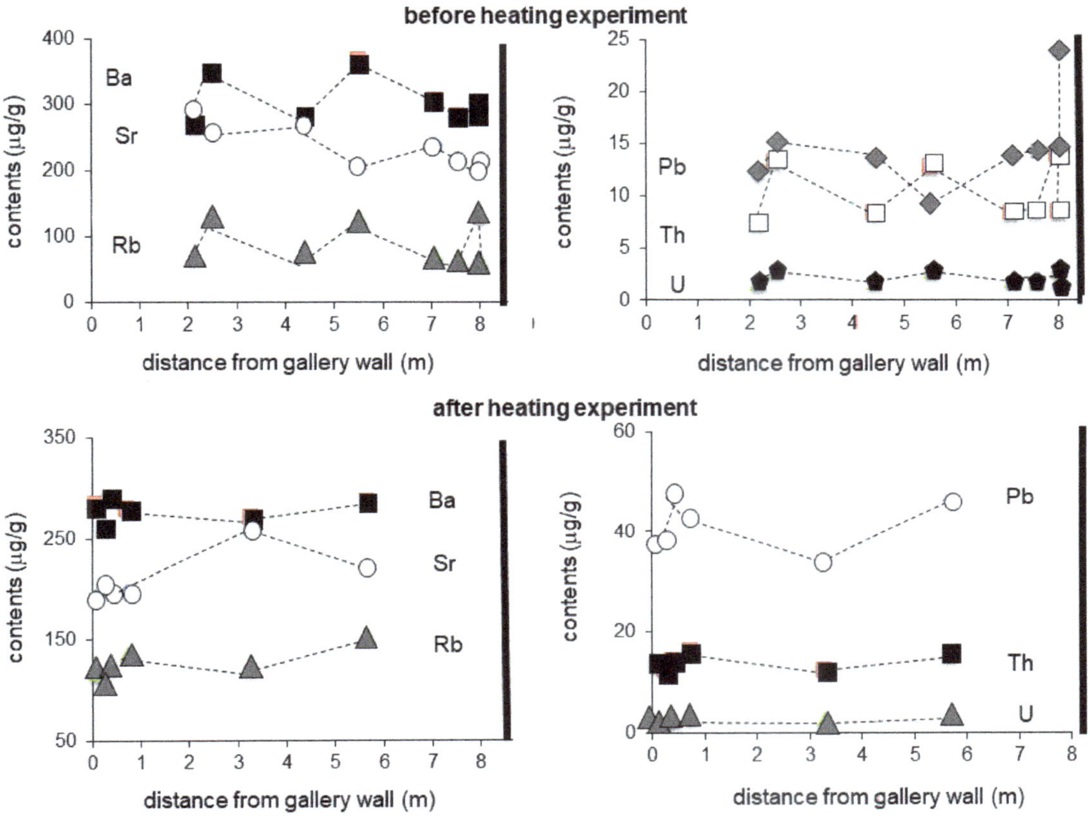

Figure 5. Variations of the trace elements from OPA whole rocks during the heating experiment. The thick bar to the right represents the heater.

The REEs yield specific characteristics including a strong ionic bond making them behave as strong acids [29]. Most occur generally with three electrons removed from three orbitals, while Eu and Ce behave differently: the former yields a half-filled orbital allowing a stability for the Eu^{2+} species and the latter yields an oxidation-reduction state allowing Ce to occur as either Ce^{3+} or Ce^{4+} [30]. As the amounts of the REEs in minerals are naturally quite dispersed among each other, their contents are often compared to those of international standards such as the North American Shale Composites (NASC) used here [30]. The widely dispersed REE contents of any kind of sample are then narrowed and easier to decrypt. Before heating, the REE patterns of the OPA rocks displayed, here, a decreasing slope from light REEs (LREEs) towards the heavy REEs (HREEs) (Table 3, Figure 6). The contents are also systematically depleted relative to the NASC reference, with an analytically significant positive Dy anomaly and a systematic increase of the Yb and Lu contents. To be also mentioned is the spectrum of the sample from EDZ/EdZ volume that is similar to those of the other samples.

Table 3. The rare-earth elemental contents of the OPA whole-rocks before and after the heating experiment.

Sample IDs	Location/ Probe (m)	La (μg/g)	Ce (μg/g)	Pr (μg/g)	Nd (μg/g)	Sm (μg/g)	Eu (μg/g)	Gd (μg/g)	Tb (μg/g)	Dy (μg/g)	Ho (μg/g)	Er (μg/g)	Tm (μg/g)	Yb (μg/g)	Lu (μg/g)	Total (μg/g)
initial																
BHE-D5/5	7.96–8.15	26.4	52.5	6.16	22.9	4.36	0.90	3.40	0.56	3.09	0.69	1.74	0.26	1.74	0.28	125.0
BHE-D5/4	7.50–7.66	27.5	52.4	5.96	20.7	3.83	0.75	2.88	0.48	2.87	0.63	1.68	0.25	1.69	0.27	121.9
BHE-D5/3	7.00–7.21	25.0	47.9	5.69	20.2	3.90	0.81	2.92	0.48	2.75	0.57	1.48	0.22	1.42	0.24	113.6
BHE-D5/2	4.36–4.50	27.1	55.8	6.57	24.4	4.81	0.95	3.65	0.60	3.25	0.67	1.70	0.24	1.59	0.26	131.6
BHE-D5/1	2.10–2.24	24.5	49.8	5.53	19.8	3.68	0.70	2.83	0.46	2.55	0.53	1.44	0.21	1.39	0.22	113.6
after heating																
BHE-D5/8	5.50–5.60	44.4	86.2	9.95	36.0	6.62	1.37	5.29	0.78	4.78	1.02	2.58	0.40	2.50	0.40	202.3
BHE-D5/6	2.45–2.60	46.3	93.0	10.5	37.7	7.11	1.51	5.48	0.84	5.04	1.04	2.74	0.43	2.62	0.42	214.7
BHE D26-14	0.4	46.6	92.7	10.5	38.3	7.16	1.48	5.78	0.85	5.16	1.10	2.79	0.44	2.86	0.45	216.2
BHE D26-12	0.1	44.7	91.2	10.7	39.7	7.68	1.57	6.38	0.89	5.39	1.11	2.89	0.45	2.78	0.44	215.9
BHE D26-10	0.27	43.4	96.3	11.8	46.3	9.51	2.01	7.63	1.09	6.10	1.23	3.01	0.45	2.67	0.42	231.9
BHE D26-8	0.75	53.0	107	12.4	45.8	8.83	1.86	7.10	1.00	6.19	1.29	3.32	0.52	3.11	0.49	251.9
BHE D26-3	3.3	42.1	81.4	9.12	33.6	6.10	1.29	4.94	0.72	4.28	0.94	2.43	0.37	2.42	0.36	190.1
BHE D26-2	5.7	48.7	96.2	10.8	39.4	7.46	1.53	5.71	0.86	5.18	1.13	2.81	0.46	2.78	0.45	223.5

ID stands for identities and m for meters.

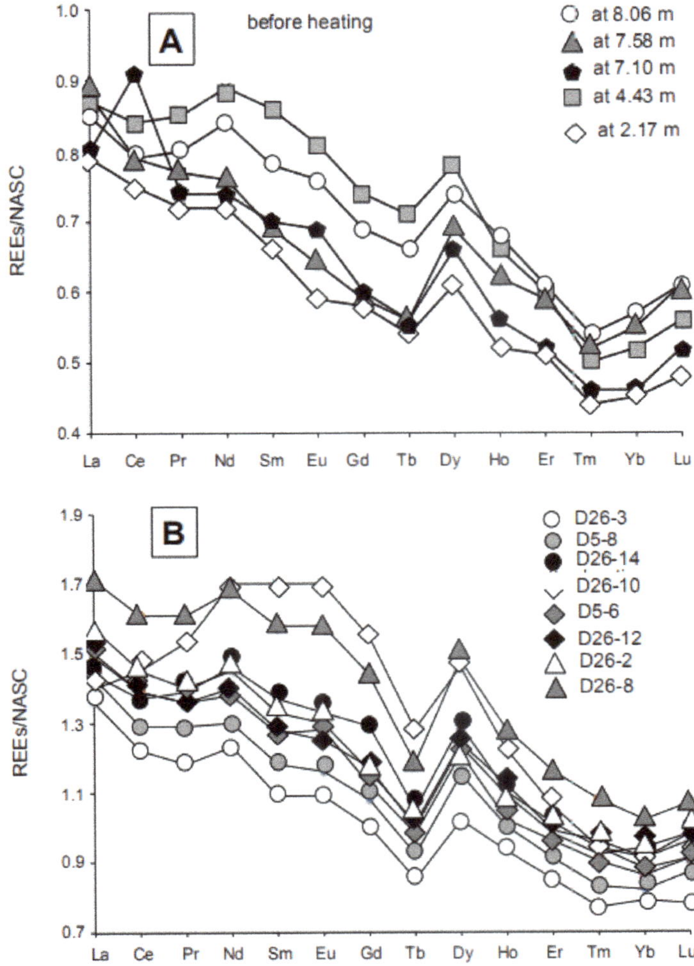

Figure 6. (**A**) The REE distribution patterns of the OPA whole rocks relative to the NASC reference at various distances from heater before the heating experiment; (**B**) the same after the heating experiment.

The K and Ar contents give K-Ar age values, with the isotopic $^{40}Ar/^{36}Ar$ compositions of the untreated rocks, from 320.2 ± 10.0 Ma at 8 m from the niche MI to 333.3 ± 11.1 Ma at 1.97 m away (Table 4). An immediate conclusion is that these age data have no stratigraphic meaning, as the sequence is of Aalenian to Toarcian (183–170 Ma) stratigraphic age. They necessarily contain detrital minerals with K-Ar ages beyond the stratigraphic deposition reference. In the detail, the K_2O contents of the analyzed samples remain quite narrow from 3.06 to 2.59%. This variation is within the 1.5% analytical uncertainty of the K determinations due a plausible interference of a natural heterogeneity of the analyzed rock volumes and no significant impact on the measured K. The variation of the K-Ar age values tends also to decrease slightly when approaching the niche, between 8 and 4.4 m, and it increases again towards the value of the farthest sample when collected closer.

Table 4. The K-Ar data of the OPA whole-rocks before and after the heating experiment.

Sample IDs	Place/MI Niche (m)	K_2O (%)	Rad. ^{40}Ar (%)	Rad. ^{40}Ar (10^{-6} cc/g)	Age (Ma +/− 2σ)
Initial					
BHE-D5 0	8.00	3.06	73.86	34.57	320.2 (10.0)
BHE D5 5B	7.96–8.15	2.63	26.46	30.88	331.7 (11.0)
BHE D5 4B	7.50–7.66	2.81	22.08	31.13	314.5 (10.2)
BHE D5 3B	7.00–7.21	2.87	22.47	31.27	309.7 (9.9)
BHE D5 2B	4.36–4.50	2.92	18.12	30.01	293.5 (9.3)
BHE D5 1B	2.10–2.24	2.76	25.77	30.32	312.0 (10.2)
BHE D5 7B	1.97	2.59	48.85	30.58	333.3 (11.1)
after heating					
BHE D5-3	7.00–7.21	2.69	51.70	30.57	345.0 (11.6)
BHE-D5 8	5.50–5.60	2.84	57.73	35.13	347.9 (11.2)
BHE-D5 6	2.45–2.60	2.88	32.10	33.70	330.7 (10.6)
BHE D5-1	2.10–2.24	2.65	37.10	32.96	326.1 (11.1)

IDs stands for identities, m for meters, and rad. for radiogenic.

3.2. *The Characteristics of the Heated Opalinus Clay Formation*

The already described morphological changes of the pyrite crystals suggest a discrete but pervasive contamination during coring, probably amplified by a bacterial pollution. After the heating test, the contents of the major elements of the rocks duplicate quite systematically those of the samples analyzed along the 8-m long cores before heating (Table 1). Among the visible changes are the Si contents that decrease towards the heater, together with those of Fe and K, whereas those of Ca and the LOI increase, also towards the heater (Figure 4). Most elemental contents scatter more widely, together with the LOI values, along the lastly drilled BHE-D26 core. A few variations concerning specifically the alkali and alkali-earth elements were observed locally. For instance, the sample located 3.3 m from heater is enriched in Ca compared to the other samples, while the sample located at 0.27 m from heater is enriched in MgO and depleted in K_2O. Again, part of these variations most probably results from discrete lithological variations expected in the studied rocks, as was the case along the BHE-D5 core. When compared to the initial unheated rocks, the chemical compositions of the heated counterparts show no significant variations in the major elements. Except for Ca and Si, the changing contents could relate to lithological variations rather than to the heating impact. Also, no chemical trend correlates with the distance to the heater (Figure 4).

The elemental contents of the rocks leached with dilute (1 M) HCl before and after heating were expected to provide information relative to their location along the drilled cores. The specific interest for the analysis of such leachates is the fact that the removed elements do not only result from mineral alteration induced by the heating impact, but more probably from amounts of soluble minerals. While quite stable away from heater, the leached Si decreases markedly by a factor of 4 when approaching the heater, the same occurring for Al and K (Table 2). These changes obviously relate to a mineralogical change in the rock collected next to the heater and were apparently induced by it. The removed elements remain quite constant away from heater, except Ca that increased continuously towards the heater before the experiment. After heating, the leached Si, Al and Mg decrease slightly towards the heater as do Na and K, while Ca remains about constant (Figure 7). This progressive decreasing contents of the former of these elements when approaching the heater are, together with an increase before heating, in favor of a discrete mineral reorganization induced by the experiment. As the amounts of leached elements are systematically lower after heating, which implies a crystallization of new soluble minerals that incorporated only part of the elements freed by the heating episode. In the detail, most of these elements are probably constitutive of chlorides, carbonates and/or sulfates dissolved by local pore fluids diffusing in the rock matrix when the temperature increased. Some were most probably trapped again by newly crystallizing soluble minerals at the end of the experiment when the temperature declined again, while the remainder of the fluids could have been driven out of the controlled rock volume.

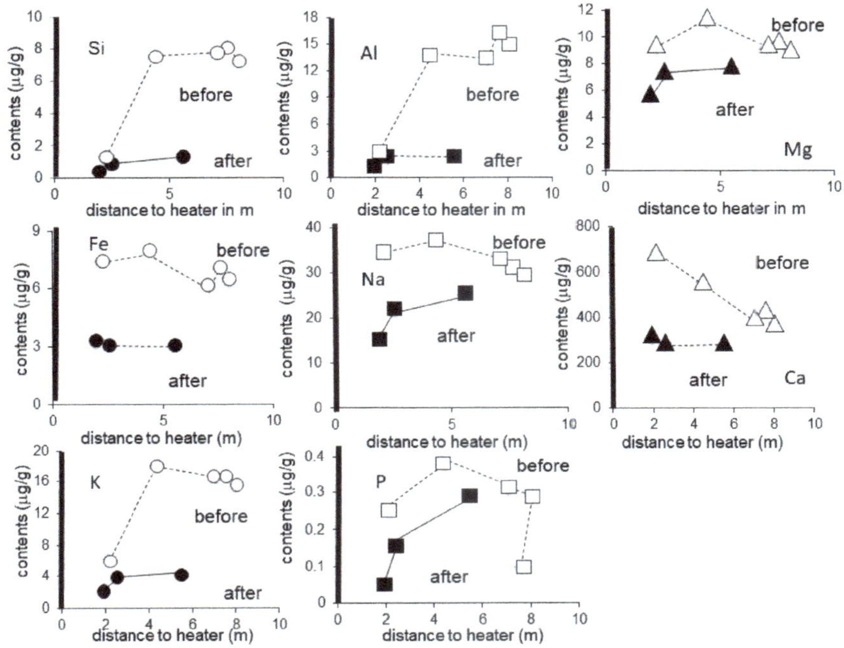

Figure 7. Variations of the elemental contents from OPA rock leachates of the OPA whole rocks before and after the heating experiment.

Before heating, the contents in REEs of the rocks collected along borehole D5 varied between 113.6 and 131.6 µg/g. They increase significantly after heating from 190.1 to 251.9 µg/g in the samples of bore hole D26 (Table 3). Therefore, the impact is easily visible by the REEs contents. On the other hand, the distribution of these REEs relative to the NASC reference is not changing significantly (Figure 6A,B). Indeed, after an incurved distribution for the LREEs La to Nd, the patterns decrease until Tb, have a short increase for Dy, decrease again until Tm and finally increase until Lu. This distribution remains similar for the heated rocks with a slight change for sample D26-8 that yields the same distribution but higher contents, and for sample D26-10 with higher Sm and Eu contents. In turn, heating impacted the contents of the rock REEs, but not really their distribution patterns. A preliminary assumption for the difference in the contents with very similar patterns could be the dissolution of homogeneous soluble minerals due to heating and fluid revival, with lower REE contents but similar distribution patterns than for the remaining insoluble counterparts.

The REE contents of the rock leachates are low and narrowly scattered along the borehole containing the heater (Table 5). Before heating, the amounts decrease from 17.9 µg/g close to the wall of the MI niche to 13.0 µg/g away from the wall. After heating, the amounts are slightly lower, between 14.7 µg/g away from heater to 12 7 µg/g next to it. However, the distribution patterns outline a decrease away from the gallery wall, even if the contents of the leached REE remain similar after heating. If resulting from an earlier excavation of the gallery, it would also explain the increase of P due to plausible atmospheric contamination. In turn, leaching with dilute acid removes constant amounts of REEs that remain about the same after heating, which suggests that the main REE carriers were apparently not altered. It also confirms that the REEs do not solubilize preferentially in diffusing fluids.

Table 5. The rare-earth elemental contents of the OPA whole-rock leachates by dilute acid before and after the heating experiment.

Sample IDs	Location/ Probe (m)	La (μg/g)	Ce (μg/g)	Pr (μg/g)	Nd (μg/g)	Sm (μg/g)	Eu (μg/g)	Gd (μg/g)	Tb (μg/g)	Dy (μg/g)	Ho (μg/g)	Er (μg/g)	Tm (μg/g)	Yb (μg/g)	Lu (μg/g)	Total (μg/g)
initial																
BHE-D5-8	5.50–5.60	1.21	1.25	1.26	1.24	1.25	1.28	1.32	1.28	1.30	1.28	1.28	1.27	1.35	1.28	17.85
BHE-D5-6	2.45–2.60	1.02	1.03	1.03	1.02	1.03	1.03	1.04	1.04	1.04	1.05	0.77	0.75	0.82	0.78	14.49
BHE-D5-7	1.90–2.00	0.95	0.94	0.92	0.92	0.91	0.89	0.91	0.89	0.90	0.91	0.91	0.97	0.97	0.94	12.93
after heating																
BHE D26-14	0.40	0.96	0.96	0.98	0.97	0.97	0.90	0.94	0.98	1.00	0.96	1.05	1.02	1.02	1.00	13.71
BHE D26-12	0.10	0.99	1.00	1.02	1.02	1.01	0.99	0.98	1.08	1.05	1.03	1.07	1.04	1.11	1.05	14.44
BHE D26-10	0.27	1.03	1.04	1.05	1.06	1.04	0.97	1.03	1.10	1.10	1.07	1.09	1.07	1.15	1.07	13.77
BHE D26-8	0.75	0.89	0.90	0.91	0.90	0.89	0.83	0.89	0.95	0.91	0.89	0.91	0.90	1.00	0.94	12.71
BHE D26-3	3.30	1.00	1.01	1.06	1.03	1.03	0.75	1.01	1.06	1.07	0.99	1.03	1.08	1.05	1.08	14.25
BHE D26-2	5.70	0.92	1.05	1.01	1.01	1.00	0.98	0.99	1.00	1.05	1.04	1.09	1.19	1.17	1.20	14.70

IDs stands for identities and m for meters.

The apparent K-Ar ages of the heated rocks measured after heating and storage are quite systematically higher than those of the samples taken before heating at about the same distance from the niche (Table 4). They range from 326.1 ± 11.1 Ma for the sample taken next to the niche to 347.9 ± 11.2 Ma for that collected 5.55 m away from the same niche. On the other hand, the K_2O contents between 2.65 and 2.88% are analytically similar to those of the rocks collected before heating. They are even identical to most of the unheated rocks except the higher 3.06% value of the unheated sample taken the farthest from the niche wall. As most data from heated rocks remain within the range of values obtained for the corresponding unheated samples, the small detected analytical scatter along the drilling profile appears to relate again mostly to slight mineralogical variations rather than to the impact of the experiment, as no clear systematic trend is observed as a function of the distance to the heater. In the detail, however, the comparison of the apparent K-Ar ages of the heated rocks with the initial references shows systematic slightly higher values after heating. This systematic increase of the K-Ar data ages is apparently related to an increase of the radiogenic ^{40}Ar, which was not really expected in the present conditions, as the K_2O contents are almost identical in the rocks before and after heating.

At this point, the changes in most elemental contents of the rocks and of their leachates after heating relate apparently more to discrete variations in the mineralogical compositions of the selected samples, rather than resulting from heating. The elements affected by heating, such as Ca, Na, Sr, the LOI and the REEs, could be constitutive of soluble minerals. On the other hand, the slight increase in radiogenic ^{40}Ar of the rocks is somehow unexpected, due to the "special" status of Ar that is not bound but squeezed within its host sheet minerals. It is considered to be most sensitive and, therefore, quite easily removable by any crystal modification (e.g., [31]).

3.3. The Results on the Boom Clay

The XRD data of the Boom rocks outline an irregular decrease of quartz until the second control after 1799 days of heating and a straight increase towards the end of the experiment. The feldspars remain at a constant amount during the whole experiment. The calcite and pyrite contents decrease slightly, as those of quartz until 1799 days, while increasing again towards the end of the experiment, the contents of dolomite and anatase remaining parochial during the whole experiment (Table 6). The contents of the 2:1 clay minerals tend to increase until the third control after 1799 days of heating and to decrease afterwards until those of the initial sample. The various clay components including illite, smectite, mixed-layers illite-smectite and smectite-chlorite, kaolinite and chlorite range also narrowly with a slight increase at the third control (Table 6). The mineral repartition of the initial reference rock and those subjected to 80 °C during 830, 1799 and 2529 days remains quite stable except for a decrease in quartz after 1799 days compensated by an increase of the 2:1 clay minerals. This mineral composition is typical for the undisturbed Boom Clay as reported earlier by Zeelmaekers et al. [19] and Frederickx et al. [20]. Mineral heterogeneity being excluded in this experiment, the observed variations resulted then, most probably, from heating. It may also be recalled that the controls had to remain limited because such batch experiment does not allow many stops and intermediate sample collections as each needs to open a container.

The decrease in the quartz content is confirmed by the variation of the SiO_2 values from 60.3% in the reference rock to about 54% after 2529 days of heating. Another significant change is the increase of Al_2O_3 from 16.1% to 19.8%, as well as that of MgO from 1.93% to 2.27%. On the other hand, K_2O increased moderately from 3.15% to 3.47%, while the other major elements remain more or less unchanged if considering the analytical uncertainty given above (Table 7, Figure 8). Also to be mentioned is a correlative increase of the same elements in the <2 μm fractions, except for P that decreases. Gentle leaching of the unheated rock and of that heated during 830 days focuses on a systematic increase of the leachate loads and, therefore, an overall elemental decrease in the rocks except for the increasing P content (Table 8). Also to be mentioned is a correlative increase of the same elements in the

<2 μm fractions, except again for P that decreases. Apparently, heating induced a transfer of major elements, probably volatile and/or soluble, from rock size into smaller crystals of the <2 μm fractions, unless smaller soluble minerals crystallized at the <2 μm size.

Table 6. The XRD mineral composition of the Boom Clay sample before and during the batch test.

Minerals	Initial Contents (%)	After 830 Days (%)	After 1799 Days (%)	After 2529 Days (%)
Quartz	23	16	12	23
Albite	7	6	6	6
K-feldspar	8	5	7	8
Calcite	4	2	1	3
Dolomite	0.6	0.7	0.0	0.1
Pyrite	4	2	2	3
Anatase	1	0.6	0.8	1
2:1 clay minerals	41	54	58	42
Smectite	12	14	15	12
Illite/Smectite	18	27	27	21
Illite	11	11	14	10
Kaolinite	4	5	2	2
KaoliniteSmectite	5	8	9	8
Chlorite	3	3	4	3

Table 7. Major elemental contents of the whole rock and <2 μm fraction of the Boom Clay during the batch test.

Sample Description	SiO_2 (%)	Al_2O_3 (%)	MgO (%)	CaO (%)	Fe_2O_3 (%)	Mn_3O_4 (%)	TiO_2 (%)	Na_2O (%)	K_2O (%)	P_2O_5 (%)	LOI (%)	Total (%)
initial bulk	60.3	16.1	1.93	0.72	5.37	0.02	0.85	0.58	3.15	0.08	10.27	99.36
initial < 2 μm	52.2	24.4	2.55	0.21	5.64	0.02	0.84	0.93	3.51	0.10	9.56	99.96
Heat1 bulk (830 d)	58.9	17.6	2.00	0.65	5.53	0.02	1.04	0.49	3.38	0.11	10.33	100.09
Heat1 < 2 μm	50.5	24.5	2.45	0.94	5.50	0.02	0.76	0.17	3.40	0.07	10.79	99.15
Heat3 bulk (1799 d)	55.0	19.2	2.19	0.67	5.60	0.02	0.93	0.46	3.36	0.09	11.27	
Heat2 bulk (2529 d)	54.2	19.8	2.27	0.67	5.60	0.04	0.84	0.58	3.47	0.13	11.71	99.34

d stands for days.

Table 8. The major elemental contents of the dilute-acid leachates of the whole rock and <2 μm fraction powders of the Boom Clay during the batch test.

Sample Description	Si (μg/g)	Al (μg/g)	Mg (μg/g)	Ca (μg/g)	Fe (μg/g)	Mn (μg/g)	Ti (μg/g)	Na (μg/g)	K (μg/g)	P (μg/g)
initial bulk	2.2	1.44	9.64	36.4	2.69	0.07	0.04	27.1	13.1	2.30
initial < 2 μm	0.8	0.26	0.21	1.30	0.42	bdl	0.01	50.7	2.56	0.63
Heat1 bulk (830 d)	1.6	1.28	7.41	27.2	1.93	0.08	0.01	19.1	10.8	3.53
Heat1 < 2 μm	4.0	1.52	1.11	31.4	1.26	0.01	0.06	1.24	6.50	0.36

d stands for days.

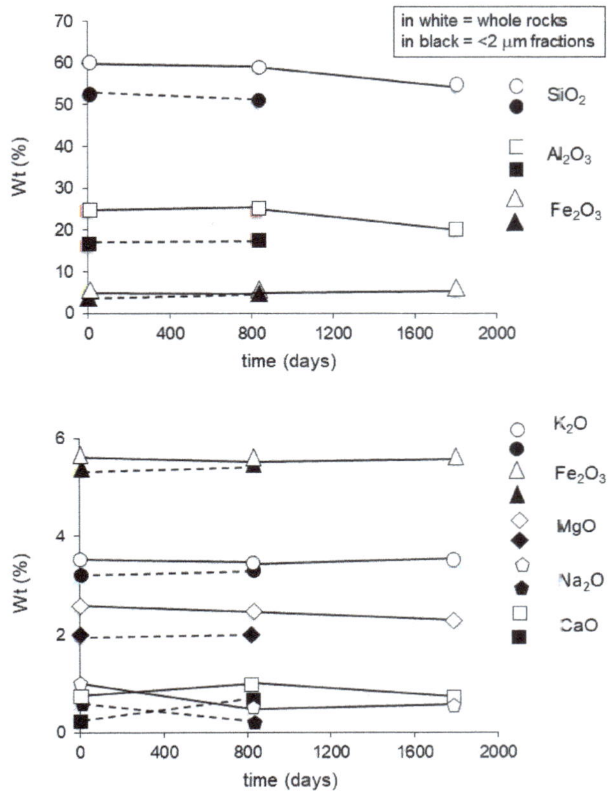

Wt stands for weight

Figure 8. Variations of the major elements from Boom Clay whole rocks before and during the heating batch experiment.

The REE distribution patterns of the Boom Clay rocks yield, again before and after heating, similar patterns relative to the NASC reference, with noticeable Nd and Tb positive anomalies. However, if the patterns are very similar, the amounts of LREEs of the sample collected at the first control are significantly higher from La to Dy than those of the untreated rock (Table 9, Figure 9A). The patterns of the corresponding <2 μm size fraction remain also similar relative to the NASC reference, before and after heating, but with higher contents after heating (Figure 9B). Furthermore, the Nd and Tb anomalies disappear after heating, while that of Dy remains visible, as it was for the OPA material. A comparison of the REE distribution in the heated rock relative to the untreated reference shows a continuous decrease from La to Lu (Figure 9C), while the <2 μm fractions yield an incurved decreasing pattern from La to Dy and increasing from Dy to Lu (Figure 9D), still relative to the NASC reference. In summary, the REE distribution patterns of both the rock and the <2 μm fraction show systematic decreases of the heavier REEs after heating.

The K-Ar apparent ages of the unheated reference rock and of its <2 μm fraction are within analytical uncertainty with those of the same rock and <2 μm fraction after 830 and 2529 days of heating (Table 10). The K-Ar values of the rocks range from a starting 258.3 ± 7.3 Ma to 256.0 ± 13.3 Ma for that heated during 830 days and to 254.4 ± 8.0 Ma after 2529 days of heating. In the case of the size fractions, the K-Ar age data range from 223.5 ± 5.1 Ma for the untreated separate to 226.2 ± 7.1 Ma after 830 days of heating and to 240.3 ± 7.4 Ma for the last step. This last value is significantly higher than the two previous ones, but not outside the analytical uncertainty. Again, the values are far higher than the

Rupelian (34–28 Ma) stratigraphic age, confirming the occurrence of detrital minerals, even at the fine <2 μm size. Moreover, the constant K-Ar values for the rocks and for the size fractions before and after heating mean that the main K carriers muscovite and illite were not altered by the experiment.

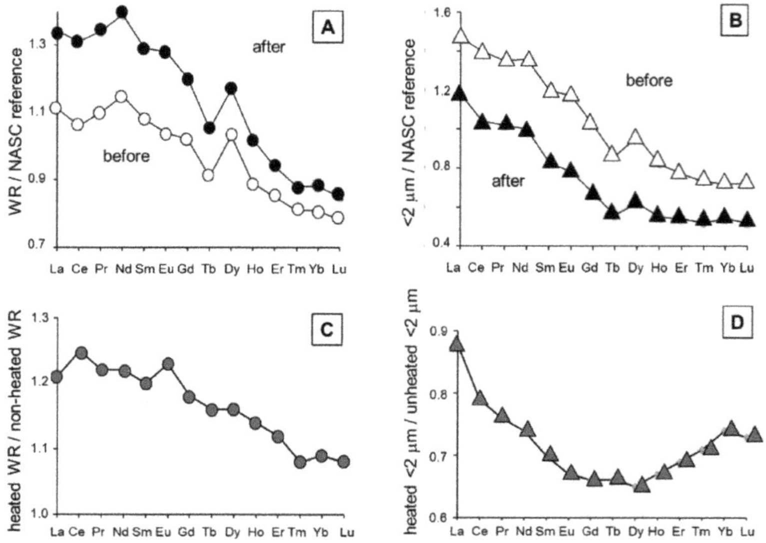

Figure 9. (**A**) REE distribution patterns of the Boom Clay rocks whole rock and (**B**) of the <2 μm fractions, before and after the heating batch experiment. (**C**) The REE contents of the heated whole rock were also compared to that of the initial contents, as well as those of the heated <2 μm relative to the initial fraction (**D**). WR stands for the whole rock heating experiment.

Table 9. The rare-earth elemental contents of the whole rock and <2 μm fraction of the Boom Clay during the batch test.

Sample Description	La (μg/g)	Ce (μg/g)	Pr (μg/g)	Nd (μg/g)	Sm (μg/g)	Eu (μg/g)	Gd (μg/g)	Tb (μg/g)	Dy (μg/g)	Ho (μg/g)	Er (μg/g)	Tm (μg/g)	Yb (μg/g)	Lu (μg/g)	Total (μg/g)
initial bulk	34.5	70.1	8.46	31.4	6.03	1.23	4.98	0.77	4.29	0.91	2.40	0.39	2.47	0.36	168.3
initial < 2 μm	45.7	93.2	10.4	37.2	6.70	1.38	5.01	0.73	4.00	0.84	2.19	0.35	2.21	0.33	210.2
Heat1 bulk (830 d)	41.6	87.7	10.3	38.2	7.23	1.51	5.88	0.89	4.89	1.04	2.68	0.42	2.68	0.39	205.4
Heat1 < 2 μm	36.7	69.3	7.86	27.5	4.66	0.92	3.30	0.48	2.59	0.56	1.52	0.25	1.64	0.24	157.5

d stands for days.

Table 10. The K-Ar data of the whole rock and <2 μm fraction of the Boom Clay during the batch test.

Sample Description	K (%)	rad ^{40}Ar (10^{-9} mol/g)	^{40}Ar (%)	Age ($\pm 2\sigma$) Ma
initial bulk	2.62	1.26	37.82	258.3 (7.3)
initial < 2 μm	2.91	1.20	82.80	223.5 (5.1)
Heat1 bulk (830 d)	2.81	1.34	15.65	256.0 13.3)
Heat1 < 2 μm	2.82	1.18	26.32	226.2 (7.1)
Heat2 bulk (2529 d)	2.74	1.30	27.56	254.4 (8.0)
Heat2 < 2 μm	2.84	1.27	30.22	240.3 (7.4)

4. Discussion

The available analyses allow discussing the chemical and isotopic, and therefore the mineralogical changes of two sedimentary formations subjected to continuous low-thermal imprints along two independent experiments. As the generated temperature increase concerns more specifically the first meter around the heater in the case of the in situ experiment, samples located in this rock volume were more detailed in the comparisons. At this point, the three OPA samples (BHE-D5-3, D5-4, D5-5) located far from the wall and the EDZ/EdZ zones of the MI niche and close to the heater are quite homogeneous in their mineral and chemical compositions.

4.1. The Changes in the Opalinus Clay

The OPA samples located away from heating devices and from EDZ/EdZ zones yield chemical and isotopic data that are slightly away from those of the samples taken closer to the heaters. These variations concern more specifically the signatures of soluble phases as their contents decrease when approaching the MI niche [2]. Apparently related to oxidizing processes due to earlier excavations of the gallery and the niche, they could be due to the longer exposition of the rocks to the atmosphere, rather than to the strict heating experiment, which agrees with the above shown SEM pictures. Very limited, they appear as representative for the 4 to 5 m wall rocks of the niche and, therefore, confirm a more pervasive impact due an earlier dehydration.

The mineralogical, chemical and isotopic data of the heated OPA samples do not outline significant changes beyond analytical uncertainty. The homogeneous K-Ar data of the samples collected far from and next to the thermal probe confirm this observation. The temperature applied to the sediments, which is considered to be representative of the potential heat produced by stored nuclear waste, did not trigger significant chemical and Ar isotopic changes. As the formation already underwent a natural burial-induced diagenetic episode during its earlier evolution with a temperature reaching at least 85 °C [32], one may expect that most or at least part of the changes occurred during this earlier natural diagenesis, especially with an impact during millions of years and not during months like for the experiment. The artificial heating did obviously not generate new altering reactions, which focuses on the maturity of the clay formation in its long-term behavior relative to any moderate thermal imprint like applied here. For a geological disposal of exothermic radioactive waste, this aspect is of importance, as the characteristics of the host formation need to be maintained during a longer period by the waste itself in a mature clay-type host sedimentary sequence, such as the OPA formation.

However, a discrete chemical evolution cannot be denied when comparing the initial and the post-heating releases into the OPA leachates, that is to say those released by the soluble minerals. Basically, the observed changes relate to elements either adsorbed on mineral surfaces or released from soluble minerals sensitive to slight temperature modifications. Such changes are asserted here by: (1) a destabilization of pyrite with a visible alteration of the crystals and the formation of blurry veils around them, probably favored by a bacterial contribution during drilling or since, and (2) a slight increase of the rock K-Ar age values. This latter change was somehow unexpected because it usually occurs only when either the K_2O content decreases and/or when that of the radiogenic ^{40}Ar increases. Here it outlines fresh information as the K_2O data of the rock samples did not change due to the heating experiment, the only explanation being then a higher content of radiogenic ^{40}Ar in some kind of mineral structures relative to the atmospheric ^{40}Ar (Table 5). The differential behavior of an inert gas, such as Ar, can only occur if the radiogenic ^{40}Ar issued from decay of the ^{40}K is retained differently than the atmospheric ^{40}Ar in the K-rich sheet-silicate minerals, such as the micas and their micrometric illite-type minerals. An immediate interpretation would then be in an increase in the radiogenic ^{40}Ar content during a degasing action, as was the case for the OPA rocks. Such behavior is analytically realistic as a preferential "degassing" of the atmospheric ^{40}Ar without any other noticeable impact on the occurring minerals, including the labile K-carrier clay minerals.

4.2. The Changes in the Boom Clay

The heated Boom Clays do, also, not outline significant mineralogical, chemical and K-Ar isotopic changes after 2529 days of heating at 80 °C. All identified mineral phases are still present in the heated samples, although with slight changes in their amounts, as suggested by the quantitative XRD and wet-chemistry analyses. In contrast, the comparison of the leachates from non-heated rock and <2 μm fraction with those from heated equivalents focuses on following changes: the contents of all major elements decrease slightly in the rocks, except that of increasing P, while the contents of the same elements from <2 μm fractions increase except, again, that of decreasing P.

The K-Ar data of the untreated rock and <2 μm fraction are strictly within those of the same rock and <2 μm separate after 830 and 2529 days of heating (Table 9). The number of analyzed samples and fractions being limited for reasons explained above, the discussion about the meaning of the K-Ar data is somehow tied. However, by comparing the data of the initial <2 μm fraction and that after 2529 days of heating, the latter yields a significantly higher K-Ar age data than the untreated reference and than that after 830 days. In fact the increase from 223.5 ± 5.1 to 240.3 ± 7.4 Ma is not due to a changing K content and has, then, to result from an "increase" of the radioactive ^{40}Ar, as was already observed for the OPA material.

The slight changes in the major elemental contents of both the rocks and the <2 μm fractions suggest some transfers of soluble elements from bulk to the <2 μm fractions. A determining aspect is provided, again, by the K-Ar data with an increase of the K content from rocks to their <2 μm fractions, which is related to the increasing amount of K-carriers in the size fractions relative to the whole rocks that contain also coarser crystals of detrital origin, such as quartz and the feldspar grains for instance. This increase is systematically accompanied by a slight decrease in the radiogenic ^{40}Ar content, both resulting in a decrease of the K-Ar age date. In turn, the probable elemental transfers from rocks to the <2 μm fractions seems not to affect the K-Ar data, except the unexpected increase of radiogenic ^{40}Ar. In turn, the K-Ar data suggest also that none of the K carriers of the rocks and the <2 μm fractions were altered by the temperature applied during the 7 years.

4.3. To Which Extent Did the Artificial Temperature Impact the Studied Material?

The data of the Boom Clay experiment build an interesting analytical basis to formulate a transfer of volatile and soluble elements from coarse to fine sized minerals, unless the coarser sized minerals lost more volatile and soluble material than the minerals of smaller size. The K-Ar data of the bulk and <2 μm fractions consolidate also the fact that all K carriers, whatever their size, remained inert during the heat experiment.

The in situ heating of the OPA sediments is close to real deep-storage conditions with the heater working at about 100 °C during 8 months and a final sampling along transversal drillings. The weak aspect is the fact that the controls after heating needed supplementary drillings that, even located very closely to the initial samples, gave samples with mineral compositions plausibly yielding mineral heterogeneity, even very limited, of the sedimentary sequence. It cannot, therefore, be excluded that this mineral heterogeneity could have introduced some minor and uncontrolled biases in the necessary comparisons. In fact, only Ca, Na, Sr, the LOI and the REEs seem to have been affected by the long-term heating with a few variations beyond analytical uncertainty. Most of these elements are major constitutive components of soluble minerals such as the carbonates or the chlorides, a relation supported in turn by the LOI trend.

The data collected during the Boom Clay experiment build an interesting analytical basis to formulate a transfer of volatile and soluble elements from coarse to fine sized minerals, unless the coarser minerals lost more volatile and soluble material than those of the <2 μm separates. The K-Ar data of the bulk and <2 μm fractions consolidate also the fact that all K carriers, whatever their size, remained inert during the heat experiment.

In fact, the most unexpected results are provided by the K-Ar data with a heating-induced increase of the radiogenic ^{40}Ar that basically results from disaggregation of the radioactive ^{40}K. Visible in both studied sequences, this increase is necessarily meaningful, as each studied sequence yields different lithologies, as well as different stratigraphic and isotopic ages. The fact that the results are similar in both sequences heated during two independent experiments constitutes a solid basis for the observation. Heating moderately sediments in their natural environment, as well as in batches, is expected to impact at least, the adsorbed gaseous content of some of the constitutive minerals. In fact, the K-Ar data suggest that the atmospheric ^{40}Ar is basically more affected by a temperature increase than the radiogenic ^{40}Ar, the former being expected to decrease and the latter to increase relatively. It is then plausible that the two Ar isotopes that are of different origins are stored at different places of the host sheet-silicates such as the micas and the micrometric argillaceous illite and illite-smectite mixed-layers. To the best of our knowledge, this is the first time such a differential behavior among Ar isotopes is observed and reported. The earlier excavations of the galleries and niches could have added fluctuations on top of those known for the Ar isotopes that are only secured in crystals by van der Waals bonding. On the other hand, these K-Ar results point towards a specific aspect of the method due to the easy escape of any of the Ar isotopes. Here, the radiogenic ^{40}Ar is surprisingly less easily removed than its atmospheric equivalent by a low heating during several months to years as its content tends to increase. Such increase only can result from a more efficient retention in the mineral structure than the other trapped gases including the atmospheric Ar. An alternative behavior could be due to atmospheric ^{40}Ar only adsorbed on the crystals, while radiogenic ^{40}Ar would be more secured in the crystals because resulting from ^{40}K disintegration.

In summary, the low heat applied to the two sedimentary sequences along two independent experiments has apparently only a very limited impact that induced: (1) a dissolution of soluble minerals, such as carbonates, sulfates and chlorides, due to the mobilization of a limited amount of fluids in the rocks, and (2) some degassing of elements rather adsorbed on than trapped in mineral structures.

5. Conclusions

Two independent heating experiments, one in situ and one in batch containers, were applied to cores samples from Aalenian-Toarcian Opalinus Clay in the Swiss Jura (Switzerland) and from Rupelian Boom Clay ground-material heated during 7 years at Mol (Belgium). These experiments were designed and completed to test the impact of 80–100 °C temperature rises on sediments currently studied as potential deep host repositories for nuclear waste, as such temperature increase is theoretically expected by stored nuclear waste on the host rocks at the disposal sites.

The Opalinus Clay rocks were heated by devices inserted into drillings during 8 months and set at about 100 °C. Some natural mineralogical heterogeneity of the rocks was an identified drawback of the design that could not be ignored. It had to be taken into account with, consequently, limited uncontrolled dispersions in some of the elemental and isotopic data. Despite this sampling aspect, the contents of most major, trace and REEs remained within analytical uncertainty, except Ca, Na, Sr and the REEs which contents, as well as the loss on ignition were affected beyond the analytical uncertainties. The results of the solid rocks match well with a systematic low removal of the same elements from heated samples after leaching with dilute hydrochloric acid. Globally, the observed releases suggest a predominant alteration of carbonates, chlorides and/or sulfates. The drawback of the batch experiment on the Boom Clay is in the necessarily limited control stops because each needs the opening of a container. Despite this aspect, the batch experiment complements well the data of the in situ experiment on the Opalinus Clay by only limited elemental transfers of only soluble minerals, the coarser minerals releasing possibly more elements than the finer. It also cannot to be excluded that some tiny minerals crystallized again in the fine fractions. In fact, no element yields changes that are clearly beyond analytical uncertainty, with the conclusion that the heat did not alter the major constitutive minerals.

The K-Ar data of the two heated rock formations provided unexpected increases of radiogenic ^{40}Ar relative to atmospheric ^{40}Ar. They suggest, in two independent experiments on two different sedimentary formations, that the latter of these isotopes is more easily removed by heating than the former. The two Ar isotopes apparently are hold differently in the K-bearing mineral space, especially of micas and argillaceous illite and illite-smectite mixed-layers, with results in favor of a preferential release of atmospheric ^{40}Ar relative to radiogenic ^{40}Ar by heating. A spontaneous interpretation is in a different holding/setting than for the other gases, the atmospheric ^{40}Ar being only adsorbed to crystals while the corresponding radiogenic ^{40}Ar is better secured in the crystals. Therefore, this differential behavior might become an interesting test for the evaluation of the potential sealing characteristics of potential host sediments.

Importantly, the different analytical protocols tested here on two sedimentary formations point to identical conclusions, namely that an extended 80–100 °C temperature increase does not alter the sealing characteristics of the studied formations. In other words, the impact by spontaneous heating does not alter the safety-related properties of the studied potential host sediments.

Author Contributions: Conceptualization, N.C., C.N.; Data curation, N.C.; Formal analysis, M.H., L.F.; Funding acquisition, N.C., C.N.; Investigation, M.H., L.F. All authors have read and agreed to the published version of the manuscript.

Funding: Mont Terri Consortium (project FI) and the NF European project.

Institutional Review Board Statement: Not applicable.

Informed Consent Statement: Not applicable.

Data Availability Statement: Some of the data are in previous publications with references and the others are available upon request.

Acknowledgments: The authors would like to thank each the Mont Terri Consortium (project FI) and the NF European project for funding support to these research projects. NC would also like to add personal thanks to the engineers and technicians of the previously named Laboratoire d'Hydrologie et de Géochimie Isotopique of the University of Strasbourg (France) for technical help in these projects of importance for the study of safety aspects that need to be carefully addressed in the prospective of deep disposal of radioactive waste.

Conflicts of Interest: The authors declare no conflict of interest.

References

1. De Craen, M.; Wang, L.; Van Geet, M.; Moors, H. Geochemistry of Boom Clay pore water at the Mol site. In *Scientific Report SCK-CEN, Blg 990*; Waste & Disposal Department SCK-CEN: Mol, Belgium, 2004; 178p.
2. Techer, I.; Clauer, N.; Liewig, N. Aging effect on the mineral and chemical composition of Opalinus Clays (Mont Terri, Switzerland) after excavation and surface storage. *Appl. Geochem.* **2009**, *24*, 2000–2014. [CrossRef]
3. Anon, 2001 Anon. In *Progress Report on Feasibility Studies and Research into Deep Geological Disposal of High-Level, Long-Lived Waste*; Agence Nationale pour la Gestion des Déchets Radioactifs (ANDRA): Châtenay-Malabry, France, 2001.
4. Bossart, P.; Meier, P.M.; Möri, A.; Trick, T. Geological and hydraulic characterisation of the excavation disturbed zone in the Opalinus Clay of the Mont Terri Rock Laboratory. *Eng. Geol.* **2002**, *66*, 19–38. [CrossRef]
5. Meier, P.M.; Trick, T.; Blümling, P.; Volckaert, G. Self-healing of fractures within the EDZ at the Mont Terri Rock Laboratory: Results after one year of experimental work. In Proceedings of the International Workshop on Geomechanics, Hydromechanical and Thermohydro-Mechanical Behaviour of Deep Argillaceous Rocks: Theory and Experiment, Paris, France, 11–12 October 2002; pp. 267–274.
6. Alheid, H.J. Lessons learned in indurated clays. In *Impact of the Excavation Disturbed or Damaged Zone (EDZ) on the Performance of Radioactive Waste Geological Repositories, Proceedings of the European Commission CLUSTER Conference, Luxembourg, 3–5 November 2003*; Publications Office of the European Union: Luxembourg, 2003.
7. Charpentier, D.; Cathelineau, M.; Mosser-Ruck, R.; Bruno, G. Evolution minéralogique des argillites en zone sous-saturée oxydée: Exemple des parois du tunnel de Tournemire (Aveyron, France). *Earth Planet. Sci. Lett.* **2001**, *332*, 601–607.
8. Matray, M.; Savoye, S.; Cabrera, J. Desaturation and structure relationships around drifts excavated in the well-compacted Tournemire's argillite (Aveyron, France). *Eng. Geol.* **2007**, *90*, 1–16. [CrossRef]

9. Mayor, J.C.; Velasco, M.; Garcia-Sineriz, J.L. Ventilation experiment in the Mont Terri underground laboratory. *Phys. Chem. Earth A/B/C* **2007**, *32*, 616–628. [CrossRef]
10. Horseman, S.T. Self-healing of fractures in argillaceous media from the geomechanical point of view. In Proceedings of the Topical Session on Self-healing, Nancy, France, 16 May 2001.
11. Pepper, D.W.; Chen, Y. Heat transfer analysis of nuclear waste casks stored in the Yucca Mountain repository. *Numer. Heat Transf.* **2005**, *47*, 671–690. [CrossRef]
12. Finsterle, S.; Muller, R.A.; Baltzer, R.; Payer, J.; Rector, J.W. Thermal evolution near heat-generating nuclear waste canisters disposed in horizontal drill-holes. *Energies* **2019**, *12*, 596–619. [CrossRef]
13. Wileveau, Y.; Rothfuchs, T. HE-D Experiment: Test Plan. In *Mont Terri Project, Technical Note 2004-20, Sept. 2003*.
14. Zhang, C.-L.; Rothfuchs, T.; Su, K.; Hoteit, N. Experimental study of the thermo-hydro-mechanical behaviour of indurated Clays. In Proceedings of the 2nd International Meeting of Clays in Natural & Engineered Barriers for Radioactive Waste Confinement, Tours, France, 14–18 March 2005.
15. Lorant, F.; Largeau, C.; Behar, F.; De Cannière, P. Improved kinetic modeling of the early generation of CO_2 from the Boom Clay kerogen. Implications for simulation of CO_2 production upon disposal of high activity nuclear waste. *Org. Geochem.* **2008**, *39*, 1294–1301. [CrossRef]
16. Deniau, I.; Behar, F.; Largeau, C.; De Canniere, P.; Beaucaire, C.; Pitsch, H. Determination of kinetic parameters and simulation of early CO_2 production from the Boom Clay kerogen under low thermal stress. *Appl. Geochem.* **2005**, *20*, 2097–2107. [CrossRef]
17. Deniau, I.; Derenne, S.; Beaucaire, C.; Pitsch, H.; Largeau, C. Simulation of thermal stress influence on the Boom Clay kerogen (Oligocene, Belgium) in relation to long-term storage of high activity nuclear waste—I. Study of generated soluble compounds. *Appl. Geochem.* **2005**, *20*, 587–597. [CrossRef]
18. Doebelin, N.; Kleeberg, R. Profex: A graphical user interface for the Rietveld refinement program BGMN. *J. Appl. Crystallogr.* **2015**, *48*, 1573–1580. [CrossRef] [PubMed]
19. Zeelmaekers, E.; Honty, M.; Derkowski, A.; Środoń, J.; De Craen, M.; Vandenberghe, N.; Adriaens, R.; Ufer, K.; Wouters, L. Qualitative and quantitative mineralogical composition of the Rupelian Boom Clay in Belgium. *Clay Miner.* **2015**, *50*, 249–272. [CrossRef]
20. Frederickx, L.; Honty, M.; De Craen, M.; Elsen, J. Evaluating the quantification of the clay mineralogy of the Rupelian Boom Clay in Belgium by a detailed study of size fractions. *Appl. Clay Sci.* **2021**, *201*, 105954. [CrossRef]
21. Samuel, J.; Rouault, R.; Besnus, Y. Analyse multi-élémentaire standardisée des matériaux géologiques en spectrométrie d'émission par plasma à couplage inductif. *Analusis* **1985**, *13*, 312–317.
22. Bonhomme, M.; Thuizat, R.; Pinault, Y.; Clauer, N.; Wendling, R.; Winkler, R. *Méthode de Datation Potassium-Argon. Appareillage et Techniques. Note Technique de l'Institut de Géologie de Strasbourg 3*; Institut de Géologie de Strasbourg: Strasbourg, France, 1978; 53p.
23. Odin, G.S.; 35 collaborators. Interlaboratory standards for dating purposes. In *Numerical Dating in Stratigraphy*; Odin, G.S., Ed.; John Wiley & Sons: Chichester, UK, 1982; pp. 123–148.
24. Lee, J.Y.; Marti, K.; Severinghaus, J.P.; Kawamura, K.; Yoo, H.S.; Lee, J.B.; Kim, J.S. A redetermination of the isotopic abundances of atmospheric Ar. *Geochim. Cosmochim. Acta* **2006**, *70*, 4507–4512. [CrossRef]
25. Steiger, R.H.; Jager, E. Subcommission on geochronology: Convention on the use of decay constants in geo- and cosmochronology. *Earth Planet. Sci. Lett.* **1977**, *36*, 359–362. [CrossRef]
26. Thury, M.; Bossart, P. The Mont Terri rock laboratory, a new international research project in a Mesozoic shale formation, in Switzerland. *Eng. Geol.* **1999**, *52*, 347–359. [CrossRef]
27. Thury, M.; Bossart, P. Results of the hydrogeological, geochemical and geotechnical experiments performed in the Opalinus Clay (1996–1997). Chapter 6.4: Organic Matter Characterisation of Rocks and Pore Waters. In *Geological Report No. 23*; Swiss Geological Survey: Bern, Switzerland, 1999.
28. Janots, E.; Brunet, F.; Goffé, B.; Poinssot, C. Thermochemistry of monazite-(La) and dissakisite-(La): Implications for monazite and allanite stability in metapelites. *Contrib. Mineral. Petrol.* **2007**, *154*, 1–14. [CrossRef]
29. Piper, D.Z.; Bau, M. Normalized rare earth elements in water, sediments, and wine: Identifying sources and environmental redox conditions. *Am. J. Anal. Chem.* **2013**, *4*, 69–83. [CrossRef]
30. Taylor, S.; McLennan, S. *The Continental Crust: Its Composition and Evolution*; Blackwell: Oxford, UK, 1985; p. 312.
31. Clauer, N.; Chaudhuri, S. *Clays in Crustal Environments: Isotope Dating and Tracing*; Springer: Heidelberg/Berlin, Germany; New York, NY, USA, 1995; 359p.
32. Mazurek, M.; Hurford, A.J.; Leu, W. Unravelling the multi-stage burial history of the Swiss Molasse Basin: Integration of apatite fission track, vitrinite reflectance and biomarker isomerisation analysis. *Basin Res.* **2006**, *13*, 27–50. [CrossRef]

sustainability

MDPI

Article

Natural Clay Minerals as a Starting Material for Matrices for the Immobilization of Radioactive Waste from Pyrochemical Processing of SNF

Anna V. Matveenko [1], Andrey P. Varlakov [2], Alexander A. Zherebtsov [3], Alexander V. Germanov [2], Ivan V. Mikheev [1], Stepan N. Kalmykov [1] and Vladimir G. Petrov [1,*]

1 Department of Chemistry, Lomonosov Moscow State University, 119991 Moscow, Russia; avd.msk11@mail.ru (A.V.M.); mikheev.ivan@gmail.com (I.V.M.); stepan@radio.chem.msu.ru (S.N.K.)
2 JSC "A.A. Bochvar High-Technology Scientific Research Institute for Inorganic Materials (VNIINM)", 123098 Moscow, Russia; varlakov_a@mail.ru (A.P.V.); avgermanov@bochvar.ru (A.V.G.)
3 JSC "Proryv", 107140 Moscow, Russia; zhala@proryv2020.ru
* Correspondence: vladimir.g.petrov@gmail.com

Abstract: Pyrochemistry is a promising technology that can provide benefits for the safe reprocessing of relatively fresh spent nuclear fuel with a short storage time (3–5 years). The radioactive waste emanating from this process is an electrolyte (LiCl–KCl) mixture with fission products included. Such wastes are rarely immobilized through common matrices such as cement and glass. In this study, samples of ceramic materials, based on natural bentonite clay, were studied as matrices for radioactive waste in the form of LiCl–KCl eutectic. The phase composition of the samples, and their mechanical, hydrolytic, and radiation resistance were characterized. The possibility of using bentonite clay as a material for immobilizing high-level waste arising from pyrochemical processing of spent nuclear fuel is further discussed in this paper.

Keywords: pyrochemical process; radioactive waste; immobilization; bentonite; ceramic matrices; spent nuclear fuel

Citation: Matveenko, A.V.; Varlakov, A.P.; Zherebtsov, A.A.; Germanov, A.V.; Mikheev, I.V.; Kalmykov, S.N.; Petrov, V.G. Natural Clay Minerals as a Starting Material for Matrices for the Immobilization of Radioactive Waste from Pyrochemical Processing of SNF. *Sustainability* **2021**, *13*, 10780. https://doi.org/10.3390/su131910780

Academic Editors: Changhyun Roh, Michael I. Ojovan, Vladislav A. Petrov and Sergey V. Yudintsev

Received: 9 August 2021
Accepted: 24 September 2021
Published: 28 September 2021

1. Introduction

Countries that use nuclear energy, including Russia, have accumulated large volumes of radioactive waste (RW) of various activity levels and aggregate states. Their main source of radioactive waste comes from the processing of spent nuclear fuel (SNF) from energetic, transport (marine), and research reactors. Converting solid RW, including slags, into a chemically and mechanically stable form suitable for final disposal, in order to isolate RW from the biosphere, is an urgent scientific, technical, and social task, and a viable solution would significantly increase the level of radiation safety for the population.

Among all types of radioactive waste, high-level waste (HLW) is the most dangerous and is dominated by ^{90}Sr, ^{137}Cs and ^{60}Co radionuclides, whose total activity can amount to 95% of the total activity of stored radioactive waste. The generally accepted approach to solving the problem of HLW management is its conversion to a solid state (for liquid waste), its inclusion into strong phases (preserving matrices) and its further disposal in underground storage facilities. At present, vitrified materials are used to immobilize HLW on an industrial scale; borosilicate glasses are still the main form used for the immobilization of high-level radioactive wastes [1].

In view of the task to close the fuel cycle, the Breakthrough (Proryv) project in Russia was launched for the reprocessing of mixed uranium–plutonium nitride (MNUP) SNF using pyrochemical or combined technology (pyrochemistry + hydrometallurgy), where the use of molten salt LiCl–KCl is assumed [2]. In contrast to technological processes in aqueous–organic systems, processes in chloride melts have advantages such as high radiation resistance of the liquid phase, internal safety of the processes, the possibility of regenerating spent nuclear fuel with a short exposure time, and a smaller amount of generated radioactive waste. However, the industrial application of pyrochemical processing technology cannot be carried out until the best methods for the disposal of the resulting radioactive wastes are determined. In this case, the LiCl–KCl eutectic with included fission products (FP) will act as the waste (Figure 1).

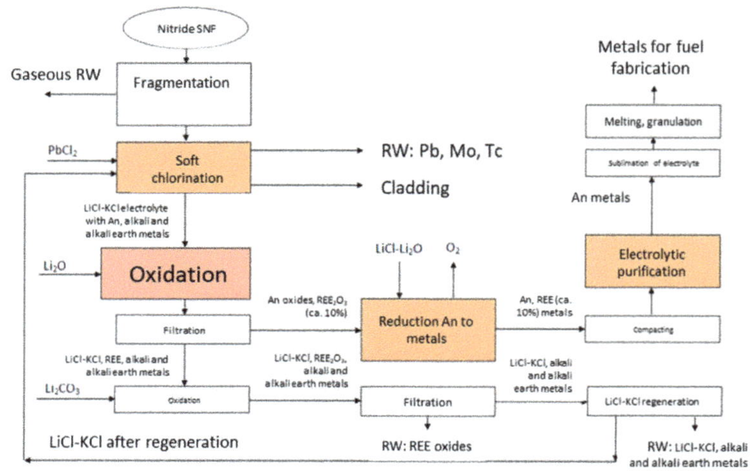

Figure 1. The technological optional scheme for pyrochemical processing of nitride SNF.

Chloride salts are difficult to immobilize in both glass matrices and cements, and in most types of ceramics. Studies of new glass compositions and glass-based composite materials show that the reliable fixation of significant amounts of alkali metals' chlorides requires preliminary treatment and their transformation to the form of oxides or phosphates, i.e., an additional technological operation is required [3–6].

In addition, the research results indicate an instability of glasses in groundwater, especially after their spontaneous devitrification (crystallization) during storage as an effect of ionizing radiation and chemical processes inside the glasses. The destruction of glass matrices due to crystallization will lead to a decrease in their chemical stability and, as a consequence, to the release of radionuclides, including long-lived ones, into the environment. Therefore, the development of alternative crystalline (mineral-like) matrices for the immobilization of high-level waste is an urgently required scientific task [7–9].

The main requirements to matrix materials are: a high isomorphic capacity for waste components (the ability to include a significant amount of waste in the bulk composition), high chemical and radiation stability (it is necessary to minimize the release of radionuclides during possible contact with groundwater), mechanical strength (it must not disintegrate during transportation, storage operations, etc.), as well as an easy industrial production process. The selection of suitable phases, with the listed properties, is conducted on the basis of the results of geochemical, mineralogical, and crystal–chemical studies of natural minerals and synthetic compounds.

Existing work on the properties of crystalline ceramics for immobilizing radioactive waste is related to mineral-like phases such as pyrochlore, perovskite, zirconolite, zirconium dioxide, garnet, hollandite, pollucite, murataite, monazite, and NZP [10–17]. The authors of [18] developed a method for immobilizing chloride salts using zeolite with sodalite as the final form of the transformed chlorides. This method involves a technologically complex and long-term process of zeolite occlusion to uniformly distribute alkali chlorides over zeolite cells. Chloride incorporation in such a material is negligible, which proves to be a problem for all chloride-containing matrices. Similar methods of incorporating salt waste (chlorides and iodides of alkali metals) into sodalite and zeolite matrices are described in [19–23].

In this work, we investigated the use of natural clay minerals as a possible matrix material for waste produced from pyrochemical processing of SNF in the form of LiCl–KCl eutectic that included fission products. The mineral-like materials will decrease the diffusion of radionuclides from the matrix into the biosphere because the matrix material, in its chemical and phase composition, will be similar to the host rocks of the waste repository.

2. Materials and Methods

The aim of this work was to develop a method for the anhydrous conversion of mono- and bivalent s-elements' chlorides into a stable, mineral-like structure. To achieve this, we used bentonite clays, that (a) are finely dispersed materials consisting of at least 70% of the layered mineral montmorillonite, (b) have a large specific surface area, (c) just as all clays, become very strong upon annealing, and (d) are able to retain their alkali metals due to their layered structure. This option may be the most promising in order to ensure the reliable long-term isolation of radionuclides from the environment.

In this study, we prepared samples of bentonite clay with a chloride mixture (ClM) content of at least 20 wt.%, of various compositions, as well as samples of bentonite clays with various silicon-containing additives and the inclusion of the at least 20 wt.% of the LiCl–KCl eutectic. The compositions of the main studied samples included the simulated spent electrolyte and are presented in Table 1. In addition to monovalent and bivalent elements, we also used analogs of trivalent and tetravalent f-elements, since such elements can contaminate the spent electrolyte.

Table 1. Composition of the samples studied in the work with different composition of the simulated spent chloride mixture and silicon-containing additives.

Sample Code	Composition of the Ceramic Base	Content, wt.%					
		LiCl	KCl	CsCl	$SrCl_2$	$CeCl_3$	$ZrOCl_2$
B20-0	Bentonite + 20 wt.% LiCl–KCl eutectic	44	56	-	-	-	-
B20-1	Bentonite + 20 wt.% ClM1	40	50	10	-	-	-
B20-2	Bentonite + 20 wt.% ClM2	40	50	5	5	-	-
B20-3	Bentonite + 20 wt.% ClM3	40	50	3	3	4	-
B20-4	Bentonite + 20 wt.% ClM4	40	50	3	3	-	4
B + mcrSi3	Bentonite + 3 wt.% microcryst.silica + 20 wt.% ClM	44	56	-	-	-	-
B + mcrSi5	Bentonite + 5 wt.% microcryst.silica + 20 wt.% ClM	44	56	-	-	-	-
B + mcrSi10	Bentonite + 10 wt.% microcryst.silica + 20 wt.% ClM	44	56	-	-	-	-
B + NaSi3	Bentonite + 3 wt.% Na_2SiO_3 + 20 wt.% ClM	44	56	-	-	-	-
B + NaSi5	Bentonite + 5 wt.% Na_2SiO_3 + 20 wt.% ClM	44	56	-	-	-	-
B + NaSi10	Bentonite + 10 wt.% Na_2SiO_3 + 20 wt.% ClM	44	56	-	-	-	-
B + A3	Bentonite + 3 wt.% AEROSIL + 20wt.% ClM	44	56	-	-	-	-
B + A5	Bentonite + 5 wt.% AEROSIL + 20wt.% ClM	44	56	-	-	-	-
B + A10	Bentonite + 10 wt.% AEROSIL + 20 wt.% ClM	44	56	-	-	-	-

The mixture of chlorides was ground, dehydrated in an oven at 250 °C for 2 h, and then stored in a desiccator over P_2O_5. Then, all the samples were prepared by mixing and grinding ClM with the main components of the matrices, after which the mixture was dried at 100–150 °C and pressed into tablets with a diameter of 10 mm and a height of 2–4 mm or 8–10 mm (for some tests) using a hand press. The last two stages of temperature preparation, the parameters of which were determined using some additional analyses, were carried out in a muffle furnace: the samples were dried at 400 °C for 4 h, ground again and pressed into tablets to obtain a more homogeneous product, and at the end they were annealed for 12 h at 900 °C (below the sintering temperature of pure bentonite).

Thermal characteristics of the prepared mixtures were investigated with a NETZSCH STA 449 C/4/G Jupiter device (Germany) with synchronous gas phase analysis using a NETZSCH 403C Aeolos quadrupole mass spectrometer (Germany). The analysis was carried out in the temperature range 40–1000 °C at a rate of 10 K/min.

The mechanical strength of each ceramic sample was determined by first measuring the maximum load that each sample could withstand without breaking, and then calculating the stresses at these given loads. A Cybertronic Cyber-Plus Evolution testing machine was used for such purposes. Three parallel samples were used for measurements; the average value calculated using the three measurements was taken as the result of the strength tests.

The phase composition of each of the samples and its change during the heating process was measured using an Empyrean, Panalytical X-ray powder diffractometer with a furnace and the ability to reach heats of up to 1200 °C (CuK_{α} radiation, two wavelengths were taken into account in the calculations: 1.5406 and 1.5444 Å with an intensity ratio in the doublet of 2:1). The phase composition of the samples was studied by heating them from room temperature to 1000 °C.

The chemical stability of the samples was studied according to the standard procedure, under static conditions, and at 25 °C in accordance with standard protocol GOST P 52126-2003. The pre-synthesized tablets were placed in plastic cups with a lid in 10 mL of distilled water. At specific time intervals, the solution was poured out, water samples were taken for chemical analysis, and a new portion of water was poured into the glass. An analysis of the leachates was performed using an ICP-AES and ionic chromatography.

The leaching rate was calculated using the following equations:

$$\mathrm{NL} = \frac{\mathrm{a}}{\mathrm{f} \times \mathrm{S}}; \quad (1)$$

$$\mathrm{R} = \frac{\mathrm{NL}}{\mathrm{t}}, \quad (2)$$

where R is the leaching rate, g/(cm^2·day);
NL is normalized weight loss, g/cm^2;
t is the time interval, in days;
a is the mass of the element that has passed into the solution, g;
f is the mass fraction of the element in the sample;
S is the sample surface in cm^2.

To determine their radiation resistance, the ceramic samples were irradiated on an electron accelerator for a month, with a maximum dose of 10^9 Gy, after which the surface morphology, phase composition, and mechanical characteristics of the samples were investigated.

3. Results and Discussion

3.1. Thermal Analysis

To determine the optimal time–temperature parameters for synthesis, we performed a thermogravimetric analysis and differential scanning calorimetry. The results of these are presented in Figure 2a. Water was released during the first stage, (200–250 °C), possibly due to the preservation of the primary layered structure of montmorillonite. During the second stage, the mineral began to lose water, represented in the hydroxyl groups of the layers, and at temperatures of 500–600 °C the mineral turned into an anhydrous modification; the transition of hydroxyl groups to oxol groups also occurred with the release of water.

Thus, the thermochemical binding of the components of pyrochemical waste reprocessing ended, and the metal cations present in the spent electrolyte were incorporated into the crystal structure. According to the analysis of off-gas mass spectrometry, at higher temperatures, weight loss was due to the release of HCl (Figure 2b). Further, CO_2 emissions were observed at temperatures of 400–600 °C due to the presence of microcline, albite, calcite, and coal fragments in the initial bentonite along with clay minerals (the measured total C content is about 0.6–0.7%) [24]. Thus, a temperature range above 700 °C was chosen for the synthesis of matrices.

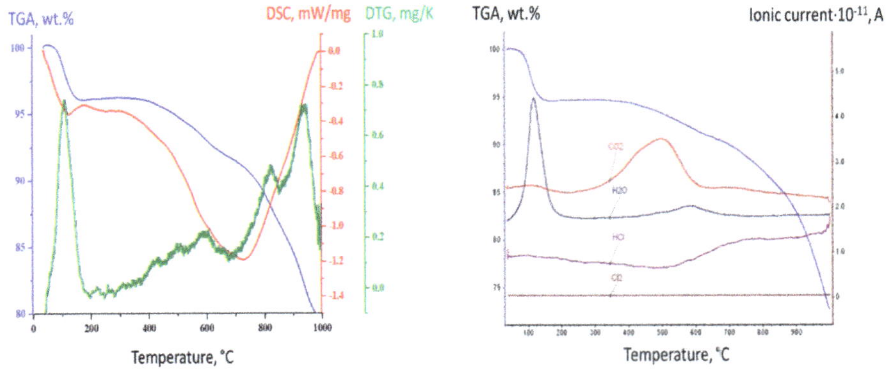

(a) (b)

Figure 2. (**a**) TGA and DSC curves for a sample *B20–0*; (**b**) TGA curves together with the composition of the released gases according to mass spectrometry data.

3.2. X-ray Diffraction Analysis

To study phase composition, the synthesized tablets were ground again and sent for XRD analysis; the data were processed using the HighScore Plus program. According to the results, the studied ceramics were polyphase samples with the following main observed phases: leucite $KAlSi_2O_6$, albite $K_{0.2}Na_{0.8}AlSi_3O_8$, sanidine $(Na,K)AlSi_3O_8$, spodumene $LiAlSi_2O_6$, strontium anorthite $SrAl_2Si_2O_8$, and strontium helenite $Sr_2Al_2SiO_7$. Small quantities of cesium kalsilit $CsAlSiO_4$ and cesium pollucite $CsAlSi_2O_6$ were also observed in some samples.

Figures 3–6 show the XRD patterns for ceramics synthesized without introducing silicon-containing additives, showing the main phases identified in the samples, as well as further possible phases.

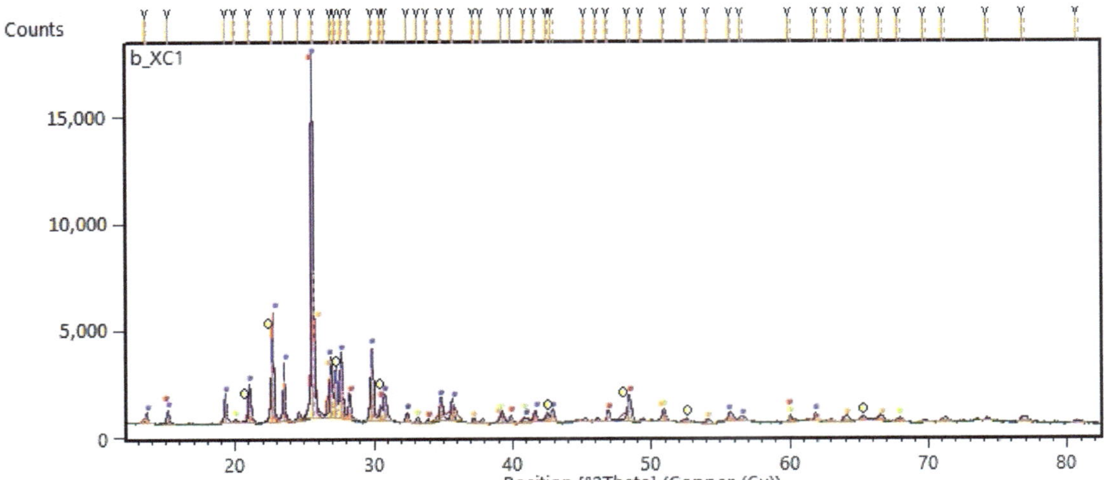

Figure 3. Diffraction pattern of a ceramic sample B20-1; phases: blue—$KAlSi_3O_8$, red—$LiAlSi_2O_6$, green—$K_{1.25}Al_{1.25}Si_{0.75}O_4$, yellow—$LiAlSi_3O_8$, orange—$Al_2SiO_5$.

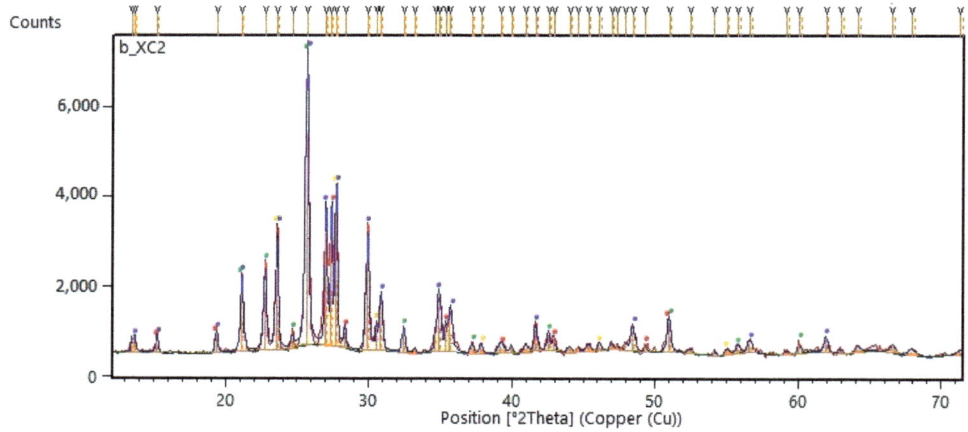

Figure 4. Diffraction pattern of a ceramic sample B20-2; phases: blue –$KAlSi_3O_8$, green—$LiAlSi_3O_8$, red—$SrAl_2Si_2O_8$, $Sr_{46}Al_{92}Si_{100}O_{384}$, orange—$CsAlSi_2O_6$; Al_2SiO_5.

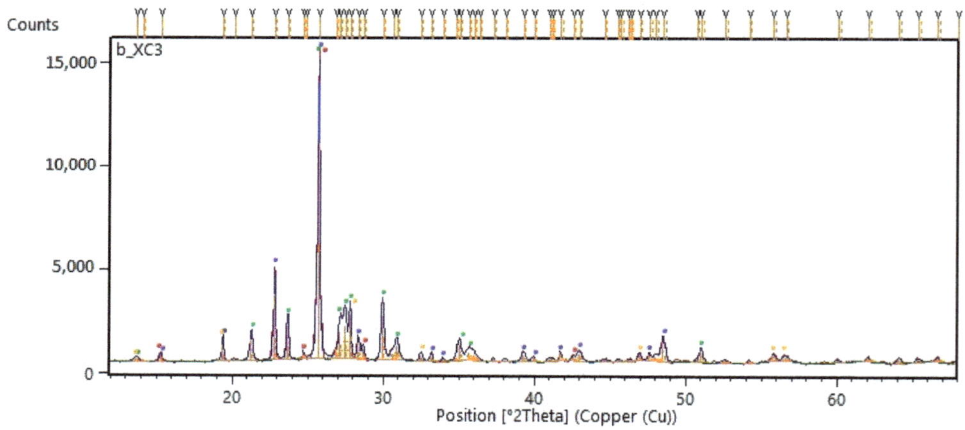

Figure 5. Diffraction pattern of a ceramic sample B20-3; phases: blue—$LiAlSi_3O_8$, green—$KAlSi_3O_8$, red—$KAlSiO_4$, orange—$SrAl_2Si_2O_8$; $K_{9.71}Al_{1.68}Si_{34.32}O_{72}$.

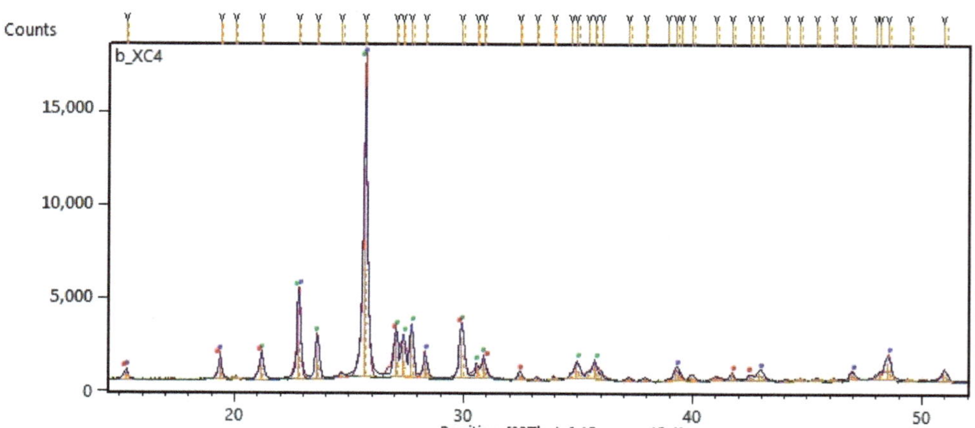

Figure 6. Diffraction pattern of a ceramic sample B20-4; phases: blue—$LiAlSi_3O_8$, green—$KAlSi_3O_8$, red—$SrAl_2Si_2O_8$.

To better understand the structure transformations that occurred through heating, but not through cooling, an X-ray phase analysis with in situ heating was carried out by heating samples from a temperature of 30 to 1000 °C and then cooling them to 35 °C. Figure 7 shows an example of the combined diffraction patterns for a sample containing 30 wt.% of eutectic. This additional sample was synthesized according to the standard procedure described above and was then investigated in order to reduce the probability of missing phases, because secondary phases with content of less than 5 wt.% could not practically be identified. It should be noted that phase changes already occur at 600 °C, but the most significant structural rearrangements occur at 750 °C. The main crystalline phases occurring at high temperatures are leucite, albite, sanidine, and spodumene, which are framework aluminosilicates capable of retaining cations. The initial phases are not observed in the final products, which means that all cations are incorporated into the new structure.

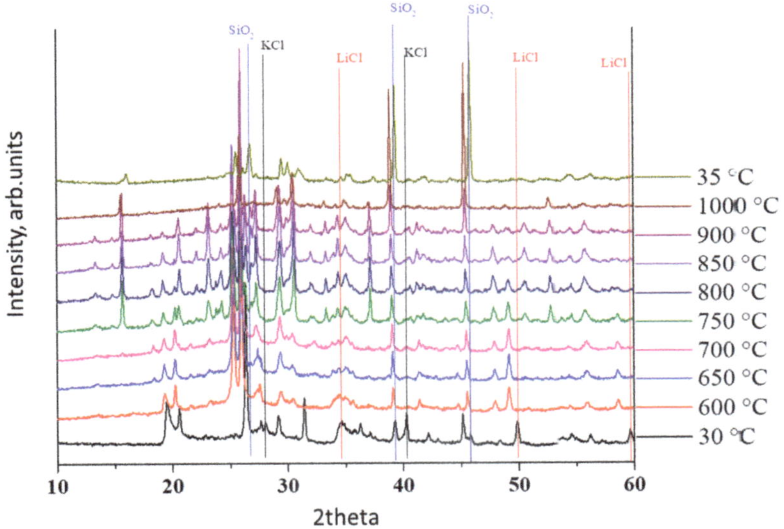

Figure 7. Diffraction patterns of a sample based on bentonite clay with the inclusion of 30 wt.% of simulated spent eutectic LiCl–KCl obtained at different temperatures.

To confirm that cesium and strontium, alongside the main components of the chloride mixture, were incorporated into the structure of aluminosilicates, additional experiments were carried out with an increased amount of chlorides of these metals both in the presence and in the absence of the LiCl–KCl eutectic, with its total amount being 20 wt.%. The XRD results (Figures 8 and 9) show that cesium kalsilit and pollucite, strontium anorthite, and gehlenite (black points, respectively) are formed even in the presence of the eutectic, and thus it can be argued that these metals may be embedded in structures, in small percentages, in the composition of simulated spent electrolyte.

Thus, based on the results of the X-ray phase analysis it can be concluded that, under the chosen synthesis conditions, the incorporation of alkali and alkali earth metals into the structure of aluminosilicates occurs.

A modification of the matrix material made by adding silicon-containing additives did not significantly affect the final major phases in the sample, which can be seen from the XRD results presented in Table 2.

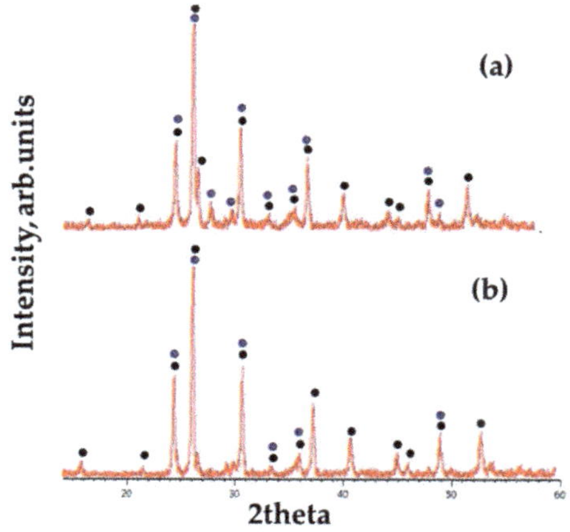

Figure 8. Diffraction patterns of auxiliary samples based on bentonite with an increased content of cesium chloride: black—$CsAlSiO_4$, $CsAlSi_2O_6$; blue—other phases ($KAlSi_3O_8$, $LiAl (SiO_3)_2$, $Al_2 (SiO_4) O$). (**a**) 1/4 part of ClM is CsCl, 3/4 parts of ClM—eutectic LiCl–KCl; (**b**) 1/2 part of ClM is CsCl, 1/2 part of ClM—eutectic LiCl–KCl.

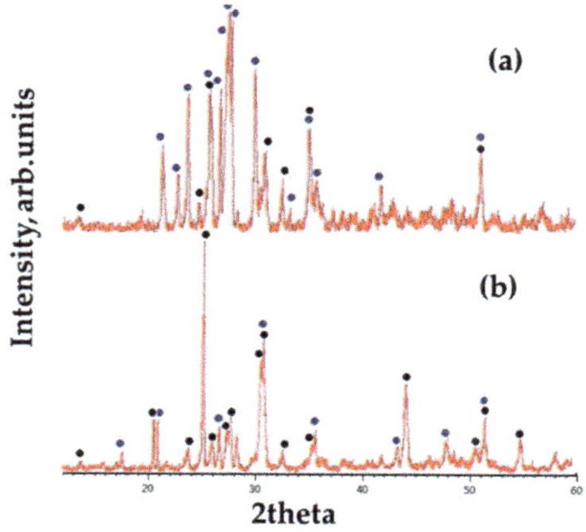

Figure 9. Diffraction patterns of auxiliary samples based on bentonite with an increased content of strontium chloride: black—$SrAl_2Si_2O_8$, $Sr_2Al_2SiO_7$; blue—other phases ($KAlSi_3O_8$, $LiAl (SiO_3)_2$, $Al_2 (SiO_4)O$). (**a**) 1/4 part of ClM is $SrCl_2$, 3/4 parts of ClM—eutectic LiCl–KCl; (**b**) 1/2 part of ClM is $SrCl_2$, 1/2 part of ClM—eutectic LiCl–KCl.

Table 2. Phase composition of the samples with various silicon-containing additives and including 20 wt.% of the simulated spent eutectic LiCl–KCl.

№	Sample Code	Phase Composition
1	B + mcrSi3	$LiAlSi_3O_8$, $KAlSi_3O_8$
2	B + mcrSi5	$LiAlSi_3O_8$, $KAlSi_3O_8$
3	B + mcrSi10	$Li_2O{\cdot}Al_2O_3{\cdot}7.5SiO_2$, $KAlSi_3O_8$, $Al_{2.806}O_{22.08}Si_{8.878}$
4	B + NaSi3	$LiAlSi_3O_8$, $LiAlSi_2O_6$, $K_{0.831}Na_{0.169}A_lSi_3O_8$, $KAlSi_3O_8$
5	B + NaSi5	$LiAlSi_2O_6$, $K_{0.831}Na_{0.169}A_lSi_3O_8$, $K_{11.7}Al_{1.8}Si_{34.2}O_{72}$
6	B + NaSi10	$LiAlSi_2O_6$, $K_{0.831}Na_{0.169}A_lSi_3O_8$, $KAlSi_3O_8$, ($Li_2O{\cdot}Al_2O_3{\cdot}7.5SiO_2$)
7	B + A3	$KAlSi_3O_8$, $LiAlSi_3O_8$
8	B + A5	$KAlSi_3O_8$, $LiAlSi_3O_8$
9	B + A10	$KAlSi_3O_8$, $LiAlSi_3O_8$

3.3. Mechanical Compressive Strength

At first the samples with a thickness of 2–4 mm were tested, but were marked as not accurately showing the mechanical properties of the synthesized materials, since such a small measurement did not allow for determining the moment of fracture, due to the rapid transition from fracturing to pressing. Due to this, samples with a greater thickness (8–10 mm) and with masses of ~1 g were prepared. The applied load was up to 5 kN at a speed of 1 mm/min. It was noted that all the tablets disintegrated gradually along the edges while maintaining the main "bulk". Table 3 shows the results of mechanical tests for some of the most interesting and important samples with regard to the obtained parameters.

Table 3. Test results for mechanical strength.

№	Sample Code	Mechanical Strength, MPa
1	B20-3	60.17
2	B20-4	61.08
3	B + mcrSi3	61.93
4	B + mcrSi10	57.29
5	B + NaSi10	17.77
6	B + A10	55.36

Similar and significantly higher values of mechanical strength are observed for all the samples that had the inclusion of a chloride mixture of various compositions. It should be noted that the addition of even 10 wt.% of microcrystalline silica or AEROSIL did not significantly change this parameter, whereas the addition of the same amount of sodium silicate resulted in the strength decreasing by almost 3 times the amount, but it still exceeded the strength threshold for glass-like materials specified in NP-019-15 (10 MPa).

3.4. Hydrolytic Stability

The results of the hydrolytic stability tests are presented in Table 4.

Table 4. Results of tests to determine hydrolytic stability.

№	Sample Code	Component Leaching Rate on the Last (28th) Day of the Study, g/cm^2·Day						
		K	Li	Sr	Cs	Al	Si	Cl$^-$
1	B20-1	2.0×10^{-6}	9.4×10^{-6}	-	7.5×10^{-8}	3.3×10^{-8}	5.4×10^{-6}	2.2×10^{-7}
2	B20-2	2.4×10^{-6}	2.7×10^{-5}	1.6×10^{-7}	6.6×10^{-8}	7.6×10^{-8}	2.8×10^{-6}	5.3×10^{-7}
3	B20-3	2.5×10^{-6}	3.7×10^{-5}	4.1×10^{-6}	6.3×10^{-7}	7.1×10^{-8}	3.6×10^{-6}	1.2×10^{-7}
4	B20-4	2.2×10^{-6}	2.3×10^{-5}	8.0×10^{-7}	8.1×10^{-8}	7.4×10^{-8}	3.0×10^{-6}	3.0×10^{-7}
5	B + mcrSi3	3.2×10^{-6}	4.7×10^{-5}	-	-	7.6×10^{-8}	3.1×10^{-6}	2.9×10^{-7}
6	B + mcrSi5	2.7×10^{-6}	2.6×10^{-5}	-	-	9.6×10^{-9}	2.0×10^{-6}	2.7×10^{-7}
7	B + mcrSi10	2.7×10^{-6}	2.0×10^{-5}	-	-	7.8×10^{-9}	1.6×10^{-6}	2.8×10^{-7}
8	B + NaSi3	2.3×10^{-6}	4.4×10^{-5}	-	-	2.4×10^{-7}	3.6×10^{-6}	1.9×10^{-7}
9	B + NaSi5	3.0×10^{-6}	7.7×10^{-5}	-	-	5.9×10^{-7}	8.5×10^{-6}	1.8×10^{-6}
10	B + NaSi10	5.6×10^{-6}	7.7×10^{-5}	-	-	1.1×10^{-7}	7.4×10^{-6}	1.8×10^{-5}
11	B + A3	3.2×10^{-6}	4.8×10^{-5}	-	-	7.5×10^{-8}	8.3×10^{-6}	2.4×10^{-7}
12	B + A5	2.8×10^{-6}	3.6×10^{-5}	-	-	3.2×10^{-8}	5.2×10^{-6}	2.6×10^{-7}
13	B + A10	5.4×10^{-6}	1.6×10^{-4}	-	-	3.9×10^{-7}	2.0×10^{-5}	2.3×10^{-7}

The leaching rate of elements from ceramic samples meets the regulatory requirements for homogeneous forms of HLW (10^{-6} g/cm^2·day). It may also be noted that the leaching rate on the 28th day of the main elements of the sample is quite low for all the samples studied, which confirms their chemical stability.

3.5. Effect of Irradiation on the Characteristics of the Samples

Figure 10 presents the data on the mechanical strength of ceramic samples before and after irradiation. The data presented in Figure 10 show that the synthesized samples do not significantly lose their mechanical characteristics for the entire range of doses up to

10^9 Gy. All the samples have a sufficient mechanical strength according to the normative document for homogeneous HLW (>10 MPa).

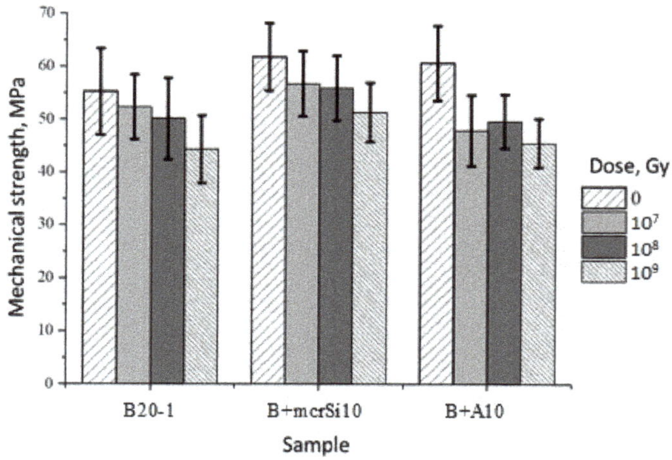

Figure 10. Dependence of the mechanical strength of samples on the dose of ionizing radiation.

Figures 11 and 12 show a comparison of the X-ray diffraction patterns of the samples B20-1 and B + mcrSi10 before and after irradiation with electrons up to a dose of 10^9 Gy. The phase composition remains practically unchanged after high doses of irradiation to ceramic samples. A slight broadening of peaks and a decrease in the intensity of individual peaks are noticeable. The relative content of minor mineral phases changes insignificantly from sample to sample, but this is most likely due to the heterogeneity of the samples.

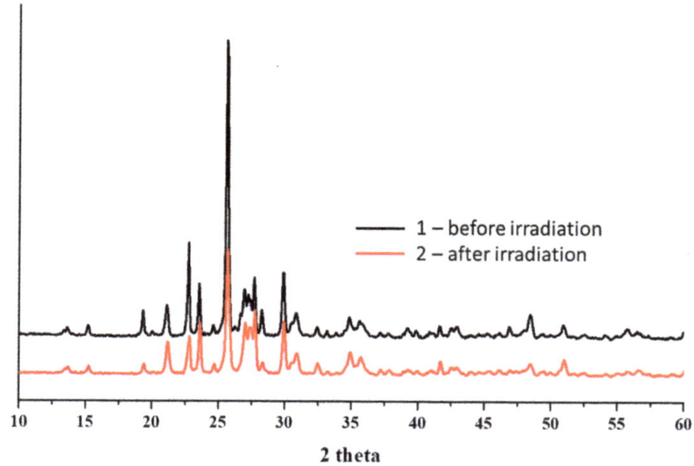

Figure 11. Comparison of the X-ray diffraction patterns of the sample B20-1 before irradiation (1) and after irradiation to the absorbed dose of 10^9 Gy (2).

The scanning electron microscope images indicate that there is practically no difference in surface morphology between the samples before and after irradiation (Figure 13). The samples are characterized by the alternation of large aggregates 5 ... 50 μm in size. The irradiated samples did not show the appearance of any new formations different from those observed in the original samples, or the development of defects in the form of microcracks and channels.

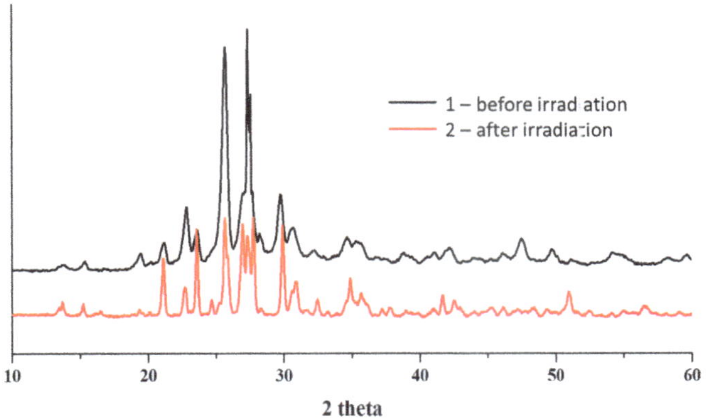

Figure 12. Comparison of the X-ray diffraction patterns of the sample B + mcrSi10 before irradiation (1) and after irradiation to the absorbed dose of 10^9 Gy (2).

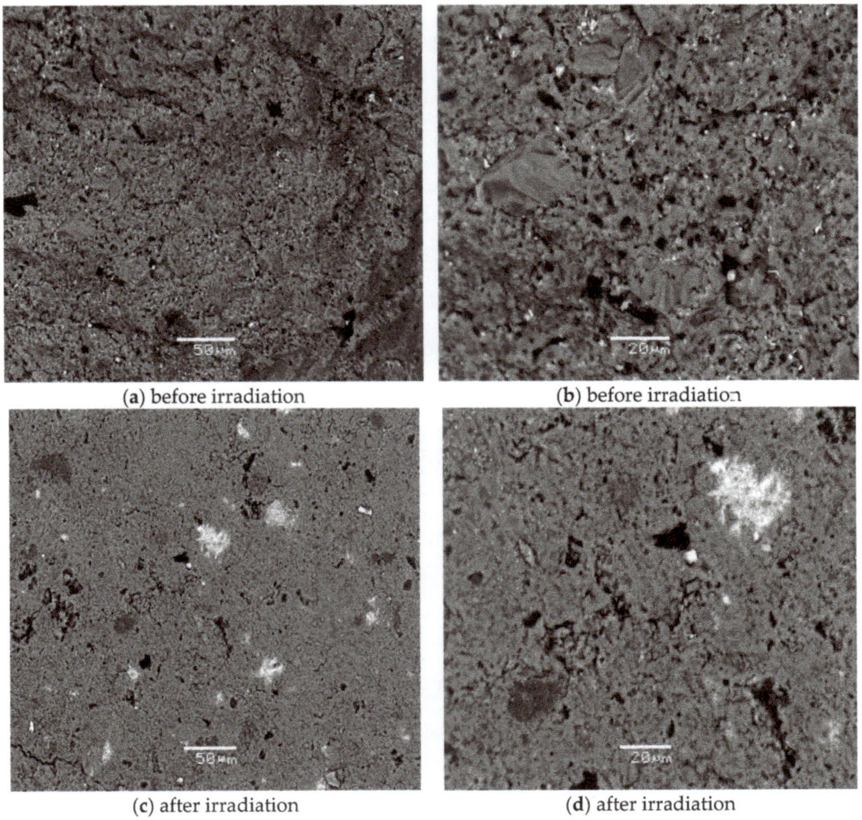

Figure 13. SEM images of the samples with composition B20-1 before and after irradiation by electrons to the absorbed dose of 10^9 Gy.

Figure 13 shows the SEM images of the sample with composition *B20-1*. Figure 13a,b show that the sample generally has low porosity, and that its composition is uniform, since there are no areas that differ greatly in shade. The shade in these SEM images reflects the average atomic number of an element—the lighter areas correspond to elements with a higher atomic number (for example, cesium). After irradiation (Figure 13c,d), individual lighter areas corresponding to the level of cesium minerals become more visible. The uneven distribution may be associated with both irradiation and with the initial uneven distribution of individual small crystallites of cesium minerals.

The morphological composition of the samples is diverse and is represented by both individual crystallites and glassy materials with crystallites scattered over the surface. Glass is similar in composition to feldspar (similar to rhyolite, but the K content is greater than Na).

Thus, we can conclude that electron irradiation doses of up to 10^9 Gy do not significantly affect the phase composition and structural characteristics of the studied ceramic samples.

4. Conclusions

We obtained samples of ceramic materials made of bentonite clay, with the addition of the 20 wt.% LiCl–KCl eutectic as a simulator of the spent electrolyte from pyrochemical reprocessing of SNF. The samples' phase composition, mechanical, hydrolytic, and radiation stabilities were studied. The results from this study indicate that the investigated matrix compositions and the method of their synthesis provide the samples with parameters that meet the regulatory requirements for solidified HLW, such as:

- Specific mechanical characteristics (compressive strength $\geq$10 MPa);
- Radiation resistance (without changes in mechanical strength and structure at an absorbed dose of up to 10^9 Gy);
- Leaching rate (for all components $\leq 10^{-6}$ g/(cm^2·day)).

Thus, it was confirmed that bentonite clays can be used as a starting material for the immobilization of HLW produced from SNF pyrochemical processing. Such HLW may consist of chlorides of alkali and alkali earth metals. In addition, the matrix may be modified by the addition of silicon-containing additives (microcrystalline silica or AEROSIL). The level of additives present may be between 10 and 20 wt.%. Such additives can help to lower the temperature of matrix synthesis.

Author Contributions: Conceptualization, V.G.P. and A.A.Z.; methodology, V.G.P. and A.P.V.; formal analysis, A.V.M. and I.V.M.; investigation, A.V.M.; resources, V.G.P., S.N.K. and A.A.Z.; writing—original draft preparation, A.V.M.; writing—review and editing, V.G.P. and A.V.G.; visualization, A.V.M. and V.G.P.; supervision, V.G.P.; project administration, A.P.V.; funding acquisition, V.G.P. All authors have read and agreed to the published version of the manuscript.

Funding: A.V.M. was funded by the Russian Foundation for Basic Research, grant number 19-33-90072.

Data Availability Statement: Not applicable.

Conflicts of Interest: The authors declare no conflict of interest.

References

1. Jantzen, C.M. Development of glass matrices for high level radioactive wastes. In *Handbook of Advanced Radioactive Waste Conditioning Technologies*; Ojovan, M.I., Ed.; Woodhead Publishing Limited: Cambridge, UK, 2011; pp. 230–292.
2. Mochalov, Y.S. Report on the Project Direction "Proryv": "Place and Advantages of the Project in the Development of the World Energy System", 7–8, June 2016. Available online: http://proryv2020.ru/en/opinion/ (accessed on 13 September 2021).
3. Sobolev, I.A.; Ozhovan, M.I.; Shcherbatova, T.D.; Batyukhnova, O.G. *Glasses for Radioactive Waste*; Energoatomizdat: Moscow, Russia, 1999.
4. Lee, K.R.; Riley, B.J.; Park, H.S.; Choi, J.H.; Han, S.Y.; Hur, J.M.; Peterson, J.A.; Zhu, Z.; Schreiber, D.K.; Kruska, K.; et al. Investigation of physical and chemical properties for upgraded SAP ($SiO_2Al_2O_3P_2O_5$) waste form to immobilize radioactive waste salt. *J. Nucl. Mat.* **2019**, *515*, 382–391. [CrossRef]
5. Lavrinovich, Y.G.; Kormilitsyn, M.V.; Konovalov, V.I.; Tselishchev, I.V.; Tomilin, S.V.; Chistyakov, V.M. Vitrification of chloride wastes in the pyroelectrochemical method of reprocessing irradiated nuclear fuel. *At. Energy* **2003**, *95*, 781–785. [CrossRef]
6. Leturcq, G.; Grandjean, A.; Rigaud, D.; Perouty, P.; Charlot, A.M. Immobilization of fission products arising from pyrometallurgical reprocessing in chloride media. *J. Nucl. Mat.* **2005**, *347*, 1–11. [CrossRef]
7. Lutze, W.; Ewing, R.C. *Radioactive Waste Forms for the Future*; Elsevier Science Pub. Co. Inc.: New York, NY, USA, 1988; p. 712.
8. Weber, W.J. Radiation and Thermal Ageing of Nuclear Waste Glass. *Proc. Mat. Sci.* **2014**, *7*, 237–246. [CrossRef]

9. Stanek, C.R.; Uberuaga, B.P.; Scott, B.L.; Feller, R.K.; Marks, N.A. Accelerated chemical aging of crystalline nuclear waste forms. *Curr. Opin. Solid State Mat. Sci.* **2012**, *16*, 126–133. [CrossRef]
10. Leinekugel-le-Cocq-Errien, A.Y.; Deniard, P.; Jobic, S.; Gautier, E.; Evain, M.; Aubin, V.; Bart, F. Structural characterization of the hollandite host lattice for the confinement of radioactive cesium: Quantification of the amorphous phase taking into account the incommensurate modulated character of the crystallized part. *J. Solid State Chem.* **2007**, *180*, 322–330. [CrossRef]
11. Marra, J.; Billings, A.; Brinkman, K.; Fox, K. Development of Ceramic Waste Forms For An Advanced Nuclear Fuel Cycle (No. SRNL-STI-2010-00622). Savannah River Site 2010. Available online: https://digital.library.unt.edu/ark:/67531/metadc834515/ (accessed on 13 September 2021).
12. Zaripov, A.R.; Orlova, V.A.; Pet'kov, V.I.; Slyunchev, O.M.; Galuzin, D.D.; Rovnyi, S.I. Synthesis and study of phosphate $Cs_2Mn_{0.5}Zr_{1.5}(PO_4)_3$. *Russ. J. Inorg. Chem.* **2009**, *54*, 45–51. [CrossRef]
13. Yudintsev, S.V. Complex Oxides of the Structural Types of Pyrochlore, Garnet and Murataite—Matrices for Immobilization of Actinide Waste from Nuclear Power. Ph.D. Thesis, Russian Academy of Sciences, Moscow. Russia, 2009.
14. Kryukova, A.I.; Kulikov, I.A.; Artemyeva, G.Y. Crystalline phosphates of the family $NaZr(PO_4)_3$—Radiation resistance. *Radiochemistry* **1992**, *34*, 194–200.
15. Oota, T.; Yamai, I. Thermal Expansion Behavior of $NaZr_2(PO4)_3$Type Compounds. *J. Am. Ceram.* **1986**, *69*, 1–6. [CrossRef]
16. Orlova, A.I.; Kemenov, D.V.; Samoilov, S.G.; Kazantsev, G.N.; Pet'kov, V.I. Thermal expansion of NZP-family alkali-metal (Na, K) zirconium phosphates. *Inorg. Mat.* **2000**, *36*, 830–834. [CrossRef]
17. Pet'kov, V.I.; Dorokhova, G.I.; Orlova, A.I. Architecture of phosphates with $\{[L_2(PO_4)_3]^{P-}\}_{3\infty}$ frameworks. *Crystallogr. Rep.* **2000**, *46*, 76–81.
18. Kim, E.; Kim, J.; Kim, J.; Yoo, J.; Lee, J. Method for Preparing Immobilization Product of Waste Chloride Salts Using Zeolite Only. Korea Patent 0089993, 7 September 2007.
19. Koyama, T. Method to Synthesize Dense Crystallized Sodalite Pellet for Immobilizing Halide Salt Radioactive Waste. U.S. Patent 5340506, 23 August 1994.
20. Lewis, M.A.; Pereira, C. Method of Preparing Sodalite from Chloride Salt Occluded Zeolite. U.S. Patent 5613240, 18 March 1997.
21. Babad, H.; Strachan, D.M. Method for Immobilizing Radioactive Iodine. U.S. Patent 4229317, 10 September 1980.
22. Toshiyuki, N.; Kenichi, O.; Shinzo, U. Production Method for Sodalite Type Waste Solid, and its Baking Device. Japan Patent 255083, 2002.
23. Hiroyasu, K.; Toshiyuki, N.; Kenichi, O.; Wataru, S.; Shinzo, U. Synthesizing Method of Sodalite Type Radioactive Waste Solidified Body. Japan Patent 346994, 1996.
24. Belousov, P.E.; Krupskaya, V.V.; Zakusin, S.V.; Zhigarev, V.V. Bentonite clays from 10th khutor deposite: Features of genesis, composition and adsorption properties. *RUDN J. Eng. Res.* **2017**, *18*, 135–143. [CrossRef]

sustainability

MDPI

Article

The Influence of Liquid Low-Radioactive Waste Repositories on the Mineral Composition of Surrounding Soils

Victoria Krupskaya [1,2,*], Anatoliy Boguslavskiy [3], Sergey Zakusin [1,4], Olga Shemelina [3], Mikhail Chernov [4], Olga Dorzhieva [1] and Ivan Morozov [1]

1 Institute of Ore Geology, Petrography, Mineralogy and Geochemistry, Russian Academy of Science (IGEM RAS), 119017 Moscow, Russia; zakusinsergey@gmail.com (S.Z.); dorzhievaov@gmail.com (O.D.); ivan.morozov@yandex.ru (I.M.)

2 Nuclear Safety Institute, Russian Academy of Science (IBRAE RAS), 115191 Moscow, Russia

3 V.S. Sobolev Institute of Geology and Mineralogy Siberian Branch Russian Academy of Sciences (IGM SB RAS), 630090 Novosibirsk, Russia; boguslav@igm.nsc.ru (A.B.); shem@igm.nsc.ru (O.S.)

4 Geological Faculty, M.V. Lomonosov Moscow State University, 119991 Moscow, Russia; miha.chernov@yandex.ru

* Correspondence: krupskaya@ruclay.com

Received: 2 August 2020; Accepted: 3 October 2020; Published: 7 October 2020

Abstract: Clay minerals may transform in various systems under the influence of geological, biological, or technogenic processes. The most active to the geological environment are technogenic and biochemical processes that, in a relatively short time, can cause transformation of the rocks' composition and structure and formation of new minerals, especially clay minerals. Isolation of radioactive waste is a complex technological problem. This work considers the influence of alkaline solutions involved in the radioactive waste (RW) disposal process. In the Russian Federation, due to historical reasons, radioactive waste has accumulated in various types of repositories and temporary storages. All these facilities are included in the federal decommissioning program. Solid radioactive wastes in cement slurries at the landfill site of the Angara Electrolysis Chemical Combine are buried in sandstones and currently suffer the influence of a highly alkaline and highly saline groundwater storage area, which leads to a considerable transformation of the sandstones. This influence results in the formation of peculiar "technogenic" illites that have smectite morphology but illite structure which was confirmed by modeling of X-ray diffraction (XRD) patterns. The described transformations will lead to the increase of porosity and permeability of the sandstones. The research results can be used in assessing the potential contamination of the areas adjacent to the disposal site and in planning the decommissioning measures of this facility.

Keywords: low-radioactive waste; waste facilities; highly concentrated solutions; clay minerals transformations

1. Introduction

The problem of radioactive waste (RW) disposal at nuclear facilities is very relevant. The main goal in this process is maintaining radiation and toxic safety during the whole lifetime of the repository [1]. At present, in Russia, there are various nuclear and radiation hazardous facilities of nuclear legacy requiring decommissioning activities. They are included in the federal program for decommissioning [2,3].

During the initial stages of the development of the nuclear industry, waste solutions with low pH were neutralized by calcium hydroxide. Then, the highly concentrated solutions with residual radionuclides were drained in subsurface ponds—sludge repositories [3]. These solutions interacted with underlying soils, changing their structural, chemical, and mineralogical composition [4].

The fate of repositories is a key problem after decommissioning. If the existing barriers are efficient enough to isolate radioactive wastes, underlying and host soils have low permeability together with high sorption characteristics, RW recovery and retrieval can carry serious environmental risks compared to the in situ preservation. If there is a risk of penetration of the contaminants in aquifers, either storage modernization or RW retrieval is needed [5]. An environmental impact assessment and study of the forms of migration and the state of barriers preventing the spread of contaminants outside the repositories should be conducted to understand the sustainability of existing facilities [6].

The best material for subsurface storage preservation is argillaceous sediments, which both prevent the groundwater flow and precipitate the dissolved radionuclides. Clay minerals form in different systems under geological, biological, or technological processes. Industrial and biochemical processes are believed to be the most sensitive towards the host geological environment. For a short period, they lead to the transformation of composition and structure of rocks and cause mineral, primarily clay, neoformations to appear [7,8].

Many studies of changes in the geological environment in the vicinity of low- and medium-level radioactive waste disposal sites are mainly aimed at the migration of radionuclides [9] and the spread of contamination prior to decommissioning operations [10,11]. Works that touch upon the issues of alteration of the mineral composition of rocks in which RW was isolated or which are located in the vicinity to the disposal sites are quite rare [12]. Thus, in works with the authors of this research, the transformations of the composition of clay minerals in sands were studied.

This research aims to determine possible changes in the composition and structure of clay minerals and to predict changes in the filtration properties of soils containing radioactive waste.

2. Geological Position and Characteristics of Uranium Recovery Facilities of AECC

The Angarsk Electrolysis Chemical Combine (AECC) is one of the oldest uranium recovery facilities. It is located in the Angarsk, Irkutsk region, Russia. The first output from this plant was issued in 1960. Until 2014, AECC had two interrelated production lines: (1) sublimation unit (production of fluorine and anhydrous hydrogen fluoride, uranium conversion, and transformation into uranium hexafluoride—UHF) and (2) separation unit (separation of uranium isotopes in multistage gas centrifuge cascades to increase the concentration of ^{235}U isotope in UHF). The sublimate line was stopped in 2014. At present, AECC receives natural raw uranium material in the form of uranium oxide U_3O_8 and tetrafluoride UF_4 with ^{235}U content about 0.7%. After enrichment, its concentration increases to 3.5%.

The low-level radioactive waste storage from the AECC is located several kilometers away from the Angarsk city border (Figure 1). It is designed for accumulation and precipitation of limewater suspension produced during the neutralization of liquid nitric acid waste from a chemical plant with $Ca(OH)_2$. The facility consists of six near-surface open reservoirs with sizes of 100×70 m and 85×90 m and 17,000 m^3 (reservoirs I-IV) and 18,000 m^3 (reservoirs V-VI) in volumes. Reservoirs I and II are filled up to the designed level and covered with a clay liner for the prevention of rainfall and melting snow infiltration. Reservoir III is at the stage of conservation and reservoirs IV-VI are still under operation [13].

At the time these repositories were designed, the technological scheme of the AECC was intended to cause precipitation of neutralized suspension and subsequent discharge of the clarified part into the Angara River. Infiltration through the reservoir bottom and walls was not considered. In the late 1980s, the production technology had been changed and the incoming amount of liquid waste sharply decreased. Nowadays, the suspension is separated into a solid precipitate and highly saline supernatant solution. Despite the waterproof measures of the storage, fluids migrate into the subsoils. As a result, a significant change in the groundwater composition is observed around the storage facilities.

The first aquifer is represented by the waters of Quaternary sediments, and the most saturated within the terraces above the floodplain. Depending on the topography, the aquifer lies at depths of 0.5 to 7.0 m. The water of the Quaternary aquifer is bicarbonate Ca > Mg > Na. Total mineralization ranges from 0.15 to 0.3 g/L. In the sampled area, the top of the technogenic altered groundwater flow occurs at a depth of 2.5 to 6.3 m. The groundwater from the solid radioactive waste (SRW) SRW construction site moves in the north–northeast direction towards the Angara river, which is about 5.5 km in the northeast direction. Technogenic waters have a carbonate–nitrate composition (Na > Ca > Mg). Directly under the waste ponds, salinity in some wells reaches 9 g/L (C70), but already at a distance of 300–500 m (C78, 79), it drops to 0.3–0.5 g/L [14].

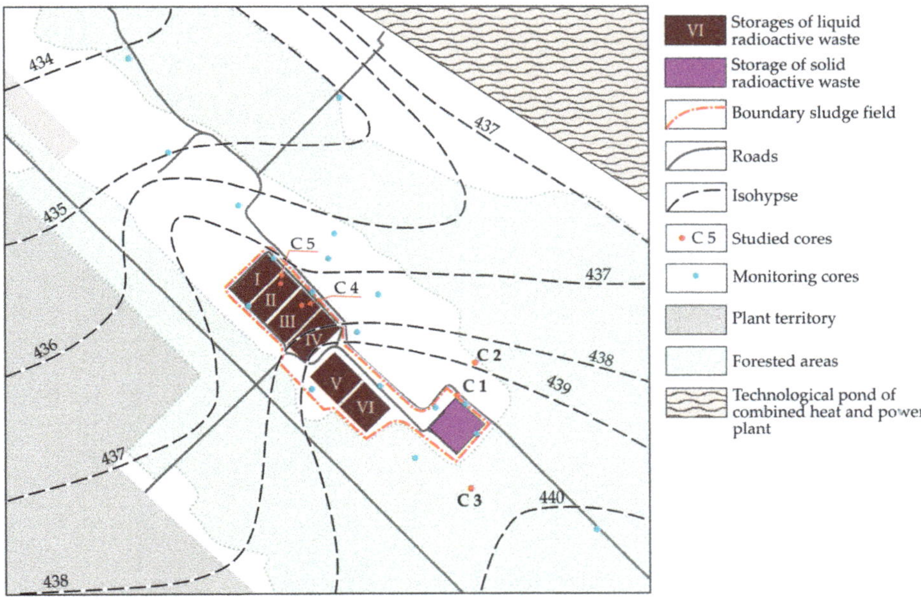

Figure 1. Overview scheme of the storage locations.

According to the research from the Irkutsk geological and ecological survey [15], at the moment of two production lines operation, the salinity of waste was not stable and varied from 13 to 31 g/L. The infiltrate was a multicomponent alkaline solution with pH 9.5–11.0. Its major components were chlorides, sulfates, carbonates, bicarbonates, nitrates, and ammonium ion at concentrations from hundreds of mg/L to a few g/L. The most common cations in solution were Na^+ (2–3 g/L), Ca^{2+} (about 0.6 g/L), and K^+ (about 0.14 g/L). The distinctive feature of the composition of the waste solution was the absence of magnesium. The foundation soil massif is subdivided into two sections. The first is the zone of aeration that lies within the alluvial sands and sandy loams. The second one is the zone of saturation, which coincides with the weathered crust of the Jurassic sandstones.

The average concentration of uranium in the solid part of the sludge is 240 ppm; the total estimated amount of uranium in the repository is about 22 tons. The Baikal region is generally characterized by a higher content of uranium in different soils than other regions of Siberia (Russia). The average content of uranium is 1.66 ppm in the bedrock and 2.4–3.6 ppm in the soil [16]. Obviously, the high background content of uranium in rocks and soils is associated with the Irkutsk coal basin. There are heavy metals and radionuclides in accessory minerals, mainly, and in the clay fraction. We found that the content of these elements in the clay fraction is 2–3.5 times higher than in bulk samples. An even higher content was detected in a sample from a carbonaceous sublayer where the concentration of uranium was 30 times higher and reached 31.5 ppm [17].

Infiltration of a pollutant from storages results in the anomalous content of uranium and other elements in the soil and groundwater; it also leads to changes in the structure and mineralogical composition of the foundation soils. The maximum concentration of uranium in the soil under the two repositories is 5.6 and 11 ppm. It is quite important that a rather low concentration of uranium has been detected in the anomalies directly under the storage bottom. The concentration of uranium in the waters directly under the waste ponds is currently in some cases 5–9 times higher than the background, but at a distance of the first hundreds of meters, it decreases to the background level of 0.2–1 μg/L [17]. However, the long-term influence of waste brines on the adjacent rocks resulted in the modification of their mineralogical composition. Waters saturated with erosion products of cement slurries with wastes are called technogenic in this work.

To estimate the risk of uranium pollution outside the repository, we conducted research on the interaction between the adjacent soils and the infiltrating solutions and the effect of this interaction on the sorption and filtration properties of soils. In the zone of direct contact between the highly saline alkaline solutions and the foundation soil, considerable transformations were revealed.

3. Materials and Methods

Figure 1 shows wells (cyan) that were used for monitoring and planning further studies and exploratory wells (C1–C5, Figure 2), which were performed by auger drilling with sampling every 0.5 m, or when the parameters of the rocks changed. The depth of the wells varied from 8 to 12 m and was determined by the depth of the top of the sandstone layer. After extraction, the core samples were dried to an air-dry state. The objects of investigation are several samples collected from the contact zone between highly saline infiltrates under the waste storage and soils.

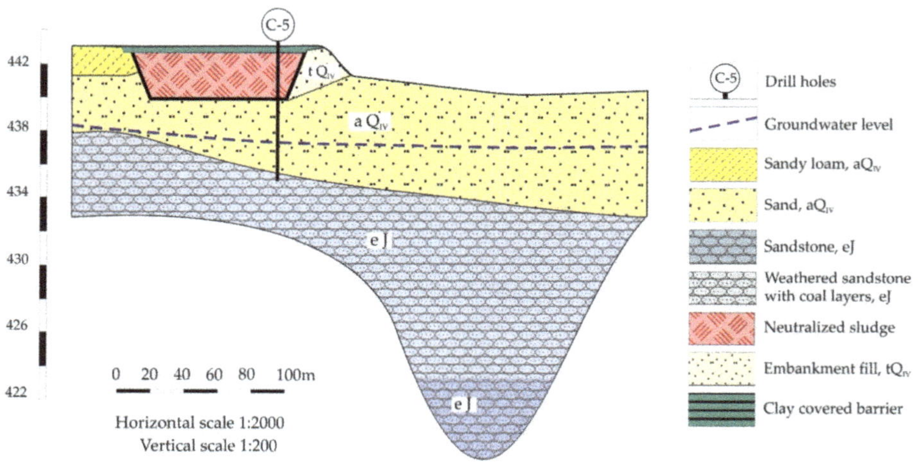

Figure 2. Cross-section through the reservoir (storage).

The mineral composition of the soils and its alteration were studied by the combination of different methods.

Mineral analysis was carried out by X-ray diffraction (XRD) for bulk samples and clay fractions (<1 μm). XRD patterns were obtained with an Ultima-IV X-ray diffractometer (Rigaku). The measurement conditions were Cu-Kα radiation, D/Tex-Ultra 1D-detector, and scan range of 3–65° 2θ. Identification of clay minerals was carried out on the XRD patterns from oriented mounts in the air-dried samples and ethylene glycol solvated. Non-oriented samples of fine fractions were used for determination of di-tri-octahedral species [18]. Quantitative analysis was carried out by the Rietveld method [19] with PROFEX GUI for BGMN [20] on the XRD patterns obtained from the random powder specimens of bulk samples after determination of the composition and structure of clay minerals in the clay fraction. The accuracy of this approach to quantitative analysis is considered about 2–3% for each individual mineral phase.

For a more detailed identification of clay minerals, a fraction < 1 μm was separated by the standard sedimentation procedure according to Stokes' law. In order to avoid modification of clay minerals, no chemical treatment was carried out; to avoid coagulation, if necessary, the samples were repeatedly washed with distilled water and dispersed by ultrasonic treatment.

Chemical analysis of soils was carried out by X-ray fluorescence analysis using synchrotron radiation with a Si (Li) detector with the VEPP-3 elemental analysis station at the Institute of Nuclear Physics, Siberian Branch, Russian Academy of Sciences.

The content of natural radionuclides was determined by scintillation gamma spectrometry (SGS) at the center for collective use of the Institute of Geology and Mineralogy, Siberian Branch, Russian Academy of Sciences (IGM SB RAS). Contents of Ra (by decay products of 222 Rn), 232 Th, 40 K were determined by the intensity of gamma radiation using a low background well-type scintillation detector NaI(Tl). The weight of the sample was 200–300 g, and the detection limit of the method is 0.4 Bq/kg.

Micromorphology study was carried out with the scanning electron microscope (SEM) LEO 1450 VP (Carl Zeiss, Germany). Samples were air-dried and then fixed on the conductive tape. To eliminate the charging effect, they were coated with gold. The samples were studied in the secondary electrons' mode, with an accelerating voltage of 30 kV.

Infrared spectroscopy analysis was carried out using FTIR spectrometer Spectrum One (PerkinElmer) equipped by LiTaO3 detector and KBr beam-splitter. The adsorption spectra recordings were performed in the 4000–400 cm^{-1} wavelength range with 100 scans for each sample and the resolution of 4 cm^{-1}. Samples were prepared as pressed KBr-pellets: 1 mg of sample was dispersed in 400 mg of KBr; this mixture was placed in a 2 cm pellet die and pressed for 20 min. The KBr pellets were then placed into a glass desiccant box with $CaCl_2$ and heated overnight in a furnace at 150 °C. Spectra manipulations were performed using the OPUS 7.1 software (Bruker Optic GmbH, Ettlingen, Germany). Baseline correction was made by Straight-Line method with 1 iteration in interactive mode.

4. Results and Discussion

The adjacent soil layer is subdivided into two main layers: (1) quaternary alluvial sands with thin layers and lenses of sandy loams and loams and (2) weathered Jurassic sandstones with inclusions of carbonaceous matter.

The minimum concentration of the most microcomponents was noted in sands and sandstones. A lower content is typical for sandy soils consisting mainly of quartz and plagioclase (albite). Heavy metals and radionuclides are mainly associated with accessory minerals, and also with the clay fraction in the adsorbed state. The amount of this fraction changes from 13% to 28%. More than 30% is represented by clay minerals: kaolinite, smectite, and illite, which is confirmed by XRD analysis and is shown below. The content of most elements is 2–3.5 times higher in the clay fraction. The background elemental composition for the main types of soil in the studied area is presented in Table 1.

The water in the zone of saturation differs by lower salinity due to the dilution of highly saline infiltrates by natural groundwater [14]. In addition, this zone is also distinguished by the soil composition. The process of technogenic mineral alteration was superimposed on the naturally altered soils of the weathered crust. Transformation of rocks by highly saline waste infiltrates is accompanied by the removal of large amounts of cations, which is revealed by the changing of the qualitative chemical composition of groundwater: natural waters of the region have a cation concentration proportion Ca > Mg > Na, while waters below the storages are characterized by the proportion Na > Ca > Mg.

The content of calcium and magnesium increases due to the ion exchange and the dissolution of solid phases. Fine-crystalline phases may be accumulating in the pore space of soils; however, their presence has not been confirmed by XRD analysis data. The chemical composition shows no accumulation of uranium in the aeration zone. This is explained by the chemical composition of highly saline solutions with a high content of nitrate ion, which determines the redox conditions [14,17,21] because in its presence uranium exists in the highly mobile form U(VI).

Table 1. Chemical composition of different types of soils.

	Jurassic Sands C-3/11.8 m	Alluvial Quaternary Sands C-3/7.5 m	Low-altered Sands C-4/6.0 m	Intermediate-Altered Sands C-5/6.0 m	Highly Altered Sands C-5/7.5 m
K, %	2.5	2.02	1.75	2.25	2.22
Ca, %	3.63	0.79	0.9	1.37	0.85
Ti, %	0.55	0.423	0.313	0.473	0.432
Mn, %	0.09	0.051	0.111	0.401	0.122
Fe, %	6.47	4.42	4.25	4.14	7.07
V, ppm	158	95	98	118	122
Cr, ppm	167	110	85	75	98
Ni, ppm	105	62	123	66	84
Cu, ppm	60	52.8	17.7	22	23.4
Zn, ppm	73	78	48.5	64	73
Rb, ppm	55	59	60	84	74
Sr, ppm	270	214	222	282	214
Y, ppm	19.9	13	13.4	21.5	14.8
Zr, ppm	170	88	156	120	102
Nb, ppm	7.01	5.59	5.03	9.02	9.48
Mo, ppm	0.36	0.7	1.19	1.07	1.41
Pb, ppm	13.8	14.2	17.4	15.7	15.8
Th, ppm	2.5	3.3	3.6	5.2	3.8
U, ppm	>1	>1	6.4	1.4	>1

The mineral composition of the studied soils is quite typical for the studied area and is represented by quartz, K-feldspar, plagioclase, carbonates, and amphiboles—also as clay minerals: smectite, chlorite, illite, and kaolinite (Figure 3). XRD patterns show a decrease of smectite content and an increase in quartz content in alluvial sands compared to Jurassic sands (Table 2, estimation was made after detailed investigation of mineral composition of bulk and clay fraction that will be shown below).

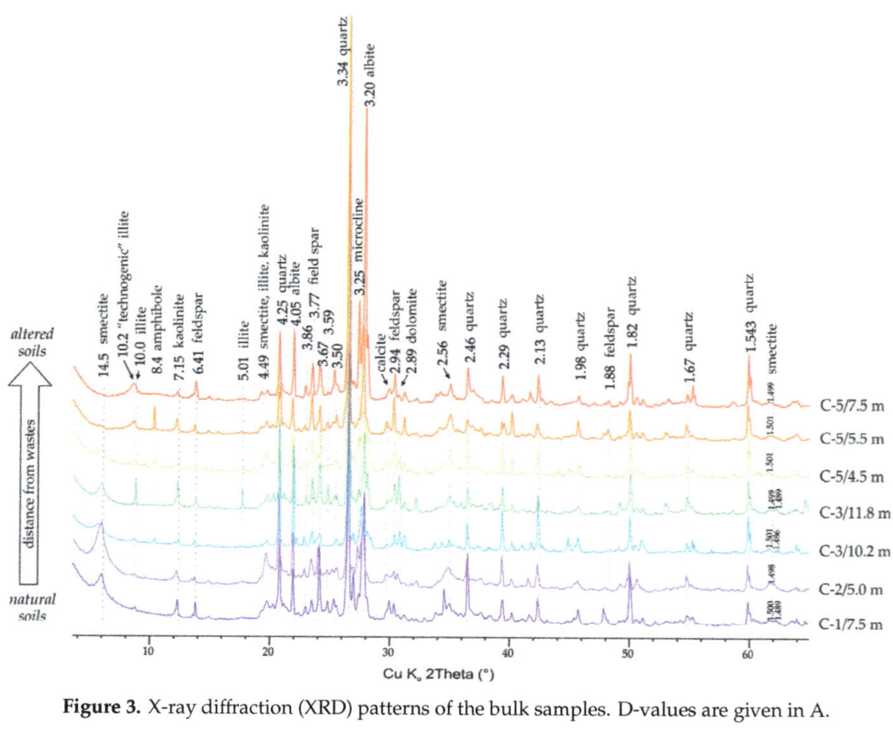

Figure 3. X-ray diffraction (XRD) patterns of the bulk samples. D-values are given in A.

Table 2. Mineral composition of the main types of soil.

Mineral Composition	Jurassic Sands C-3/11.8 m	Alluvial Quaternary Sands C-3/7.5 m	Low-Altered Sands C-4/6.0 m	Intermediate-Altered Sands C-5/6.0 m	Highly Altered Sands C-5/7.5 m
smectite	31.1	10.7	10.7	0.0	0.0
illite and TI*	4.8	5.8	13.5	28.7	33.9
kaolinite	4.6	5.1	1.1	5.7	3.0
chlorite	4.8	3.0	5.6	4.1	3.9
carbonates	0.6	4.8	4.5	0.8	0.9
quartz	15.4	32.2	29.5	26.3	15.3
microcline	14.3	10.7	8.8	8.6	11.4
albite	23.0	25.2	23.5	23.1	30.3
amphibole	1.4	2.5	2.7	2.8	1.3

TI*: "technogenic illite", explanation in the text below.

There is a clear change in the diffraction patterns of the transformed soils, in the range of 3–15°2θ. These changes are observed quite clearly in the samples of the lower horizons of well C5 (5.5–7.5 m), which was drilled directly through the maps and sedimentary rocks, which are affected by technogenic waters.

For more confident identification of technogenic changes in soils, the clay fraction of soils was analyzed. XRD patterns from oriented mounts and fragments of patterns from non-oriented mounts in the region of (060) peaks are shown in Figure 4. The presence of di- and tri-smectite varieties was noted in all the studied samples. At the same time, the peculiarities in XRD patterns from oriented specimens suggest that the swelling component is represented not only by di- and tri-smectites but also by mixed-layer minerals of the chlorite–smectite series with a predominance of smectite interlayers, which requires further, more detailed studies.

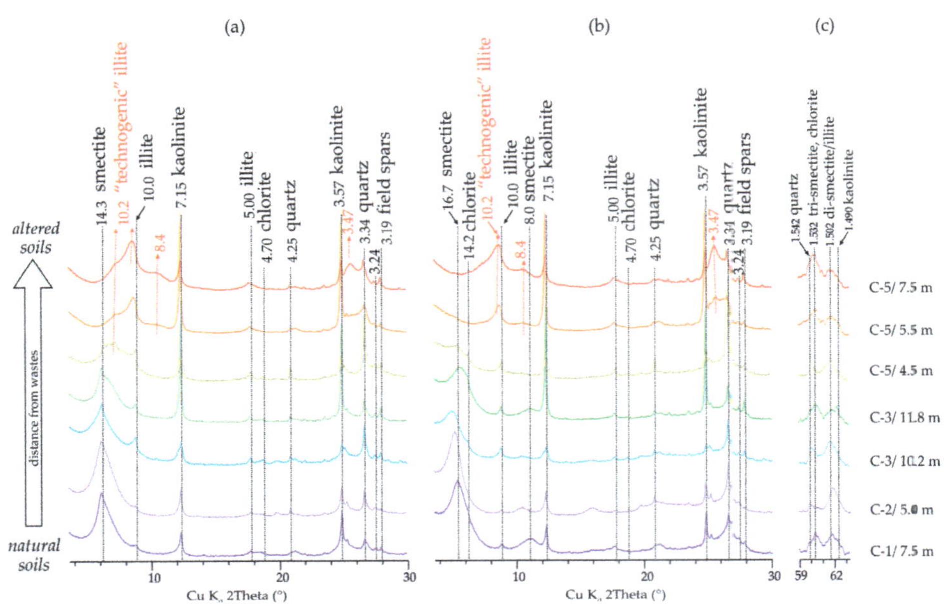

Figure 4. Fragments of the XRD patterns of oriented specimens (**a**,**b**) and non-oriented specimens (**c**). a—air-dried state, b—ethylene glycol solvated, c—(060) area. D-values are given in angstroms. TI peaks are marked by red color.

XRD patterns in the (060) part indicate that the visible changes in the composition of di- and tri-octahedral clay minerals in the transformed sands in comparison with the background soils may be caused by changes in the composition of the original soils and cannot be attributed to technogenic changes. At the same time, changes in the XRD patterns from clay fractions (Figure 4) in the small-angle region are obvious and can be observed not only in the deeper horizons (5.5–7.5 m) of the well 5, as noted above, but also in the horizon 4.5 m of the well C5 and in the lower horizon (11.8) of the well C3. Samples C1/7.5, C3/5.0, and C3/10.2 can be classified as sands that have not undergone visible alterations due to the impact of industrial waters.

Changes in samples C3/11.8 and C5/4.5 are expressed by the appearance of peaks at 12.2 and 10.2 Å. In samples C/5.5 and C/7.5, the peak at 12.2 Å disappears, while the 10.2 Å remains and becomes more intensive; also, a peak 8.4 Å appears and both of them do not shift in XRD patterns from ethylene glycol solvated specimens.

The infra-red (IR) spectroscopy data of the fine fractions of non-altered C-3/11.8 and altered C-5/7.5 soil samples show the polymineral composition: predominantly dioctahedral smectite, chlorite (or mixed-layer minerals with chlorite layers), and kaolinite. Identification of minerals was performed in accordance with recommendations of [22,23]. Wavenumber values and profiles of the absorption bands in the IR spectra (Figure 5) for both samples are similar; the noticeable difference is caused only by a higher content of kaolinite in the altered soil specimen. The IR-spectroscopy data show that there are no "illitic" phases in the altered sample, otherwise, there would be a small band at ≈ 420 cm^{-1} on the spectra [24].

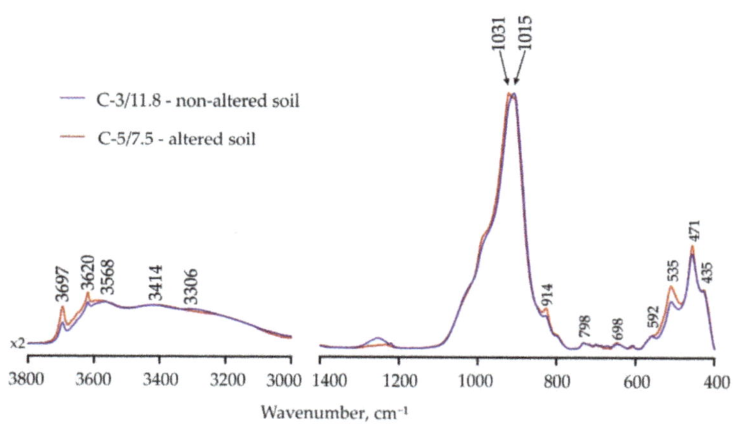

Figure 5. Fragments of the IR-spectra of non-altered and altered soil samples.

Electron microscopy reveals that the host rock was transformed significantly. Leached feldspars and neoformed (altered) clay minerals with an "openwork" structure can be observed (Figure 6c,d). The relict "openwork" structure of the technogenic illite (TI) indicates a rather fast transformation of smectite into illite without change of micromorphology and interaction between particles.

Among other clay minerals, the formation of chlorite on the surface of plagioclase grains was observed. At a depth of about 5 m under the storage bottom where the pH is close to neutral, neoformed opals were found (Figure 6e,f). We assume that the observed processes take place in the soils with a lack of silica caused by weathering, thus, the authigenic opal was formed as a result of a decrease of pH and precipitation of SiO_2 from the supersaturated solution.

Microstructures observed with a scanning electronic microscope are quite typical for weathered sandstones. Quite large isometric particles and aggregates of dense clay particles (probably kaolinite, illite, and smectite) and thin openwork smectite domains are clearly visible (Figure 6 a,b).

Thus, in samples of soil that were subjected to the filtration of technogenic waters, a specific clay phase is noted. It is most likely di-octahedral, which, according to IR spectroscopy and scanning electron microscopy data, is close to smectite but, according to XRD data, loses the ability to swelling and behaves more like illite, while not being illite exactly. Since this phase was found in technogenically altered soils, it was named "technogenic illite"—TI, as previously suggested [25]. The loss of the swelling ability can significantly affect the insulating properties of soils, therefore, the phase was assigned to "illite" and not to "smectite", like the "technogenic smectites" previously described by the authors in the sands-collectors of liquid radioactive waste at the facility of Siberian Chemical Combine [12].

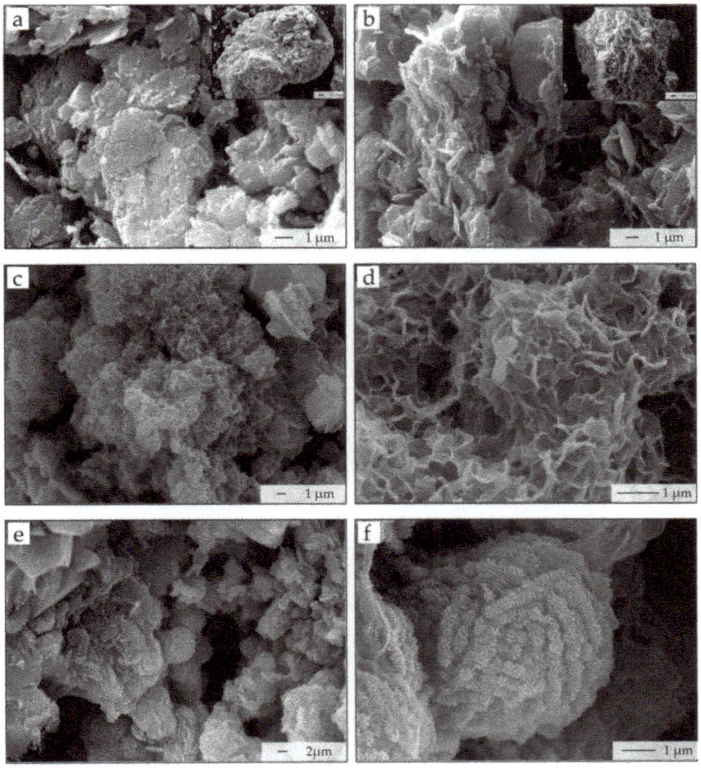

Figure 6. The microstructure of soils from different depths. Background soils: (**a**) loam from the depth of 7.5 m (Core C-3, aQIV); (**b**) weathered sandstone from the depth of 11.8 m (Core C-3 eJ). Underlying soils: (**c**) and (**d**) weathered sandstone from the depth of 7.5 m (Core C-5, aQIV); (**e**) and (**f**) neoformed opal from the depth of 7.5 m (Core C-5, aQIV).

Since the structural parameters of the new phase are unknown at the moment, a modified illite phase with modified parameters of the unit cell was incorporated into the model for calculating the quantitative content. The mineral composition is represented in the Table. 2, from which it is possible to trace the gradual increase of the TI content in weakly transformed soils and its predominance in technologically modified soils.

Based on the research results, it is possible to recommend conducting detailed studies of the composition of soils and, in particular, the composition of clay minerals when monitoring nuclear and radiation hazardous facilities and when implementing measures for their decommissioning.

5. Conclusions

Significant changes of the soils in the area of a long-term impact of the solutions from the low-radioactive liquid waste storages were revealed in the mineral composition of soils under the influence of waste infiltrates compared to the background samples. The transformation of minerals especially clay minerals and the appearance of new mineral phases causes a change of the soil properties: permeability and sorption capacity, mainly.

Thus, directly under the bottom of storage throughout about two meters to groundwater level, there are no favorable conditions for the formation of a geochemical barrier that could prevent the migration of uranium. Although, below the groundwater level, the hydrochemical environment changes, which significantly reduces the effect of nitrate ion on the mobility of uranium, it also could decrease the sorption capacity of the soil at the observed area to a depth of about 3 m.

In the zone of influence of technogenic waters, "technogenic illite" was found. Many of its properties are similar to dioctahedral smectite, however, it does not swell, which primarily distinguishes smectites from illites. The loss of the swelling capacity of "technogenic illites" in the clay fraction of soils can lead to a decrease in the waterproofing properties of rocks and potentially increase the risk of pollution by technogenic waters.

Author Contributions: Conceptualization, V.K. and A.B.; data curation, V.K. and S.Z., methodology, V.K. and S.Z.; formal analysis, S.Z., O.S., M.C., O.D., and I.M.; investigation, V.K., S.Z., M.C., and O.D.; writing—original draft preparation, V.K., A.B., O.S., and S.Z.; writing—review and editing, V.K., A.B., O.S., O.D., and S.Z. All authors have read and agreed to the published version of the manuscript.

Funding: This research was done on state assignment of IGM SB RAS. Investigations of clay minerals transformation were carried out within the framework of the project IGEM RAS AAAA-A18-118021590167-1. Experimental studies were partially performed on the equipment acquired with the funding of the Moscow State University Development Program (X-ray diffractometer Ultima-IV, Rigaku and scanning electron microscope LEO 1450VP, Carl Zeiss).

Acknowledgments: The authors are grateful to Svetlana Garanina (Lomonosov Moscow State University) for assistance in XRD measurements.

Conflicts of Interest: The authors declare no conflict of interest.

References

1. Handbook of Parameter Values for the Prediction of Radionuclide Transfer in Terrestrial and Freshwater Environments: Technical Reports Series 2010. IAEA, Vienna, 472. Available online: https://www.iaea.org/publications/8201/handbook-of-parameter-values-for-the-prediction-of-radionuclide-transfer-in-terrestrial-and-freshwater-environments (accessed on 3 October 2020).
2. Abramov, A.A.; Dorofeev, A.N.; Deryabin, S.A. Development of USS RW in the Framework of Federal Targeted Program of Nuclear and Radiation Safety Assurance. *Radioact. Waste* **2019**, *1*, 6–20. [CrossRef]
3. Danilovich, A.S.; Pavlenko, V.I.; Potapov, V.N.; Semenov, S.G.; Chesnokov, A.V.; Shisha, A. DTechnologies of radwaste management at a decommissioning of the MR and RFT research reactors. *Radioact. Waste* **2018**, *2*, 63–72.
4. Hosseini, A.; Thørring, H.; Brown, J.E.; Saxén, R.; Ilus, E. Transfer of radionuclides in aquatic ecosystems–default concentration ratios for aquatic biota in the Erica Tool. *J. Environ. Radioact.* **2018**, *99*, 1408–1429. [CrossRef] [PubMed]
5. Rahn, F.J.; Adamantiades, A.G.; Kenton, J.E.; Braun, C. *A Guide to Nuclear Power Technology: A Resource for Decision Making*; Wiley-Interscience Publication John Wiley and Sons: New York, NY, USA, 1984.
6. Choppin, G.; Liljenzin, J.O.; Rydberg, J.; Ekberg, C. *Radiochemistry and Nuclear Chemistry*; Elsevier: Amsterdam, The Netherlands; Academic Press: Cambridge, MA, USA, 2013.
7. Gaskova, O.L.; Boguslavskiy, A.E.; Shemelina, O.V. Uranium release from contaminated sludge materials and uptake by subsurface sediments: Experimental study and thermodynamic modeling. *Appl. Geochem.* **2015**, *55*, 152–159. [CrossRef]
8. Mergelov, N.S.; Shorkunov, I.G.; Targulian, V.O.; Dolgikh, A.V.; Abrosimov, K.N.; Zazovskaya, E.P.; Goryachkin, S.V. Soil-like patterns inside the rocks: Structure, genesis, and research techniques. In *Biogenic-Abiogenic Interactions in Natural and Anthropogenic Systems*; Frank-Kamenetskaya, O.V., Panova, E.G., Vlasov, D.Y., Eds.; Springer International Publishing: Cham, Switzerland, 2016; pp. 205–222.
9. Danilov, V.V.; Istomin, A.D.; Noskov, M.D. Multilevel digital model of sedimentary layer of the Siberian Chemical Combine area. *Tomsk State Univ. Bull.* **2008**, *329*, 256–261.

10. Gavrilov, P.M.; Antonenko, M.V.; Druz, D.V.; Chubreev, D.O. Monitoring the Points of Special RAW Placement at FSUE "MCC". *Radioact. Waste* **2018**, *3*, 50–59.
11. Orlova, N.A.; Kropotkin, M.P.; Il'ina, O.A.; Prasolov, A.A.; Krupskaya, V.V. Geoecological risks arising from the disposal of toxic chemical and radioactive waste in Kolomenskoe (Moscow) and the options for territory rehabilitation. *Environ. Geosci.* **2020**, *1*, 57–63.
12. Krupskaya, V.V.; Zakusin, S.V.; Zhuchlistov, A.P.; Dorzhieva, O.V.; Sudin, V.V.; Kruchkova, I.U.; Zubkov, A.A. Newly formed smectite as an indicator for geological environment transformation under the high-reactive solutions, which accompany liquid radioactive wastes. Geoecology. Engineering geology. *Hydrogeol. Geocryol.* **2016**, *5*, 412–419.
13. Environmental Audits of the Angarsk Electrolysis Chemical Combine 2007; Limnological Institute SB RAS: Irkutsk, Russia. 2007. Available online: https://www.aiche.org/resources/proceedings/aiche-annual-meeting/2007 (accessed on 3 October 2020).
14. Boguslavskiy, A.E.; Gaskova, O.L.; Shemelina, O.V. Uranium Migration in the Ground Water of the Region of Sludge Dumps of the Angarsk Electrolysis Chemical Combine. *Chem. Sustain. Dev.* **2012**, *20*, 465–478.
15. Matveeva, I.V.; Shenkman, B.M.; Sakharov, N.V. *Evaluation of Filtration Properties of the Containing Rocks and Complex Hydrogeological Assessment of Slurry Field AECC to Justify the Regulation of Hydrogeological Monitoring*; Irkutsk State Technical University, Center of Geological and Ecological Researches: Irkutsk, Russia, 2009.
16. Grebenshchikova, V.I.; Kitaev, N.A.; Lustenberg, E.E.; Medvedev, V.I.; Lomonosov, I.S.; Karchevsky, A.N. Radioactive Elements Scattering in the Environment in Pribaikalia (Communication 1. Uranium). *Contemp. Probl. Ecol.* **2009**, *1*, 17–28.
17. Shemelina, O.V.; Boguslavskiy, A.E.; Yurkevich, N.V. Determination of soil immobilization characteristics around influence of the fuel and nuclear cycle plant on an example AECC radioactive waste storages. In Proceedings of the VII International Scientific-Practical Conference: Heavy Metals and Radionuclides in the Environment, Semey, Kazakhstan, 4–8 October 2012; pp. 373–380.
18. Moore, D.M.; Reynolds, R.C., Jr. *X-ray Diffraction and the Identification and Analysis of Clay Minerals*; Oxford University Press: Oxford, UK, 1997.
19. Post, J.E.; Bish, D.L. Rietveld refinement of crystal structures using powder X-ray diffraction data. *Rev. Mineral. Geochem.* **1989**, *20*, 277–308.
20. Doebelin, N.; Kleeberg, R. Profex: A graphical user interface for the Rietveld refinement program BGMN. *J. Appl. Crystallogr.* **2015**, *48*, 1573–1580. [CrossRef] [PubMed]
21. Shemelina, O.V.; Boguslavskiy, A.E.; Kolmogorov, Y.P. Measuring the amount of radioactive elements in slime pits and enveloping soils. *Bull. Russ. Acad. Sci. Phys.* **2013**, *77*, 199–202. [CrossRef]
22. Russell, J.D.; Fraser, A.R. Infrared methods. In *Clay Mineralogy: Spectroscopic and Chemical Determinative Methods*; Chapter 2; Wilson, M.J., Ed.; Chapman & Hall: London, UK, 1994; pp. 11–67.
23. Madejova, J.; Gates, W.P.; Petit, S. IR Spectra of Clay Minerals. In *Infrared and Raman Spectroscopies of Clay Minerals*; Chapter 5; Gates, W.P., Kloprogge, J.T., Madejova, J., Bergaya, F., Eds.; Elsevier Ltd.: Amsterdam, The Netherlands, 2017; Volume 8.
24. Madejova, J.; Komadel, P. Information available from infrared spectra of the fine fractions of bentonites. In *The Application of Vibrational Spectroscopy to Clay Minerals and Layered Double Hydroxides*; CMS Workshop Lectures; Kloprogge, J.T., Ed.; The Clay Mineral Society: Aurora, CO, USA, 2005; Volume 3.
25. Krupskaya, V.; Zakusin, S.; Dorzhieva, O.; Boguslavskiy, A.; Shemelina, O.; Chernov, M.; Zubkov, A., II. *International symposium "Clays and ceramics". Book of abstracts. Transformation of Clay Minerals Due to Technogenic Processes Associated with the Disposal of Radioactive Waste*; University of Latvia: Riga, Latvia, 2018; pp. 67–68.

sustainability

MDPI

Article

Calculation of Potential Radiation Doses Associated with Predisposal Management of Dismantled Steam Generators from Nuclear Power Plants

Ga Hyun Chun, Jin-ho Park and Jae Hak Cheong *

Department of Nuclear Engineering, Kyung Hee University, 1732 Deokyoung-daero, Giheung-gu, Yongin-si, Gyeonggi-do 17104, Korea; gahyunchun@khu.ac.kr (G.H.C.); jinhp@khu.ac.kr (J.-h.P.)

* Correspondence: jhcheong@khu.ac.kr; Tel.: +82-031-201-3689

Received: 14 May 2020; Accepted: 22 June 2020; Published: 24 June 2020

Abstract: Although the generation of large components from nuclear power plants is expected to gradually increase in the future, comprehensive studies on the radiological risks of the predisposal management of large components have been rarely reported in open literature. With a view to generalizing the assessment framework for the radiological risks of the processing and transport of a representative large component—a steam generator—12 scenarios were modeled in this study based on past experiences and practices. In addition, the general pathway dose factors normalized to the unit activity concentration of radionuclides for processing and transportation were derived. Using the general pathway dose factors, as derived using the approach established in this study, a specific assessment was conducted for steam generators from a pressurized water reactor (PWR) or a pressurized heavy water reactor (PHWR) in Korea. In order to demonstrate the applicability of the developed approach, radiation doses reported from actual experiences and studies are compared to the calculated values in this study. The applicability of special arrangement transportation of steam generators assumed in this study is evaluated in accordance with international guidance. The generalized approach to assessing the radiation doses can be used to support optimizing the predisposal management of large components in terms of radiological risk.

Keywords: large components; steam generator; dose calculation; predisposal management; processing; transport; special arrangement

1. Introduction

Various types of large components which are mainly made of metal, including reactor pressure vessels (RPVs), pressurizers (PZRs), and steam generators (SGs), are generated from the decommissioning of nuclear power plants (NPPs); these can even be produced during the operation period of pressurized water reactors (PWRs) and pressurized heavy water reactors (PHWRs), as well from replacement projects of degraded large metal components such as SGs and others [1]. In 2009, the International Atomic Energy Agency (IAEA) reported that SGs in 175 units of NPPs have been replaced in total worldwide [2]. Furthermore, the numbers of NPPs approaching their designed lifetimes or which have permanently ceased operation have increased continually, which supports the forecasting of a future gradual increase in replaced and/or dismantled large components [3].

Special considerations are needed when managing large components from NPPs because of their bulky size, heavy weight, and high cost of handling, in addition to other factors. In this respect, the Nuclear Energy Agency (NEA) under the Organization for Economic Co-operation and Development (OECD) published a specific issue report on the management of large components to be generated from the decommissioning of NPPs in 2012 [4]. In the issue report, the OECD/NEA suggests that multiple considerations of technical, regulatory, and economic aspects should be taken at "all stages" including the predisposal management (i.e., transport, treatment, and storage) of dismantled large components onsite or offsite of NPPs, and further pointed out the significance of radiological assessment in deciding the management options of large components. In addition, the IAEA stated that the hazards induced by the cutting and handling of large components should be assessed and managed so that the potential consequences of such hazards can be prevented or mitigated [5]. Furthermore, the IAEA's requirements of interdependences among all steps in the predisposal management of radioactive waste are also applicable to the management of radioactive large components from NPPs [6].

In practice, replaced or dismantled large components have been managed in various ways according to the relevant country or site. In many cases, replaced large components such as SGs and reactor heads have been temporarily stored onsite without segmentation [7,8]. Dismantled RPVs and SGs have been transported in one piece without segmentation from NPPs to the low-level radioactive waste repository, by barge, through an inland waterway in the United States [9,10]. Segmentation operations of the head, pressure vessel, and internal parts of the reactor have been reported [11]. It has also been reported that replaced large components have been transported to a domestic storage facility via road and inland waterways and to a smelter in Sweden through the overseas waterway [12,13].

However, no comprehensive studies have been reported on the radiological risk assessments for all possible predisposal management options of large components in an integrated manner. For instance, a preliminary study has estimated the radiation doses for the decommissioning workers involved in the cutting and smelting of an SG, with neither assessment of the handling of the SG in one piece nor its transportation to offsite [14]. Other studies on radiological risk assessment for the transportation—but not the processing—of two actual SGs from German NPPs to offsite facilities have been conducted, such as (a) the overseas transport of four SGs from Stade NPP to Studsvik processing plant in Sweden and (b) the transport of SGs in one piece from Obrigheim NPP to the Lubmin offsite interim storage facility through road and inland waterways [13]. In addition, one study taking a wider scope has reported on the expected direct exposure from the in situ cutting and handling of an SG at Bohunice NPP Unit 1 and the offsite transportation of segmented and conditioned waste packages to Mochovce waste repository in Slovakia [15]. Although this study is more comprehensive than previous works, the following aspects can be noted as limitations: (a) the handling of the SG in one piece was not considered, (b) there was a simple assumption of the unit activity (1 Bq) of ^{60}Co only as a radioactive source term, (c) inhalation and ingestion pathways were not considered, and (d) alternative waterway transportation was not assessed.

Accordingly, this study proposes an integrated radiation dose calculation framework for various predisposal management scenarios of SGs to support the decision making of the effective management of large components. In order to attain this goal, radiological dose calculation models have been structured, and their applicability has been demonstrated through case studies.

2. Methodology

2.1. Target Large Component and Its Characteristics

Among the various large and heavy components installed at NPPs, SGs are known to be larger and heavier than other components. Hundreds of experiences of replacement and dismantling of SGs and their subsequent processing, storage, and disposal at operating or decommissioned NPPs have been reported [2]. Accordingly, for this study, the SG was chosen as a target large component for modeling the management options and evaluate the radiological impacts. Although the weights and dimensions of SGs vary widely depending on the capacities of nuclear reactors and the specific design of the SGs, SGs are typically cylindrical metal objects weighing hundreds of tons that are a few meters in diameter and over 10 m in height [16].

A representative U-tube SG consists of an external shell (or body) that is usually made of a metal alloy, a bottom side chamber, and a tube bundle through which primary coolant passes. Therefore, the shell side of an SG is usually not significantly radioactive under normal operation conditions because it does not contact the radioactive primary coolant, whereas the water chamber is slightly contaminated and the inner surface of the tube bundles is contaminated with radionuclides present in the primary coolant. In this study, the SG is assumed to be a cylindrical metal solid having the same mass as the SG for the simplification and generalization of the problem. In addition, publicly reported actual radioactive source terms for SGs including activated corrosion products such as ^{60}Co, fission products such as ^{137}Cs and ^{90}Sr, actinides such as ^{244}Cm, tritium (^{3}H), or radiocarbon (^{14}C) characterized through radiological surveys and measurements have been assumed and used in this study (see Section 3.1).

2.2. Scenarios for Predisposal Management of Dismantled Steam Generators

Large metal components such as SGs dismantled from NPPs have been managed in various ways. Based on a literature review of past experiences, 12 generalized potential management scenarios of SGs have been derived according to the various processing methods (i.e., segmentation, smelting, or handling in one piece without processing), places of processing (i.e., onsite, offsite, or overseas), and transportation means (i.e., road, inland, or overseas waterway) as depicted in Figure 1 and Table 1 [7–13,17].

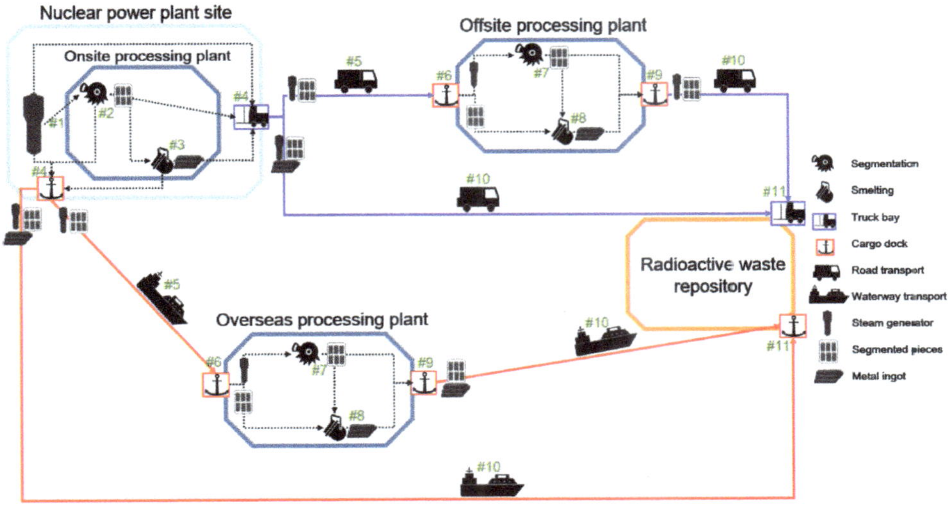

Figure 1. Schematic view of the generalized potential scenarios and the respective locations for the predisposal management of steam generators from nuclear power plants.

The detailed description of the symbol in Table 1 is as follow; S_j (j = 1 to 4) represents the different segmentation workers; scrap cutter, scrap loader, scrap transfer worker, and scrap processor, respectively. M_k (k = 1 to 7) represents the different smelting workers; smelter loader, furnace operator, baghouse processor, slag worker, ingot caster, ingot loader, and ingot transfer worker, respectively. T1 indicates a trailer driver who transports a steam generator in one piece. T2 indicates a truck driver who transports the processed objects of a steam generator. T3 indicates a ship crew who transports objects by waterway. The receptor H represents the package handler who loads or unloads a package (both the one-piece steam generator and the processed objects). Furthermore, the subscript after H and T means the location number in Figure 1.

As summarized in Table 1, Scenarios 1 to 6 refer to the road transportation and Scenarios 7 to 12 refer to waterway transportation. Scenarios 1 and 7 refer to the handling of the SG in one piece without processing, whereas the others consider processing such as segmentation only or smelting after segmentation. Scenarios 2, 3, 8, and 9 represent processing at the NPP site and transportation to the repository, but Scenarios 5, 6, 11, and 12 assume offsite (domestic or overseas) processing and transportation to the repository. It is worth noting that a possible option for onsite segmentation at an NPP and subsequent offsite smelting is reflected in Scenarios 4 and 10. Taking into consideration the Korean situation in which all NPPs and radioactive waste repositories are located at coastal regions, it is assumed that offsite transportation to a domestic processing plant is conducted using public roads or through overseas waterways in the case of an overseas processing plant.

Table 1. Generalized potential scenarios for the management of large metal components from nuclear power plants (NPPs) categorized by the measures for their processing and transportation at each stage and the considered receptors.

Scenario No.	Stage of Predisposal Management										
	Nuclear Power Plant Site				On/Offsite Processing Plant						Repository
	#1	#2	#3	#4	#5	#6	#7	#8	#9	#10	#11
1	-	-	-	H_4	-	-	-	-	-	$T1_{10}$	H_{11}
2	H_1	S_j	-	H_4	-	-	-	-	-	$T2_{10}$	H_{11}
3	H_1	S_j	M_k	H_4	-	-	-	-	-	$T2_{10}$	H_{11}
4	H_1	S_j	-	H_4	$T2_5$	H_6	-	M_k	H_9	$T2_{10}$	H_{11}
5	-	-	-	H_4	$T1_5$	H_6	S_j	-	H_9	$T2_{10}$	H_{11}
6	-	-	-	H_4	$T1_5$	H_6	S_j	M_k	H_9	$T2_{10}$	H_{11}
7	-	-	-	H_4	-	-	-	-	-	$T3_{10}$	H_{11}
8	H_1	S_j	-	H_4	-	-	-	-	-	$T3_{10}$	H_{11}
9	H_1	S_j	M_k	H_4	-	-	-	-	-	$T3_{10}$	H_{11}
10	H_1	S_j	-	H_4	$T3_5$	H_6	-	M_k	H_9	$T3_{10}$	H_{11}
11	-	-	-	H_4	$T3_5$	H_6	S_j	-	H_9	$T3_{10}$	H_{11}
12	-	-	-	H_4	$T3_5$	H_6	S_j	M_k	H_9	$T3_{10}$	H_{11}

2.3. Basic Equations to Calculate Radiation Dose

In order to estimate the radiation dose to each receptor (see Table 1) exposed to radiation for each stage of the predisposal management of SGs, a set of potential exposure pathways are assumed, and the appropriate dose calculation models applied. In the segmentation and smelting processes, direct exposure from radioactive metal is expected, and the inhalation of radioactive materials suspended in the air of the workplace and inadvertent secondary ingestion of radioactive materials can be assumed as potential exposure pathways inducing the radiation exposure to the receptors [18]. The total radiation dose (D_{tot} in mSv/year) to the receptors participating in the processing of SGs can be calculated by summing the estimated radiation doses from all the possible pathways, which are generally categorized into direct radiation and internal exposure due to the inhalation or ingestion of radionuclides as follows:

$$D_{tot} = D_{ext} + D_{inh} + D_{ing} \tag{1}$$

where D_{ext} is the annual external dose to receptors from direct radiation (mSv/y), and D_{inh} and D_{ing} are the annual internal doses from the inhalation and ingestion pathways, respectively (mSv/y).

The annual effective dose to the receptors involved in segmentation and smelting (i.e., S_1 to S_4, and M_1 to M_7) from direct radiation D_{ext} can be calculated by

$$D_{ext} = \sum_{i}^{N} D_{ext,i} = \sum_{i}^{N} C_{S,i} \cdot DCF_{ext,i} \cdot t = \sum_{i}^{N} C_{S,i} \cdot DCF_{ext,i} \cdot \left(\frac{W}{TP}\right) \tag{2}$$

where $C_{S,i}$ is the activity concentration of radionuclide i in metal scrap (Bq/g), $DCF_{ext,i}$ is the external dose conversion factor of radionuclide i (mSv/h per Bq/g) which is determined by the source geometry, dimension, and distance between the source and the receptor, t is the annual exposure time (h/y). W is the weight of SG annually processed (ton/y), TP is the throughput of processing (ton/h), and N is the number of radionuclides.

The annual effective dose to the receptors (i.e., S_1 to S_4, and M_1 to M_7) from the inhalation of radioactive materials in the air D_{inh} can be calculated by [19]

$$D_{inh} = \sum_{i}^{N} D_{inh,i} = C_D \cdot \varepsilon \cdot BR \cdot f_R \cdot \left(\frac{W}{TP}\right) \cdot \sum_{i}^{N} C_{S,i} \cdot DCF_{inh,i} \tag{3}$$

where C_D is the concentration of respirable dust in the air (g/m^3), $DCF_{inh,i}$ is the inhalation dose conversion factor of radionuclide i (mSv/Bq), ε is the efficiency of the respiratory protection equipment, BR is the breathing rate of the receptor (m^3/h), and f_R is the respirable fraction of airborne dust.

In addition, the annual radiation dose to the receptors (i.e., S_1 to S_4, and M_1 to M_7) from the ingestion of radioactive materials D_{ing} can be estimated as [19]

$$D_{ing} = \sum_{i}^{N} D_{ing,i} = (C_D \cdot BR \cdot (1 - f_R) + IR) \cdot \left(\frac{W}{TP}\right) \cdot \sum_{i}^{N} C_{S,i} \cdot DCF_{ing,i} \tag{4}$$

where $DCF_{ing,i}$ is the ingestion dose conversion factor of radionuclide i (mSv/Bq) and IR is the inadvertent ingestion rate (g/h).

It is noted that the radionuclides present in metal scrap are redistributed into resulting matrices such as ingot, slag, and dust if the metal scrap is melted in a smelter such as an electric arc furnace [18]. Accordingly, $C_{S,i}$ in Equations (1) to (3) should be replaced with an adjusted activity concentration of radionuclide i in each matrix (i.e., ingot, slag, or dust) taking into consideration the respective matrix involved in a specific scenario using Equation (5):

$$C_{P,i} = C_{S,i} \cdot \frac{f_{E,P,i}}{f_{M,P}}, \tag{5}$$

where $C_{P,i}$ is the adjusted activity concentration of radionuclide i in a resulting matrix P (i.e., ingot, slag, or dust) in Bq/g, $f_{E,P,i}$ is the element partitioning factor of radionuclide i in matrix P, and $f_{M,P}$ is the mass partitioning factor of metal scrap into matrix P.

On the other hand, receptors involved in transport operations (i.e., T1 to T3 and H) are also exposed mainly due to the direct radiation from the radionuclides present in SGs or transport packaging containing processed objects under normal transport conditions. During normal transport operation, the radionuclides are assumed to be contained in the package, and the inhalation and ingestion pathways can be ignored. The annual radiation dose for receptors T1 to T3 caused by direct radiation from transport packaging D_{Trans} (in mSv/y) can be calculated by

$$D_{Trans} = \sum_{i}^{N} D_{Trans,i} = \sum_{i}^{N} C_{S,i} \cdot N_{trans} \cdot DR_i \cdot \frac{L}{v} \tag{6}$$

where DR_i is the normalized dose rate at the receptor's location per unit activity concentration of radionuclide i (mSv/h per Bq/g), N_{Trans} is the number of transport operations per year (y^{-1}), L is the distance between the origin and the destination of the packaging to be transported (km), and v is the average speed of the transport carrier (i.e., vehicle for road transport and vessel for waterway transport) in km/h. After the smelting process, the dose to receptor T1 to T3 can be calculated by the following equation:

$$D_{Trans} = \sum_{i}^{N} D_{Trans,i} = \sum_{i}^{N} C_{P,i} \cdot N_{Trans} \cdot DR_i \cdot \frac{L}{v} \quad (7)$$

For receptor H, the annual radiation dose caused by direct radiation from the transport packaging during the handling operation $D_{handle,i}$ (in mSv/y) can be calculated by replacing $DCF_{ext,i}$ with DR_i in Equation (2).

It is noted that only the potential radiological impact from scenarios representing normal conditions in the processing and transportation of SGs have been considered in this comparative study, whereas off-normal or accidental scenarios with low probabilities are subject to preparedness and responses to emergencies [13,14].

2.4. Calculation Tools

Here, we report a few models which have been developed to estimate the radiological impacts from the recycling of scrap metals in which numerical models similar to Equations (1) to (4) are used as basic formulas for calculation [18–20]. Among these models, RESRAD-RECYCLE, developed by the Argonne National Laboratory as a computation tool to calculate radiation doses and risks resulting from the recycling of radioactive scrap metal, was adopted in this study in order to facilitate numerical calculation [19]. The RESRAD-RECYCLE code has been validated and widely applied to the prediction of radiation doses from the recycling of scrap metals containing radionuclides through a series of processes including cutting, smelting, and fabrication [21,22].

On the other hand, the potential radiological consequences to receptors T1 to T3 and H from the transportation and handling of SGs and/or the resultant processed objects have been modeled and calculated in this study using the RADTRAN 6 computer code developed by Sandia National Laboratories under the funding of the U.S. Nuclear Regulatory Commission and the U.S. Department of Energy [23]. RADTRAN 6, which simulates the radiation risk based on the measured or calculated dose rate at 1 m from the package, has also been well validated and is frequently used for environmental impact assessments of nuclear installations and transportation risk analysis worldwide [24].

3. Results and Discussions

3.1. General Assessment for the Predisposal Management of Dismantled Steam Generators

In this section, a set of general pathway dose factors (PDFs) was derived for each radionuclide at its unit activity concentration based upon the processing and transportation of a reference SG of the Optimized Power Reactor (OPR 1000) with a weight of 540 tons and reported physical dimensions of 21 m in height and 5.7 m in outer diameter [25,26]. In total, 26 radionuclides are referenced from publicly available inventory data for SGs from a PWR (Kori Unit 1) and a PHWR (Bruce A Unit 1) and considered in the general assessment [14,27]: eight actinides (^{237}Np, ^{238}Pu, ^{239}Pu, ^{240}Pu, ^{241}Am, ^{242}Pu, ^{243}Am, and ^{244}Cm), six fission products (^{90}Sr, ^{99}Tc, ^{106}Ru, ^{129}I, ^{137}Cs, and ^{144}Ce), 10 activated corrosion products (^{54}Mn, ^{55}Fe, ^{57}Co, ^{59}Ni, ^{60}Co, ^{63}Ni, ^{65}Zn, ^{94}Nb, ^{125}Sb, and ^{154}Eu), and two others (^{3}H and ^{14}C).

3.1.1. General Pathway Dose Factors for Processing of Dismantled Steam Generators

The potential radiation dose to each receptor involved in the processing of an SG does not depend on the specific site or scenario but instead relies on the activity concentrations of existing radionuclides and weight of the SG, as implied in Equations (1) to (4). In order to derive general PDFs for the processing of an SG, anticipated doses to all potential receptors (i.e., S_1 to S_4, and M_1 to M_7, as shown in Table 1) were calculated using RESRAD-RECYCLE code by assuming the unit activity concentration of each radionuclide.

The general PDF for each radionuclide and each receptor can be expressed in terms of (mSv/y) per $(Bq/g){\cdot}ton$, as below, from Equations (1) to (4):

$$General\ PDF_i = \frac{D_{ext,i}+D_{inh,i}+D_{ing,i}}{W{\cdot}C_{S,i}} = \frac{DCF_{ext,i}+C_D{\cdot}\varepsilon{\cdot}BR{\cdot}f_R{\cdot}DCF_{inh,i}+(C_D{\cdot}BR{\cdot}(1-f_R)+IR)\ DCF_{ing,i}}{TP} \quad (8)$$

where the value of TP is the reciprocal of time for processing 1 ton of steel scrap which can be derived from the default exposure time for processing 100 ton of steel scrap as proposed in the RESRAD-RECYCLE model [19]. In addition, the values of $f_{M,P}$ for SG as steel scrap in Equation (5) are assumed to be 90% for ingot, 10% for slag, and 1% for dust using the default values in RESRAD-RECYCLE [19]. The elements with low boiling points, such as cesium, typically concentrate in the dust, and the elements that easily oxidize tend to concentrate in the slag [19]. It is noted that default dose coefficients for inhalation and ingestion (see Equations (3) and (4)) in RESRAD-RECYCLE, which are based upon Federal Guidance Report No. 11, have been replaced with those recently introduced in the International Commission on Radiological Protection (ICRP) Publication 119, in order to calculate the effective dose in accordance with the radiation protection recommendations of ICRP Publication 60 [28–30]. The geometry and dimensions of objects handled by receptors participating in SG processing and the distance from the receptors are assumed to be the same as the reference values as proposed in the RESRAD-RECYCLE manual [19]. Thus, the assumed $DCF_{ext,i}$ is using a default value in the RESRAD-RECYCLE [19]. Other parameters (C_D, ε, BR, f_R, IR) are also assumed to be the default value in the RESRAD-RECYCLE [19].

^{3}H, ^{14}C, and ^{129}I are not taken into account in RESRAD-RECYCLE, which may be due to the fact that ^{3}H, ^{14}C, and ^{129}I emit very weak photons, and direct exposure from them is negligible [31]. However, internal exposure from the inhalation and ingestion of these radionuclides may be of concern; radiological impacts from ^{3}H, ^{14}C, and ^{129}I have been frequently considered in the assessment of radioactive waste management [20,32]. In this study, internal exposure from the inhalation and ingestion of these radionuclides has therefore been separately calculated, whereas direct radiation has not been assessed for these three radionuclides.

Values of other parameters used in this study with regard to the processing of a SG are assumed to be the same as the default values in RESRAD-RECYCLE, as mentioned above. Figure 2 shows the general PDF for each actinide and for each receptor involved in the processing of the SG, which is calculated using Equation (8).

It is worth noting that the general PDFs for receptors handling metal ingot (i.e., M_5 to M_7) are calculated to be zero due to there being no elemental partitioning of actinides into ingot through the smelting process (see Table 2 and Equation (5)) [19].

Significant differences are not found in the general PDFs for receptors S_1, S_2, S_4, M_1, M_2, and M_3, while the PDFs for M_4 show much higher values than for other receptors. Furthermore, the variability in PDF values among the eight actinides for a receptor is generally small (e.g., the highest ratio of the maximum to minimum PDF is 2.27), except for receptor S_3, for which the significant differences in the PDFs of actinides are observed (e.g., the ratio of the maximum to minimum PDF is about 50,000).

The small differences in the general PDFs for radionuclides (except receptor S_3) can be ascribed to the comparable dose coefficients for intake among actinides (i.e., 1.1×10^{-7} to 2.5×10^{-7} Sv/Bq for ingestion and 2.1×10^{-5} to 4.7×10^{-5} Sv/Bq for inhalation) and the dominance of internal exposure for the respective receptors. In addition, irregularities of the general PDFs among the actinides observed for receptor S_3 (i.e., scrap transfer worker) result from the fact that direct radiation becomes the only applicable exposure pathway due to the general assumption of negligible portions of releasable radionuclides under normal transfer conditions [33]. On the other hand, the higher values of general

PDFs for receptor M_4 (i.e., slag worker) than other receptors, can be attributed to the longer exposure time, higher external dose conversion factor due to larger dimensions of objects, shorter distance from the receptor, and the higher slag-partitioning factor of actinides, as shown in Table 2 [19].

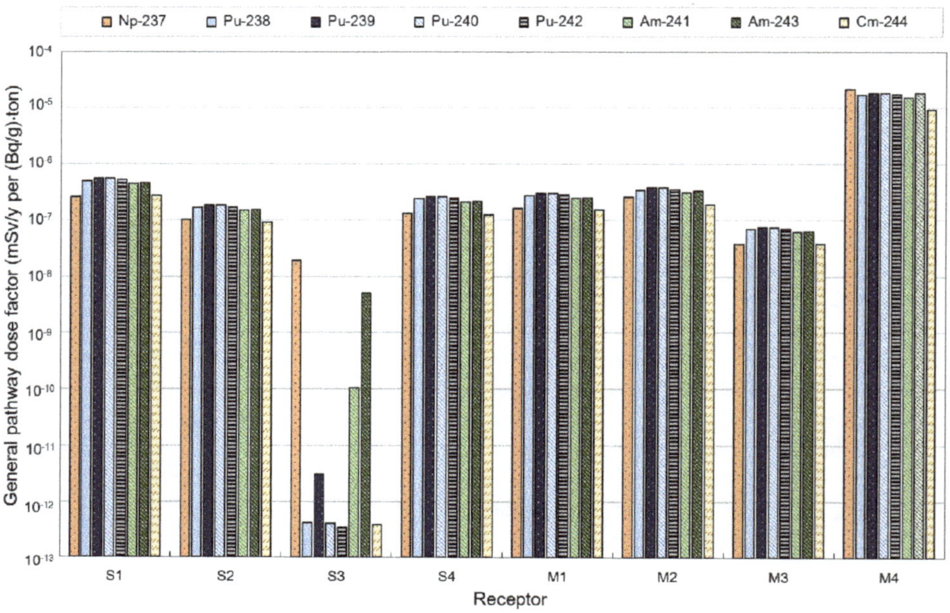

Figure 2. General pathway dose factor for each actinide and for each receptor in the processing of a steam generator as calculated in this study.

Table 2. Element partitioning factor of each radionuclide in dismantled steam generators [19].

Nuclide	Element Partitioning Factor (%)			Nuclide	Element Partitioning Factor (%)		
	Ingot	Slag	Dust		Ingot	Slag	Dust
^{3}H [1]	10	0	0	^{125}Sb	80	20	0
^{14}C [1]	63.5	0	0	^{129}I [1]	0	25	25
^{54}Mn	49	50	1	^{137}Cs	0	3	97
^{55}Fe	97	2	1	^{144}Ce	0	99	1
^{57}Co	99	0	1	^{154}Eu	0	99	1
^{60}Co	99	0	1	^{237}Np	0	99	1
^{59}Ni	99	0	1	^{238}Pu	0	99	1
^{63}Ni	99	0	1	^{239}Pu	0	99	1
^{65}Zn	1	0	99	^{240}Pu	0	99	1
^{90}Sr	0	99	1	^{242}Pu	0	99	1
^{94}Nb	99	0	1	^{241}Am	0	99	1
^{99}Tc	99	0	1	^{243}Am	0	99	1
^{106}Ru	99	0	1	^{244}Cm	0	99	1

[1] Note: element-partitioning factors are as given in Table 2 and balanced average atmospheric release fractions of volatile elements ^{3}H, ^{14}C, and ^{129}I (i.e., 90%, 36.5%, and 50%, respectively) are taken from NUREG-1640 [20].

Figure 3 shows the calculated general PDF for each non-actinide and for each receptor involved in the processing of the SG, in accordance with Equation (8).

For receptors S_1 to S_4, which are involved prior to the smelting process, where the mass of metal scrap and the constituent elements are redistributed into resulting matrices, the external exposure pathway is dominant, and thus the radiological impacts from gamma-emitting radionuclides ^{60}Co, ^{94}Nb, ^{154}Eu, ^{137}Cs, ^{129}I, and ^{54}Mn are higher than for other radionuclides.

In the beginning of the smelting process, in which the receptors M_1 and M_2 are involved, both direct radiation from scrap metal and internal exposure from the intake of radioactive dust at the smelter or furnace are in effect. Therefore, the general PDF values of radionuclides emitting high-energy gamma rays (e.g., ^{60}Co, ^{94}Nb, and ^{154}Eu) and partitioned into dust (e.g., ^{137}Cs, ^{129}I and ^{65}Zn) are remarkably high (see Table 2). For receptor M_3, which handles the baghouse filter, the general PDF values of radionuclides that preferably partition into the dust phase (e.g., ^{137}Cs and ^{65}Zn) are higher compared to the others. Likewise, the general PDFs of ^{94}Nb, ^{154}Eu, ^{129}I, ^{54}Mn, and ^{125}Sb, which tend to be redistributed into slag, turn out to be dominant for slag workers (receptor M_4). It is also noted that the general PDFs for receptor M_4 are much higher than other receptors, as shown in Figure 3, which can be ascribed to the same arguments already addressed to interpret the similar trend observed in Figure 2. In addition, gamma-emitting radionuclides preferably partitioned into ingot (i.e., ^{60}Co, ^{54}Mn, ^{125}Sb, ^{106}Ru, ^{65}Zn, etc.) induce higher PDF values for receptors handling ingot (i.e., M_5 to M_7).

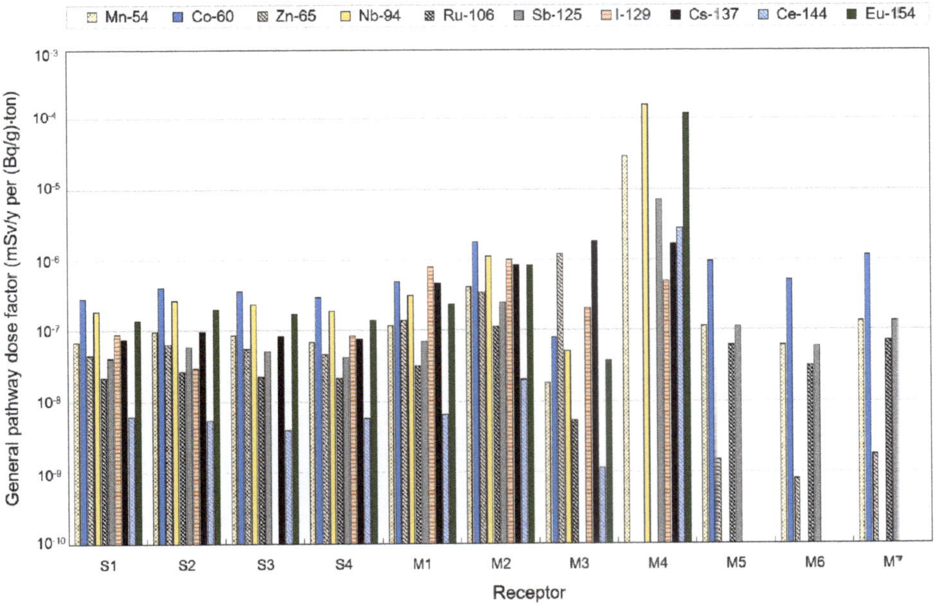

Figure 3. General pathway dose factor calculated in this study for each non-actinide and for each receptor in the processing of a steam generator.

Due to the volatile characteristics of ^{3}H, ^{14}C and ^{129}I, however, the radionuclides may not be trapped by the baghouse filter and may ultimately be dispersed into the atmosphere [20]. As implied in the footnote of Table 2, significant portions of volatile elements are released into the atmosphere from the processing facility and the radiological impacts from them may require additional care regarding public exposure, due to the release of airborne radionuclides. In this regard, the activity concentration of radionuclide i at the boundary of the processing facility, $C_{B,i}$ (Bq/m^3), can be calculated as [34]

$$C_{B,i} = \frac{(1000)}{(31,536,000)} \cdot C_{S,i} \cdot W \cdot f_{D,i} \cdot (X/Q) \tag{9}$$

where $f_{D,i}$ is the dischargeable fraction of the volatile element i, 1000 is the factor used to convert tons into grams, 31,536,000 is the factor used to convert seconds to years, and X/Q is the atmospheric dispersion factor (s/m^3). By applying $C_{S,i}$ of 1 Bq/g, W of 1 ton/year and $f_{D,i}$ of 90%, 36.5%, and 50% for ^{3}H, ^{14}C and ^{129}I, respectively, as well as X/Q of 4.605 × 10^{-6} s/m^3 (at 1000 m downwind distance) as suggested in comparable studies, $C_{B,i}$ for ^{3}H, ^{14}C, and ^{129}I were calculated and compared to the effluent concentration limit for airborne radionuclide ($ECL_{A,i}$) set forth in 10 CFR Part 20 Appendix B, based upon the annual radiation dose limit (i.e., 1 mSv/year) for the members of the public, as shown in Table 3 [20,35].

As shown in Table 3, the ratio of $C_{B,i}$ to $ECL_{A,i}$ lies within a range from an order of approximately 10^{-14} to 10^{-9}, and the activity concentrations equivalent to the ratio of $C_{B,i}$ to $ECL_{A,i}$ to unity are calculated to be 2.03 × 10^8 to 2.81× 10^{13} Bq/g, which conforms to a total activity of each radionuclide of 1.09 × 10^{11} to 1.52 × 10^{16} MBq in the SG (i.e., 540 tons). It is noted that the above estimated total activity for each volatile radionuclide equivalent to the public dose limit is much lower than the actual radioactive source terms of SGs in Section 3.2.1. Thus, the potential exposure due to volatile radionuclides released into the atmosphere from a processing plant are not further taken into account in this study, since their contributions to radiological impacts turn out to be negligible.

Table 3. Estimated concentration of the volatile radionuclides at the boundary of the processing facility, the ratio with regard to the applicable effluent concentration limits for airborne radionuclides, assuming the unit activity concentration of each radionuclide in the steam generator, and activity concentration equivalent to the unit ratio.

Radionuclide	$C_{B,i}$ (Bq/m^3 per (Bq/g)·ton)	$ECL_{A,i}$ (Bq/m^3)	Ratio of $\frac{C_{B,i}}{ECL_{A,i}}$	Activity Concentration Equivalent to $\frac{C_{B,i}}{ECL_{A,i}}=1$ (Bq/g)
^{3}H	1.31 × 10^{-10}	3.70 × 10^3	3.55 × 10^{-14}	2.81 × 10^{13}
^{14}C	5.33 × 10^{-9}	1.11 × 10^2	4.80 × 10^{-11}	2.08 × 10^{10}
^{129}I	7.30 × 10^{-9}	1.48 × 10^0	4.93 × 10^{-9}	2.03 × 10^8

3.1.2. General Pathway Dose Factors for Transportation and Handling of Dismantled Steam Generators

Using Equation (6), the general PDF for each radionuclide and each receptor (i.e., T1 to T3) involved in the transport operations can be defined in terms of (mSv/y) per $(Bq/g){\cdot}km$, as below:

$$General\ PDF_{Trans,i} = \frac{D_{Trans,i}}{C_{S,i}{\cdot}L} = \frac{DR_i{\cdot}N_{Trans}}{v} \tag{10}$$

and receptor H, involved in the handling operations, can be defined in terms of (mSv/y) per $(Bq/g){\cdot}ton$ as follows:

$$General\ PDF_{Handle,i} = \frac{D_{Handle,i}}{C_{S,i}{\cdot}W} = \frac{DR_i}{TP} \tag{11}$$

where $D_{handle,i}$ (in mSv/year) can be calculated by replacing $DCF_{ext,i}$ with DR_i in Equation (2).

In order to calculate the radiation dose from transportation using RADTRAN 6, the dose rate 1 m from the package should be provided as an input [32]. Thus, the dose rate 1 m from a package containing 1 Bq/g of each radionuclide listed in Table 4 has been derived using the MicroShield® computer code with regard to the SG in one piece, and a container containing processed (i.e., segmented or smelted) objects [36]. For simplification, each radionuclide is assumed to be homogeneously distributed in the total volume of the SG weighing 540 tons in one piece, defined in Section 3.1, and in the International Organization for Standardization (ISO) 1496/1 container (length 12 m, width 2.4 m, and height 2.5 m), which is widely used in the transportation of low and intermediate level radioactive waste (LILW), containing 20 tons of processed objects [23]. As such, the dose rate 1 m from each package containing each of the 11 key radionuclides for SG transportation at unit activity concentration was derived as shown in Table 4, ranging in the order of 10^{-7}–10^{-4} (mSv/h) per (Bq/g). It should be noted that other radionuclides showing negligibly low dose rates ranging in the order of 10^{-25}–10^{-15} (mSv/y) per (Bq/g) are not given in Table 4.

Table 4. The dose rate 1 m from packages per unit of radioactivity concentration.

Radionuclide	Dose Rate (mSv/h per Bq/g)		
	One Piece	**Segmented**	**Smelted**
^{54}Mn	1.34×10^{-4}	8.96×10^{-5}	4.88×10^{-5}
^{57}Co	1.18×10^{-5}	2.78×10^{-6}	3.06×10^{-6}
^{60}Co	4.04×10^{-4}	2.75×10^{-4}	3.02×10^{-4}
^{65}Zn	9.38×10^{-5}	6.35×10^{-5}	7.06×10^{-7}
^{94}Nb	2.53×10^{-4}	1.68×10^{-4}	1.84×10^{-4}
^{106}Ru	3.33×10^{-5}	2.10×10^{-5}	2.31×10^{-5}
^{125}Sb	6.66×10^{-5}	4.00×10^{-5}	3.55×10^{-5}
^{137}Cs	9.06×10^{-5}	5.84×10^{-5}	0
^{144}Ce	6.54×10^{-6}	3.70×10^{-6}	0
^{154}Eu	2.07×10^{-4}	1.37×10^{-7}	0
^{237}Np	1.42×10^{-6}	3.27×10^{-7}	0

In Table 4, the ratio of the dose rate 1 m from the SG to that from the ISO 1496/1 container that emplaced segmented objects is about 1.6, on average, except for ^{57}Co and ^{237}Np. The higher dose rate 1 m from the SG than from the container can be attributed to the difference in total radioactivity present in the whole SG and in one container; that is, there is 27 times higher radioactivity in the SG than in the single container for segmented SG. Due to the redistribution of radionuclides after smelting, for some radionuclides that are concentrated to ingot after smelting as the concentration of these radionuclides can be calculated by Equation (5) (i.e., Co^{57}, Co^{60}, Nb^{94} and Ru^{106}) using the values in Table 2, the dose rate at 1 m from the container containing smelted SG (i.e., ingot) is higher than that for segmented SG. However, for other nuclides that are not redistributed to ingot after smelting (i.e., Cs^{137}, Ce^{144}, Eu^{154} and Np^{237}), no radiological impacts of these nuclides for smelted SG transportation were observed. The much lower ratios for two low-energy photon emitters ^{57}Co and ^{237}Np (a few keV of average photon energy) can be ascribed to the fact that the low-energy photon is very susceptible to self-absorption [37].

As shown in Figure 4, the general PDFs for T1–T3 and H have been derived using the RADTRAN 6 code in accordance with Equations (10) and (11), using the assumed values of parameters referenced from the open literature, as given in Table 5.

Table 5. Assumed conditions and input parameter values to derive general pathway dose factors using RADTRAN 6 code.

Transport Means	Processing Methods	Speed (km/h) [38–40]	Vehicle	Weight, W (ton/y)	Throughput, TP (ton/h)	N_{trans} (y^{-1})
Road	One piece	20	Trailer	540	100	1
	Segmented	60	Truck	20	100	27
	Smelted	60	Truck	20	100	24.3
Waterway [41]	One piece	22	Ship [42]	540	100	1
	Segmented	22	Ship	540	100	1
	Smelted	22	Ship	486	100	1

NOTE: The assumed distance from the receptor to the object transported is 4 m for T1, T2 and T3. In addition, it is assumed that the receptor H handles the object 10 m away with a crane. The total weight of the smelted objects transported is 486 ton which is partitioned to ingot mass.

As shown in Figure 4, the general PDF values for receptors T1 and T2 which use road transportation are 766 to 1952 times higher than those for the waterway transport workers, T3, and the general PDFs for the receptor involved in the transportation/handling of the processed SG are about two to six times higher than those for SG in one piece. The higher PDF values for the road transportation than for the waterway transportation result from an inspector checking the packaging for only two minutes per day being considered a crew member for waterway transport in the RADTRAN 6 model, while a driver for road transportation, exposed during the whole transport operation, is assumed to be a crew member [32].

On the other hand, the higher general PDF values for the processed SG transportation/handling for receptors T1, T2 and H than those for SG in one piece can be attributed to N_{trans} which have values of 27 y^{-1} for the segmented SG transportation and 24.3 y^{-1} for smelted SG because the loading limit of the container is 20 ton so that 540 ton of segmented objects are transported in 27 times, and 486 ton that are partitioned to the ingot mass (90%) of smelted objects are transported in 24.3 times. Furthermore, although the dose rate at 1 m from the container containing smelted SG is higher than that containing segmented SG for Co^{57}, Co^{60}, Nb^{94} and Ru^{106} given in Table 4, the general PDF values for segmented SG are higher than those for smelted SG due to the N_{trans} mentioned in the previous paragraph. Thus, the effect of N_{trans} is greater than that of the concentrated radioactivity concentration of the ingot.

However, for waterway receptor, T3, N_{trans} is the same whether the transportation of SG in one piece or in the container including processed objects. Therefore, the general PDFs for SG in one piece is higher than that for the containers with processed objects, which conforms to the relative magnitudes of the dose rate at 1 m from package calculated in this study (see Table 4).

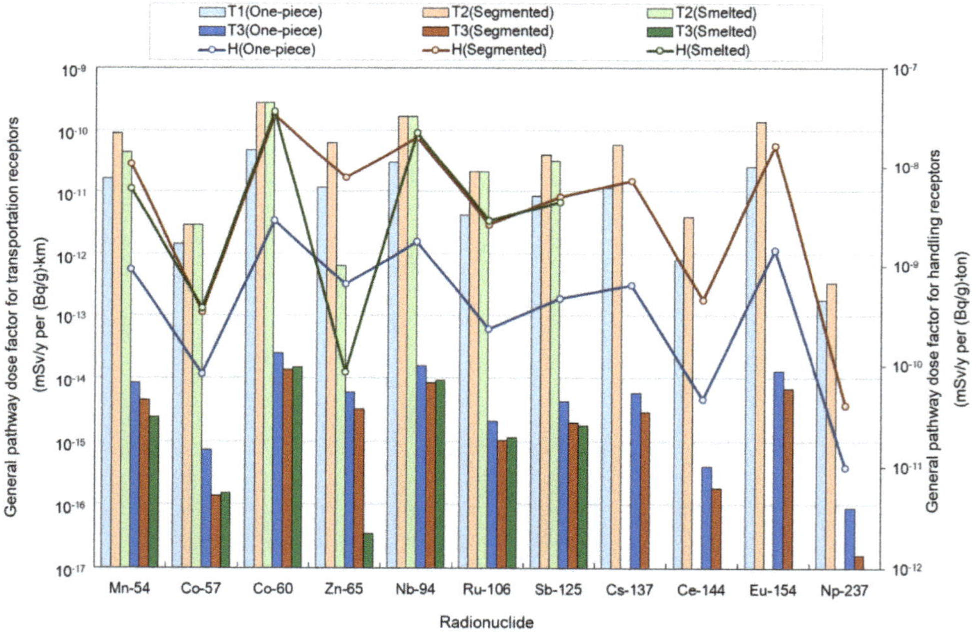

Figure 4. General pathway dose factor for each radionuclide and for each receptor, from T1 to T3 (read left Y axis) and H (read right Y axis), involved in the transportation of the steam generator as calculated in this study.

3.2. Specific Assessment for Predisposal Management of Dismantled Steam Generators

In order to estimate the potential radiological impacts from the management of actually dismantled SGs, a set of assessments have been conducted for two types of SGs from PWRs and PHWRs to be managed in Korea. One of the SGs was replaced from Kori Unit 1 in 1998 and the other from Bruce A Unit 1 in 1997; these are assumed to be representative large components from PWRs and PHWRs, respectively.

3.2.1. Characteristics and Source Terms of Dismantled Steam Generators

Kori Unit 1 is a two-loop 576 MWe PWR in Korea under permanent shutdown since 2017, and its replaced SG has a dimension of about 20 m in height and 4.88 m in outer diameter, and weighs 300 tons [42]. On other hand, Bruce A Unit 1 is a one-loop 840 MWe Canadian PHWR, and its replaced SG has a dimension of about 11.7 m in height and 2.6 m in outer diameter and weighs 100 tons [27,43]. Table 6 shows the actual radionuclide-specific inventories of the two types of SGs obtained from the open literature [38,44].

The composition and inventory of each radionuclide in the PWR SG were reported to be characterized by a smear test for radioactive deposits onto the surface of the SG chamber and the measurement of the dose rates from the SG tubes in 1998; however, non-gamma emitters were not included in the source terms [44]. Ten short-lived radionuclides with half-lives of less than 180 days were excluded from the 16 reported radionuclides, resulting in six radionuclides (e.g., ^{54}Mn, ^{57}Co, ^{60}Co, ^{65}Zn, ^{106}Ru, and ^{144}Ce) being assessed for the PWR SG. On the other hand, the radioactive source terms for the PHWR SG were reported to be determined by multiple measures, not only including direct measurements by in situ gamma spectrometry, but also the application of scaling factors and neutron activation calculations to estimate the hard-to-detect radionuclides in 2010 [27]. From the reported 22 radionuclides, short-lived radionuclides were screened out as well, and 21 radionuclides in total were retained for assessment in this study.

Table 6. Radioactive source terms of the assumed dismantled steam generators for a pressurized water reactor (PWR) and a pressurized heavy water reactor (PHWR).

PWR Steam Generator from Kori Unit 1 [44]			**PHWR Steam Generator from Bruce A Unit 1 [38]**		
Radionuclide	**Half-Life (year)**	**Total Activity (MBq)**	**Radionuclide**	**Half-Life (year)**	**Total Activity (MBq)**
^{51}Cr	7.59×10^{-2}	1.71×10^{5}	^{3}H	1.20×10^{1}	2.15×10^{4}
^{54}Mn	8.57×10^{-1}	3.50×10^{4}	^{14}C	5.70×10^{3}	1.50×10^{3}
^{59}Fe	1.22×10^{-1}	2.59×10^{4}	^{55}Fe	2.70×10^{0}	2.43×10^{4}
^{57}Co	7.42×10^{-1}	3.06×10^{3}	^{60}Co	5.27×10^{0}	8.47×10^{4}
^{58}Co	1.94×10^{-1}	1.13×10^{6}	^{59}Ni	7.50×10^{4}	5.20×10^{2}
^{60}Co	5.27×10^{0}	6.36×10^{5}	^{63}Ni	9.60×10^{1}	6.69×10^{4}
^{65}Zn	6.70×10^{-1}	1.97×10^{4}	^{90}Sr	2.90×10^{1}	4.71×10^{4}
^{85}Sr	1.78×10^{-1}	1.51×10^{5}	^{94}Nb	2.00×10^{4}	1.50×10^{1}
^{95}Zr	1.76×10^{-1}	6.40×10^{4}	^{125}Sb	2.80×10^{0}	4.11×10^{1}
^{95}Nb	9.61×10^{-2}	1.25×10^{5}	^{99}Tc	2.10×10^{5}	9.00×10^{-2}
^{103}Ru	1.08×10^{-1}	1.93×10^{5}	^{129}I	1.60×10^{7}	3.90×10^{-4}
^{106}Ru	1.01×10^{0}	1.85×10^{5}	^{137}Cs	3.00×10^{1}	8.21×10^{2}
^{113}Sn	3.15×10^{-1}	5.67×10^{3}	^{181}Hf	1.16×10^{-1}	6.74×10^{2}
^{136}Cs	3.62×10^{-2}	5.13×10^{5}	^{154}Eu	8.80×10^{0}	2.91×10^{2}
^{141}Ce	8.90×10^{-2}	3.91×10^{4}	^{237}Np	2.10×10^{6}	7.50×10^{-1}
^{144}Ce	7.79×10^{-1}	3.84×10^{4}	^{238}Pu	8.80×10^{1}	4.80×10^{3}
			^{239}Pu	2.40×10^{4}	4.90×10^{3}
			^{240}Pu	6.50×10^{3}	6.99×10^{3}
			^{242}Pu	3.80×10^{5}	7.10×10^{0}
			^{241}Am	4.30×10^{2}	1.61×10^{4}
			^{243}Am	7.40×10^{3}	1.60×10^{1}
			^{244}Cm	7.80×10^{2}	8.24×10^{3}
Total activity		3.3×10^{6}	Total activity		2.89×10^{5}

3.2.2. Specific Scenarios for Predisposal Management of Dismantled Steam Generators

One PWR SG is assumed to be produced from one of the six PWRs in the Hanul NPPs where PWRs are in operation, then processed or unprocessed and ultimately transported using a road or waterway to the final radioactive waste repository, Wolsong LILW Disposal Center (WLDC), in accordance with the 12 scenarios given in Table 1. In addition, it is assumed that one PHWR SG is generated from one of the four PHWRs in the Wolsong NPPs site and then transported to WLDC with/without processing via road transport only; waterway transport is screened out due to the proximity of WLDC to the Wolsong NPPs site.

The distance from each NPP site to the overseas processing plant and WLDC is estimated using publicly available geographic information, as shown in Table 7. The distance from each NPP site to the offsite domestic processing plant is simply assumed to be 100 km for road transportation, while the distance to the overseas processing plant (21,500 km, as shown in Table 7) is estimated by assuming transportation from the Wolsong NPP site to the Studsvik radioactive metal processing plant in Sweden [45].

As addressed in Section 3.1.2 and Table 4, the dose rate 1 m from each packaging (i.e., two types of SGs and containers including processed objects) was calculated using the radioactive source terms of each SG given in Table 6 as follows: 0.805 and 0.237 mSv/h for the one-piece SGs from PWR and PHWR, respectively, and 0.604 and 0.193 mSv/h for the containers with segmented objects produced from the processing of the SG from each NPP, and 0.654, 0.213 mSv/h for the containers with smelted objects of the SG from each NPP. Moreover, the values of N_{Trans} are 15 and 5 y^{-1} for the PWR and PHWR segmented SGs, respectively, and 13.5 and 4.5 y^{-1} for the PWR and PHWR smelted SGs.

Table 7. Lengths of the transport routes from the designated origins to the destinations assumed for specific assessment. LILW: low and intermediate-level radioactive waste.

Origin	Distance to Destination (km)			
	Wolsong LILW Disposal Center		Processing Plant	
	Road [46]	Waterway [47]	Road	Waterway [45]
Kori	80	96	100	21,500
Hanul	170	207	100	21,500

3.2.3. Calculation of Radiation Dose from Processing of Dismantled Steam Generators

The total dose of the receptors processing SGs can be rewritten from Equations (1) and (8) as

$$D_{tot} = W \cdot \sum_{i=1}^{N} C_{s.i} \cdot General\ PDF_i \tag{12}$$

where N is the number of radionuclides. Accordingly, the total dose of each receptor processing the PWR SG and PHWR SG shown in Figure 5 was calculated by the general PDF for each radionuclide i (see Figures 2 and 3), the weight of PWR SG and PHWR SG (300 and 100 ton, respectively), and the concentration of each radionuclide i (see Table 6).

Although it is not specified in Figure 5, ^{60}Co turns out to be the most dominant for all receptors except for M_4 (slag worker), which is attributed to it having the highest activity concentration among the 26 radionuclides, as given in Table 6, and the high general PDF value for ^{60}Co, except for M_4, due to the very high element partitioning of ^{60}Co into ingot and no partitioning into the slag phase, as shown in Table 2.

As shown in Figure 5, all the radiation doses of the receptors S_1 to S_4 and M_1 to M_7 for PWR SG are higher than those for the PHWR SG. This can be explained by the higher inventory of radionuclides in the PWR SG (see Table 6) and the three times heavier weight of the PWR SG compared to the PHWR SG, as implied in Equation (12). In addition, two receptors—M_2 (furnace operator) and M_4 (slag worker)—received the highest radiation doses in the processing of both types of SGs. The high exposure of the receptor M_2 can be attributed to the high dust loadings (i.e., C_D in Equation (3)) in the smelting process compared to the segmentation process [19]. On the other hand, the high radiation dose for receptor M_4 can be ascribed to the fact that more radionuclides tend to redistribute into slag than into ingot and dust phases in the smelting process (see Table 2). Moreover, the relatively high radiation dose of M_4 among others involved in the processing of PHWR SGs results from the selective partitioning (i.e., 99%) of all actinides into the slag phase. Finally, the lower radiation dose of receptor M_3 (baghouse processor) than the other receptors in the processing of PWR and PHWR SGs is explained by the very low contents of ^{65}Zn and ^{137}Cs (i.e., 0.3% and 0.6%, respectively, as given in Table 6), whereas only two out of the 26 radionuclides of concern tend to distribute selectively into the dust phase, which are the radionuclides that most affect the receptor M_3.

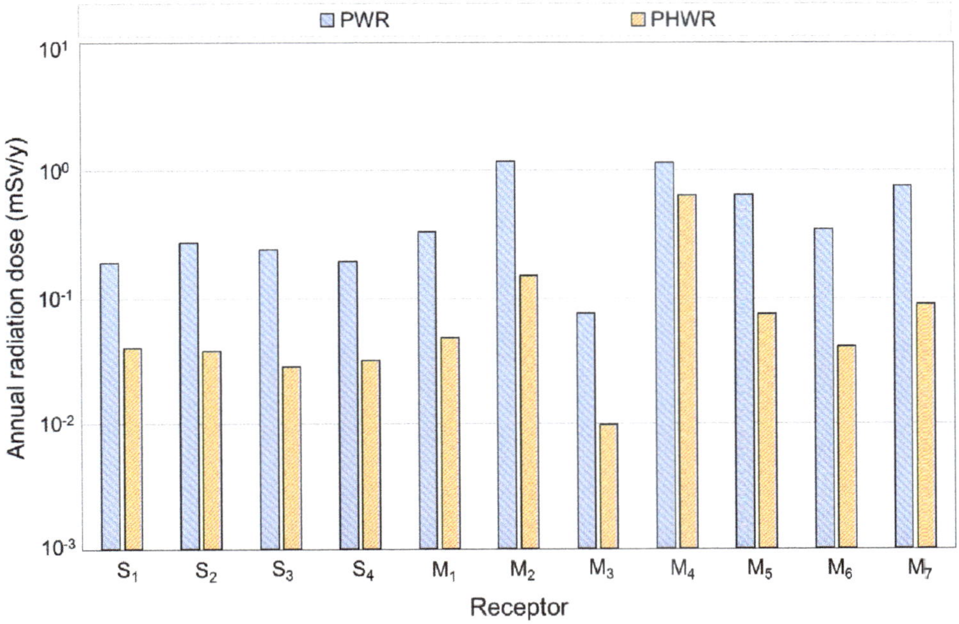

Figure 5. Radiation dose of receptors S_1 to M_7 involved in the processing of the PWR or PHWR steam generator calculated in the specific assessments.

3.2.4. Calculation of Radiation Dose from Transportation and Handling of Dismantled Steam Generators

In contrast to the specific dose calculation for the processing receptors that can be directly calculated using the general PDFs and the given input parameters (see Section 3.2.3), the radiation dose of the receptors involved in the transportation or handling of SGs in a specific case should be calculated in a different way. That is, the total dose of the transportation receptor can be calculated using Equation (6) as below:

$$D_{trans} = \frac{L \cdot N_{Trans}}{v} \cdot \sum_{i=1}^{N} DR_i \cdot C_{s.i} \quad (13)$$

On the other hand, the total dose of receptor-handling SG can be derived from Equation (2) as follows:

$$D_{Handle} = \frac{W}{TP} \cdot \sum_{i=1}^{N} DR_i \cdot C_{s.i} \quad (14)$$

Radiation doses for the receptors (T1 to T3 and H) participating in the transportation of SGs from the PWRs NPP site in Korea (i.e., Hanbit) or from PHWRs NPP site (i.e., Wolsong) are calculated using Equations (13) and (14) and the values of parameters v given in Table 5, $C_{S,i}$ in Table 6, L in Table 7, and N_{Trans} in Section 3.2.2. No plots for the Scenarios 4–9 in Figure 6b result from the screening out of the inland waterway transport of the PHWR SG from Wolsong Site to WLDC (see Section 3.2.2).

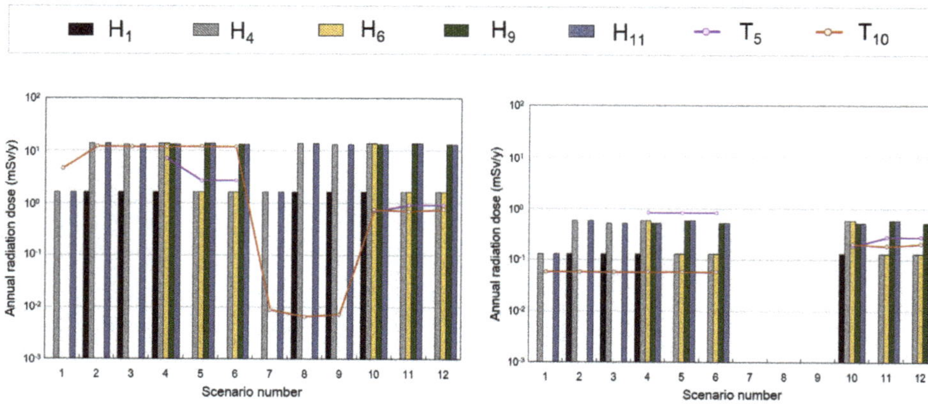

(**a**) Hanbit site (**b**) Wolsong site

Figure 6. Calculated radiation dose of the receptors T1–T3 and H involved in the transportation of the steam generators from two NPP sites.

The radiation doses for the package handlers (H) in all scenarios in Figure 6a,b are calculated to be the same, as the target package is same (e.g., $H_4 = H_6$ and $H_9 = H_{11}$ for Scenarios 4–6 and Scenarios 10–12). The lower doses for H in Figure 6b than Figure 6a can be explained by the differences in the radionuclides inventories (Table 6) and the general PDFs derived in this study (Figure 4) between SGs from PWR and PHWR.

The calculated doses for $T2_{10}$ in Scenarios 2 and 5 (i.e., the transportation of segmented SG) show the highest level among the road transport workers (T1 and T2) in Scenarios 1–6 in Figure 6a, which conforms to the relative magnitudes of the general PDFs calculated in this study (see Figure 4). However, the higher doses for $T3_{10}$ in Scenarios 11 and 12 (i.e., the transportation of the SG in one piece), among waterway transport workers in Figure 6a, can also attributed to the highest general PDF values for the transportation of the SG in one piece (see Figure 4). Likewise, the higher radiation dose for $T3_{10}$ (i.e., the transportation of segmented SG) than that for $T3_5$ (i.e., the transportation of smelted SG) in Scenario 10 can be explained by the higher general PDF for the transportation of segmented SG than for the transportation of smelted SG (see Figure 4).

The lower radiation doses for the receptors involved in the inland or overseas waterway transportation of the PWR SGs in Scenarios 7–12 in Figure 6a compared to the road transport drivers in Scenarios 1–6 were already predicted in the respective general PDFs, as shown in Figure 4. On the other hand, the higher doses for the overseas waterway transportation of the PHWR SGs in Scenarios 10–12 in Figure 6b compared with road transportation in Scenarios 1–3 are contrary to the general PDFs in Figure 4, which can be ascribed to the much longer distance for the overseas transportation (21, 500 km) than for the road transportation (7 km), as shown in Table 7.

3.3. Comparison with Actual Experiences in and Studies on Predisposal Management of Steam Generators

In order to test the applicability of the models proposed in this study, a comparison with other modeling studies or actual experiences is helpful. Comprehensive studies covering every stage in the predisposal management of SGs are not available in the open literature; however, the radiological impact assessment models established in this study are partly compared to a reference study only on the onsite processing of an SG in Korea and another practical reference study only on the transportation of SGs in Germany [13,14].

3.3.1. Comparative Studies for Processing of Steam Generators

Conditions and input parameters for comparison with the reference study on the onsite processing of an SG are shown in Table 8, where the conditions or parameter values are assumed to be the same as or equivalent to the reference as much as practicable [14]. Receptors S_1, S_2, M_2, and M_4 (see Table 2), which were common to both studies, were selected for comparison. The specific considerations assumed in the reference study, such as the separation of the tubes and chambers of the SG, the decontamination of segmented pieces, and the shielding of radioactive objects from workers, could not be fully reflected due to limitations in the known information or inherent differences in the basic models and scenarios between the two studies.

Table 8. Conditions and input parameters assumed for the comparison with the reference study on the onsite processing of steam generators in Korea.

Parameter	This Study	Reference Study [14]
Component and weight	Kori Unit 1 SG (300 ton)	
Parts	Assumed to be in one piece	Separated into tubes and chambers
Radionuclide	^{54}Mn, ^{60}Co, ^{65}Zn, ^{106}Ru, and ^{144}Ce	
Activity concentration	Homogeneous (see PWR SG in Table 6)	Chamber: 15.9 and 0.03 Bq/g for ^{60}Co and ^{106}Ru, respectively Tube: 0.53, 4309, 0.04, 8.47, and 0.28 Bq/g for ^{54}Mn, ^{60}Co, ^{65}Zn, ^{106}Ru, and ^{144}Ce, respectively
Radioactive decay	Decay prior to processing not considered	Decay for 28 years prior to processing
Shielding	Not considered	Considered
Decontamination	Not considered	Considered
Mass partitioning	90% for ingot, 10% for slag, and 1% for dust (see Section 3.1)	98.35% for ingot, 1.64% for slag, and 0.01% for dust
Exposure duration	See footnote [1]	S_1: 40 h (chamber), 100 h (tube) S_2: 43 h (chamber), 15 h (tube) M_2: 73 h (chamber), 60 h (tube) M_4: 3 h (chamber), 2.5 h (tube)

[1] The exposure duration (t) was not directly used in the dose calculation of this study. However, the exposure duration can be estimated from the weight and throughput in accordance with the relation in Equations (2)–(4).

The radiation doses for the four designated receptors were calculated using Equation (12), the general PDFs in Figure 3, and the values of parameters in Table 8, and then compared with the respective results in the reference study as shown in Figure 7.

As shown in Figure 7, the radiation doses for the receptors S_1 and S_2 calculated in this study lie between those for the receptors involved in the processing of chambers and tubes given in the reference study, which is mainly attributed to the fact that more radioactivity is distributed to the tube sides (reported to be about 95% of total activity) rather than to the chambers (about 5%) in the SG, as reported in the actual experience report on the SG replacement [42]. Regarding receptors M_2 and M_4, on the other hand, the estimated radiation doses in this study are higher than those reported for the respective receptors in the reference study [14]. The reference study underestimates the radiation doses for M_2 (furnace operator) and M_4 (slag worker), which can be explained by the multiple factors considered in the reference study, such as decontamination prior to smelting, the shorter exposure duration for M_4, lower mass partitioning into slag, additional shielding, and the 28-year-long radioactive decay (low-element partitioning of the relatively long lived ^{60}Co into slag, as shown in Table 6) prior to processing.

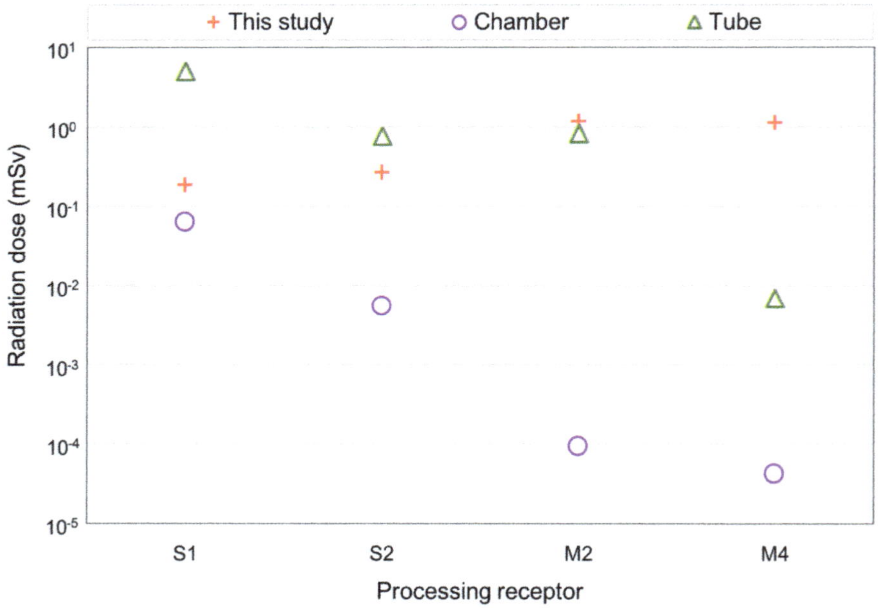

Figure 7. Comparison of the radiation doses for the receptors S_1, S_2, M_2, and M_4 calculated in this study and presented in the comparable reference study [14]. Plots for the "Camber" and "Tube" represent the estimated radiation doses for the workers processing the chambers and tubes of the steam generator, respectively, in the reference study.

3.3.2. Comparative Studies for Transportation of Steam Generators

The conditions and values of the input parameters, for comparison with the reference study on the offsite transportation of SGs in one piece, are shown in Table 9 [13].

Table 9. Conditions and input parameters assumed for comparison with the reference study on the offsite transportation of steam generators in Germany.

Parameter	Case I		Case II	
	This Study	**Reference Study [13]**	**This Study**	**Reference Study [13]**
Component	Two SGs (16.5 m height, 3.6 m outer diameter, and 177 ton weight)		Four SGs (16.5 m height, 3.5 m outer diameter, and 160 ton weight)	
Route	From Obrigheim NPP to Lubmin interim storage		From Stade NPP to Studsvik in Sweden	
Transport measure and distance	Road: 1 km [48] Waterway: 1399 km	Road (onsite) and waterway 1400 km total	Road: 1 km [49] Waterway: 920 km [50]	Road (onsite) and waterway Distances not given
Transport time [1]	Road: 1 h Waterway: 63.6 h or 15 days	15 days	Road: 1 h Waterway: 50 h or 4 days	4 days
Dose rate from SG	0.1 mSv/h 2 m from surface [51]	Not given	0.1 mSv/h 2 m from surface [51]	Not given

[1] Road transport time is assumed to be 1 h considering the practical arrangements needed for onsite transport. On the other hand, transport time for the waterway is either derived from the transport distance and respective speed (see Table 5) or assumed to be the same period reported in the reference study.

Due to the limited information presented in the reference study, a few conditions and parameters should be assumed based on regulatory limits or reasonable inference. The distance of onsite road transportation in each case, which is not given in the reference study, has been assumed to be about 1 km by measuring the length of the routes using a publicly available map [48,49]. Likewise, the distance from the Stade NPP to Studsvik through the overseas waterway (see Case II) was assumed to be 920 km [50]. Furthermore, the dose rates from the packages, which are not reported in the reference study, are assumed to be 0.1 mSv/h at 2 m from the surface of the package, in accordance with international transport regulations [51].

As stated in Section 3.2.4, the total doses of the representative transportation receptors T1 and T3 are calculated using Equation (13) and the assumed conditions in Table 9, and then compared with the results in the reference study (see Figure 8).

As shown in Figure 8, the calculated radiation doses for the designated receptors are quite comparable to those reported in the reference study, which implies that the assessment models developed in this study and the assumed conditions for comparison (see Table 9) are reasonable. The range of radiation doses for receptor T3 in Case I calculated in this study lies between those calculated and actually measured (i.e., below the detection limit) in the reference study. For Case II, the calculated dose for T3 is much closer to the measured value rather than the higher estimated value in the reference study [13]. This study underestimates the radiation dose for receptor T1 in Case I compared to the reference study (i.e., 1.4% to 25% of the reported values), but further analysis could not be conducted due to the limited information (e.g., actual radioactive source terms) in the reference case.

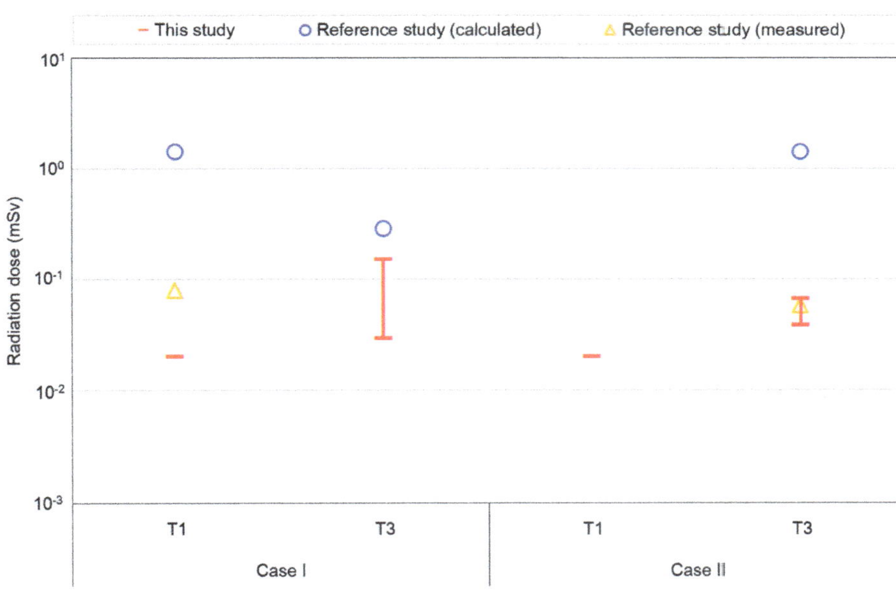

Figure 8. The comparison of the radiation doses for receptors T1 and T3 calculated in this study and presented in the comparable reference study for Cases 1 and 2 [13]. Plots for the "Reference study (calculated)" and "Reference study (measured)" represent the radiation doses for the transport workers as calculated and actually measured, respectively, in the reference study [13]. The plot for T3 in Case I is not shown since the measured radiation dose is reported to be below the detection level. The reference study does not provide the value for T1 in Case II, and therefore, it is not plotted in Figure 8.

3.4. Application of Regulations to Transport of One-Piece Steam Generators

Segmented pieces of SG or byproducts (e.g., ingot, slag, baghouse dust filter) from the processing of SGs can be transported using designated standard packages (e.g., Type IP (industrial package), Type A package) under the full-scope transport regulations for radioactive materials [51]. Due to the bulk size and heavy weight of SGs, however, an SG in one piece without segmentation cannot be placed into any available standard packages and the transport regulations may not be fully applicable [4]. Special arrangement for the transportation of consignments which cannot satisfy all applicable requirements has been already adopted in the IAEA Transport Regulations; however, the IAEA provides much more specific guidance on the transport of "large components" under special arrangements, as shown in Table 10 [51,52].

Table 10. Recommended criteria to approve the special arrangement transport of steam generators in one piece suggested by the International Atomic Energy Agency (IAEA) [52].

	Quantitative Criterion	Qualitative Criterion
(a)	Conveyance activity limit ≤ $10A_2$ for inland waterway or $100A_2$ for other modes [51]	- Non-fissile or fissile excepted - No unnecessary extraneous material in interior void space of the component - Negligible liquid content - Satisfying Type IP (Industrial Package)-2 requirements for the component including any unpackaged penetrations, openings and crevices, and additional shielding - Consigned as exclusive use of the component - Excluded from air transport due to size and mass of the component
(b)	External radiation level 3 m from unshielded component ≤ 10 mSv/h	
(c)	Maximum radiation level on outside shell of and at plane formed by opening/penetration on the component < 2 mSv/h	
(d)	Accidental intake of radionuclide by a person ≤ ~$10^{-6}A_2$ or corresponding inhalation dose of 50 mSv	
(e)	Non-fixed contamination of the component's accessible surface ≤ Limiting value for surface contaminated objects (SCOs) [51] [1]	

[1] Limiting the value for the non-fixed contamination on the inaccessible surface for beta and gamma emitters and low-toxicity alpha emitters is 40×10^3 Bq/cm^2, and 4×10^3 Bq/cm^2 for all other alpha emitters.

In order to evaluate if the SGs from the PWRs and PHWRs assumed in this study (see Section 3.2) can meet all the criteria for the approval of transportation under special arrangements, the quantitative criteria were assessed for each SG, while a set of qualitative criteria which should be confirmed case by case are assumed to be satisfied in this study. In accordance with qualitative criterion (a), firstly, the sum of the fraction for activity contents in each SG (see Table 3) is calculated using the following equation:

$$Sum\ of\ fraction = \sum_{i}^{N} \frac{A_i}{A_{2,i}} \tag{15}$$

where A_i is the activity of each radionuclide i in SG (TBq) and $A_{2,i}$ is the A_2 value defined in Table 2 of the IAEA Specific Safety Requirements No. SSR-6 (TBq) [51].

The sum of fractions is calculated to be 5.6 (≤10) and 37.3 (>10 but ≤100) for the PWR SG and the PHWR SG, respectively, which can be interpreted to mean that the PWR SG assumed in this study meets qualitative criterion (a) for all the transport measures but the PHWR SG is not appropriate for inland waterway transport under special arrangement.

Using the MicroShield® computer code and specifications of the SGs (see Section 3.1.2), the dose rates 3 m from the PWR SG and the PHWR SG were calculated to be lower than the limiting value of 10 mSv/h, at 0.46 and 0.12 mSv/h, respectively. Accordingly, the qualitative criterion (b) is demonstrated to be satisfied for both SGs. At the same time, the maximum radiation levels on the outside shell of the PWR SG and the PHWR SG are 1.38 and 0.50 mSv/h, respectively, which are both lower than the 2 mSv/h specified as a limiting value for the qualitative criterion (c).

With regard to the qualitative criterion (d), finally, the potential intake of radionuclide i by a person $Q_{INT,i}$ (TBq) can be simply calculated by Equation (16):

$$Q_{INT,i} = Q_{IV,i} \cdot F_{REL} \cdot F_{RSUS} \cdot F_{INT} \tag{16}$$

where $Q_{IV,i}$ is the inventory in the package (TBq) of radionuclide i, F_{REL} is the releasable fraction of the activity which is to be released from the package in an accident, F_{RSUS} is the fraction of the released activity which is in respirable aerosol, and F_{INT} is the fraction of the respirable released activity to be inhaled by a person in the vicinity of the accident [52]. By simply adopting the reference values for F_{REL} (0.1), F_{RSUS} (0.01), and F_{INT} (10^{-4}) suggested by the IAEA and assuming the radioactivity in Table 6 for $Q_{IV,i}$, the sum of fraction for $Q_{INT,i}$ is calculated by the following equation:

$$Sum\ of\ fraction = \sum_{i}^{N} \frac{Q_{INT,i}}{A_{2,i}} \tag{17}$$

where the calculated values of the sums of fractions are 0.56×10^{-6} for the PWR SG and 3.73×10^{-6} for the PHWR SG, respectively, which can be interpreted to mean that the PWR SG assumed in this study meets the qualitative criterion (d), but the PHWR SG is not appropriate for transport under special arrangement. Through more specific assessments for F_{REL}, F_{RSUS}, and F_{INT}, however, even higher levels of the total activity content could be justified [52].

Qualitative criterion (e) is subject to a specific SG's non-fixed contamination levels; therefore, the criteria cannot be evaluated for the two SGs assumed in this study. Therefore, it is assumed that both SGs are demonstrated to meet the qualitative criterion (e) and to be defined as surface contaminated objects (SCOs) with the provided information regarding the surface contamination. Under this assumption, the SGs should meet Type IP package requirements for transportation under special arrangement and furthermore, the SGs should be categorized into SCO-I and SCO-II subject to Type IP-1 and IP-2 package requirements, respectively [51].

Based upon the set of assumptions made and the evaluation conducted in this study, the PWR SG meets the criteria for transportation under special arrangement through all the transport modes, while the PHWR SG is not to be applicable for inland waterway transportation under special arrangement. The practical applicability of the special arrangement transportation of SGs is also shown in Table 11, which introduces a few respective cases reported in the open literature [13,38,53–55].

Table 11. Reported cases of the suggestion and application of special arrangements for the transportation of steam generators.

Country	Applied Standards	Type of Object and Package
United States [53,54]	- Dose rate ≤ 10 mSv/h (3 m from SG) - Conveyance activity ≤ $100A_2$. - Surface contamination for beta, gamma and low toxicity alpha ≤ 40×10^3 Bq/cm^2 - Surface contamination for other alpha ≤ 4×10^3 Bq/cm^2	SCO Type IP-2
Germany [13]	- Dose rate ≤ 10 mSv/h (at 3 m from SG) - $10A_2$ for inland waterway and $100A_2$ for other modes - Intake by inhalation ≤ $10^{-6}A_2$	SCO-II Type IP-2
Canada [38]	- Surface contamination for beta, gamma and low toxicity alpha ≤ 40×10^3 Bq/cm^2 - Surface contamination for other alpha ≤ 4×10^3 Bq/cm^2 - Speed limit for road transport vehicle ≤ 20 km/h	SCO-I Type IP-1
Japan [55]	- Surface dose rate < 10 mSv/h (at 3 m from SG) - Total activity ≤ $100A_2$ - Surface contamination for beta and gamma ≤ 40×10^3 Bq/cm^2 - Activity concentration limit for sea transport ≤ 74 Bq/g [1]	SCO-I Type IP-1

[1] It is reported that compliance with this criterion facilitates the process of the sea transportation of an SG which is not considered a "dangerous material" [55].

4. Conclusions

Twelve comprehensive scenarios including every stage in the predisposal management of a representative large component from NPPs (i.e., SG) were modeled based upon processing methods, places of processing, and transportation means, and an integrated framework was established to assess the radiological risk for 15 receptors involved in the processing and transport of SG in each scenario. Assuming a unit activity concentration of each radionuclide reported to be present in SGs, a set of normalized general PDFs were derived for 26 radionuclides.

It was found that the normalized potential radiation dose is greatly affected by the selective partitioning of mass and element in the smelting of metal component SGs. The general PDF value for slag workers is higher than for the other receptors for most radionuclides except ^{129}I, due to multiple factors such as the relatively long exposure time and selective partitioning of all actinides into slag. In addition, the representative high-energy gamma emitter ^{60}Co selectively partitioned into ingot is the most dominant radionuclide for the receptors involved in the processing and handling of scrap or ingots. With regard to transportation operations, it is shown that the general PDFs for road transport are much higher (766 to 1952 times) than for waterway transport, and those for the transport of processed objects from SG are two to six times higher than the transport SG in one piece.

Assuming two types of SGs are generated from PWRs and PHWRs at two nuclear sites, processed and ultimately disposed of in Korea, a set of specific assessments was conducted by using the derived general PDFs for directly processing and separately derived equations for transport, according to the methodology established in this study and additional specific data such as the actually measured source terms. The estimated radiation doses for the processing workers were affected by the weight and radioactive source terms of the SG rather than its origin. The higher radiation dose calculated for the workers processing SGs from PWRs than those from PHWRs can be attributed to the heavier weight and differences in the characterized radioactive source terms. Under the conditions assumed in this study, the maximum annual individual doses for the receptors involved in the predisposal management of SGs lies between 13.7 mSv (for a PWR SG) and 1.14 mSv (for a PHWR SG) for each generated and processed SG. It is worth noting that the highest calculated individual dose was about 27.4% of the effective dose limit for the radiation workers (50 mSv/year), whereas the calculated dose for the workers involved in long-distance transportation was about 68.5% of the 5 year average effective dose limit for a radiation worker (20 mSv/year), which suggests additional shielding for rotating transport workers should be considered.

Compared with reference cases (i.e., actual experience in the transportation of SGs and studies on the processing of SGs from PWRs), the estimated maximum radiation doses calculated in this study are comparable to those in the reference cases, from 1.44% to 165% for processing and from 1.4% to 25% for the transportation of the doses measured or estimated in the reference cases.

Finally, the feasibility of the special arrangement transport for SGs in one piece that do not meet international transportation regulations has been partly demonstrated in terms of both qualitative and quantitative manners. It was shown that the PWR SG may satisfy all the quantitative criteria for special arrangement transportation, while the assumed PHWR SG is unsuitable for inland waterway transport and are even not applicable for special arrangement itself unless a further detailed evaluation of the accidental intake of radionuclides is conducted.

The normalized general PDFs derived in this study can be used for the preliminary estimation of radiological risk in each stage of the predisposal management of SGs. Furthermore, the comprehensive safety assessment framework, together with the developed scenarios, can be used for a more detailed assessment with site-specific data and conditions. It is expected that the radiological risk assessment framework, together with the general PDFs developed in this study, may contribute to finding an optimal management option for large components to be generated from the decommissioning of NPPs, taking into account multiple attributes including both radiological and non-radiological factors.

Author Contributions: Writing—original draft preparation, G.H.C.; writing—review and editing, J.H.C.; methodology, J.-h.P. All authors have read and agreed to the published version of the manuscript.

Funding: This work was supported by the Korea Institute of Energy Technology Evaluation and Planning (KETEP) and the Ministry of the Trade, Industry & Energy (MOTIE) of the Republic of Korea (No.20191510301290). This work was supported by the Nuclear Safety Research Program through the Korea Foundation of Nuclear Safety (KoFONS), granted financial from the Nuclear Safety and Security Commission (NSSC), Republic of Korea (No.2003017).

Conflicts of Interest: The authors declare no conflict of interest.

References

1. Nuclear Energy Agency (NEA). *The Economics of Long-term Operation of Nuclear Power Plants*; NEA No 7054; NEA: Paris, France, 2012.
2. International Atomic Energy Agency (IAEA). *Heavy Component Replacement in Nuclear Power Plants: Experience and Guidelines*; Nuclear Energy Series No.NP-T-3.2; IAEA: Vienna, Austria, 2008.
3. International Atomic Energy Agency (IAEA). Power Reactor Information System (PRIS). Available online: https://pris.iaea.org/PRIS/WorldStatistics/OperationalByAge.aspx (accessed on 17 April 2020).
4. Nuclear Energy Agency (NEA). *The Management of Large Components from Decommissioning to Storage and Disposal*; NEA/RWM/R(2012)8; NEA: Paris, France, 2012.
5. International Atomic Energy Agency (IAEA). *Decommissioning of Facilities*; Safety Standards Series No. GSR Part 6; IAEA: Vienna, Austria, 2014.
6. International Atomic Energy Agency (IAEA). *Predisposal of Management of Radioactive Waste*; Safety Standards Series No. GSR Part 5; IAEA: Vienna, Austria, 2009.
7. Korea Hydro & Nuclear Power Co., Ltd. Replacement of Large Components for Operation Nuclear Power Plants. Available online: https://npp.khnp.co.kr/index.khnp?menuCd=DOM_000000105002002000 (accessed on 15 June 2020).
8. Nuclear Regulatory Agency (NRA). *Radiation Management Report on Nuclear Facilities in 2017*; NRA: Tokyo, Japan, 2018.
9. Maheras, S.J.; Best, R.E.; Ross, S.B.; Buxton, K.A.; England, J.L.; McConnell, P.E.; Massaro, L.M.; Jensen, P.J. *Preliminary Evaluation of Removing Used Nuclear Fuel from Shutdown Sites*; PNNL-22676 Rev. 10; PNNL: Richland, WA, USA, 2017.
10. McGrath, R.; Reid, R. Waste Management for Decommissioning of Nuclear Power Plants: An EPRI Decommissioning Program Report. In Proceedings of the Symposium on Recycling Metals Arising from Operation and Decommissioning of Nuclear Facilities, Nyköping, Sweden, 8–10 April 2014; EPRI: California, CA, USA, 2014.
11. Westinghouse Electric Company. *PWR Reactor Vessel and Reactor Internals Segmentation and Packaging*; DDR-0004; Westinghouse: New York city, NY, USA, 2018.
12. Ken, S. *Safe and Secure Transport and Storage of Radioactive Materials*, 1st ed.; Woodhead Publishing: Cambridge, UK, 2015; p. 277.
13. Nitsche, F.O.; Fasten, C. Transport of large components in Germany-some experiences and regulatory aspects. *RAMTRANS* **2010**, *22*, 54–58. [CrossRef]
14. Son, Y.J.; Park, S.J.; Byon, J.; Ahn, S. The assessment and Reduction Plan of Radiation Exposure during Decommissioning of the Steam Generator in Kori Unit 1. *JNFCWT* **2018**, *16*, 377–387. [CrossRef]
15. Hornáček, M.; Nečas, V. The analysis of management of radioactive waste arisen from steam generator's dismantling from radiological and economical point of view. *Prog. Nucl. Energy* **2017**, *100*, 406–418. [CrossRef]
16. Riznic, J. *Steam Generators for Nuclear Power Plants*; Woodhead: Duxford, UK, 2017; p. 53.
17. Park, J.H.; Cheong, J.H. Potential Management Options of Radioactive Large Metallic Components from NPPs in Operation and Decommissioning. In Proceedings of the Korean Radioactive Waste Society Spring 2018, Busan, Korea, 30 May–1 June 2018.
18. International Atomic Energy Agency (IAEA). *Application of Exemption Principles to the Recycle and Reuse of Materials from Nuclear Facilities*; Safety Series No.111-p-1.1; IAEA: Vienna, Austria, 1992.

19. Cheng, J.J.; Kassas, B.; Yu, C.; Lepoire, D.; Arnish, J.; Dovel, E.S.; Chen, S.Y.; Williams, W.A.; Wallo, A.; Peterson, H. *RESRAD-RECYCLE: A Computer Model For Analyzing the Radiological Doses and Risks Resulting from the Recycling of Radioactive Scrap Metal and the Reuse of Surface-Contaminated Material and Equipment*; ANL/EAD-3; Argonne National Laboratory: Illinois, IL, USA, 2000.
20. United States Nuclear Regulatory Commission (US NRC). *Radiological Assessment for Clearance of Materials from Nuclear Facilities*; NUREG-1640; US NRC: Washington, DC, USA, 2003.
21. Cheng, J.J.; Yu, C.; Williams, W.A.; Murphie, W. Validation of the RESRAD-RECYCLE computer code. In Proceedings of the WM'02 Conference, Tucson, AZ, USA, 24–28 February 2002.
22. Song, J.S.; Kim, D.M.; Lee, S.H. A Study on the Application of Standards for Clearance of Metal Waste Generated During the Decommissioning of NPP by Using the RESRAD-RECYCLE. *JNFCWT* **2016**, *14*, 305–320. [CrossRef]
23. Weiner, R.F.; Hinojosa, D.; Heames, T.J.; Farnum, C.O.; Kalinina, E.A. *RADTRAN 6/RadCat 6 User Guide*; SAND2013-8095; Sandia National Laboratories: Albuquerque, NM, USA, 2013.
24. Obsborn, D.M.; Weiner, R.F.; Mills, G.S.; Hamp, S.C. *Verification and Validation of RADTRAN 5.5*; SAND2005-12741274; Sandia National Laboratories: Albuquerque, NM, USA, 2005.
25. Yoon, J.; Kim, W.; Park, K.T.; Kang, W.M.; Kang, B.S. Development of Temporary Lifting Device for Steam Generator in Containment Vessel at Nuclear Power Plant, Transaction of the Korean Society of Mechanical Engineers. 2020. Available online: https://doi.org/10.3795/KSME-A.2020.44.4.291 (accessed on 23 June 2020).
26. Korea Institute of Nuclear Safety (KINS). *Technical Review Report on Hanul Nuclear Power Plant Units 3 and 4 Operation Change Permit (Steam Generator Replacement)*; 12-060, U-AM-57(12); KINS: Daejeon, Korea, 2013.
27. Ontario Power Generation (OPG), Inc. *Reference Low and Intermediate Level Waste Inventory for the Deep Geological Repository*; 00216-REP-03902-00003-R003; OPG, Inc.: Ontario, ON, Canada, 2010.
28. U.S. Environmental Protection Agency (EPA). *Limiting Values of Radionuclide Intake and Air Concentration and Dose Conversion Factors for Inhalation, Submersion, and Ingestion*; Federal Guidance Report No. 11; U.S. EPA: Washington, DC, USA, 1988.
29. International Commission on Radiological Protection (ICRP). *ICRP Publication 119: Compendium of Dose Coefficients Based on ICRP Publication 60*; Annals of the ICRP; ICRP: Stockholm, Sweden, 2013; Volume 42, pp. 1–130.
30. International Commission on Radiological Protection (ICRP). *ICRP Publication 60: 1990 Recommendations of the International Commission on Radiological Protection*; Ann. ICRP 21 (1-3); ICRP: Stockholm, Sweden, 1991.
31. U.S. Environmental Protection Agency (EPA). *Potential Recycling of Scrap Metal from Nuclear Facilities, Part I: Radiological Assessment of Exposed Individuals Volume 1*; U.S. EPA: Washington, DC, USA, 2001.
32. Alexander, W.R.; McKinley, L.E. *Deep Geological Disposal of Radioactive Waste*; Elsevier Radioactivity in the Environment Volume 9; Elsevier: Amsterdam, The Netherlands, 2011; p. 30.
33. Weiner, R.F.; Neuhauser, K.S.; Heames, T.J.; O'Donell, B.M.; Dennis, M.L. *RADTRAN 6 Technical Manual*; SAND2014-0780; Sandia National Laboratories: Albuquerque, NM, USA, 2014.
34. United States Nuclear Regulatory Commission (US NRC). *Calculation of Annual Doses to Man from Routine Releases of Reactor Effluents for the Purpose of Evaluating Compliance with 10 CFR 50, Appendix I*; Regulatory Guide 1.109, Revision 1; US NRC: Washington, DC, USA, 1977.
35. United States Nuclear Regulatory Commission (US NRC). *Standards for Protection against Radiation*; Code of Federal Regulations Title 10, Part 20, Appendix B—Annual Limits on Intake (ALIs) and Derived Air Concentrations (DACs) of Radionuclides for Occupational Exposure; Effluent Concentrations; Concentrations for Release to Sewerage; US NRC: Washington, DC, USA, 2015.
36. Grove Software, Inc. *MicroShield®User's Manual*; Grove Software Inc.: Lynchburg, USA, 2011.
37. Michael, F.L. *Handbook of Radioactivity Analysis (Third Edition)*; Academic Press: Oxford, UK, 2012; pp. 1051–1052.
38. Canadian Nuclear Safety Commission (CNSC). *Special Form Certificates*; CDN/5255/X-96 (Rev. 0); CNSC: Ottawa, ON, Canada, 2010.
39. Seo, M.H.; Hong, S.W.; Park, J.B. Radiological Impact Assessment for the Domestic On-road Transportation of Radioactive Isotopes Wastes. *JNFCWT* **2016**, *14*, 279–287. [CrossRef]
40. Swedish Nuclear Fuel and Waste Management (SKB). Available online: https://web.archive.org/web/20120506092817/http://www.skb.se/Templates/Standard____15149.aspx (accessed on 11 May 2020).

41. Walberg, M.; Viermann, J.; Beverungen, M.; Kemp, L.; Lindström, A. Disposal of Steam Generators from Decommissioning of PWR Nuclear Power Plants. In Proceedings of the International Youth Nuclear Congress, Interlaken, Switzerland, 20–26 September 2008.
42. Korea Electric Power Corporation (KEPCO). *Kori Unit 1 Steam Generator Replacing Experience Report*; '99-0328-DAN-38; KEPCO: Naju, Korea, 1999.
43. Bruce Power. *Bruce A Refurbishment for Life Extension and Continued Operation Project*; Bruce Power: Ontario, ON, Canada, 2004.
44. Shin, S.W.; Son, J.K.; Cho, C.H.; Song, M.J. Analysis of Dose Rates from Steam Generators to be Replaced from Kori Unit 1. *JRPR* **1998**, *23*, 175–184.
45. SEARATES. Available online: https://www.searates.com/services/distances-time/ (accessed on 11 May 2020).
46. Google Map. Available online: https://goo.gl/maps/jMEeWeMBeKjYwR6i8 (accessed on 13 May 2020).
47. Kim, J.K.; Jin, H.H.; Lee, Y.C. A Legal Study on the Transportation of the Radioactive Waste by Sea under Act on the Arrival, Departure, etc. of Ships. *Marit. Law Rev.* **2017**, *29*, 201–228.
48. Google Map. Available online: https://goo.gl/maps/KYBDENWaqc3W8asn7 (accessed on 13 May 2020).
49. Google Map. Available online: https://goo.gl/maps/XnLYcbSJ1ZJFTeWj6 (accessed on 13 May 2020).
50. Google Map. Available online: https://goo.gl/maps/ah4FA1nWxu7j2vTTA (accessed on 11 May 2020).
51. International Atomic Energy Agency (IAEA). *Regulations for the Safe Transport of Radioactive Materials*; Safety Standards Series No. SSR-6 (Rev.1); IAEA: Vienna, Austria, 2018.
52. International Atomic Energy Agency (IAEA). *Advisory Material for the IAEA Regulations for the Safe Transport of Radioactive Material (2012 Edition)*; Safety Standards Series No. SSG-26; IAEA: Vienna, Austria, 2014.
53. United States Nuclear Regulatory Commission (US NRC). *Interim Guidance on Transportation of Steam Generators (Generic Letter 96-07)*; US NRC: Washington, DC, USA, 1996.
54. James, L.W.; Rick, W.B. Large Component Regulatory Relief in the United States. In Proceedings of the 14th International Symposium on the Packaging and Transportation of Radioactive Materials (PATRAM (2004)), Berlin, Germany, 20–24 September 2004.
55. Shirai, K.; Ozaki, S. Evaluation of transport experience for large-scale radioactive wastes produced from decommissioning of nuclear plant and re-use of metallic radioactive wastes to the transport container. *Denryoku Chuo Kenkyusho Hokoku* **2003**, *12*, 1–4.

sustainability

MDPI

Article

Effect of Gamma Irradiation on Structural Features and Dissolution of Nuclear Waste Na–Al–P Glasses in Water

Alexey V. Luzhetsky [1], Vladislav A. Petrov [2], Sergey V. Yudintsev [2], Viktor L Malkovsky [2], Michael I. Ojovan [2,3,*], Maximilian S. Nickolsky [2], Andrey A. Shiryaev [2,4], Sergey S. Danilov [5] and Elizaveta E. Ostashkina [1]

1 Federal State Unitary Enterprise-United Ecological, Scientific and Research Centre of Decontamination of Radioactive Waste (RAW) and Environmental Protection, 119121 Moscow, Russia; sia-radon@radon.ru (A.V.L.); ostash-eliz@mail.ru (E.E.O.)

2 Institute of Geology of Ore deposits, Petrography, Mineralogy and Geochemistry, 119017 Moscow, Russia; vlad243@igem.ru (V.A.P.); yudintsevsv@gmail.com (S.V.Y.); malkovsky@inbox.ru (V.I.M.); mnickolsky@gmail.com (M.S.N.); a_shiryaev@mail.ru (A.A.S.)

3 Department of Materials, Faculty of Engineering, The University of Sheffield, Sheffield S1 3JD, UK

4 Institute of Physical Chemistry and Electrochemistry, 119071 Moscow, Russia

5 Vernadsky Institute of Geochemistry and Analytical Chemistry, 119991 Moscow, Russia; danilov070992@gmail.com

* Correspondence: m.ojovan@sheffield.ac.uk; Tel.: +7-929-652-1757

Received: 17 April 2020; Accepted: 12 May 2020; Published: 19 May 2020

Abstract: Structural properties and water dissolution of six sodium–aluminum–phosphate (NAP) glasses have been investigated before and after irradiation by a gamma-ray source based on ^{60}Co. Two of these samples were of simple composition, and four samples had a complex composition with radionuclide simulants representing actinides, fission, and activated corrosion products. Samples of the simple composition are fully vitreous, whereas samples of the complex composition contained up to 10 vol.% of aluminum–phosphate, $AlPO_4$, and traces of ruthenium dioxide, RuO_2. Based on the study of pristine and irradiated glasses, it was established that the radiation dose of 62 million Gray had practically no effect on the phase composition and structure of samples. At the same time, the rate of leaching of elements from the irradiated samples by water was decreased by about two times.

Keywords: highly radioactive waste; immobilization; matrix; sodium–aluminum–phosphate glass; properties; irradiation; leaching resistance.

1. Introduction

Nuclear energy is an important source of electricity and a key component of low-carbon energy. In 2018, 450 nuclear power reactors with a total installed capacity of 396 GW€ generated 10% of all electricity in the world, including one-third of low-carbon power generation [1]. In OECD countries, the USA, and Russia, 18–20% of electric energy is produced at nuclear power plants. In France, the amount increased to over 70% [2]. Nuclear energy also has other applications—production of isotopes for various purposes [3], heat generation, desalination, etc. Its negative side is associated with the accumulation of spent nuclear fuel (SNF) and radioactive waste (RW). Up to 12,000 tons of SNF are produced in the world each year, whereas a PWR-VVER type reactor generates annually 25–30 tons of SNF [1,4]. By 2014, about 370,000 tons of SNF had already been produced, of which almost a third was reprocessed [5]. A closed nuclear fuel cycle with SNF reprocessing is being implemented in Russia, France, United Kingdom, China, whereas in Canada, Sweden and Finland, South Korea,

Germany, and some other countries, the emphasis is on an open cycle with SNF disposal without reprocessing [6–8].

The volumes of SNF and RW resulting from the SNF reprocessing will inevitably grow over time [9]. The reprocessing is accompanied by the generation of liquid radioactive waste, including high-level waste (HLW) in an amount of 13–31 m^3 per ton of SNF [10]; after evaporation, its volume is reduced to 250 L [3]. Liquid HLW are isolated in vitreous matrices: Sodium–aluminum–phosphate (Na–Al–P) glasses in Russia [11–15], and alkali-borosilicate (A-B-Si) glasses elsewhere [16–24]. By 2013, about 30,000 tons of vitrified HLW had been accumulated [6,8,25,26]. Taking into account the current rates of SNF reprocessing and solidification of liquid HLW, at present, the total volume of nuclear waste glass can be estimated as 35,000 tons with an approximate weight ratio between Na–Al–P and A-B-Si glasses of 20% to 80%. SNF and RW management in Russia is described in national reports on fulfilling obligations arising from the Joint Convention on the Safety of Spent Fuel and Radioactive Waste Management [27–29]. For SNF reprocessing PUREX technology developed at the end of the 1940s in the USA to extract fissile Pu isotopes for military applications is used [30]. The amount of SNF in the Russian Federation at the end of 2016 reached approximately 22500 tons, of which about 2/3 were from RBMK and 1/3 was from VVER reactors. The reprocessing of SNF of VVER-440, BN-600 and research reactors is carried out at the RT-1 "Mayak" reprocessing plant; about 6000 tons of SNF have been processed to date. Up to 200 tons of the SNF per year is reprocessed now. Recently reprocessing SNF of VVER-1000 and defective SNF of RBMK-1000 have also been started at this plant. Federal State Unitary Enterprise (FSUE) "MCC" has a facility for reprocessing of 5 tons VVER-1000 SNF per year and in 2021 its second stage is expected to launch its productivity to 250 tons of SNF per year [29]. Given the import of SNF from Russian-designed reactors from abroad, the rate of accumulation of spent nuclear fuel in Russia is much higher than the rate of reprocessing. However, according to the planned upgrade of the reprocessing plants, it is assumed that by 2030 the rate of reprocessing of SNF in Russia will exceed the rate of its formation. At the same time, this will greatly increase the volume of vitrified HLW.

The composition of SNF depends on the type of reactor, the composition of the initial fuel, its burnup and storage time [5,6,9]. Radionuclides in SNF are represented by fission products, actinides (uranium and newly formed radioisotopes), activated elements of fuel assemblies and construction materials involved [6,22,23]. The main fission products include rare earth elements (REE), Zr, Mo, Tc, Ru, Pd, Cs, Sr, Rh, Te, Xe, Kr, I. Among them, REE accounts for 25 wt.%, Platinoids—16%, Zr—15%, Mo—12%, Cs—6% [3]. After three years of storage of SNF with a burnup of 33 GWd/t (at 3.5% initial enrichment of ^{235}U), the content, in kg per ton of SNF, [23] is: alkalis (Cs, Rb)—3, alkaline earths (Sr, Ba)—2.4, rare earths (mainly light, Ce groups)—10.2, transition 4d-metals (Mo, Zr, Tc)—7.7, noble metals of the platinum group (Ru, Rh, Pd)—3.9, as well as 0.5 kg of Se and Te, 0.2 kg of I and Br; Ag, Cd, Sn, Sb make another 0.1 kg. These proportions are largely inherited into liquid HLW after SNF reprocessing [3,10,17], where various technological impurities are also present. In RW, in particular, there are [23]—19 wt.% of transition metals (Mo, Zr, Tc, etc.), 18% of REE (La, Ce, Pr, Nd, Sm), 10% alkaline (Gs, Rb) and alkaline-earth (Sr, Ba) elements, 7% platinoids (Ru, Rh, Pd), 2% of minor actinides (Np, Am, Cm), up to 1% Se and Te, the remaining 43 wt.% on Fe, Ni, Cr, Na, P, and a number of less significant elements.

Investigations are underway to support available vitrification technologies and develop improved borosilicate and aluminum-iron phosphate glasses in order to prove the possibility of reliable immobilization of HLW. The compositions of the modernized glasses differ from the traditional ones in the ratio of the main components. They additionally include HLW components such as oxides of Cs, Sr, Ba, Nd, Cr, Zr, Ru, and U. To determine the suitability of such glasses for immobilization and disposal of HLW, it is necessary to study their thermal and radiation resistance and influence of radiation on durability at interaction with water. There are many works on this topic for borosilicate glass matrices, whereas for sodium–aluminum–phosphate glasses, such data are contained in very few publications [11].

The main dose rate from HLW is alpha, beta and gamma rays emitted from the decay of fission products. To study the resistance of the glasses to gamma radiation, high activity ^{60}Co sources are used. In this paper, we studied the effect of γ-radiation on properties of nuclear waste sodium–aluminum–phosphate (NAP) glasses—their structure and resistance to leaching in the water at 25 and 90 °C. Effects of irradiation on properties of NAP glasses of various compositions are compared with previous studies reported in the literature. The aim of this work is to study the effect of radiation (γ-irradiation from ^{60}Co source) on the structure and properties of two groups of nuclear waste NAP glasses. Samples of the first group—"simple composition"—comprise only principal components of the glass matrix (oxides of Na, Al, Fe and P), serving as reference material. Glasses of the second group—"complex composition"—contain simulants of HLW with respective changes in the content of the major elements (Table 1).

Table 1. Compositions of the NAP glasses investigated, wt.% (SEM/EDS)*. NAPas0 (NAPcm0)—pristine glass of simple and complex (in brackets) composition. NAPas1 (NAPcm, NAP1, NAP2)—the same samples after gamma irradiation up to a dose of 6.2×10^7 Gy by ^{60}Co source.

No.	Na_2O	Al_2O_3	Fe_2O_3	P_2O_5	CaO	NiO	La_2O_3	Cr_2O_3	MnO	RuO_2	UO_2
NAPas0	22.3	9.2	15.6	52.1	-	-	-	-	-	-	-
NAPas1	22.1	9.2	15.2	52.4	-	-	-	-	-	-	-
NAPcm0	26.2	15.6	1.3	53.4	0.9	0.4	0.3	0.3	0.3	0.5 *	1.4
NAPcm	26.4	14.4	1.3	54.2	0.9	0.3	0.5	0.4	0.2	0.5 *	1.3
NAP1	26.7	14.7	1.4	53.8	0.9	0.3	0.4	0.2	0.3	0.5 *	1.3
NAP2	24.9	16.2	1.5	53.6	0.9	0.4	0.4	0.4	0.4	0.5 *	1.2
Mayak [1]	23.5	15.8	1.6	53.3	1.0	0.3	0.1	0.4	-	-	-

* 0.4–0.7 wt% of SiO_2 and 0.4–0.5 wt.% K_2O are also present in all samples. Dash—not added, the ruthenium content in the samples according to SEM/EDS is below the detection limit (0.2 wt.%). The value of 0.5 wt.% is estimated from the data on the area fraction of the RuO_2 phase in SEM images, see text. [1] model composition of vitrified HLW of FSUE "Mayak" [13]. Other elements, in wt.%: 0.5 Cs_2O, 0.5 ZrO_2, 0.6 Ce_2O_3, 0.5 Nd_2O_3, 0.5 MoO_3, 0.4 SO_2.

2. Materials and Methods

The NAP glasses studied in this work, potentially suitable for isolation of historical (legacy) HLW of FSUE "Mayak", were synthesized in corundum crucibles as follows. The glass batch mixture was heated to 1100 °C at a rate of 10–30 °C/h; for complete homogenization, the melt was held for 2 hours and quenched. The glass was cooled in the isothermal annealing mode. The compositions of glasses of simple and complex composition are summarized in Table 1. The composition of simulant sodium-aluminum-phosphate glass of FSUE "Mayak" [13] is also presented. Glass samples (Figure 1) are visually translucent, opaque, possess green or dark green color and a conchoidal fracture.

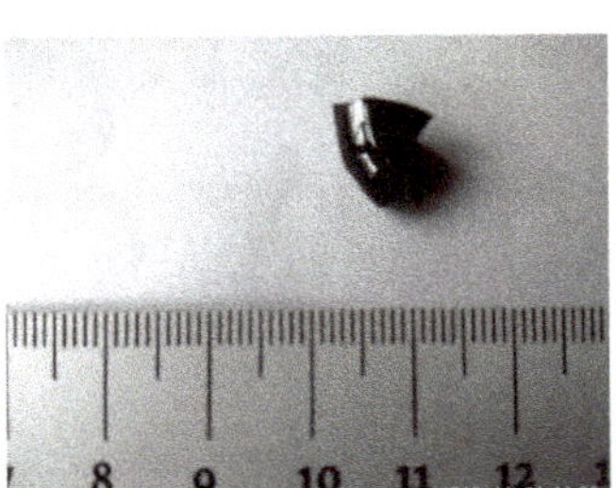

(NAPas0)

(NAPas1)

Figure 1. *Cont.*

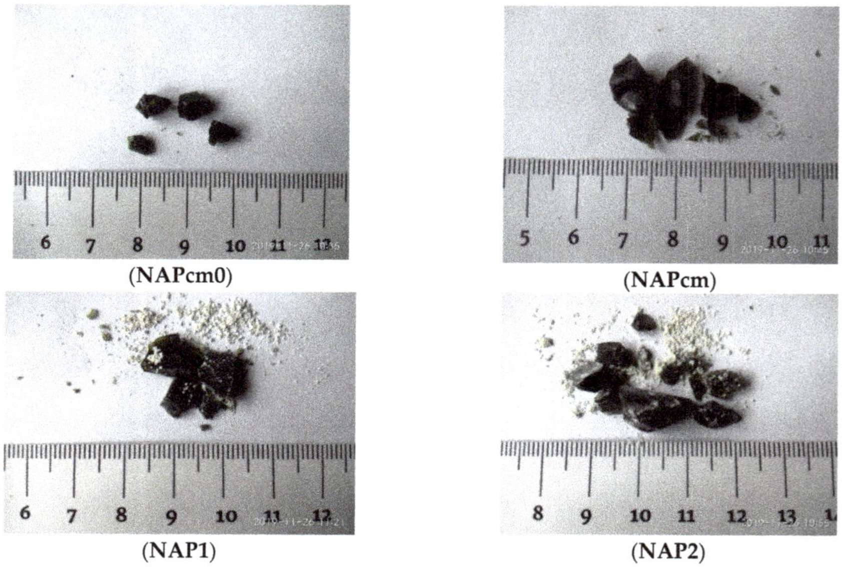

Figure 1. Photos of samples investigated (monolithic and powdered for analysis): NAPas0, NAPas1, NAPcm0—pristine, and NAPcm1, NAP1, NAP2—irradiated. The scale is in cm.

The radiation durability of HLW glasses is one of the important criteria for acceptability of high-level radioactive waste for geological disposal (Class 1 in Russian classification, vitrified HLW of FSUE "Mayak" and Mining and Chemical Combine). Radiation durability is the maximum dose at which there is practically no change in the structure, chemical and mechanical strength of the immobilized RW. For instance, after gamma irradiation with a dose of 10^8 Gy, the glass uniformity should be maintained in the volume of the glass block as controlled by X-ray phase analysis, and changes in the content of the main components in the glass composition should be less than 10% [31].

In order to verify compliance with regulatory requirements, the glass samples were irradiated with a dose of $6.2 \cdot 10^7$ Gy, which is close to that specified by regulatory requirements [31], at the RHM-gamma installation of FSUE "RADON". The dose rate of irradiation was $1 \cdot 10^4$ Gy/h. No thermal (overheating) effects were observed during or after irradiation. Analyses of both unirradiated (pristine) and irradiated samples were performed on a GSM 5610LV scanning electron microscope (SEM) equipped with energy dispersive (EDS) detector.

The phase composition of unirradiated and irradiated NAP glasses were studied by the X-ray diffraction method, infrared and Raman spectroscopy and scanning electron microscopy. X-ray diffraction measurements were performed in the Bragg—Brentano geometry ("reflection") on an EMPYREAN diffractometer using Ni filtered CuK_α radiation. An X'Celerator linear semiconductor detector was used. Samples were mechanically grounded and placed on a zero-background holder made of single-crystal silicon. No binders were not used; the samples were wetted with ethanol to prevent loss during the measurements. HighScore software and PDF2 database (International diffraction data center, ICDD) were used for data analysis.

Raman spectra of the samples investigated were recorded on a Senterra (Bruker) spectrometer; for excitation laser with a wavelength of 532 nm was used. Infrared spectra were obtained on a SpectrumOne IR® spectrometer (Perkin Elmer). For some samples, spectra were also recorded using an AutoImage IR microscope. To record the absorption spectra in the infrared region, the powders were pressed into KBr tablets.

Chemical resistance of NAP glasses was analyzed using standard Russian test protocols [32,33].

3. Structural Properties of the Glasses before and after Irradiation

3.1. SEM/EDS Data

Samples of simple composition (NAPas0, NAPas1) before and after irradiation are composed only of the glassy phase, the composition of which does not change as a result of irradiation. According to SEM/EDS analysis, they comprise (in descending order): P_2O_5, Na_2O, Fe_2O_3, Al_2O_3, the amount of SiO_2 and K_2O oxides is at the detection limit, that is, their presence in the samples shall be checked by more accurate methods. Samples of NAPas0 and NAPas1 glasses are homogeneous with only vitreous phase seen (Figure 2), whereas in samples of the complex composition (NAPcm0, NAPcm, NAP1, and NAP2) in addition to glassy phase, there are crystalline phases: aluminum orthophosphate and ruthenium dioxide (Figures 3 and 4).

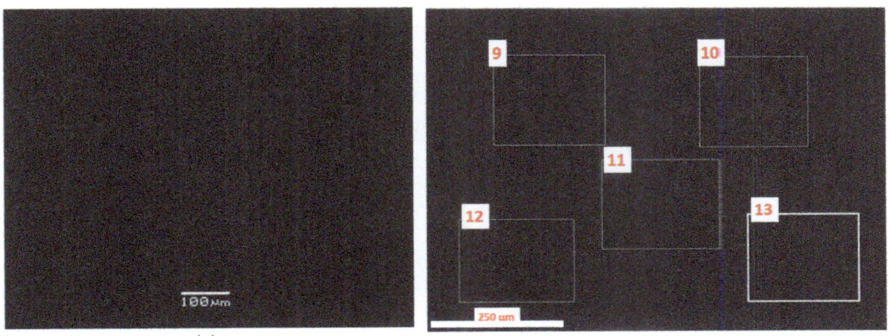

(**a**) (**b**)

Figure 2. SEM images of NAPas0 glasses: (**a**) Image showing the homogeneity of glass; (**b**) sections of glass where the composition is analyzed (Table 2).

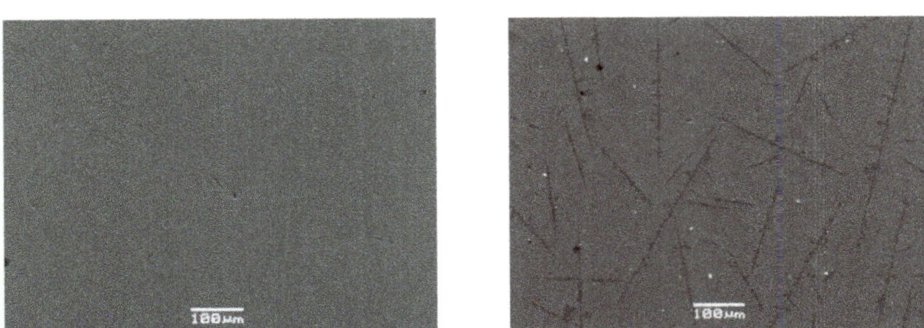

(**a**) (**b**)

Figure 3. SEM images of the samples: (**a**) NAPas1 glass that is seen practically homogeneous; (**b**) NAPcm glass with gray areas that belong to vitreous phase, dark gray—to aluminum orthophosphate ($AlPO_4$, phosphotridimite), and small white particles—to ruthenium dioxide, RuO_2. These phases are confirmed by joint data of SEM/EDS analysis and XRD investigation.

The compositions of the glassy phase of samples NAPas0 and NAPas1, measured in rectangular areas like those, as shown in Figure 2, are given in Tables 2–4. The detection limit of elements is 0.3–0.5 wt.% in light elements (Na, K, Al, Si) and 0.1–0.2 wt.% in heavy elements (Ru, Ln, U). No ruthenium was detected in the glass matrix (its amount is below the detection limit, 0.2 wt.%). Due to very low solubility in the glass [11,20], the main fraction of this element formed oxide precipitates. These particles, as well as Mo-based grains, can merge into large aggregates and settle to the bottom of the furnace, eventually leading to failure of the electrodes of the industrial vitrification furnace EP-500 at FSUE "Mayak". They can also serve as centers for phosphate phases (monazite) crystallization in the glass matrix of HLW during storage and heating, due to the decay of short-lived radionuclides of fission products (Cs, Sr, Ln, etc.).

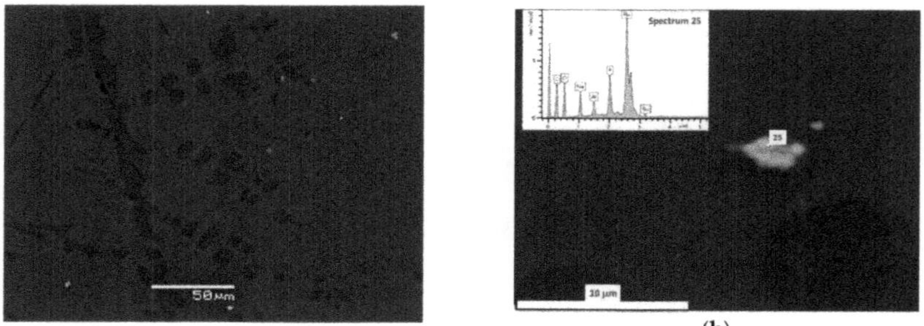

(**a**) (**b**)

Figure 4. SEM images of NAPcm0 glass: (**a**) $AlPO_4$ grains are dark areas in the glass (grey color), whereas white fine grains are ruthenium dioxide; (**b**) particles of RuO_2 in the glass matrix confirmed by joint data of SEM/EDS analysis. Scale bar is equal to 10 μm.

Table 2. Composition of NAPas0 glass (before irradiation), SEM-EDS data (Σ = 100 wt.%). The accuracy of the determinations is 3–5 rel.%.

Oxide, wt.%	Spectrum 9	Spectrum 10	Spectrum 11	Spectrum 12	Spectrum 13
Na_2O	22.3	22.4	22.4	22.3	22.6
Al_2O_3	9.2	9.5	9.2	9.4	9.2
SiO_2	0.4	0.4	0.4	0.4	0.4
P_2O_5	52.1	51.8	51.7	51.9	51.7
K_2O	0.5	0.5	0.5	0.5	0.5
Fe_2O_3	15.6	15.4	15.8	15.5	15.7

Table 3. Composition of NAPas1 glass (after irradiation), SEM-EDS data (Σ = 100 wt.%).

Oxide, wt.%	Spectrum 1	Spectrum 2	Spectrum 3	Spectrum 5	Spectrum 6
Na_2O	22.1	22.4	22.4	22.3	22.8
Al_2O_3	9.2	9.4	9.3	9.3	9.3
SiO_2	0.7	1.2	0.3	0.4	0.2
P_2O_5	52.4	51.8	51.8	51.9	51.5
K_2O	0.5	0.4	0.5	0.5	0.6
Fe_2O_3	15.2	14.8	15.7	15.7	15.6

Table 4. Composition of NAPcm0 glass (before irradiation), SEM-EDS data (Σ = 100 wt.%).

Oxide, wt.%	Spectrum 15	Spectrum 16	Spectrum 17	Spectrum 18	Spectrum 19
Na_2O	26.2	26.7	26.1	26.7	26.7
Al_2O_3	15.6	15.0	15.2	15.1	14.9
SiO_2	bdl *	bdl	bdl	0.3	bdl
P_2O_5	53.4	53.7	53.9	53.4	53.2
CaO	0.9	0.9	0.8	1.0	1.1
Cr_2O_3	0.3	0.2	0.3	0.4	0.3
MnO	0.3	0.4	0.3	0.4	0.3
Fe_2O_3	1.3	1.2	1.2	1.2	1.4
NiO	0.4	0.4	0.4	0.4	0.4
La_2O_3	0.3	0.3	0.5	0.3	0.6
UO_2	1.4	1.2	1.3	1.0	1.1

* bdlbelow detection limit (0.3 wt.%).

The composition of aluminum orthophosphate phase in the initial (unirradiated) glass NAPcm0, as shown in Figure 4a, is given in Table 5.

Table 5. Composition of orthophosphate phase particles in NAPcm0 glass (before irradiation).

Oxide, wt.%	Spectrum 20	Spectrum 21	Spectrum 22	Spectrum 23	Spectrum 24
Na_2O	0.4	0.3	0.2	0.2	0.4
Al_2O_3	41.8	41.4	42.1	42.0	41.5
P_2O_5	57.4	58.2	57.5	57.8	57.8
Fe_2O_3	0.3	0.2	0.2	0.0	0.3

On the EDS spectra of ruthenium-rich particles (Figure 4b), low-intensity peaks of other elements of—C, Na, Al, P—are observed. Carbon is associated with its deliberate deposition on the surface of samples for analysis, whereas the presence of other elements is associated with a partial capture of the surrounding glass matrix by the electron beam. Oxygen is a component of these particles; that is, the ruthenium phase is represented by its oxide. This conclusion is confirmed by the XRD results. SEM investigations have not revealed any structural (Figure 5) and compositional (Table 6) difference between unirradiated and irradiated glasses.

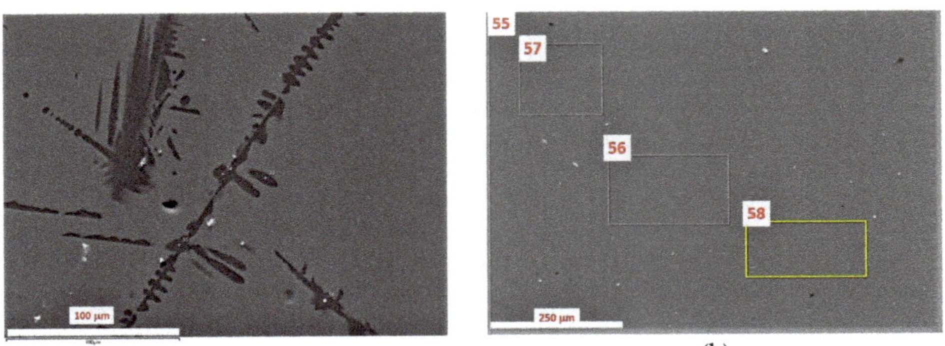

(**a**) (**b**)

Figure 5. SEM images of the irradiated NAP1 and NAP2 glasses: (**a**) NAP1 glass, where gray—glass, dark—$AlPO_4$ (phosphotridimite), small white particles—ruthenium dioxide; (**b**) NAP2 glass with white colored RuO_2 particles in gray glass. Rectangles indicate measurements areas.

Table 6. Compositions of irradiated NAP1 (Spectra 27–31) and NAP2 (Spectra 55–58) glasses, SEM/EDS data (Σ. = 100 wt.%).

Oxide, wt.%	Sp. 27	Sp. 28	Sp. 29	Sp. 30	Sp. 31	Sp. 55	Sp. 56	Sp. 57	Sp. 58
Na_2O	26.7	26.8	26.7	26.9	26.5	24.9	25.2	25.0	24.9
Al_2O_3	14.7	14.8	14.9	14.6	14.9	16.2	16.7	16.0	16.5
P_2O_5	53.8	53.7	53.9	53.4	53.2	53.2	53.2	53.2	53.2
CaO	0.9	0.9	0.9	0.9	0.9	0.9	0.9	0.9	1.0
Cr_2O_3	0.2	0.4	0.4	0.3	0.3	0.4	0.3	0.4	0.3
MnO	0.3	0.3	0.3	0.2	0.3	0.4	0.3	0.3	0.3
Fe_2O_3	1.4	1.3	1.2	1.5	1.2	1.5	1.4	1.5	1.2
NiO	0.3	0.3	0.3	0.5	0.4	0.4	0.4	0.4	0.5
La_2O_3	0.4	0.7	0.6	0.5	0.5	0.4	0.5	0.2	0.5
UO_2	1.3	1.3	1.3	0.9	1.4	1.2	1.4	1.3	1.2

Thus, the main difference in the structure of the initial (unirradiated) glasses of simple (NAPas0) and modified (NAPcm0) compositions is that the latter, in addition to the glass matrix, contains crystalline phases: aluminum-phosphate and ruthenium-oxide. The crystalline phases are clearly visible on the cleaved surface of the glass.

According to SEM studies, both samples of simple composition—pristine (NAPas0) and irradiated (NAPas1) are composed of homogeneous glass phase only. The modified compositions are characterized by the appearance of a small amount (<10 vol.%) of two crystalline phases—phosphotridmite (dominant) and ruthenium dioxide. The appearance of the latter is due to the extremely low solubility of this element in the studied aluminum–phosphate glass. The study of the evolution of the molten ternary system $Na_2O–Al_2O_3–P_2O_5$ on cooling [34] showed the possibility of the formation of several polymorphs of the $AlPO_4$ phase with different structures: P-cristobalite, P-tridimite, berlinite. Crystallization of $AlPO_4$ can begin in the melt in the furnace with a Na_2O content of less than 5 mol% and then continue when the melt (glass) cools down after pouring into canisters. In general, phosphotridimite and alkaline aluminum (iron) phosphates are typical products arising from the partial or complete crystallization of phosphate glass matrices [35–45]. At high contents of light rare earths (the Ce group), lanthanide phosphate with the structure of monazite appears, REE solubility in the aluminum–phosphate melt is determined as few wt.% [11,42,43].

Differences in the structure of irradiated (NAPas1, NAPcm) and unirradiated (NAPas0, NAPcm0) glasses of the same composition were not observed by scanning electron microscopy. The appearance of aluminum–phosphate and RuO_2 is also characteristic of these glasses of the complex composition, both pristine (initial) and irradiated as well.

3.2. X-ray Diffraction

Despite adequate measurements duration and averaging over two or more independent experiments, the signal-to-noise ratio for most samples is relatively poor, which is explained by the small volume fraction of crystalline phases. High background signal level increasing with the diffraction angle is due to fluorescence of the iron admixture.

From the XRD data, the samples studied can be confidently divided into two groups. The first group includes glasses of simple composition—NAPas0 and NAPas1, in which no crystalline phases are observed. There is no noticeable difference between the irradiated and initial samples. All other samples with complex composition contain both an amorphous component (glass) and crystalline phases. Composition of the crystalline phases is the same for all studied samples (Figure 6). The predominant phase is orthorhombic $AlPO_4$ (phosphotridimite) (ICDD card 00-048-0652); the second crystalline phase is ruthenium oxide (RuO_2) (card 01-070-2662).

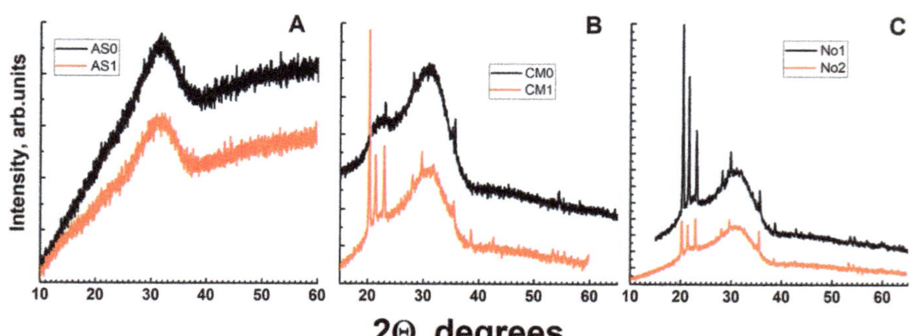

Figure 6. XRD patterns of samples: (**a**) With a simple composition (AS0—initial, AS1—after irradiation); (**b**) with a complex composition (CM0—initial, CM1—irradiated); (**c**) No1 and No2—irradiated).

In general, according to the XRD data, no significant differences between glasses of the same composition before and after irradiation were revealed.

3.3. *Raman and Infrared Spectroscopy*

Both Raman and infrared spectra have revealed no differences between the spectra obtained on nonirradiated and irradiated samples, which indirectly indicates sufficiently high radiation stability of the material. Details are given in Appendix A.

4. Leaching and Dissolution

4.1. *Hydrolytic Durability*

Hydrolytic stability of aluminum–phosphate glass matrices was reported in References [11,44–47]. However, very few publications are devoted to the effect of self-irradiation of the glasses on structure and properties [11]. Therefore, analysis of the effect of irradiation on the leaching of glass matrices is the most important part of this work. To assess the chemical (hydrolytic) stability of glasses of different compositions, leaching experiments were performed. The leaching of glass samples (NAPas0, NAPas1, NAPcm0, NAPcm1, NAP1 and NAP2) was carried out in the water in autoclaves at 90°C; the solution was changed after 1, 3, 7, 10, 14, 21, and 28 days. Besides, similar leaching tests were carried out for samples NAPcm0 and NAPcm1 at 25°C. Due to the small amount of material and difficulties in the preparation of monolithic cubic or parallelepiped specimen, the samples were crushed. The particle size fraction in each sample was selected using sieves with mesh sizes of 0.16 mm and 0.071 mm. From average particles size, the surface area of the powdered glass (S) in runs was estimated to be 200 cm^2.

Analysis of solutions after experiments on interaction with glass was carried out by inductively coupled plasma mass spectrometry on an X-Series instrument at the GEOKHI RAS (Vernadsky Institute of Geochemistry and Analytical Chemistry of Russian Academyof Sciences).

Let us denote the time at which the water was replaced (and the aliquot of the solution for analysis was taken) as $\{t_i, i = 1, ..., 7\}$; $t_0 = 0$. Then, for the time interval between t_{i-1} and t_i, the differential dissolution rate of the glass, normalized by one of its main structural elements, is determined as

$$J_{El} = \frac{C_{El}V}{S(t_i - t_{i-1})m_{El}}, \quad (1)$$

where C_{El} is the concentration of the glass element, which determines the normalized dissolution rate; V is the volume of the autoclave; S is the total surface area of the particles, m_{El} is the mass fraction of the element El in the glass. Since J_{El} is the average value of the normalized dissolution rate in the time interval between t_{i-1} and t_i, when determining the function $J_{El}(t)$, it is approximately attributed to the middle of this interval, i.e., at time $\tau_i = (t_{i-1} + t_i)/2$, since the smooth function is assumed.

Data on integral dissolution rates of the glasses investigated are presented in Appendix B. A typical dependence of the normalized dissolution rate of glasses on time is shown in Figure 7.

In general, all values of J_{El} decrease with time, which is consistent with the simplest diffusion models for the advancement of the leach front. However, at τ_i = 8.5 days, a local maximum occurs. Such jumps in J_{El} values were noted in many studies on the leaching of borosilicate glasses [26,48–52]; they are termed as leaching resumption stage. The presence of these events is typically explained by redeposition of sparingly soluble matrix components as secondary phases. The formation of these phases circumvents the protective effect of the altered glass layer, which acts as a barrier against leaching [53,54]. However, this explanation cannot be considered exhaustive, since in many works, the effect of leaching recovery was not observed, although secondary phases were also deposited on the glass surface. An alternative explanation is the formation of micro-cracks on glass surface [26]. In the case of investigated NAP glasses, the presence of the local maximum is neither an artifact nor a random deviation, as proved by maxima of $J_{El}(t)$ in experiments with other glasses (Figure 8).

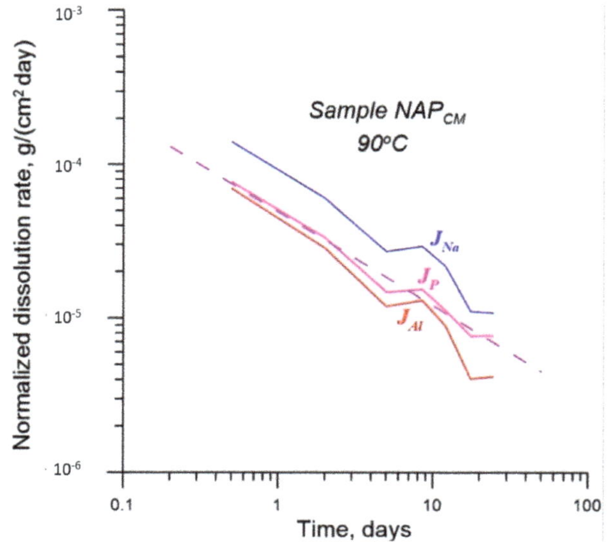

Figure 7. Time dependences of normalized dissolution rate of irradiated glass NAPcm at 90 °C. An approximation of time dependence of dissolution rate normalized by phosphorus by power function $0.489 \cdot 10^{-4}/t^{0.61}$ (values of parameters of the power function were obtained from least-squares fit) is shown by the dashed purple line.

In all the samples studied the mass fraction of phosphorus is approximately the same. In addition to the presence of modifiers in samples NAPcm, NAPcm0, NAP1 and NAP2, significant differences of these four types of glasses from samples NAPas0 and NAPas1 relate only to the ratio of mass fractions of sodium and aluminum. Analysis of values of J_{El} (t) in these two groups of glasses shows that glasses with a higher mass content of aluminum dissolve more slowly in water, i.e., have smaller $J_{El}(t)$ (Figure 8).

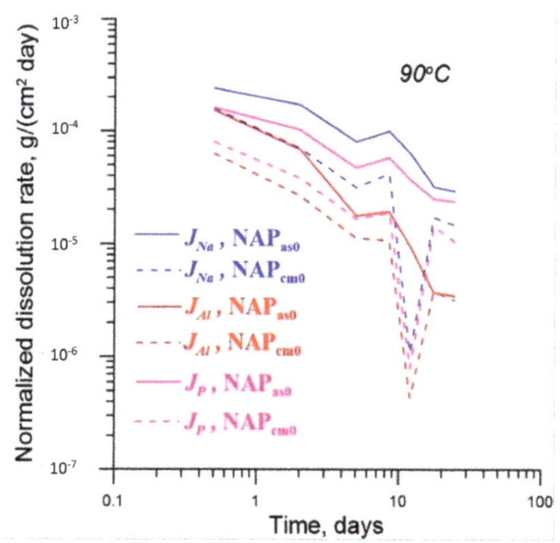

Figure 8. Dependences $J_{El}(t)$ for the pristine (unirradiated) glasses with high (NAPas0) and low (NAPcm0) values of the ratios of the mass fractions of sodium and aluminum in the glass composition.

Analysis of $J_{El}(t)$ over the entire range of time values shows that for all t, the ratio $J_{El}(t)$ for sample NAPas1 and samples NAPcm, NAPcm0, NAP1, and NAP2 for all elements for which the normalized dissolution rate was calculated, has close values apart from the unreliable value of $J_{El}(t)$ for NAPcm0. The dependence of $J_{El}(t)$ on the ratio of sodium and aluminum in phosphate glasses has the form

$$J_{El}(t) = J^0_{El}(t)f(m_{Na}/m_{Al}), \quad (2)$$

where $J^0_{El}(t)$ is a function that does not depend on the ratio m_{Na}, m_{Al}, e.g., mass fractions of sodium and aluminum in glasses.

Dependences $J_{El}(t)$ are satisfactorily approximated by power functions of the form (see Figure 7):

$$J_{El}(t) = \frac{A}{t^{0.6}}, \quad (3)$$

where A is a coefficient depending on a certain element by which the normalized dissolution rate of the glass is calculated, and on the ratio of the mass fractions of sodium and aluminum in the glass composition. Values of normalized dissolution rates calculated for different elements satisfy the inequalities:

$$J_{Al}(t) < J_P(t) < J_{Na}(t). \quad (4)$$

This seems to be well justified: Sodium passes into solution more easily than phosphorus, and that, in turn, is easier leached compared to aluminum. If values of $J_{Na}(t)$ are used in the estimate of the safety of repositories to calculate leaching rates, then this estimate will be conservative (i.e., pessimistic). In this regard, it seems appropriate to find such a form of the analytic function $J^{appr}_{Na}(t)$ that approximates with satisfactory accuracy (with the mean-squared error of about 30%) the dependences obtained from experiments in the entire range of ratios m_{Na} and m_{Al} studied. Accounting for Equation (2), we seek this analytical approximating dependence in the form

$$J^{calc}_{Na}(t, m_{Na}/m_{Al}, a, b) = \frac{a}{t^{0.6}} \exp\{b(m_{Na}/m_{Al} - 2.5)\}, \quad (5)$$

where a, b are unknown parameters that we will determine by the least-squares method [55] from the condition

$$\Phi(a,b) = \sum_{m_{Na}/m_{Al}} \sum_{n=1}^{5} \left[J^{calc}_{Na}(t_n, m_{Na}/m_{Al}, a, b)/J^{exper}_{Na}(t_n, m_{Na}/m_{Al}) - 1\right]^2 \to \min \quad (6)$$

The procedure for finding the minimum of the function Φ was carried out by the gradient method with the control of convergence as follows [53]. The initial approximation was set: $a_0 = 9 \cdot 10^{-5}$, $b_0 = 0.8$ Then, with the known n-th approximation—a_n, b_n—the (n+1)-th approximation was determined by the formulae

$$a_{n+1} = a_n - \frac{\partial\Phi}{\partial a}\delta_n, \qquad b_{n+1} = b_n - \frac{\partial\Phi}{\partial b}\delta_n \quad (7)$$

Here we have

$$\delta_n = \frac{1}{2^m}\min\left\{\frac{|a_n|}{|\partial\Phi/\partial a|}, \frac{|b_n|}{|\partial\Phi/\partial b|}\right\} \quad (8)$$

where m is the minimum positive integer under which the condition for convergence of the method is satisfied:

$$\Phi(a_{n+1}, b_{n+1}) < \Phi(a_n, b_n) \quad (9)$$

The minimum function Φ (a,b) found in this way from the data on the leaching rate of Al is reached at $a = 1.183 \cdot 10^{-4}$, $b = 0.75$. Therefore, if we substitute these values of the parameters into dependence (5), it will have the form

$$J_{Na}^{calc}(t, m_{Na}/m_{Al}, a, b) = \frac{0.0001183}{t^{0.6}} F_{Na,Al} \quad (10)$$

where $[J_{Na}] = g/cm^2 \cdot day$, $[t] = day$, $F_{Na,Al} = \exp\{0.75 m_{Na}/m_{Al} - 1.875\}$. The calculated dependence (9) is compared with the experimental data on the leaching of glasses (Figure 9) and shows a satisfactory accuracy of the approximation of experimental data by power functions.

The temperature has a significant effect on the leaching of sodium–aluminum–phosphate glasses [54]. A comparison of the $J_{El}(t)$ dependences obtained on the NAPcm0 sample in experiments at 90 °C and 25 °C for different structural elements of glass is shown in Figure 10.

The dependence of leaching rates on temperature, as well as the intensity of many chemical reactions, can be described by the Arrhenius formula:

$$J_{El}(T) = J \cdot \exp\left\{-\frac{E_a}{RT}\right\}, \quad (11)$$

where E_a is the activation energy, and R is the gas constant (R = 8.3 J/(mol · K)).

As noted above the final portion of $J_{El}(t)$ dependences of NAPcm0 can be ignored. Thus, for calculating the $J_{El}(t)$ ratios at different temperatures, only two time points remain, e.g., 5 and 8.5 days. The values of the relations $J_{El}(t)$ at different temperatures T at these points are given in Table 7.

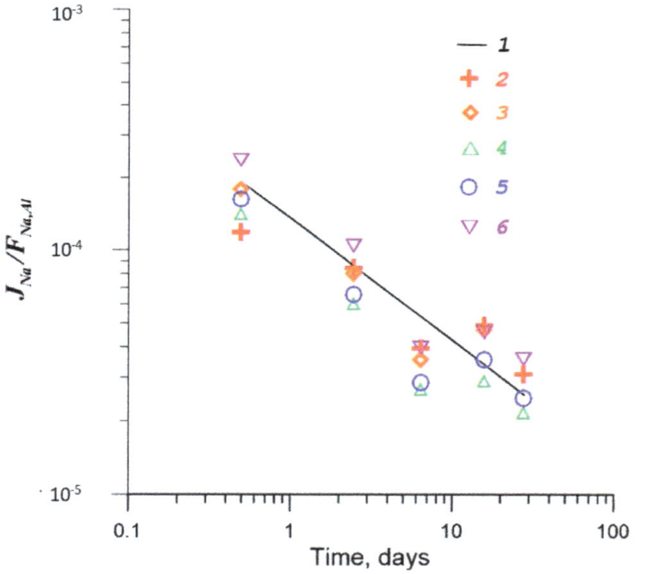

Figure 9. Comparison of the calculated approximating line by Equation (9) and experimental data of glasses studied in this work. (1) $1.183 \cdot 10^{-4}/t^{0.6}$, (2) NAPas0, (3) NAPcm0, (4) NAPcm1, (5) NAP1, (6) NAP2.

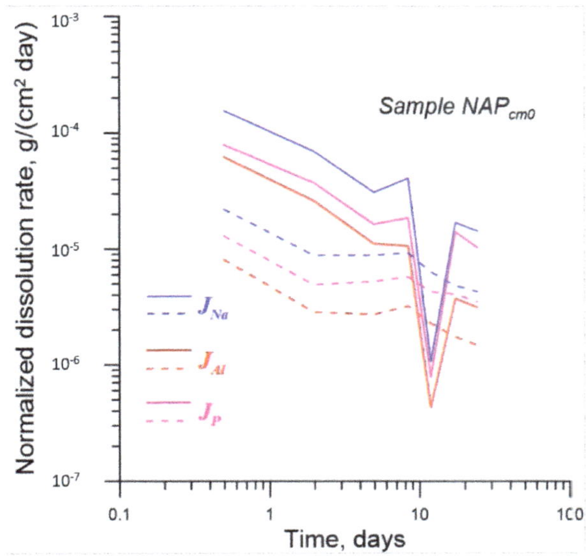

Figure 10. Normalized leaching rates of glass NAPcm0 at 25 °C (dashed lines) and 90 °C (solid lines).

Table 7. $J_{El}(t)$ ratios at temperatures of 90 °C and 25 °C for NAPcm0 glass.

Element	t = 5 Days	t = 8.5 Days
Al	4.07	3.30
Na	3.51	4.50
P	3.12	3.23

The average value over Table 7 is 3.62. Then we get from expression (11) $\ln 3.62 = \frac{E_a}{R}\left(-\frac{1}{363} + \frac{1}{298}\right)$, therefore for glass NAPcm0 we obtain $E_a = 17.7$ kJ/(mol · K). In a similar manner, we consider the $J_{El}(t)$ ratios at temperatures of 90 °C and 25 °C of NAPcm1 glass. These data are given in Table 8. The average over Table 8 value of the ratio of leaching rates at 90 °C and 25 °C, in this case, is 3.92. Hence, the activation energy of dissolution of glass NAPcm1 is $E_a = 18.9$ kJ/(mol · K). This value is close to the previously calculated value of 17.7 kJ/(mol · K) for NAPcm0, which supports the correctness of the results. Note, however, that these activation energies are significantly lower compared with those in borosilicate nuclear waste glasses, which are more durable in water [56].

Table 8. $J_{El}(t)$ ratios at temperatures of 90 °C and 25 °C for NAPcm1 glass.

Element	t = 5 Days	t = 8.5 Days	t = 12 Days
Al	4.83	4.85	5.04
Na	3.80	3.75	3.28
P	3.35	3.06	3.30

4.2. *Effect of γ-Radiation on Dissolution of Glass*

Effects of irradiation of glasses on their dissolution behavior remain insufficiently studied, but recent works on International Simple (borosilicate) Glass showed that it is indeed necessary to take into account radiation damage effects in the prediction of water-glass interaction in HLW repository [57]. The absolute majority of irradiation experiments employ much higher dose rates than is expected in real HLW. Whereas, the dose-rate effect is difficult to estimate consistently, one can reasonably expect that higher rates may alter the mechanism of radiation damage (e.g., cascade overlap vs. defects accumulation) and influence recovery kinetics. Presumably, in most cases, the influence of high dose rates on dissolution in irradiation experiments will be larger than in real HLW glasses.

Only very few studies addressed the effects of irradiation on properties of aluminum–phosphate glasses. In Reference [58] it was shown that during electron irradiation of aluminum–phosphate glass gas bubbles were formed even at low doses ($0.8 \cdot 10^{23}$ electrons/m^2). As the absorbed dose increases, the bubbles grow, and they move towards the glass surface. At a dose of $2.2 \cdot 10^{26}$ electrons/m^2 ($6.6 \cdot 10^{11}$ Gy) all bubbles leave the glass sample. With further exposure of the glass to a dose of $4.5 \cdot 10^{26}$ electrons/m^2, areas enriched in Al and P appeared. Sodium–aluminum–phosphate glasses were irradiated to a dose of 10^8 Gy (electrons and γ-photons), and up to $2 \cdot 10^{18}$ α-decays/cm^3 [11]. These values are close to the radiation dose that real vitrified HLW will receive in 10,000 years. Study of the irradiated glass samples by X-ray diffraction analysis, nuclear magnetic resonance and electron paramagnetic resonance, showed absence of changes in the matrix structure [11]. It was also found that in experiments lasting one day, the rate of Na leaching at room temperature increases with the rise of radiation dose up to 10^8 Gy, but its value remains constant or slightly decreases after 30 days of interaction, amounting to $4 \cdot 10^{-7}$ g/(cm^2·day). The damage during gamma-irradiation is mostly due to Compton electrons, thus, the studies employing electrons mentioned above are fully relevant.

A comparison of data on the dissolution of NAPas0 and NAPas1 glass samples is shown in Figure 11. On a logarithmic scale, the dependency graphs $J_{El}(t)$ for the same structural elements obtained on samples NAPas0 and NAPasa1 run approximately parallel to each other. This suggests that over the entire time range, the ratio $J_{El}(t)$ obtained for the initial and irradiated glasses remains roughly the same. The ratios of values of $J_{El}(t)$ obtained on samples of the pristine NAPas0 and irradiated NAPas1 glasses at 90 °C are given in Table 9. The average value over the whole table is 2.03; the data of Tables 8 and 9 compared in Figure 11.

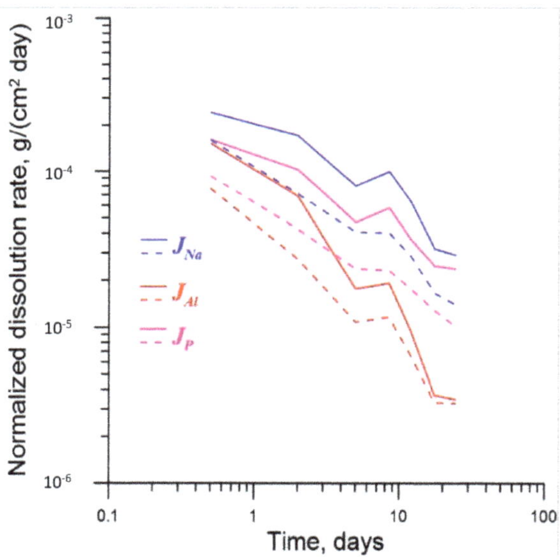

Figure 11. Comparison of the normalized dissolution rates of unirradiated glass NAPas0 (solid lines) with irradiated glass NAPas1 (dashed lines) at 90 °C.

Table 9. Ratios of dissolution rates $J_{El}(t)$ of nonirradiated NAPas0 and irradiated NAPas1 glasses.

Element	t = 0.5 Days	t =2 Days	t = 5 Days	t = 8.5 Days	t = 12 Days
Al	1.98	2.51	1.63	1.64	1.44
Na	1.51	2.41	1.98	2.48	2.22
P	1.74	2.41	1.98	2.51	2.08

Figure 12 shows that the J_{El}/J_{El0} ratios obtained on samples NAPas0 and NAPas1 remain close to 2 through the entire studied range of duration of runs.

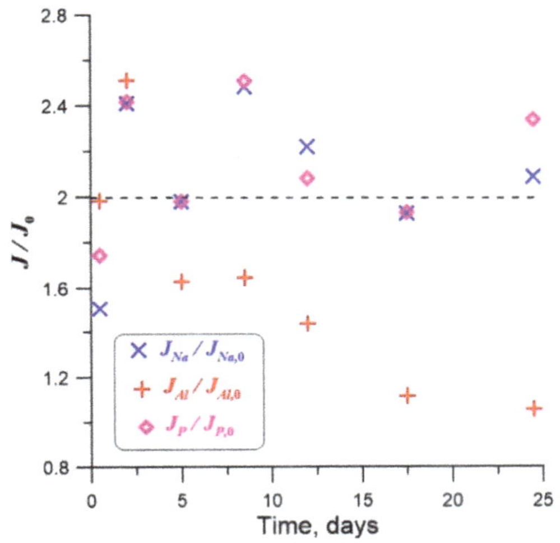

Figure 12. Time dependence of ratios J_{El}/J_{El0} obtained on samples NAPas0 and NAPas1.

Thus, as a result of γ-irradiation, the leaching rate of the sodium–aluminum–phosphate glass has averagely decreased by about 2 times. This can be caused by undetectable structural changes such as radiation-induced annealing that occurs in some other type of glasses [22]. Much higher effect on the deterioration of properties of the aluminum–phosphate vitreous waste form, including their solubility in hot water, is observed due to glass crystallization in temporary storage and after ultimate disposal in deep underground repository [11,14,15,36–40,43–45].

5. Conclusions

Six samples of glasses used in Russia for HLW immobilization were investigated before and after irradiation. The studied sodium–aluminum–phosphate glasses studied are divided into two compositional groups: Group 1—samples NAPas0 and NAPas1, which are glasses of simple composition (i.e., only principal oxides); group 2—samples NAPcm0, NAPcm, NAP1, NAP2—glasses of a modified composition with additives of radionuclide simulants. Each group contains one unirradiated sample and the same samples after their irradiation to a dose of 62 MGy at the gamma source of FSUE RADON.

Using SEM/EDS, XRD, Raman, and infrared spectroscopy, it was established that samples of the simple composition of group 1 are composed only of the glass matrix. In samples of group 2, in addition to dominating glass matrix, precipitates of the crystalline phases (up to 10 vol.%) of aluminum–phosphate, and a small amount (a fraction of vol.%) of ruthenium dioxide are observed. Dissolution behavior of glasses was analyzed at 25 and 90 °C. According to the results of comparison of pristine and irradiated glasses, it was found that irradiation with a dose of $6.2 \cdot 10^7$ Gy had practically no important effect on the structure and composition of samples. Simultaneously, it was found that the rate of leaching of elements from irradiated samples decreased by approximately two times.

Author Contributions: Conceptualization, A.V.L., V.A.P. and S.V.Y.; methodology, S.V.Y. and V.I.M.; software, V.I.M.; validation, S.V.Y., V.I.M. and M.I.O.; investigation, M.S.N., A.A.S., S.S.D. and E.E.O.; data curation, S.V.Y., V.I.M. and M.I.O.; writing—original draft preparation, S.V.Y. and V.I.M.; writing—review and editing, S.V.Y., V.I.M. and M.I.O.; visualization, S.V.Y., V.I.M. and M.I.O.; supervision, S.V.Y. and M.I.O.; project administration, S.V.Y.; funding acquisition, A.V.L., V.A.P. and S.V.Y. All authors have read and agreed to the published version of the manuscript.

Funding: The study was funded by the FSUE "Radon" under Agreement with State Corporation "Rosatom" and from the Federal budget within the framework of the state tasks of scientific investigations for the IGEM RAS.

Acknowledgments: The research was carried out under Agreement between FSUE "Radon" and State Corporation "Rosatom" and in according with topic of scientific research within the framework of the state task of the IGEM RAS funded from the Federal budget. We are indebted to O.I. Stefanovsky for assistance at synthesis of the samples investigated; to B.S. Nikonov for participation at SEM/EDS study and to A.A. Averin for help in Raman measurements. Authors are also grateful to anonymous reviewers whose valuable comments helped to improve this paper.

Conflicts of Interest: The authors declare no conflict of interest.

Appendix A. Raman and Infrared Spectra of Sodium–Aluminum–Phosphate Glasses Investigated

Appendix A.1. Raman Spectroscopy

Raman spectra are shown in Figure A1. The measurements were performed in quasi-backscattering geometry without sample preparation on macroscopic glass samples. The size of the laser spot on the sample was 2–3 micrometers.

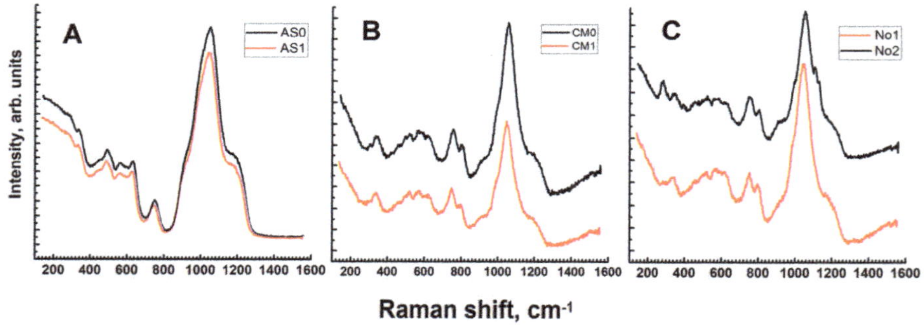

Figure A1. Raman spectra of the initial and irradiated samples: (**A**) With a simple composition (AS0—initial, AS1—after irradiation); (**B**) with a complex composition (CM0—initial, CM1—irradiated); (**C**) No1 and No2)irradiated).

The spectra contain bands because of the vibrations of the aluminum–phosphate groups. Narrow peaks due to crystalline phases are observed in many spectra, but their amount is small. There is no difference between the spectra obtained in these measurements, which indirectly indicates sufficiently high radiation stability of the material.

Appendix A.2. Infrared Spectroscopy

Infrared spectra of samples are shown in Figure 8. The rise in the background with increasing wave number is due to light scattering by glass particles in the salt matrix. As for the Raman spectra, vibrational bands of Al–P groups are observed at the IR spectra of the glasses. The data obtained allow us to conclude that difference between the initial and irradiated glasses is not observed (Figure A2).

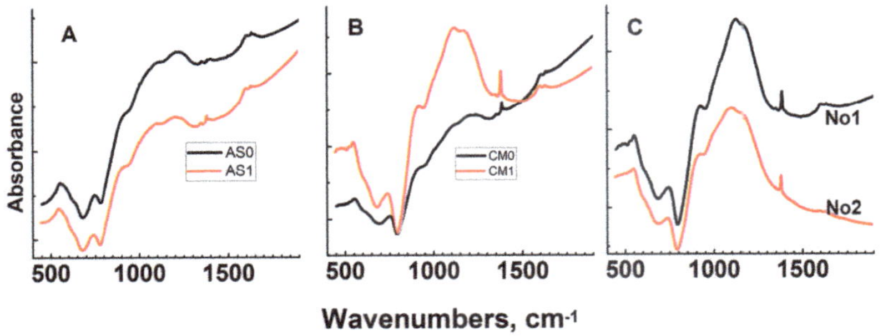

Figure A2. Infrared absorption spectra: (**A**) NAPas0—AS0 initial, AS1 after irradiation; (**B**) NAPcm0—CM0 initial and NAPcm1 (CM1)—irradiated; (**C**) NAP1 (No1) and NAP2 (No2)—irradiated. The narrow peak at 1384 cm^{-1} is due to KBr matrix.

Appendix B. Integral Dissolution Rates of Sodium–Aluminum–Phosphate Glasses Investigated

Table A1 presents the integral leaching rate (LR) of elements from samples NAPas0, NAPas1, NAPcm0, NAPcm, NAPcm1, NAP1, and NAP2. LR for contents below the detection limit were calculated for a value equal to the detection limit (Table A2). Ru was not detected.

Table A1. The integral leaching rate of elements from NAP glasses investigated, g/(cm^2 · day).

Sample	Time, d	Na	Al	Fe	P	Ca	Cr	Mn	Ni	La	U
NAPas0 (90°C)	1	2.41×10^{-4}	1.52×10^{-4}	1.75×10^{-6}	1.61×10^{-4}						
	3	1.94×10^{-4}	9.64×10^{-5}	7.09×10^{-7}	1.22×10^{-4}						
	7	1.29×10^{-4}	5.14×10^{-5}	3.49×10^{-7}	7.91×10^{-5}						
	10	1.20×10^{-4}	4.17×10^{-5}	2.89×10^{-7}	7.28×10^{-5}						
	14	1.04×10^{-4}	3.25×10^{-5}	2.83×10^{-7}	6.24×10^{-5}						
	21	7.99×10^{-5}	2.29×10^{-5}	2.05×10^{-7}	4.98×10^{-5}						
	28	6.72×10^{-5}	1.80×10^{-5}	1.55×10^{-7}	4.32×10^{-5}						
NAPas1 (90°C)	1	1.60×10^{-4}	7.67×10^{-5}	1.34×10^{-5}	9.24×10^{-5}						
	3	1.01×10^{-4}	4.37×10^{-5}	6.12×10^{-6}	5.89×10^{-5}						
	7	6.63×10^{-5}	2.49×10^{-5}	3.03×10^{-6}	3.88×10^{-5}						
	10	5.84×10^{-5}	2.09×10^{-5}	2.40×10^{-6}	3.41×10^{-5}						
	14	4.99×10^{-5}	1.68×10^{-5}	1.83×10^{-6}	2.93×10^{-5}						
	21	3.88×10^{-5}	1.23×10^{-5}	1.32×10^{-6}	2.38×10^{-5}						
	28	3.25×10^{-5}	1.01×10^{-5}	1.13×10^{-6}	2.04×10^{-5}						
NAPcm0 (90°C)	1	1.57×10^{-4}	6.28×10^{-5}	4.53×10^{-5}	7.98×10^{-5}	5.90×10^{-5}	6.30×10^{-5}	6.98×10^{-5}	5.40×10^{-5}	2.50×10^{-5}	8.07×10^{-5}
	3	9.93×10^{-5}	3.86×10^{-5}	2.43×10^{-5}	5.17×10^{-5}	2.31×10^{-5}	3.83×10^{-5}	3.13×10^{-5}	3.37×10^{-5}	1.42×10^{-5}	5.11×10^{-5}
	7	6.04×10^{-5}	2.29×10^{-5}	1.36×10^{-5}	3.16×10^{-5}	2.88×10^{-5}	2.25×10^{-5}	1.50×10^{-5}	1.99×10^{-5}	7.57×10^{-6}	3.15×10^{-5}
	10	5.47×10^{-5}	1.93×10^{-5}	1.02×10^{-5}	2.78×10^{-5}	6.81×10^{-5}	1.74×10^{-5}	1.24×10^{-5}	1.84×10^{-5}	6.10×10^{-6}	2.74×10^{-5}
	14	3.94×10^{-5}	1.39×10^{-5}	7.44×10^{-6}	2.01×10^{-5}	6.42×10^{-5}	1.25×10^{-5}	9.02×10^{-6}	1.32×10^{-5}	4.42×10^{-6}	2.04×10^{-5}
	21	3.20×10^{-5}	1.05×10^{-5}	5.23×10^{-6}	1.81×10^{-5}	4.28×10^{-5}	8.40×10^{-6}	6.10×10^{-6}	9.43×10^{-6}	2.99×10^{-6}	1.52×10^{-5}
	28	2.76×10^{-5}	8.67×10^{-6}	4.21×10^{-6}	1.62×10^{-5}	3.21×10^{-5}	6.34×10^{-6}	4.64×10^{-6}	7.51×10^{-6}	2.41×10^{-6}	1.25×10^{-5}

Table A1. *Cont.*

Sample	Time, d	Na	Al	Fe	P	Ca	Cr	Mn	Ni	La	U
	1	2.23×10^{-5}	8.16×10^{-6}	7.37×10^{-6}	1.31×10^{-5}	9.72×10^{-5}	1.14×10^{-6}	7.86×10^{-6}	7.03×10^{-6}	6.25×10^{-6}	1.19×10^{-5}
	3	1.34×10^{-5}	4.64×10^{-6}	3.96×10^{-6}	7.67×10^{-6}	5.09×10^{-5}	7.58×10^{-7}	4.82×10^{-6}	4.21×10^{-6}	2.40×10^{-6}	7.94×10^{-6}
NAPcm0 (25°C)	7	1.08×10^{-5}	3.56×10^{-6}	2.93×10^{-6}	6.30×10^{-6}	2.83×10^{-5}	4.87×10^{-7}	4.90×10^{-6}	4.40×10^{-6}	1.17×10^{-6}	5.10×10^{-6}
	10	1.04×10^{-5}	3.46×10^{-6}	2.86×10^{-6}	6.16×10^{-6}	2.67×10^{-5}	4.55×10^{-7}	6.34×10^{-6}	5.20×10^{-6}	1.48×10^{-6}	4.76×10^{-6}
	14	9.24×10^{-6}	3.13×10^{-6}	2.56×10^{-6}	5.63×10^{-6}	2.75×10^{-5}	4.06×10^{-7}	6.14×10^{-6}	4.91×10^{-6}	1.41×10^{-6}	4.25×10^{-6}
	21	7.78×10^{-6}	2.67×10^{-6}	2.08×10^{-6}	5.13×10^{-6}	1.85×10^{-5}	3.25×10^{-7}	5.84×10^{-6}	4.50×10^{-6}	9.84×10^{-7}	3.40×10^{-6}
	28	6.93×10^{-6}	2.38×10^{-6}	1.80×10^{-6}	4.73×10^{-6}	1.40×10^{-5}	2.84×10^{-7}	5.77×10^{-6}	4.24×10^{-6}	7.73×10^{-7}	2.98×10^{-6}
	1	1.40×10^{-4}	6.96×10^{-5}	3.85×10^{-5}	7.67×10^{-5}	4.31×10^{-6}	6.26×10^{-5}	3.40×10^{-5}	8.89×10^{-5}	2.89×10^{-5}	6.01×10^{-5}
	3	8.68×10^{-5}	4.21×10^{-5}	2.05×10^{-5}	4.76×10^{-5}	3.36×10^{-4}	3.79×10^{-5}	1.67×10^{-5}	5.49×10^{-5}	1.49×10^{-5}	3.89×10^{-5}
NAPcm (90°C)	7	5.26×10^{-5}	2.48×10^{-5}	1.10×10^{-5}	2.88×10^{-5}	1.48×10^{-4}	2.16×10^{-5}	8.14×10^{-6}	3.20×10^{-5}	8.14×10^{-6}	2.38×10^{-5}
	10	4.55×10^{-5}	2.13×10^{-5}	8.70×10^{-6}	2.48×10^{-5}	1.04×10^{-4}	1.83×10^{-5}	6.17×10^{-6}	2.74×10^{-5}	6.61×10^{-6}	2.06×10^{-5}
	14	3.87×10^{-5}	1.77×10^{-5}	6.78×10^{-6}	2.09×10^{-5}	1.01×10^{-4}	1.51×10^{-5}	4.40×10^{-6}	2.26×10^{-5}	5.13×10^{-6}	1.73×10^{-5}
	21	2.95×10^{-5}	1.32×10^{-5}	4.73×10^{-6}	1.65×10^{-5}	6.73×10^{-5}	1.09×10^{-5}	3.04×10^{-6}	1.63×10^{-5}	3.48×10^{-6}	1.25×10^{-5}
	28	2.48×10^{-5}	1.09×10^{-5}	3.82×10^{-6}	1.43×10^{-5}	5.06×10^{-5}	8.96×10^{-6}	2.31×10^{-6}	1.34×10^{-5}	3.03×10^{-6}	1.02×10^{-5}
	1	2.13×10^{-5}	8.53×10^{-6}	7.96×10^{-6}	1.26×10^{-5}	5.56×10^{-5}	1.09×10^{-6}	6.06×10^{-6}	1.04×10^{-5}	2.76×10^{-6}	9.92×10^{-6}
	3	1.29×10^{-5}	5.02×10^{-6}	4.63×10^{-6}	7.75×10^{-6}	5.29×10^{-5}	7.25×10^{-7}	3.88×10^{-6}	5.98×10^{-6}	1.09×10^{-6}	6.61×10^{-6}
NAPcm1 (25°C)	7	9.57×10^{-6}	3.55×10^{-6}	3.13×10^{-6}	5.83×10^{-6}	2.38×10^{-5}	4.66×10^{-7}	2.73×10^{-6}	4.60×10^{-6}	5.43×10^{-7}	4.25×10^{-6}
	10	9.02×10^{-6}	3.29×10^{-6}	2.82×10^{-6}	5.57×10^{-6}	1.71×10^{-5}	4.35×10^{-7}	2.85×10^{-6}	5.08×10^{-6}	4.32×10^{-7}	3.97×10^{-6}
	14	8.32×10^{-6}	2.86×10^{-6}	2.38×10^{-6}	4.95×10^{-6}	1.73×10^{-5}	3.88×10^{-7}	3.03×10^{-6}	5.13×10^{-6}	3.46×10^{-7}	3.54×10^{-6}
	21	6.62×10^{-6}	2.21×10^{-6}	1.79×10^{-6}	4.19×10^{-6}	1.17×10^{-5}	3.11×10^{-7}	2.55×10^{-6}	4.12×10^{-6}	2.55×10^{-7}	2.83×10^{-6}
	28	5.74×10^{-6}	1.87×10^{-6}	1.35×10^{-6}	3.80×10^{-6}	8.93×10^{-6}	2.72×10^{-7}	2.31×10^{-6}	3.65×10^{-6}	2.10×10^{-7}	2.48×10^{-6}

Table A1. *Cont.*

Sample	Time, d	Na	Al	Fe	P	Ca	Cr	Mn	Ni	La	U
NAP1	1	1.67×10^{-4}	7.31×10^{-5}	5.49×10^{-5}	9.11×10^{-5}	4.17×10^{-6}	9.61×10^{-5}	5.40×10^{-5}	7.45×10^{-5}	3.34×10^{-5}	1.13×10^{-4}
	3	1.01×10^{-4}	4.25×10^{-5}	2.90×10^{-5}	5.58×10^{-5}	2.78×10^{-6}	5.73×10^{-5}	2.54×10^{-5}	4.42×10^{-5}	1.44×10^{-5}	6.40×10^{-5}
	7	6.02×10^{-5}	2.43×10^{-5}	1.58×10^{-5}	3.33×10^{-5}	3.75×10^{-5}	3.35×10^{-5}	1.22×10^{-5}	2.52×10^{-5}	7.11×10^{-6}	3.72×10^{-5}
	10	5.31×10^{-5}	2.10×10^{-5}	1.25×10^{-5}	2.92×10^{-5}	3.28×10^{-5}	2.85×10^{-5}	8.95×10^{-6}	2.08×10^{-5}	5.42×10^{-6}	3.21×10^{-5}
	14	4.53×10^{-5}	1.76×10^{-5}	1.05×10^{-5}	2.52×10^{-5}	3.21×10^{-5}	2.42×10^{-5}	7.34×10^{-6}	1.67×10^{-5}	4.93×10^{-6}	2.66×10^{-5}
	21	3.44×10^{-5}	1.26×10^{-5}	7.20×10^{-6}	2.02×10^{-5}	2.16×10^{-5}	1.72×10^{-5}	4.95×10^{-6}	1.17×10^{-5}	3.31×10^{-6}	1.90×10^{-5}
	28	2.91×10^{-5}	1.00×10^{-5}	5.41×10^{-6}	1.78×10^{-5}	1.64×10^{-5}	1.35×10^{-5}	3.75×10^{-6}	9.10×10^{-6}	2.51×10^{-6}	1.50×10^{-5}
NAP2 (90°C)	1	1.78×10^{-4}	8.75×10^{-5}	5.01×10^{-5}	9.77×10^{-5}	1.21×10^{-5}	7.95×10^{-5}	4.58×10^{-5}	6.98×10^{-5}	1.68×10^{-5}	7.76×10^{-5}
	3	1.12×10^{-4}	5.59×10^{-5}	3.12×10^{-5}	6.34×10^{-5}	4.03×10^{-6}	5.18×10^{-5}	2.90×10^{-5}	4.69×10^{-5}	1.56×10^{-5}	5.05×10^{-5}
	7	6.51×10^{-5}	3.15×10^{-5}	1.60×10^{-5}	3.68×10^{-5}	1.73×10^{-6}	2.85×10^{-5}	1.44×10^{-5}	2.64×10^{-5}	7.78×10^{-6}	2.94×10^{-5}
	10	5.59×10^{-5}	2.67×10^{-5}	1.27×10^{-5}	3.15×10^{-5}	1.69×10^{-5}	2.40×10^{-5}	1.10×10^{-5}	2.24×10^{-5}	6.01×10^{-6}	2.55×10^{-5}
	14	4.77×10^{-5}	2.25×10^{-5}	9.82×10^{-6}	2.67×10^{-5}	2.13×10^{-5}	1.97×10^{-5}	8.38×10^{-6}	1.85×10^{-5}	4.69×10^{-6}	2.14×10^{-5}
	21	3.55×10^{-5}	1.63×10^{-5}	6.81×10^{-6}	2.04×10^{-5}	1.48×10^{-5}	1.41×10^{-5}	5.64×10^{-6}	1.33×10^{-5}	3.16×10^{-6}	1.54×10^{-5}
	28	2.93×10^{-5}	1.33×10^{-5}	5.37×10^{-6}	1.72×10^{-5}	1.12×10^{-5}	1.12×10^{-5}	4.27×10^{-6}	1.06×10^{-5}	2.55×10^{-6}	1.23×10^{-5}

Table A2. The limits of detection of elements by ICP-AS and ICP-MS, in mg/kg of water (10^{-4} wt.%).

Al	Ca	Cr	Fe	La	Mn	Na	Ni	P	Ru	U
0.01	0.01	0.001	0.01	0.001	0.001	0.01	0.001	0.02	0.05	0.05

References

1. IAEA. *Energy, Electricity and Nuclear Power Estimates for the Period. up to 2050*, 2019 ed.; IAEA: Vienna, Austria, 2019; p. 139.
2. NEA. *Nuclear Energy Data*; OECD, NEA: Paris, France, 2019; p. 117.
3. Batorshin, G.S.; Kirillov, S.N.; Smirnov, I.V.; Sarychev, G.A.; Tananaev, I.G.; Fedorova, O.V.; Myasoedov, B.F. Integrated separation of valuable components from technogenic radioactive waste as an option to create a cost-effective nuclear fuel cycle. *Radiat. Saf. Issues* **2015**, *3*, 30–36. (in Russian).
4. IAEA PRIS. The Power Reactor Information System. Available online: https://pris.iaea.org/PRIS/home.aspx (accessed on 6 April 2020).
5. IAEA. *Status and Trends in Spent Fuel and Radioactive Waste Management*; IAEA: Vienna, Austria, 2018; p. 59.
6. Ojovan, M.I.; Lee, W.E.; Kalmykov, S.N. *An. Introduction to Nuclear Waste Immobilisation*, 3rd ed.; Elsevier: Amsterdam, The Netherlands, 2005; p. 497.
7. Phillips, J.R.; Korinny, A.; Depisch, F.; Drace, Z. Update of the INPRO methodology in the area of waste management. *Sustainability* **2018**, *10*, 9. [CrossRef]
8. Diaz-Maurin, F.; Ewing, R.C. Mission Impossible? Socio-technical integration of nuclear waste geological disposal systems. *Sustainability* **2018**, *10*, 39. [CrossRef]
9. Fedorov, Y.S.; Zil'berman, B.Y.; Aloi, A.S.; Puzikov, E.A.; Shadrin, A.Y.; Alyapyshev, M.Y. Problems of modernization of spent nuclear fuel extraction processing. *Russ. J. Gen. Chem.* **2011**, *81*, 1932–1948. [CrossRef]
10. Kopyrin, A.A.; Karelin, A.I.; Karelin, V.A. *Technology for the Production and Radio-Chemical Processing of Nuclear Fuel*; Atomizdat Publishing House CJSC: Moscow, Russia, 2008; p. 576.
11. Vashman, A.A.; Demine, A.V.; Krylova, N.V.; Kushnikov, V.V.; Matyunin, Y.I.; Poluektov, P.P.; Polyakov, A.S.; Teterin, E.G. *Phosphate Glasses with Radioactive Waste*; CNIIatominform: Moscow, Russia, 1997; p. 172.
12. Remizov, M.B.; Kozlov, P.V.; Logunov, M.V.; Koltyshev, V.K.; Korchenkin, K.K. Conceptual and technical solutions for the creation of Mayak plants for vitrification of current and accumulated liquid HLW. *Radiat. Saf. Issues* **2014**, *3*, 17–25.
13. Kozlov, P.V.; Remizov, M.B.; Belanova, E.A.; Vlasova, N.V.; Orlova, V.A.; Martynov, K.V. Modification of the composition of aluminophosphate glasses with HLW simulators to increase their stability. 1. The effect of modifiers on the viscosity and crystallization ability of melts. *Probl. Radiat. Saf.* **2019**, *1*, 3–15. (in Russian).
14. Laverov, N.P.; Yudintsev, S.V.; Kochkin, B.T.; Malkovsky, V.I. The Russian strategy of using crystalline rock as a repository for nuclear waste. *Elements* **2016**, *12*, 253–256. [CrossRef]
15. Kochkin, B.T.; Malkovsky, V.I.; Yudintsev, S.V. *Scientific Basis for Assessing the Safety of the Geological Isolation of Long-Lived Radioactive Waste (Yenisei Project)*; IGEM RAS: Moscow, Russia, 2017; p. 384.
16. Lutze, W. *Radioactive Waste Forms for the Future*; Elsevier Science Publishers, B.V.: Amsterdam, The Netherlands, 1988; pp. 1–160.
17. Donald, I.W. *Waste Immobilisation in Glass and Ceramic Based Hosts*; Wiley: Chichester, UK, 2010.
18. Jantzen, C.M. Development of glass matrices for HLW radioactive wastes. In *Handbook of Advanced Radioactive Waste Conditioning Technologies*; Woodhead: Cambridge, UK, 2011; pp. 230–292.
19. Sobolev, I.A.; Ojovan, M.I.; Scherbatova, T.D.; Batyukhnova, O.G. *Glasses for Radioactive Waste*; Energoatomizdat: Moscow, Russia, 1999.
20. Gin, S.; Jollivet, P.; Tribet, M.; Peuget, S.; Schuller, S. Radionuclides containment in nuclear glasses: An overview. *Radiochim. Acta* **2017**, *105*, 927–959. [CrossRef]
21. Laverov, N.P.; Velichkin, V.I.; Omelyanenko, B.I.; Yudintsev, S.V.; Petrov, V.A.; Bychkov, A.V. *Isolation of Spent Nuclear Materials: Geological and Geochemical Foundations*; IFZ RAS: Moscow, Russia, 2008; p. 280.
22. Ojovan, M.I.; Lee, W.E. *New Developments in Glassy Nuclear Wasteforms*; Nova: New York, NY, USA, 2007; p. 131.
23. Caurant, D.; Loiseau, P.; Majérus, O.; Aubin-Cheval-donnet, V.; Bardez, I.; Quintas, A. *Glasses, Glass-Ceramics and Ceramics for Immobilization of Highly Radioactive Nuclear Wastes*; Nova: New York, NY, USA, 2009; p. 445.

24. Jantzen, C.M.; Ojovan, M.I. On selection of matrix (wasteform) material for higher activity nuclear waste immobilization (Review). *Russ. J. Inorg. Chem.* **2019**, *64*, 1611–1624. [CrossRef]
25. Gin, S.; Abdelouas, A.; Criscenti, L.J.; Ebert, W.L.; Ferrand, K.; Geisler, T.; Harrison, M.T.; Inagaki, Y.; Mitsui, S.; Mueller, K.T.; et al. An international initiative on long-term behavior of high-level nuclear waste glass. *Mater. Today* **2013**, *16*, 243–248. [CrossRef]
26. Ojovan, M.I. On alteration rate renewal stage of nuclear waste glass corrosion. *MRS Adv.* **2020**, *5*, 111–120. [CrossRef]
27. The Fifth National Report of the Russian Federation on the Fulfillment of Obligations Arising from the Joint Convention on the Safety of Spent Fuel Management and on the Safety of Radioactive Waste Management. Moscow. 2017. Available online: http://www.gosnadzor.ru/activity/international/national%20reports/Russian_Federation_rus.pdf (accessed on 6 April 2020). (In Russian)
28. Dorogov, V.I.; Ponizov, A.V.; Khaperskaya, A.V. On the preparation of the fifth national report of the Russian Federation on the fulfillment of obligations arising from the Joint Convention on the Safety of Spent Fuel Management and on the Safety of Radioactive Waste Management. *Radioact. Waste* **2017**, *1*, 100–107. (In Russian)
29. Kryukov, O.V.; Khaperskaya, A.V.; Dorofeev, A.N.; Ferapontov, A.V.; Kudryavtsev, E.G.; Linge, I.I.; Utkin, S.S.; Dorogov, V.I.; Sharafutdinov, R.B.; Ponizov, A.V.; et al. Fulfillment of Russia's obligations under the Joint Convention on the Safety of Spent Fuel Management and on the Safety of Radioactive Waste Management. *Radioact. Waste* **2019**, *1*, 25–36. (In Russian)
30. Kenneth, L.N.; Gregg, J.L. *Advanced Separation Techniques for Nuclear Fuel Reprocessing and Radioactive Waste Treatment*; Woodhead Publishing Limited: Cambridge, UK, 2011; p. 492.
31. State Standard of Russian Federation. *Waste Highly Radioactive Solidified. General Technical Requirements*; GOST P50926-96; Standards Publishing: Moscow, Russia, 2013; p. 5. Available online: http://docs.cntd.ru/document/gost-r-50926-96 (accessed on 6 April 2020).
32. State Standard of Russian Federation. *Radioactive Waste. Determination of Chemical Resistance of Solidified High-Level Waste by Continuous Leaching*; GOST P52126-2003; Standards Publishing: Moscow, Russia, 2003. p. 6. Available online: http://docs.cntd.ru/document/gost-r-52126-2003 (accessed on 6 April 2020).
33. American Society for Testing and Materials. *Standard Test. Method for Static Leaching of Monolithic Waste Forms for Disposal of Radioactive Waste*; ASTM C 1220—98; ASTM International: West Conshohocken, PA, USA, 2004; p. 16.
34. Martynov, K.V.; Nekrasov, A.N.; Kotelnikov, A.R.; Shiryaev, A.A.; Stefanovsky, S.V. Subliquidus phase relations in the low-alumina section of the $Na_2O–Al_2O_3–P_2O_5$ system and the radioactive waste vitrification. *Glas. Phys. Chem.* **2018**, *44*, 591–600. [CrossRef]
35. Remizov, M.B.; Myasoedov, B.F.; Nikonov, B.S.; Stefanovsky, S.V.; Belanova, E.A. Phase composition and structure of molybdenum-, copper- and cesium-containing sodium-aluminum-phosphate glass materials for immobilization of highly active waste from nuclear reactors. *Phys. Chem. Glass* **2014**, *40*, 707–717. [CrossRef]
36. Yudintsev, S.V.; Malkovsky, V.I.; Mokhov, A.V. Hydrothermal transformations in an aluminophosphate glass matrix containing simulators of high-level radioactive wastes. *Dokl. Earth Sci.* **2016**, *468*, 503–506. [CrossRef]
37. Stefanovsky, S.V.; Stefanovsky, O.; Remizov, M.; Kozlov, P.V.; Belanova, E.; Makarovsky, R.; Myasoedov, B. Sodium–aluminum–iron phosphate glasses as legacy high level waste forms. *Prog. Nucl. Energy* **2017**, *94*, 229–234. [CrossRef]
38. Stefanovsky, S.V.; Stefanovsky, O.; Myasoedov, B.; Vinikurov, S.; Danilov, S.S.; Nikonov, B.; Maslakov, K.I.; Teterin, Y.A. The phase composition, structure, and hydrolytic durability of sodium-aluminum-(iron)-phosphate glassy materials doped with lanthanum, cerium, europium, and gadolinium oxides. *J. Non-Cryst. Solids* **2017**, *471*, 421–428. [CrossRef]
39. Malkovsky, V.I.; Yudintsev, S.V.; Mokhov, A.V.; Pervukhina, A.M. Leaching of degraded matrices—Preservatives of highly radioactive waste based on Na-Al-P-glasses. *At. Energy* **2018**, *123*, 3–177.
40. Malkovsky, V.I.; Yudintsev, S.V.; Aleksandrova, E.V. Influence of Na-Al-Fe-P glass alteration in hot non-saturated vapor on leaching of vitrified radioactive wastes in water. *J. Nucl. Mater.* **2018**, *508*, 212–218. [CrossRef]
41. Stefanovsky, S.V.; Prusakov, I.L.; Stefanovsky, O.I.; Kadyko, M.I.; Averin, A.A.; Makarenkov, V.I.; Trigub, A.L.; Nikonov, B.S. The structure of rhenium-containing sodium alumino (iron) phosphate glasses. *Int. J. Appl. Glass Sci.* **2019**, *10*, 479–487. [CrossRef]

42. Stefanovsky, S.V.; Stefanovsky, O.I.; Danilov, S.S.; Kadyko, M.I. Phosphate-based glasses and glass ceramics for immobilization of lanthanides and actinides. *Ceram. Int.* **2019**, *45*, 9331–9338. [CrossRef]
43. Matyunin, Y.I. *The Dissertation for the Degree of Candidate of Chemical Sciences*; VNIINM: Moscow, Russia, 2000.
44. Stefanovsky, S.V.; Stefanovskaya, O.I.; Vinokurov, S.E.; Danilov, S.S.; Myasoedov, B.F. Phase composition, structure, and hydrolytic durability of glasses in the Na_2O-Al_2O_3-(Fe_2O_3)-P_2O_5 system at replacement of Al_2O_3 by Fe_2O_3. *Radiochemistry* **2015**, *57*, 348–355. [CrossRef]
45. Danilov, S.S.; Vinokurov, S.E.; Stefanovsky, S.V.; Myasoedov, B.F. Hydrolytic stability of uranium-containing sodium aluminum (iron) phosphate glasses. *Radiochemistry* **2017**, *59*, 259–263. [CrossRef]
46. Poluektov, P.P.; Schmidt, O.V.; Kascheev, V.A.; Ojovan, M.I. Modelling aqueous corrosion of nuclear waste phosphate glass. *J. Nucl. Mater.* **2017**, *484*, 357–366. [CrossRef]
47. Boldyrev, K.A.; Martynov, K.V.; Kryuchkov, D.V.; Zakharova, E.V.; Ermolaev, V.M. Numerical modeling of leaching of aluminophosphate glass in the batch mode in the presence of bentonite. *Radiochemistry* **2019**, *61*, 612–618. [CrossRef]
48. Van Iseghem, P.; Grambow, B. The long-term corrosion and modeling of two simulated Belgian reference high-level waste glasses. *Res. Soc. Symp. Proc.* **1988**, *176*, 631–639.
49. Frugier, P.; Chave, T.; Gin, S.; Lartigue, J.-E. Application of the GRAAL model to leaching experiments with SON68 nuclear glass in initially pure water. *J. Nucl. Mater.* **2009**, *392*, 552–567. [CrossRef]
50. Ojovan, M.I. Mass spectrometric evidencing on modified random network microstructure and medium range order in silicate glasses. *J. Non-Cryst. Solids* **2016**, *434*, 71–78. [CrossRef]
51. Gin, S.; Ribet, I.; Peuget, S.; Delaye, J.-M. Long-term behavior of glasses. In *Nuclear Waste Conditioning*; CEA: Paris, France, 2009; pp. 51–64.
52. Fournier, M.; Gin, S.; Frugier, P. Resumption of nuclear glass alteration: State of the art. *J. Nucl. Mater.* **2014**, *448*, 348–363. [CrossRef]
53. Gill, P.E.; Murray, W.; Wright, M.H. *Practical Optimization*; Academic Press Inc.: London, UK, 1981; p. 401.
54. Zotov, A.V.; Levin, K.A.; Magazina, L.O.; Mukhamet-Galeev, A.P.; Omelianenko, B.I.; Samotoin, N.D.; Shapovalov, Y.B. Interaction of aluminophosphate glass with water at elevated temperatures. *Geochem. Int.* **1996**, *34*, 805–817.
55. Charnes, A.; Frome, E.L.; Yu, P.L. The equivalence of generalized least squares and maximum likelihood estimates in the exponential family. *J. Am. Stat. Assoc.* **1976**, *71*, 169–171. [CrossRef]
56. Ojovan, M.I.; Pankov, A.S.; Lee, W.E. The ion exchange phase in corrosion of nuclear waste glasses. *J. Nucl. Mater.* **2006**, *358*, 57–68. [CrossRef]
57. Peuget, S.; Tribet, M.; Mougnaud, S.; Miro, S.; Jégou, C. Radiations effects in ISG glass: From structural changes to long-term aqueous behavior. *NPJ Mater. Degrad.* **2018**, *2*, 23. [CrossRef]
58. Sun, K.; Wang, L.M.; Ewing, R.C. Microstructure and chemistry of an aluminophosphate glass waste form under electron beam irradiation. *Mater. Res. Soc. Symp. Proc.* **2004**, *807*, 121–126. [CrossRef]

sustainability

MDPI

Review

On the Sustainable Utilization of Geopolymers for Safe Management of Radioactive Waste: A Review

Esther Phillip [1], Thye Foo Choo [1], Nurul Wahida Ahmad Khairuddin [1] and Rehab O. Abdel Rahman [2,*]

[1] Malaysian Nuclear Agency (Nuclear Malaysia) Bangi, Kajang 43000, Selangor, Malaysia
[2] Hot Laboratory Center, Atomic Energy Authority of Egypt, Cairo 13759, Egypt
[*] Correspondence: alaarehab@yahoo.com; Tel.: +20-10-6140-4462

Abstract: The application of geopolymers for the safe management of radioactive waste has not been implemented on a large scale, where they are tirelessly examined with the purpose of facilitating the practicality and feasibility of the actual application towards the sustainable performance of these materials. This review therefore compiles the findings of the utilization of geopolymers as sorbents for removal of radio-contaminants from aqueous waste streams and as immobilization matrices for the containment of different radioactive wastes. The investigated geopolymer base materials encompass a wide range of reactive aluminosilicate precursor sources that include natural materials, industrial wastes, and chemicals. This work introduces to the reader the scientific interest in the field of geopolymer studies, their sustainability analysis, and their application in the nuclear industry, in particular in radioactive waste treatment and immobilization. The geopolymer classification, radiation stability, and structural characterizations were summarized with special reference to the characterization of the structure alteration due to the inclusion of functional materials or radioactive wastes. The effect of the application of metakaolin-based materials, fly ash-based materials and other base materials, and their blend on radio-contaminant removal from aqueous solutions and the immobilization of different problematic radioactive waste streams were reviewed and analyzed to identify the gaps in the sustainable performance of these materials. Finally, perspectives on geopolymer sustainability are presented, and the identified gaps in sustainable application included the need to investigate new areas of application, e.g., in pretreatment and membrane separation. The reusability and the regeneration of the geopolymer sorbents/exchangers need to be addressed to reduce the material footprints of this application. Moreover, there is a need to develop durability tests and standards based on the record of the application of the geopolymers.

Keywords: geopolymers; radioactive waste; sorbent; immobilization matrices; performance measures

Citation: Phillip, E.; Choo, T.F.; Khairuddin, N.W.A.; Abdel Rahman, R.O. On the Sustainable Utilization of Geopolymers for Safe Management of Radioactive Waste: A Review. *Sustainability* **2023**, *15*, 1117. https://doi.org/10.3390/su15021117

Academic Editor: Changhyun Roh

Received: 17 November 2022
Revised: 27 December 2022
Accepted: 3 January 2023
Published: 6 January 2023

1. Introduction

In nuclear industry, large quantities of radioactive effluents are generated by nuclear power plants [1], hospitals and medical and research laboratories [2], as well as nuclear accidents [2,3].The volumes of the stored effluents are dependent on the size and nature of the national nuclear program, which was reported to fall in the range of few tens to ten of thousands of cubic meters [4]. The main contaminants of concerns in these effluents include, ^{3}H, ^{60}Co, ^{85}Kr, ^{131}I, ^{133}Xe, ^{134}Cs, ^{137}Cs and ^{90}Sr [1].These radioactive effluents need to be managed safely to ensure the protection of human health and to minimize their environmental impacts. The radiological, chemical, physical, and biological characteristics of these effluents are dependent on the generating process, and these characteristics determine the selected waste treatment route that usually necessitates the use of combined treatment technologies. Various chemical treatment technologies are widely employed to reduce the volume of these effluents. Examples of these technologies include membrane separation [5,6], coagulation [7], electrochemical precipitation [8–11], and sorption/ion exchange [12–14]. After the treatment, the treated effluents are discharged or reused according to the adopted procedure at the treatment facility and the exhausted materials and/or produced sludge are managed as radioactive wastes. Among these technologies, sorption/ion exchange is more common and widely used. The studies in this field are directed to develop assorted sorbent materials such as modified clay [15], zeolite [16], synthetic polymer [17], cellulose [18], and geopolymer [19–28] to ensure the effective removal of radionuclides from aqueous solutions. In addition to these effluents, organic liquid wastes are generated due to the operation of nuclear fuel cycle and some medical research facilities and research laboratories. The volumes of these organic liquid wastes are relatively small compared to those of the effluents [29]. These wastes have varying radiological, biological, and chemical characteristics and require special treatment that includes the application of non-destructive, direct immobilization or destructive methods, refs. [29–33]. In non-destructive treatment, the organic content remains intact, and physical changes in the waste properties are targeted. Absorption is widely applied as a non-destructive method to treat spent lubricants and solvents; this process does not aim to reduce the volume of the wastes, but it aims to improve the subsequent immobilization practice [30,31]. Drying and evaporation are non-destructive methods that are applied to reduce the volume of the generated organic solvents, and the resultant concentrates are managed as radioactive wastes. Direct immobilization is used to manage the spent ion-exchangers and organic liquids [30,31,33]. Finally, the destructive treatment of the organic wastes involves chemical changes in the waste that leads to considerable volume reduction in comparison with the previously mentioned methods. These methods include thermal, chemical, and biological treatments; the first, e.g., incineration and plasma treatment, are employed to treat spent solvents and lubricants. The ashes produced from these methods are sent for immobilization, and the off-gases are treated [29–31]. The chemical and biological treatments are also applied for the treatment of the organic liquid wastes, and the resultant sludge is managed as radioactive wastes, and the off gases are treated [29–31].

After treatment, hazardous wastes, including radioactive wastes, are recommended to be immobilized in a suitable matrix to ensure their containment [34].This method involves confining the radioactive waste within a binder material to produce a stable wasteform that complies with specified requirements on their durability. The wasteforms can effectively reduce the mobility of radionuclides by physical encapsulation, sorption, or chemical interaction processes [35]. At present, the immobilization of different radioactive wastes has been achieved using several types of binders, including ceramics [36–39], glasses [40,41], conventional and innovative cement-based materials [31–33,42–63], bitumens [64,65], and polymers [66,67]. Ceramic wasteforms include variable chemical structures, e.g., simple oxides with a fluorite structure; complex oxides; simple silicates; and silicate, phosphate, and aluminate frameworks [36]. They are used to immobilize nuclear wastes. Phosphate and borosilicate glasses are widely employed to immobilize nuclear wastes, and in some cases, they were applied to low- and intermediate-level radioactive waste immobilization [39,40]. Conventional cement-based wasteforms relied on the use of ordinary Portland cement (OPC), with or without additives, as an immobilization matrix, whereas the innovative cements include calcium aluminate cements (CAC), calcium sulfo-aluminate cements (CSAC), magnesium phosphate cements (MPC), and alkali-activated cements (AAC) [31,32]. Cement-based wasteforms are widely used to immobilize different types of low- and intermediate-level radioactive wastes. Bitumens and polymers have been used as immobilizing media for the encapsulation of low- and intermediate-level radioactive wastes; they can be optimized to allow high waste loading and good retention characteristics [31,42].

Geopolymers are inorganic materials that are produced by low-temperature polymerization of an aluminosilicate precursor in an alkaline solution [68]. It was known as soil cement, inorganic polymer, then was named as geopolymer by Joseph Davidovits in 1978 [31]. This relatively innovative class of materials received wide scientific interest to develop various applications in a vast array of industrial sectors. This interest has tremendously increased within the last decade; an unrestricted analysis of Scopus database using the keyword "Geopolymer" was recently conducted on 8 October 2022, showing that there were39,115 published items in that database; nearly two thirds of these works were published in the prior 4 years (Figure 1a). According to the analysis of the data, review papers published in this field represents only 5.98% of the published work. Geopolymers find their applications as civil engineering materials [69], insulation materials [70], coating materials [71], ceramic materials [72], fire-resistance materials [73], catalysts [74], municipal-waste immobilization matrix [75], etc. These applications are supported by the reduced environmental impacts of these materials compared to conventional cement-based materials in terms of reduced energy consumption and CO_2 emissions. In addition, the economy of a geopolymer prepared from waste materials supports the transition from a linear to a circular economy. Subsequently, the sustainability of these materials received considerable scientific attention. Figure 1b illustrates the increasing scientific interest in addressing the sustainability of the geopolymer materials that represents nearly 19.11% of the total research in geopolymer. Most of the published works are research papers that represent the main contributing publication type for evaluating the geopolymers and their sustainability (Figure 1c,d). The review papers have increased contributions to the "geopolymer sustainability" publications compared to that of the "geopolymer" publications.

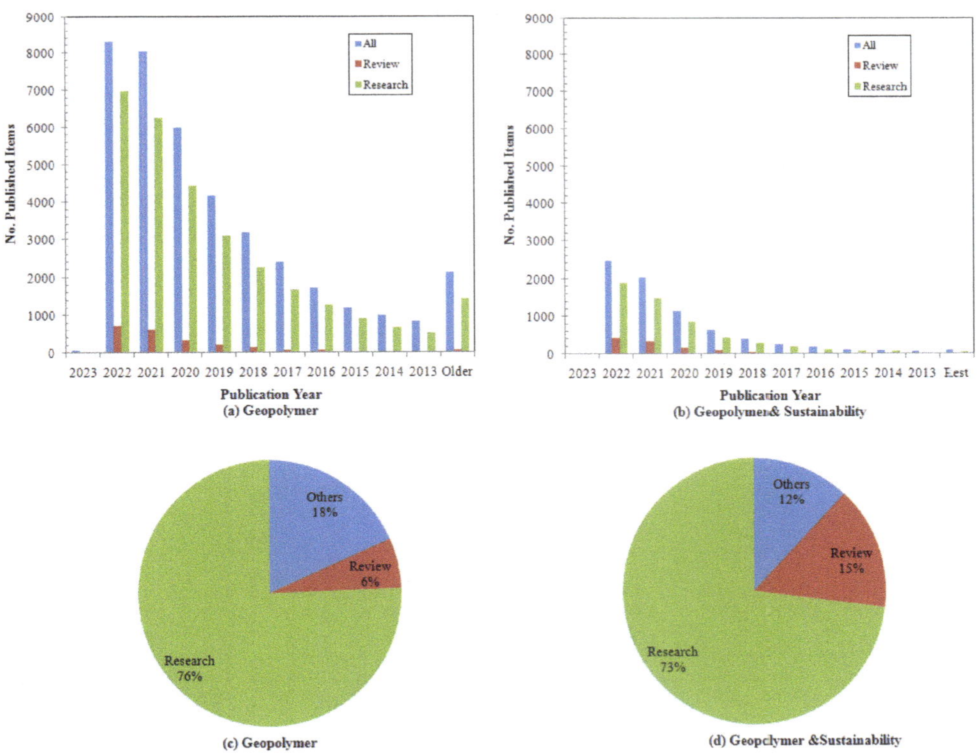

Figure 1. Bibliometric data analysis (**a**,**b**) number of annual scientific publications, (**c**,**d**) relative contribution of the research and review papers to the total published items that addressed geopolymers and geopolymer sustainability, respectively.

In the radioactive-waste-management field, geopolymers have also attracted considerable attention for their applications in the treatment and immobilization of these wastes. This is due to their many advantages, including low cost and simple preparation process, as well as good mechanical properties and thermal and chemical stability. By analyzing the bibliometric data in Scopus database, there are 4460 publications that motioned the word "Geopolymer" AND "Nuclear". Figure 2a (insert) shows the annual distribution of the published work that included both words in comparison with that related to "geopolymers" over that last decade. Both publication fields have exponential growing trends with a relatively higher exponential constant for the general field. The number of the publications that mentioned "Removal" was lower than those mentioned "Immobilization" until 2018, then this trend was reversed. The publications that mentioned "Nuclear" represent nearly 11% of the total geopolymers publications (Figure 2b). Overall of the analyzed data, there is a fairly equal contribution from the publications that mentioned "Removal" and "Immobilization" and from the total publications that motioned "Geopolymer" And "Nuclear" (Figure 2c). The addition of the word "Sustainability" reduces the numbers of the published work by nearly 15%, with a noted exponential increasing trend and with no clear pattern on the relative contribution of "Removal" and "Immobilization" to the annual published items (Figure 2d).

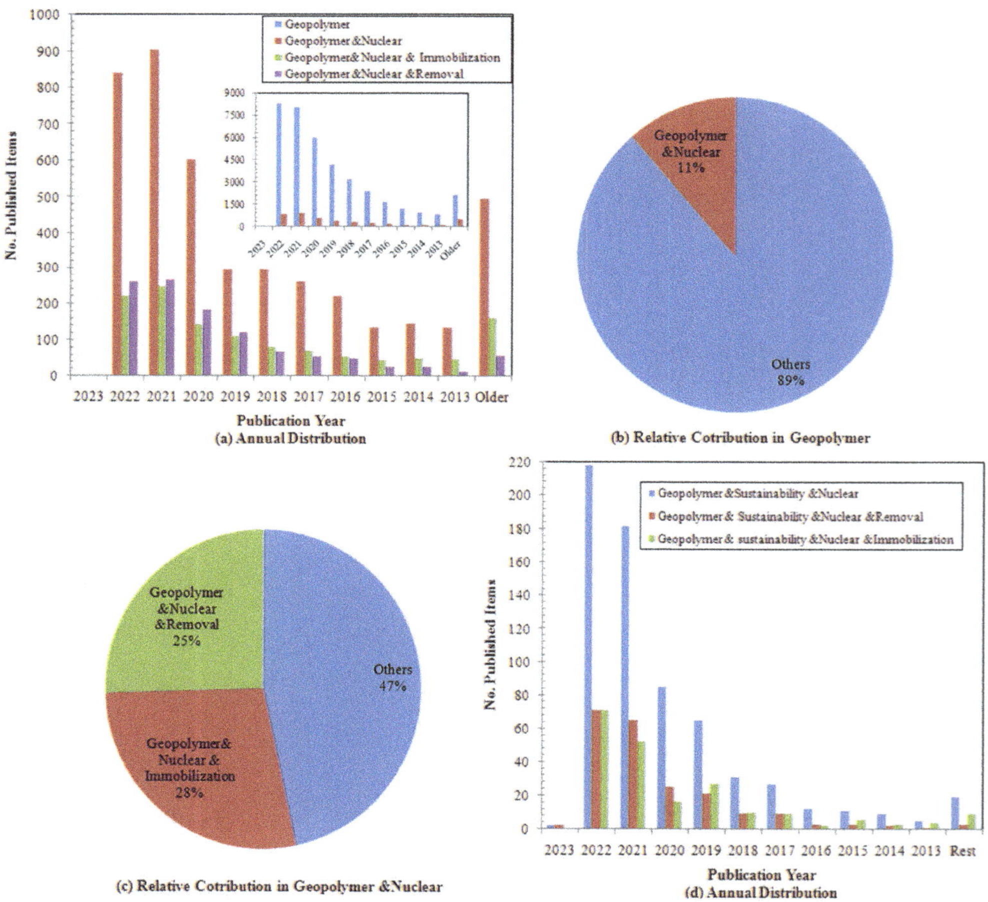

Figure 2. Bibliometric data analysis for publications that mentioned a combination of the words geopolymer, nuclear, removal, and immobilization. (**a**) Annual distribution during the last decade. (**b**) Relative contribution of "Geopolymer" and "Nuclear" to the geopolymer publications. (**c**) Relative contribution of the paper type to the total published work on nuclear and geopolymers. (**d**) Annual distribution of the papers that addressed the sustainability in this field.

The development of geopolymers for various applications has been widely reported in literature, and several comprehensive reviews have been published recently. Recent review papers were directed to assess the geopolymer synthesis, applications, and challenges, but only one addressed the sustainability of the geopolymer in environmental remediation [76–82]. Of these review papers, four papers were directed to cover the use of geopolymer in water and wastewater treatment [76,78–80]. Review papers that addressed the application of geopolymer materials in immobilization covered both the solidification/stabilization and removal applications for heavy metal [83–89], whereas only two review articles addressed it in the context of radioactive wastes [87,89], and one addressed the immobilization of organic liquid wastes in general [90]. However, none of these reviews was directed at discussing the integrated applications of geopolymers in radioactive waste management from the sustainability point of view. Thus, the objective of this paper is to compile and review various geopolymers, aiming to identify the gaps in the current knowledge about the sustainable performance of these materials. In particular, gaps towards the large-scale application of these materials as sorbents and gaps in identifying the factors that affect the durability of the geopolymer immobilization matrices will be addressed. In addition, areas that have not been addressed in radioactive waste management will be identified with reference to similar applications in the non-nuclear wastewater treatment field. In this context, the classification of geopolymers, their radiation stability, and their structure characterizations will be summarized with special reference to the geopolymers that have been tested as sorbent/immobilization matrix for radioactive wastes. Recent advances in testing different geopolymers for their applications as sorbents will be reviewed with a focus on the studied precursors and analyzing their performance in the removal of radio-contaminants from an aqueous solution. Similarly, the recent advances in testing the geopolymer immobilization matrices will be reviewed with a special focus on analyzing the sustainability of their safety function. Finally, a perspective on the sustainability of these materials will be presented.

2. Geopolymers

Geopolymers are a relatively new class of inorganic amorphous materials, which has recently been used in large-scale applications. The main ingredients to prepare geopolymers are reactive aluminosilicate precursor (named here as base material) and activating alkali solutions, e.g., NaOH, KOH, waterglass. The kinetics of geopolymerization to form a three-dimensional network structure of aluminates and silicates tetrahedrons are complex and include: dissolution, speciation equilibrium, gelation, reorganization, and polymerization and hardening [31,91]. The main binding phase in geopolymers is the aluminosilicate gel and is classified, based on its Si/Al ratio, into; poly(si-alate) (Si/Al = 1), poly(sialate-siloxo) (Si/Al = 2), and poly(sialate-disiloxo)(Si/Al = 3) [92–94]. In comparison with zeolite structure, these Si/Al ratios correspond to low (≤ 2) and intermediate ($2 < Si/Al \leq 5$) silica zeolite [95]. The final properties of geopolymers are related to their microstructures, which are strongly dependent on the formulation and the nature of the base materials and the preparation and curing conditions [68]. General properties for these classes are as follows [31,96,97]:

- Si/Al < 1 noted zeolite crystallization is observed in geopolymers;
- 1 < Si/Al < 2 increased polymerization degree, with reduced porosity;
- 2 < Si/Al the polymerization extent is dependent on the solubility of the Si source.

Based on the alkaline nature of the geopolymers and the ability of their structures to be tailored, this class of materials can be employed in various radioactive-waste-management activities as follows:

- Pre-treatment activity: Due to the high buffer capacity of these materials, geopolymers can be used to regulate the pH of the aqueous radioactive waste streams;
- Aqueous effluents treatment activity: Porous geopolymers composites can be used in membrane separation, sorption/ion exchange, filtration, and photocatalytic degradation;

- Immobilization activity: Impermeable geopolymers can be used in the direct immobilization of problematic operational waste streams, e.g., organic liquid wastes, spent ion-exchangers, and evaporated concentrates.

Up to now, scientific efforts have not covered all these applications in radioactive waste management. Most of the conducted work focused on the ability of the geopolymers to remove some radionuclides of concern via sorption/ion exchange or to immobilize the radioactive wastes in geopolymers [20–28,31,45–63,68–74,98–110]. In this section, an overview of the used materials in geopolymers preparation, the effect of the radiation on these materials, and techniques to characterize their structure are presented.

2.1. Base Materials

A large number of materials have been used as base materials (source for silica and alumina) to synthesize geopolymers; these materials can be classified into three groups:

- Natural minerals: These are the most popular structural elements sources for geopolymer synthesis. Calcined kaolin (CK)/metakaolin (MK) have been extensively tested to prepare sorbents [20,24,26–28,100,111] and immobilization matrices [35,45,46,48–50, 52,53,55–62,101,103]. Limited research investigated other minerals, including feldspar (F), bentonite (B), and mordenite (M), for the same purposes [54,62,98,104,105].
- Industrial wastes: Fly ash (FA) is the most widely used waste in the preparation of geopolymer sorbents [21–23,25] and immobilization matrices [47,55,61,98,102,104,106]. Some research used blast furnace slag (BFS) with other materials to prepare sorbents [23–25] and geopolymeric immobilization matrices [50,51,55,61]. Manganese slag (MS) was employed to prepare immobilization matrices [45]. Prior to the utilization of these materials, they should be tested using toxicity characteristics leaching test (TCLP) to ensure that their heavy metal content, if any, is in stable form. Additionally, the amount of the natural occurring radioactive materials in these wastes should be quantified, if suspected.
- Synthetic materials: Chemical sodium silicate and aluminum nitrate solutions have been used to prepare sorbent material to test its potential application in radioactive metal removal from aqueous solutions [110]. In addition, Betol 39T was investigated to prepare geopolymer immobilization matrix [59].

2.2. Effect of Radiation on Geopolymers

Exposure to radiation can lead to various changes in the materials depending on their structures and exposure doses. Chemical changes can occur in materials due to the radiolysis reactions; the radiation chemical yield (G, μmol/J) is used to quantify the extent of these reactions and subsequently, the radiological stability of these materials. It is defined as the number of formed species due to the absorption of 100 eV. Physical changes can occur on the macro- and/or micro-scales, e.g., change in the volume [13], pore number, and structure [51]. These changes can affect the performance of the material and its life time. In order to assess the suitability of geopolymers for their applications in radioactive waste management, their radiological stability and durability should be assessed [112]. Several studies have closely examined the changes in geopolymers under gamma (γ) irradiation. The studied parameters included the radiation chemical yield compressive strength (σ, MPa) as an indication for the changes in mechanical properties and cumulative leach fraction (CLF, cm^{-1}) as an indicator for the stabilization performance of radionuclides within the immobilization matrix; a summary of these studies is shown in Table 1. Limited studies presented the effect of γ-radiation on MK and FA geopolymers by measuring the hydrogen chemical radiation yield under a wide range of exposure doses [101–103]. The hydrogen radiolytic yields were reported in the range of 2.1 to 9.0×10^{-3} μmol/J (Table 1); these values are lower than those reported for OPC cements (~1.0×10^{-2} μmol/J) and pure bulk water (4.7×10^{-2} μmol/J) [105,106]. It should be noted that the yield is highly dependent on the water saturation in the sample and the exposure doses, which should be considered as a factor in these studies [113].

Table 1. Effect of gamma (γ) irradiation on geopolymers: Hydrogen radiolytic yield, compressive strength, and leaching.

Geopolymer	Dose, kGy	Property	Effect	Ref.
GGBFS*/Wollastonite	1000	Compressive strength	Δσ increased by 35%	[51]
MK	50	Hydrogen radiolytic	$G = 6.1 \times 10^{-3}$ µmol/J	[101]
FA	700	Hydrogen radiolytic	$G = 2.1 \times 10^{-3}$ µmol/J	[102]
MK	750	Hydrogen radiolytic	$G = 9.0 \times 10^{-3}$ µmol/J	[103]
	50–1000	Compressive strength	Δσ~10%	
FA	100	Compressive strength	Δσ 7.8%	[106]
		Radionuclide leaching	ΔCLF 5, 22.3 and 47.3%, in DIW **, GW *** & SW ****	

* GGBFS granulated grounded BFS. ** DIW deionized water. *** GW ground water. **** SW seawater.

The results of investigating the effect of γ-radiation on the geopolymer compressive strength confirmed increased compressive strength after irradiation for doses in the range of 50 to 1000 kGy (Table 1) [51,103,106]. A change of about 10% after irradiation was reported at different doses in the range 50 to 1000 kGy for MK-based geopolymer; it was attributed to the densification in the geopolymer network structure [103].For BFS-based geopolymers, a higher increase in the compressive strength of 35% was noted at an irradiation dose equal to 1000 kGy; this value was reduced progressively as the waste loading increased [51]. The increase in compressive strength under γ-radiation was attributed to the decrease in the mean Si–O–Si angle and the decrease in the average pore size of the geopolymer [51]. It should be noted that no significant change occurred due to the irradiation of the Fe-rich geopolymer prepared from synthetic plasma slag up to 5 kGy [114]. The irradiation of the geopolymer prepared from BSF/FA blend, up to 10,214 kGy, revealed that two competing mechanisms are responsible for the changes that occur in the geopolymer during the irradiation. The first is beneficial, enhancing the gelation and cross-linking of the geopolymer, and is dominant at lower irradiation doses. The second is detrimental; it causes structural and micro-structural destabilization and is dominant at higher irradiation doses [115]. The reported improvement in the compressive strength in Table 1 was associated with increases in the CLF, which is a negative change for the stabilization performance. This change in the CLF is highly dependent on the type of the leaching solution (Table 1). However, this increase in the CLF was reported as still complying with the acceptance criteria set by the American Nuclear Society Standards committee [106,116].

2.3. *GeopolymerStructure Characterization*

By default, geopolymers are amorphous aluminosilicate materials that can contain some crystalline phases embedded in it. These phases might be residual from the base material or due to the preparation of the geopolymer composite. In general, there are various techniques that can be employed to quantify the morphology, and the pore and chemical structure of the materials. The morphology is usually identified using microscopy techniques, e.g., optical, secondary electron microscope (SEM), and transmission electron microscope (TEM). Moreover, SEM and TEM can give information about the elemental distribution in the material and its crystal structure, respectively. Figure 3a illustrates the elemental distribution of the structural elements, i.e., Si, O, Al, and the immobilized contaminant, i.e., Cd, in a BFS-based geopolymer [117]. The uniform distribution of the contaminant in the geopolymer is an indication of the effective immobilization [117]. Figure 3b,c illustrates the inclusion of crystalline nano-particles in a MK-based geopolymer, wherein the dimensions of these inclusions can be measured from the TEM image, i.e., Figure 3b, and the d-spacing of the crystalline material can be determined from the electron diffraction pattern, i.e., Figure 3c [118].

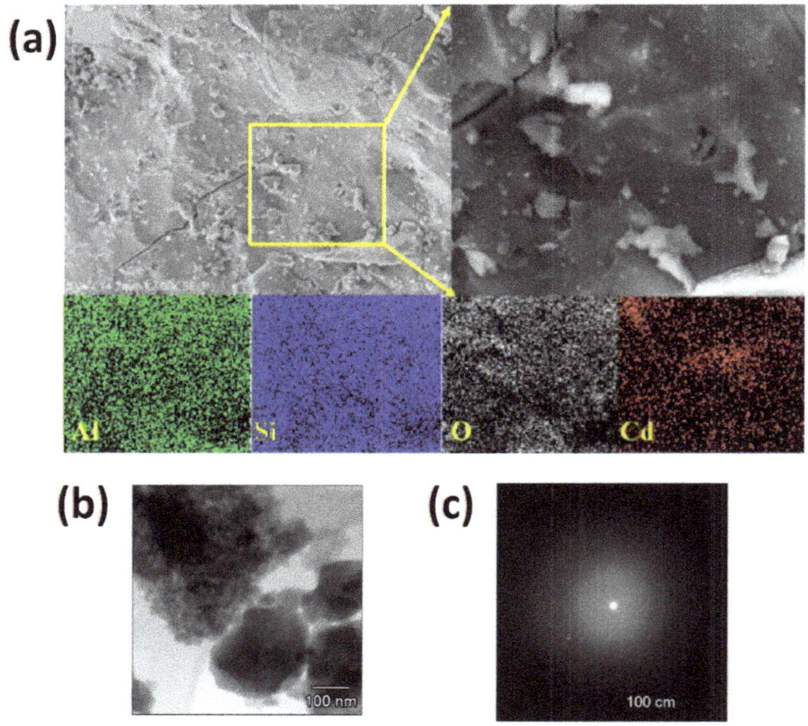

Figure 3. (**a**) SEM micrograph and elemental mapping of the geopolymers with immobilized Cd. (copyrighted, from [117]).(**b**) TEM micrograph of the metakaolin-based geopolymer and (**c**) the electron diffraction showing evidence of crystallinity (copyrighted, from [118] (CC BY 4.0)).

The pore structure is identified using gravimetric and absorption tests, e.g., MIP and BET, and the chemical structure is usually identified via spectroscopic analysis, e.g., XRF, Raman, FTIR, and NMR [112]. The most simple and versatile technique that can be employed for this purpose is Fourier-transform infrared (FTIR) spectroscopy analysis, which is widely used to identify the chemical structure and the changes that occur in it, in a qualitative way, by identifying the function groups and the shifts that occur in them. Typical geopolymer structure is confirmed by detecting the characteristics peaks of OH and M-O, where M is the metal, i.e., Si and/or Al. Figure 4 shows the results of the FTIR and Raman spectroscopy analysis for geopolymers prepared from chemicals (Figure 4a,b) and MK geopolymers with different prefabricated foam ratios (Figure 4c) [100,110]. FTIR spectra show the stretching vibration (near 3500 cm^{-1}) and bending vibration (near 1650 cm^{-1}) of the OH group. The characteristic M–O peaks appear near 1050–1020 cm^{-1} and 450–400 cm^{-1} represents the internal stretching vibration of the SiO_2 tetrahedral and the bending of Si–O–Al. The shift of the peaks between the base material and the geopolymers can give insights into the inclusion of the Al in the Si tetrahedrons and the dissolution of the base material (Figure 4c). In addition, FTIR can clarify the changes in the geopolymers' composite structure in a comparative way, wherein the relative changes in the peak position and magnitude give indications on the changes that occur in the structure (Figure 4a). Raman spectroscopy can also be used in identifying the changes that occur in the crystalline inclusion within the geopolymers matrix. Figure 4b shows the characteristic peak for geopolymers at 500 cm^{-1} and the changes in the crystalline layer houble hydroxide (LDH) due to its inclusion in geopolymers [110].

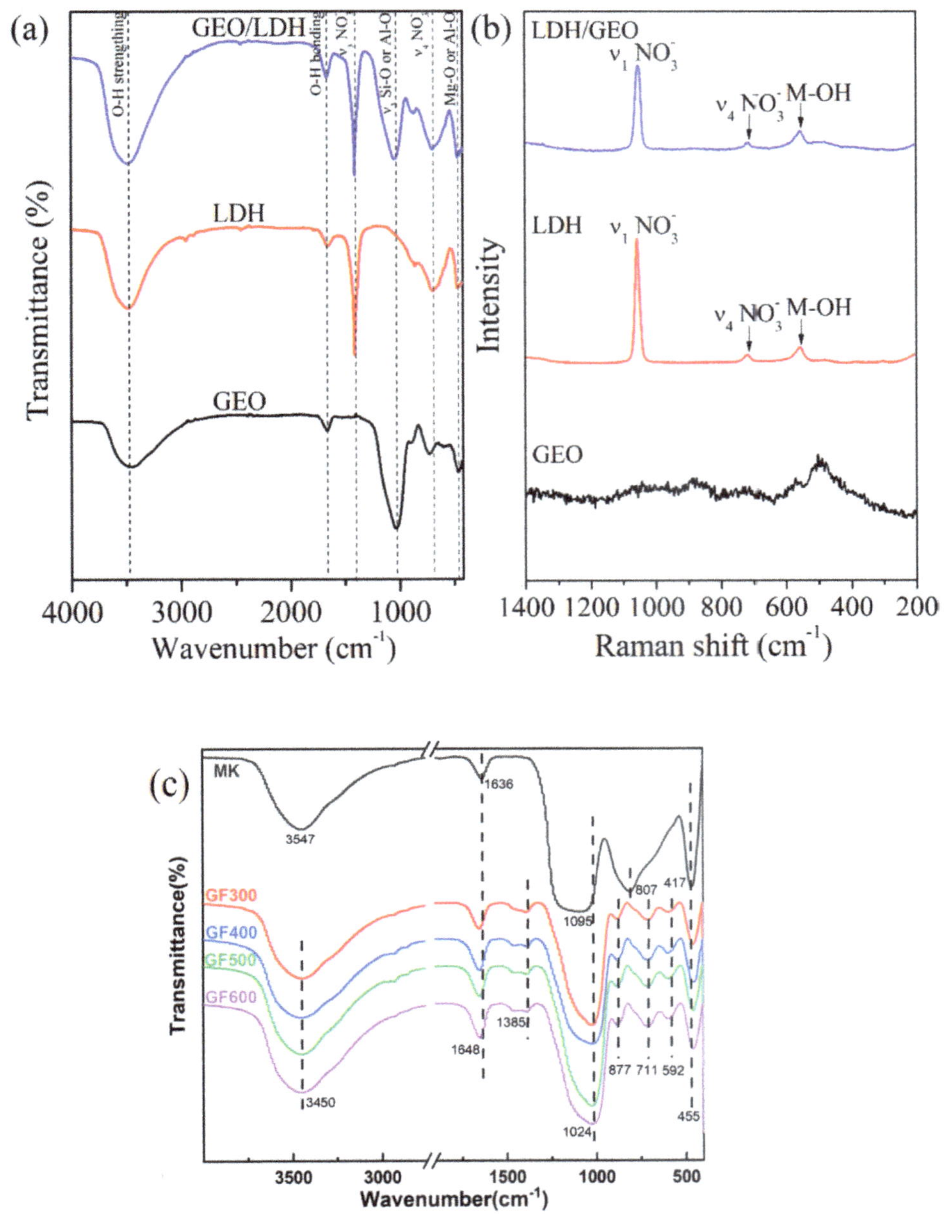

Figure 4. Geopolymer characterization using FTIR and Raman spectroscopy. (**a**,**b**) FTIR and Raman characterization for geopolymers prepared from chemicals and its modification with LDH (copyrighted, from [110]). (**c**) FTIR spectra for MK and geopolymers with prefabricated foam (copyrighted, from [100]).

To check if the prepared geopolymer contain any crystalline phases, either due to the incomplete dissolution of the crystalline base material or due to composite formation, XRD is considered a powerful tool. In general, the XRD pattern of amorphous silicate-based materials shows a diffuse hump that extends over the range 15–35°, and the inclusion of crystalline phases in these materials appears as characteristic Bragg peaks superimposed on the hump [119]. Figure 5 presents the XRD results for geopolymers prepared from different sources and their composites [100,110,120]. The characteristic geopolymer diffuse hump reaches its peak near the quartz characteristic peak (≈27°) (Figure 5a–d). For geopolymers prepared from crystalline-based materials, the shift of the hump peak to higher 2θ values indicates the complete dissolution of the base materials. The characterization of geopolymers composites using XRD gives information on the inclusion of the crystalline phases within the geopolymers matrix and their crystallographic structure. Figure 5a,c shows the inclusion of LDH and sewage sludge ash (SSA) in geopolymers. The comparison between the geopolymer without SSA (Figure 5c) and with 30% SSA (Figure 5d) revealed that the crystalline waste, i.e., SSA, transformed to an amorphous structure during its reaction with the alkali solution [120]. That work concluded that the Cs leachability from the geopolymer matrix is improved over that of the OPC due to the following mechanisms [120]: (1) Cs is librated during the transformation of the crystalline SSA into amorphous structure; (2) the librated Cs is sorbed into the geopolymer matrix.

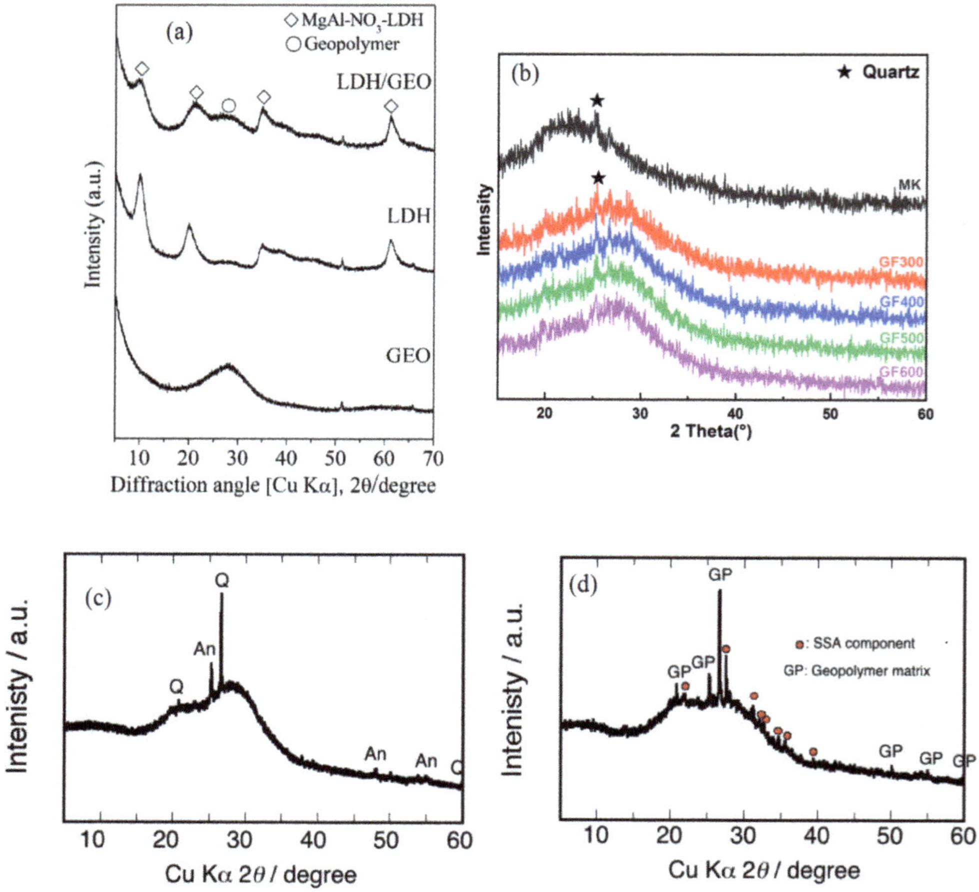

Figure 5. Geopolymer characterization using XRD (**a**) geopolymers prepared from chemicals (GEO), layer double hydroxide (LDH) and LDH—geopolymers composite (LDH/GEO) (Copyrighted, from [110]), (**b**) MK and geopolymers prepared with different inclusions of pre-fabricated foam(copyrighted, from [100]), (**c**) MK Geopolymer (copyrighted, from [120]) (**d**) MK—geopolymers with 30% sewage sludge ash (SSA) (copyrighted from [120]).

3. Geopolymers as Sorbents/Ion Exchangers for Radio-Contaminant Removal

Recent advancements in testing geopolymers for their potential applications in contaminant removal covered a wide spectrum of organic, inorganic, and biological contaminants, e.g., heavy metals, dyes, ammonium, sulfates, and microorganisms, etc, [75,77–79]. These efforts were motivated by the chemical and mechanical stability and the low energy and material footprints of these materials. In addition, the structure of the alumina silicate gel resembles the low and intermediate Si zeolites, which have a good electrostatic field that supports its application in the sorption of polar contaminants [95]. Subsequently, this class of material attracted workers in the field of radioactive waste treatment to test it.

Sorption/ion-exchange techniques are applied in nuclear reactors to control the system chemistry, minimize corrosion or degradation of system components, remove radioactive contaminants, and to clean and decontaminate aqueous streams during the operation of the power plant. Thus, this technique is used to decontaminate the primary coolant, primary effluents, wet storage waters, etc. [121]. In addition, it is employed in the treatment of liquid radioactive wastes generated from research centers, radioisotope production laboratories, uranium mining, and decommissioning activities of nuclear/radioactive facilities. Cs, Sr, Co, and Eu are used widely to test the performance of the cationic sorbents/exchangers for their potential applications in radioactive waste treatment; these elements are selected as models for alkali, alkaline earth, transition, and rare earth elements. In addition, Se, As, and I are used to assess the performance of anionic sorbents/exchangers.

3.1. Types of Studied Geopolymers

The development of geopolymer technology revealed promising applications of geopolymers in the removal of radio-contaminants from aqueous waste streams by means of sorption/ion exchange. Of the many types of geopolymers, MK- and FA-based geopolymers were employed [20–28] since geopolymers were first introduced. The emergence of other types of base materials, e.g., slag and silica fume, has inspired researchers to use blended base materials instead of a single base material. In several studies, the successful testing of geopolymers formulated from blending different types of base materials were reported [20,24,25]. Lei et al. [24] fabricated MK/slag-based zeolite microsphere geopolymers to be used in the removal of Cs^{+} and Sr^{2+} from radioactively contaminated wastewater, and their findings revealed maximum adsorption capacities in batch sorption experiments equal to 103.74 and 54.90 mg/g, respectively. Earlier, Lee et al. [25] reported successful Cs^{+} adsorption onto a hybrid mesoporous geopolymer containing zeolites formulated from FA and slag. This hybrid geopolymer was able to remove more than 90% of Cs^{+}, and the maximum sorption capacity recorded was higher than most other materials including ceiling tiles, walnut shell, and chabazite pellets. The retention of Cs^{+} by geopolymers is benchmarked against ordinary Portland cement (OPC) and frequently shows higher retention compared to OPC. For instance, Jang et al. [22] reported higher Cs adsorption capacity for a FA-slag-based geopolymer (29.87–35.23 mg/g) compared to cement (19.09 mg/g). This result was attributed to the presence of aluminosilicate gel and the C-S-H with a low Ca/Si ratio in the FA/slag-based geopolymer [22,122,123]. These phases contain active sites that boost Cs sorption. In particular, the C-S-H with a low Ca/Si ratio was reported to have increased de-protonated weak sites in the silanol group compared to that in the C-S-H with a high Ca/Si ratio, which favors Cs sorption [124]. In comparison with the studied OPC in that work, the slag-based geopolymer was reported to have C-S-H with a higher Ca/Si ratio [22]. It is remarkable in that work that geopolymers prepared from FA only had better removal performance in terms of adsorption capacity compared to that of FA-slag geopolymers, which was attributed to the effect of increasing the slag on the extent of aluminosilicate gel formation [22].

The incorporation of other materials can provide additional sorption sites, as adopted by Chen et al. [26]; their study showed the increased removal of radioactive iodide (I^{-}) from wastewater when hexa-decyl-tri-methyl-ammonium (HDTMA)-incorporated MK geopolymers were used. Petlitckaia et al. [28] developed a hybrid material with functionalized sorbent (potassium copper hexacyanoferrate, $K_2CuFe(CN)_6$) grafted onto a lightweight MK geopolymer foam to selectively remove Cs^+ from radioactive aqueous waste. Their findings revealed that the grafted geopolymer showed very high selectivity for Cs^+ in the presence of other cations compared to the un-grafted geopolymer. Additional material grafted onto the geopolymer is also useful for the purpose of the co-immobilization of more than one type of radionuclide, as shown by Tian and Sasaki [109], wherein a composite of layered double hydroxide/geopolymer (LDH/GEO) was employed for the adsorption of Cs^+ and ${SeO_4}^{2-}$.

Alkali cations (Na^+, K^+) originating from the hydroxide or silicate activator may affect the sorption performance of geopolymers with favorable radionuclide sorption by a Na-based geopolymer, as indicated by Lee et al. [25] and El-Naggar and Amin [27]. The retention of radionuclides by geopolymers is also affected by the Si/Al ratio, as well as the temperature. A decrease in the pore volume with increasing Si/Al ratio was reported to result in the lower retention of radionuclides [23], whereas the presence of crystalline phases, i.e., nepheline or pollucite, that were formed during the high-temperature treatment of the geopolymer were reported to improve the retention of the radionuclides [23]. In another paper, the calcination of the MK/SA geopolymer was reported to reduce its ability to remove Cs from its aqueous solution [125]. This reduced sorption capacity is explained by the effect of the calcination temperature on the Al- coordination in the geopolymer, wherein a part of the Al tetrahedrons were proven to be transformed to penta- and hexahedrons with the increasing temperature. Moreover, it was found that approximately 10% of the aluminum on the surface of the calcined geopolymer decreased compared to the un-calcinated one.

3.2. Testing Techniques

There are two widely employed methods to test the performance of sorbents/ion exchangers, which are batch and column tests, to simulate the practical conditions during the operation. Batch tests are used to study the effect of the variation in the sorbent mass to the contaminated liquid volume (m/V, g/L), and the contaminated solution characteristics, i.e., the initial contamination level (C_o, ppm), the acidity or alkalinity (pH), and temperature (T, °C), under static conditions on the radio-contaminant removal behavior. Based on the field of application, either removal or separation, different performance measures can be obtained that include percentage removal, distribution coefficient, sorbed amount, and sorption capacity (Q_o, mg/g). In particular, the latter can be used as a base of comparison between the performances of different materials. Batch tests can be run to investigate the unsteady state behavior of the sorption process, i.e., kinetics and the equilibrium behavior. On one hand, the kinetic investigations are conducted to determine the time to reach equilibrium and rate constant and to have insights into the controlling removal mechanism. On the other hand, equilibrium investigations are employed to determine the sorption capacity, have information about the nature of the energy sites, and determine the reaction thermodynamic parameters. Several kinetics and equilibrium models are used to analyze these experimental data, a summary of these models and their features are found elsewhere [95]. Geopolymers prepared from chemicals, MK, FA, MK/slag, and FA/BFS (as indicated above), were tested to check their feasibility to be used in the removal of radio-contaminants; the researchers focused their efforts on studying the batch sorption behavior toward Cs and Sr (cations) and Se and As (anions) in aqueous solutions. Moreover, geopolymer composites with LDH, potassium copper hexacyanoferrate ($K_2CuFe(CN)_6$]), hexa decyl-tri-methyl-ammonium bromide (HDTMA), and iron (Fe) were tested to remove Cs, Sr, Se, I, and As. Tables 2 and 3 list the main studied experimental conditions, the specific surface area (SSA) of the geopolymers, and the main findings of these studies [20–28,98,100,109,110].

Table 2. Feasibility assessment of the use of geopolymers for radio-contaminant removal from aqueous solutions.

Base Material	Studied Experimental Conditions				SSA, m^2/g	Main Findings				Ref.
	Cont.	C_o, ppm	T, °C	m/V, g/L		t_{eq}, min	pH_{Opt}	Capacity, mg/g	Comments	
Chemical	Cs	100	-	1	121	60	8	-	2nd order kinetics Monolayer sorption	[110]
	SeO_4^{2-}	100	-			30	8	-		
F/perlite	Cs	50	24.85–49.85	10	-	60	8	4.28**	2nd order kinetics	[99]
	Eu	50	24.85–49.85			60	4	1.45**		
MK	Cs	100–1000	24.85–59.85	10	18.72	-	-	74.95*	Spontaneous endothermic reaction	[27]
	Cs	20–1000	24.85	1	37.77	120	7	216.1*	2nd order kinetics Monolayer sorption Spontaneous endothermic reaction Reused for 2 cycles	[100]
FA	Cs	85–150	RT	1	-	10	>7	281.74*	2nd order kinetics Monolayer sorption	[98]
	Sr					60		169.07*		
	Cs	100	RT	1	215	10	7	92.63	2nd order kinetics Low Si/Al ratio result in better sorption of Cs+	[23]
MK/Slag	Cs	1000	-	1	77.6	10	-	59.56**	2nd order kinetics Mixture of fly ash and slag reduces the removal performance	[22]
	Sr	1000	-			30	-	54.52**		
	Cs	10–170	25	1.2	23.22	30	>4	103.74*	2nd order kinetics Monolayer sorption Regeneration for 4 and 2 cycles for Cs and Sr, respectively without significant loss***	[24]
	Sr	10–170	25	1.23		60	>4	54.91*		
FA& BFS	Cs	1000	-	1	12.72	30	-	29.22**	2nd order kinetics Mixture of fly ash and slag reduces the removal performance	[22]
	Sr	1000	-			30	-	44.64**		
	Cs	10–150	-	10	114.16	40	4	15.24*	Multilayer sorption	[25]

* Langmuir mono sorption capacity. ** Highest sorbed amount from kinetic only. *** studied column.

Table 3. Feasibility assessment of the use of geopolymers composites for radio- contaminant removal from aqueous solutions.

Base Material	Studied Experimental Conditions				SSA, m^2/g	Main Findings				Ref.
	Cont.	C_o, ppm	T, °C	m/V, g/L		t_{eq}, min	pH_{Opt}	Capacity, mg/g	Comments	
Chemical/LDH	Cs	100	-	1	134.1	120	8	84.14	2nd order kinetics Monolayer sorption	[110]
	SeO_4^{2-}	100	-			5	8	71.3		
MK/$K_2CuFe(CN)_6$]	Cs	3–1000	RT	1	35	4–5	-	250–175***	The material is very selective for Cs.	[28]
MK/HDTMA	I	250	24.85	1		180	>7	36.1**	2nd order kinetics Multilayer spontaneous and exothermic process	[26]
FA/Fe	Cs	100	RT	1	107.9	10	7	111.9*	2nd order kinetics Cs^+ & Sr^{2+} monolayer sorption AsO_4 is multilayer sorption.	[21]
	Sr	100				30	7	14.19*		
	AsO_4^{2-}	50				150	5	21.51*		
MK-FA	Cs	27	25	2	-	120–240	7–7.5	113.3	The adsorption of geopolymer on Sr^{2+}, Co^{2+}, and Cs^+ is mainly chemical adsorption.	[20]
	Sr	18			-			85.7		
	Co	12			-			58.8		

* Langmuir mono sorption capacity. ** experimental. *** studied desorption.

Column tests are used to investigate the sorption dynamics, optimize the column parameters, i.e., bed depth, flow rate, and identify the breakthrough curve characteristics. The application of the column technique in studying the removal performance of the radio-contaminants using geopolymer was limitedly addressed [24]. Cs and Sr removal using a MK/slag geopolymer was tested in a fixed-bed column with bed depth equal to 0.5 and 1 cm at two flow rates equal to1 and 4 mL/min [24]. The breakthrough curve characteristics for the highest bed-depth and slowest flow-rate experiments were as follows: (1) the breakthrough points were 12 and 3.3 h; (2) the saturation points were 26 and 18 h; (3) the column adsorption capacities were 121.1 and 58.73 mg/g, respectively [24]. A recent paper studied Cs removal using a MK/sodium lauryl sulfate (SLS) geopolymer cylinder with diameter and height equal 0.5 and 0.7 cm, respectively, under varying flow rate of 20–50 mL/s [126]. The maximum adsorption capacities were reported to be 10.3–13 mg/g. In addition, the reusability and regeneration ability of the ion-exchanger/sorbent is an important topic to be identified to ensure the economic feasibility of the materials and to reduce the environmental footprint by reducing the material requirements for the treatment process [95]. The reusability and regeneration studies are very limited for geopolymer sorbents/ion exchangers. From the presented data herein on geopolymer testing for applications in radio-contaminant removal, the following remarks can be drawn:

- In terms of the number of conducted batch experiments, these experiments can provide a basis for evaluating the performance of the studied geopolymers, whereas the column and reusability and regeneration studies are lacking;
- The studied batch experimental data covered a wide range of initial contaminant concentrations in the range (3–1000 ppm) tested mostly at room temperature, except in two papers [27,99], and the sorbent-mass-to-liquid-volume ratio was in the range 1–10 g/L;
- The sorption data follow the pseudo-second-order reaction model, which shows that the reaction has a chemisorption nature that involves electron-sharing between the contaminants and the sorbent;
- For most of the sorption equilibrium batch tests, it was found that the sorption occurs on sites of equal energy, i.e., monolayer sorption, with exceptions for Cs removal using (FA/BFS) geopolymers and I and AsO_4^{2-}removal by MK/HDTMA and Fa/Fe geopolymers, respectively;
- The conducted thermodynamic studies indicated that the reactions were mainly spontaneous and endothermic, except for the removal of I using MK/HDTMA geopolymers.

4. Geopolymers for the Immobilization of Radioactive Wastes

Using a geopolymer as a containment matrix for the immobilization of radio-contaminants involves the mixing of radioactive waste with a reactive base material (such as MK, FA) and/or an activating solution containing alkali (Na^+ or K^+) hydroxides and silicates, then applying suitable curing conditions [127]. This process aims, as in the case of other immobilization matrices, to produce an acceptable wasteform that can comply with the regulatory requirements on radionuclide retention, water ingression, and structural stability provision in near-surface disposal facilities. The immobilization is achieved by solidification, embedding, or encapsulation [31]. The first is usually used to describe the immobilization of liquid and liquid-like wastes, and it is achieved through the chemical incorporation of the waste components into the structure of a suitable matrix. The latter is achieved by physically surrounding the waste in the immobilization matrix [108]. Basically, the radionuclides leachability is used as a performance measure to quantify the retention safety function. It is affected by the characteristics of the containment matrix, the radionuclide being leached, and the leaching environment [128]. Compressive strength and permeability are used as performance measures to quantify the provision of structural stability and for the prevention of water ingression, respectively.

4.1. MK Based Geopolymer

Geopolymers have been investigated for the containment of radionuclides in low- and intermediate-level wastes (LILW) [33,129] that usually contain long-lived radionuclides, such as ^{137}Cs, ^{90}Sr, ^{226}Ra, ^{232}Th, and ^{238}U. Among the many radionuclides, Cs and Sr have been expansively studied due to their unfavorable immobilization by conventional cement-based materials. As for removal applications, MK geopolymers are frequently investigated by numerous researchers for radioactive waste immobilization [47,49,57,130–136]. Examples of these investigations include the immobilization of solid wastes generated from the nuclear fuel cycle that encompasses operational and decommissioning wastes from operating and decommissioned nuclear reactors, respectively [33]. Operational wastes such as graphite containing ^{14}C has been successfully conditioned using geopolymers to produce a wasteform with acceptable compressive strength and structural stability [60]. Another study addressed the immobilization of fuel cladding in MK geopolymers [101]. Besides solid waste, organic liquid wastes generated from nuclear reactors, such as lubricating oil contaminated with ^{60}Co and ^{137}Cs, have also been conditioned using MK geopolymers [62]. Moreover, secondary wastes, which are generated from the treatment of primary wastes, e.g., exhausted filters and ion exchangers/sorbents, have been tested for their potential immobilization in MK geopolymers. As in the case of the Fukushima Daiichi Accident, the feasibility of immobilizing spent ion-exchange resins containing ^{137}Cs and ^{90}Sr that resulted from the treatment of contaminated cooling water in a MK geopolymer was investigated. The immobilization of Sr-loaded titanate ion-exchangers in a MK geopolymer has proven the potential of the geopolymer as confinement matrix for this exhausted ion exchanger [134]. Similar findings were observed by Kuenzel et al. [135] for the immobilization of zeolite clinoptilolite ion exchangers contaminated with Cs and Sr. Walkley et al. [56] also observed the immobilization of ^{90}Sr in ion-exchange resins by using MK geopolymers. One of the primary challenges in the immobilization of radioactive waste is the high sulfate content in some waste streams. Ahn et al. [35], in their study, have proven the applicability of using a MK geopolymer in immobilizing a high-sulfate hybrid sludge from a Hydrazine Based Reductive Metal Ion Decontamination (HyBRID) process that contained Fe, Ni, Cr, and Co ions with increased waste loading up to 53.8 wt%.

4.2. Other Geopolymers

Other than MK geopolymers, FA geopolymers have also been investigated for the immobilization of different radio-contaminants, such as Cs^+ [104]. Slag-based geopolymers have also been investigated for the immobilization of radio-contaminants in many radioactive waste streams. The advantages of geopolymers not only lie in their favorable mechanical strength and radionuclides stabilization potential but also in their waste loading capacity. For instance, a geopolymer prepared from GGBFS was applied successfully in the immobilization of ion-exchange resins contaminated with Cs^+ and Sr^{2+} with maximum solidified wet resin content of about 45 wt% [51].

Blended materials were tested for their potential application to enhance the radio-contaminant retention, waste loading capacity, and mechanical strength of an immobilization matrix. As an example, the findings of Lin et al. [50] indicated the superior performance of geopolymer compared to cement when a blended MK-slag-based geopolymer was used for the immobilization of reactor spent resins containing Cs^+ and Sr^{2+} with a loading capacity of ion-exchange resins up to 12 wt% (wet base). Similar findings were reported by El-Naggar [109] for the immobilization of ^{60}Co using blended slag-seeded Egyptian Sinai kaolin geopolymer. As in the case of the well-known Hanford Project in the USA, the performance of the BFS-MK DuraLith geopolymer was enhanced, e.g., improved workability, reduced hydration heat, and higher waste loading, by the addition of fly ash [61].

Frequently, additional materials are incorporated into a geopolymer's formulation to either increase its radionuclides immobilization performance or to provide selective containment. These materials provide structure that creates extra fixation sites for radionuclides [137] or phases that favor the attachment of ions [138]. Besides the containment of radionuclides, they may contribute to the enhancement of the mechanical strength of the geopolymer [51]. Moreover, these materials can cause changes to the chemical state of the radionuclides that further affect the performance of the geopolymers. Yu et al. [45] mentioned that, in the presence of Mn, Co was transformed from divalent to trivalent in an oxidation environment.

4.3. Effect of Alkali Activator and Thermal Treatment

In terms of material types, another consideration is the type of the used alkali activators. In many studies, Na-based geopolymers showed better performance than K-based geopolymers [49,132,134]. This preference was attributed to the hard and soft acids and bases (HSAB) principles, in which high-charge-density Na^+ resulted in Na^+ being the stronger Lewis acid that favored reaction with Cs^+ [134]. The incorporation of radionuclides into the geopolymer framework occurs through several mechanisms. One of the important mechanisms involved the replacement of alkali ions (Na^+ and/or K^+) by the radionuclide [56,134] with preference towards a Na-based geopolymer compared to a K-based geopolymer [48]. This replacement can cause structural changes to the geopolymers.

Apart from the types of used materials, the radionuclide containment performance of geopolymers is also affected by the thermal treatment of the geopolymer's matrix. Generally, geopolymers are thermally stable in a wide range of temperatures up to approximately 800 °C [139]. By exceeding this point up to around 1100 °C, crystalline phases are formed; nepheline forms in Na-based geopolymers; leucite forms in K-based geopolymers, and pollucite forms in the presence of Cs [48]. These crystalline phases have been shown to immobilize this radionuclide. The investigation of MK-based geopolymers formulated from the Na-alkali activator for Sr immobilization at different temperatures revealed lower Sr^{2+}leaching in the presence of nepheline structures formed by calcination in comparison to the uncalcinated geopolymer [135]. Few publications, however, noted the formation of these crystalline phases at lower temperatures [46,48], and the crystalline phases formed at these temperatures demonstrated better performance than the ones produced at high temperature. For instance, pollucite obtained at lower temperature via alkali metal ions doping and optimizing the Na/Cs ratio demonstrated high Cs immobilization compared to pollucite formed by a high-temperature hydrothermal treatment [140].

As much as geopolymer applications can, in certain cases, lead to better mechanical strength and durability compared to Portland cement, its dry shrinkage is relatively higher and therefore prone to cracking. Additionally, the geopolymer has also been found to have low affinity for anions, such as $SeO_4{}^{2-}$ and $Cr_2O_7{}^{2-}$ [141],which eventually hinders its application in immobilizing waste streams containing these contaminants. In order to rectify these shortcomings, the optimization of the alkali activator and the use of additive material can provide successful solutions for this problem [58,110].

Even though the application of a geopolymer in the immobilization of radioactive waste still remains largely in research stage, nevertheless the actual implementation has been adopted by the Slovak Republic. The immobilization of intermediate-level waste containing ^{137}Cs from nuclear power reactors (sludge, resin, liquid wastes, and their mixtures) in Slovakia and Czech Republic was successfully implemented using a proprietary MK-based geopolymer matrix called SIAL [142].

4.4. Geopolymers Performance

As indicated above, the radioactive wasteform should be designed to comply with the regulatory requirements on its safety functions, so they should be durable and able to mitigate the impact of anticipated accidents that can occur during its life time. The durability of the wasteform is affected by the waste compositions, amount of free water content, and the presence of environmental stressors that subsequently affect the characteristics of the wasteform (i.e., porosity, density, thermal and radiation stability, compressive strength, leaching resistance, chemical attack resistance, and freeze and thaw) [57,112,113,143–148]. It should be noted that the durability of the geopolymer under chemical attack is very much dependent on the calcium content in the base material. It was reported that the high-calcium-content base material, i.e., class F fly ash, will produce a geopolymer more vulnerable to sulfate attack and carbonation than that withlow calcium content [148]. Several durability tests were developed to test the sustainability of the safety function of the wasteforms throughout their life cycle [31,112,145–147]. In this section, the testing techniques and the analysis of the leaching behavior and compressive strength of the geopolymer wasteforms are presented.

4.4.1. Testing Techniques

During the design of the wasteforms, there are several options can be considered for testing these forms to ensure their sustainable performance. These options include the selection of the testing procedure, the factors that affect the wasteform performance, and the optimization technique. Details of these factors are found elsewhere [33].Geopolymer studies addressed the sustainable performance of these wasteforms by evaluating the effect of geopolymer formulation on the radionuclide retention and compressive strength. In addition, the effects of the leaching solution on the wasteform stability, in terms of radionuclide leachability, were investigated. The effect of the freeze–thaw cycles, irradiation, and water immersion on the leaching and the compressive strength were quantified. The following subsections present the results of these investigations for different geopolymer types.

4.4.2. Leaching Behavior of Geopolymer Wasteforms

The high pH environment of the cement-based material favors the stabilization of lanthanides and actinides within the immobilization matrix, yet alkali and alkaline metal remain substantially soluble depending on the waste constituents and the presence of additives [33].Subsequently, most of the geopolymer studies were directed at assessing the potential of these materials to immobilize different primary and secondary waste streams contaminated with Cs and Sr and, to a lesser extent, Co, as mentioned in the previous sections [35,45–49,53–55,57,58,62,76,106]. In addition, the stabilization of some anions was addressed [58]. Most of the studied geopolymer were MK-based, with few tests for other geopolymers.

Several types of leaching procedures were employed in this respect, including the standardized ANSI/ANS, TCLP, and ASTM. The duration of the leaching-test applications varied from very-short-duration tests, i.e., 5 days [51], to longer-duration tests, i.e., 42 days [47,49,57]. The obtained leaching parameters were also very variable, including Cumulative Leach Fractions (CLF, cm^{-1}), Normalized Leach Fraction (NLR, $g.m^{-2}.d^{-1}$), Leach Rate (LR, cm. d^{-1}), Leaching index (Li), and Inhabitation grade (I, %). Tables 4 and 5 list the compositions of the studied immobilization matrices and the performed leaching tests and their results for Mk-based and other geopolymers, respectively. Despite the reported values of the leaching index providing insights into the acceptability of the wasteform performance, the variability of the reported leaching measures and the conducted leaching procedures inhibit the determination of a suitable geopolymer formulation to immobilize certain type of radioactive wastes.

Over the years, the utilization of geopolymers as a containment matrix has gained lots of attention, especially due to their improved performance in terms of thermal stability (at high temperature and during freeze–thaw cycles), acid resistance, mechanical strength, and radionuclide containment performance [47]. Findings from the study carried out by Liu et al. [55] suggested that the immobilization performance of the blended FA/slag/MK-based geopolymer exceeded that of cement as shown by the higher cumulative fraction leaching rate of cement compared to that of the geopolymer. In addition, the findings of Jang et al. [144] for the FA/slag-based geopolymer also showed better Cs- and Sr-containing radioactive-waste-immobilization performance in comparison to Portland cement and therefore proposing the potential of these geopolymers as promising barrier materials to retard the migration of radio-contaminants.

Table 4. Containment performance for metakaolin and metakaolin-blend geopolymers.

Studied Immobilization Matrix				Matrix Leaching Studies and Results			Ref.
Base Material	Activators	Radionuclides Simulant/Waste	Waste Loading,%	Leach Test	Leaching Measure	Findings	
MK	SiO_2, NaOH, or KOH, or LiOH	$CsOH{\cdot}H_2O$	14–20	ANSI/ANS 16.1	Li = 8.93–12.66 [a] NLR = 2.51×10^{-4} gm^{-2}/d	Cs effectively immobilized in pollucite at ≤1000 °C.	[46]
	Na_2SiO_3, SiO_2 and NaOH		2–18		Li = 8.93–12.66 [a]		
	Silica sol gel and NaOH	Cs OH	6–30	ANSI/ANS-16.1	NLR = 1.14×10^{-3} gm^{-2}/d	Hydrothermal treatment increases the performance and compressive strength.	[48]
	Sol gel NaOH	$CsNO_3$	3.52	Leaching for 42 d	CLR < 1%	The Na-based geopolymer showed a lower leaching rate than the K-based geopolymer.	[49]
		$Sr(NO_3)_2$	5.82		CLR < 1%		
	KOH	^{152}Eu	-	Leaching for 24 d	I = 98.9%	The radionuclides were not leached in water, even after the fine pulverization of samples, but remained in the geopolymer matrices.	[53]
		^{134}Cs	-		I = 97.7%		
		^{60}Co	-		I = 99.0%		
		^{59}Fe	-		I = 99.0%		
	Water glass & H_2O 17.6	Zeolite-loaded Sr	29.4	Leaching for 42	CLF = 1.8×10^{-3} cm^{-1}	Has better leaching resistance than those of cement in different leaching solutions	[57]
	NaOH, KOH, Fumed silica, DIW	Sulfate ions in sludge	0–40%	-	CFL < 1.0%	-	[35]
	Sodium silicate NaOH	Heavy metals (Th(IV), U(VI), Pb(II), Cd(II), Cu(II))	-	Leaching for 24 h	LC: Deionized water = 11% 1 M HCl = 8% 0.1 M NaCl = 4.6% 1.0 M NaCl = 3.4% 0.1 M NaOH = 5.7%	The MK-based geopolymer is very effective in the stabilization of heavy metal ions.	[132]
	Na silicate	Sr	-	TCLP	Leaching rate: Deionized water at 1200 °C = 5.82×10^{-7} gm^{-2}/d Simulated seawater at 1200 °C = 4.64×10^{-6} gm^{-2}/d	Low leaching is achieved at higher temperature (1200 °C) due to the immobilization of Sr in nepheline structures.	[134]

Table 4. *Cont.*

Studied Immobilization Matrix				Matrix Leaching Studies and Results			Ref.
Base Material	Activators	Radionuclides Simulant/Waste	Waste Loading,%	Leach Test	Leaching Measure	Findings	
MK/MS	Water glass & NaOH	$CoCl_2$	5.56	TCLP	LC = 0.20%	Has higher acid-leaching resistance compared to the MK geopolymer	[45]
MK/ Hydrotalcite/ SF	Sodium silicate NaOH	SeO_3^{2-}	2	TCLP	LC = 10%	Na2SiO3-activated geopolymers have better leaching performance than those of NaOH-activated geopolymers.	[58]
		SeO_4^{2-}			-		
FA/ BFS/ MK/Sand/SF	The waste and NaOH KOH	Re	-	TCLP	LC = 0.65 mg/L	-	[61]
MK/B	Sodium silicate NaOH	Oil contaminated by Co	15.5–25	ASTM C130A for 12 day	LR = 8.5×10^{-5} cm/day	Leaching rate complied with the Brazilian regulations	[62]

[a] after thermal treatment to 1000C. SF = silica fume.

Table 5. Containment performance for fly ash and fly ash blend geopolymers.

Studied Immobilization Matrix				Matrix Leaching Studies and Results			Ref.
Base Material	Activators	Radionuclides Simulant/Waste	Waste Loading,%	Leach Test	Leaching Measure	Findings	
FA	Sodium silicate and NaOH	$CsNO_3$	1.46	Leaching for 42 d	CFL = ~9×10^{-3} cm^{-1}	-	[47]
GGBFS/ Wollastonite	NaOH	Cs*	32%	Leaching for 5 days	CFL = 0.152 cm^{-1}	-	[51]
		Sr*			CFL = 9.72×10^{-4} cm^{-1}		
B/wood ash	NaOH	Sr	-	Leaching for 28 Day	NLR = 10−6 gm^{-2}/d	The clay-based geopolymer shows better Sr immobilization than that of OPC.	[54]
FA/Slag/MK	-	Sr	-	-	CFL = 1.1×10^{-3} cm^{-1}	FA/slag/MK has improved immobilization performance over that of OPC.	[55]
FA	Sodium silicate and aOH	$CsNO_3$	2	ANSI/ANS 16 for 40 days	Li = 8.7–10.7	-	[105]

* ion-exchange resin purolite NRW-10 and purolite NRW-4004.4.3 compressive strength of geopolymer wasteforms.

The compressive strength of the wasteforms is another key parameter that must meet the minimum criteria set by various regulatory bodies. The U.S. Nuclear Regulatory Commission (NRC) recommends a mean compressive strength of at least 500 psi (3.45 MPa) for wasteform specimens cured for a minimum of 28 days (ASTM C39/C39M-01). According to the standard GB 14569.1-2011 set by the National Standards of the People's Republic of China, which regulates the performance requirements for solidified wasteforms of low- and intermediate-level radioactive wastes, the compressive strength of the solidified sample should not be less than 7 MPa. Meanwhile, a minimum compressive strength of 4.9 MPa for low- and intermediate-level radioactive cementitious wasteforms are set by the Russian Federation (GOST R 51883–2002). Table 6 shows that the compressive strength of the geopolymer wasteforms reported in this review not only met the minimum criteria specified in the standards but also most of them exceeded the criteria by multiple folds.

Table 6. The compressive strength of the geopolymer wasteforms reported in the literature.

Geopolymer	Curing Conditions		Type of Simulant/Waste	Waste Loading (wt%)	σ (MPa)	Ref.
	Temp. (°C)	Duration (Day)				
MK	25	7	Ba-loaded sludge waste	40.0	49.6	[35]
	60	2	Cs(OH) solution	47.7	65.8	[46]
	25	28	Sr-loaded zeolite	29.4	37.6	[57]
	25	28	Na_2SeO_3powder	2.45	30.0	[58]
	20	30	Nuclear graphite	10.0	22.0	[60]
	RT	28	Co-loaded bentonite	15.5–25	9.5 ± 0.9	[62]
Clay-based	RT	28	Sr-loaded wood ash	57.0	12.7	[54]
BFS	RT	28	Cs,Sr-loaded ion-exchange resins	5–45	10.2–22	[51]
FA/SF	60	28	$^{133}Cs^+$solution	2.0	57.2	[47]
MK/BFS	RT	28	Cs,Sr-loaded ion-exchange resins	12.0	13.6	[50]
	25	28	$Sr(NO_3)_2$ powder	9.0	24.5	[55]
FA/slag/Mk	RT	28	Re-loaded waste solution	26.8	57.5–121.7	[61]

The compressive strength of a wasteform can be affected by the chemical and physical properties, as well as the proportions of the radioactive wastes. Table 6 shows that the geopolymer wasteforms produced from liquid wastes [46,47,61] have higher compressive strength than the others. This better performance of liquid waste solidification is attributed to the ease of liquid-waste incorporation into the geopolymer slurries to form homogenous wasteforms after setting. In contrast, the insoluble solid wastes are encapsulated in the geopolymer slurries to form heterogeneous wasteforms. For example, it can be seen that the geopolymer wasteforms produced from spent ion-exchange resins [50,51] have relatively low compressive strength. The insoluble ion-exchange resins have weak contact with the geopolymer matrix, which, when under load, can easily lead to waste–matrix debonding, thus weakening the strength of the wasteforms.

The effect of the waste loading on the compressive strength of wasteforms was briefly investigated in the studies of Ahn et al. [35], Lin et al. [50], Lee et al. [51], and Liu et al. [55]. As for a comparison, the results from these studies are replotted with correlation coefficient values (R^2) in Figure 6. In general, the waste loading influences the compressive strength of the wasteforms inversely. For example, strong and intermediate negative correlations are shown in the geopolymer wasteforms produced with spent ion-exchange resins [50,51] and $Sr(NO_3)_2$ powder [55]. However, a weak correlation is shown in the geopolymer wasteforms with sludge waste, wherein the compressive strength initially decreased but increased at waste loading of 30 wt% and 40 wt%. [35]. This behavior was attributed, according to Ahn et al. [35],to the effect of the increased sludge waste loading on the

H_2O/Al ratio in the geopolymer matrix, which led to a recovery of the compressive strength.

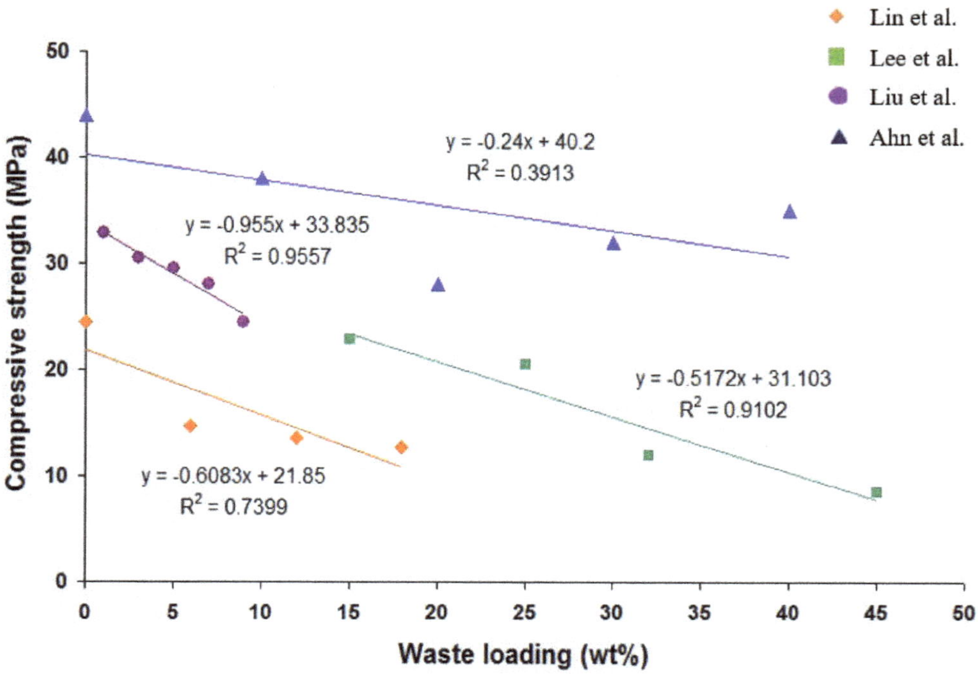

Figure 6. Compressive strength correlations with waste loading [35,50,51,55].

As for the type of binder matrix, there are no significant relationships between the geopolymer raw materials and wasteforms' compressive strength that can be observed in Table 6. However, Li et al. [47], Lee et al. [51], and Xu et al. [57], in their studies, have compared the compressive strength of geopolymer wasteforms with cement wasteforms. These studies found that, under the same conditions, wasteforms produced using geopolymers have significantly higher compressive strength than cement, as shown in Figure 7.

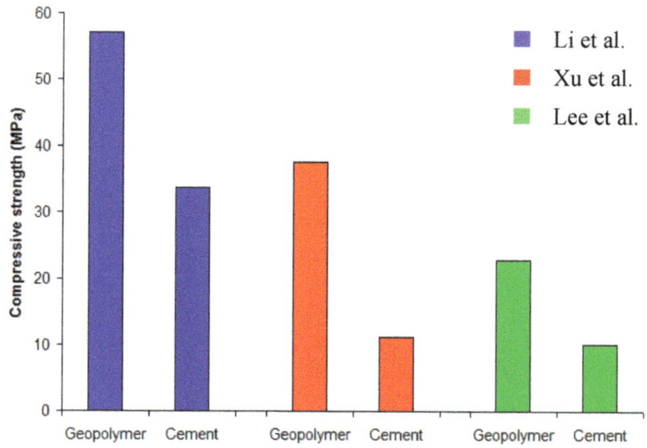

Figure 7. Compressive strength of geopolymer and cement wasteforms [47,51,57].

Besides being used as the basic criterion for solidified radioactive wasteforms, the compressive strength of a wasteform after calcination or freeze–thaw cycles are often used to assess its stability and sustainable performance. Li et al. [47] reported that their studied geopolymer matrix maintained an adequate compressive strength after calcination at 1000 °C. In comparison, the compressive strength of their CPC matrix was reduced to just 5 MPa, as shown in Table 7. Similarly, Xu et al. [57] reported that effect of the temperature on the compressive strength in OPC is more significant than that in the geopolymer. It should be noted that the study of such a very-high-temperature effect is not required for low- and intermediate-level wastes that are not associated with heat generation and is not required for scenarios that do not include volcanic eruptions. These studies also showed that the losses of compressive strength due to freeze–thaw cycles ranged from 3.5 to 10.5%, with higher losses observed for cement wasteforms ranging from 14.7 to 18.2%, as shown in Table 8. Both studies concluded that geopolymer wasteforms are superior to cement wasteforms in thermal and freeze–thaw stability.

Table 7. Compressive strength of the geopolymer/cement wasteforms after thermal exposure.

Wasteforms	Compressive Strength (MPa)					Loss (%) after 1000 °C Calcination	Ref.
	Initial	400 °C	600 °C	800 °C	1000 °C		
Geopolymer	57	52	45	38	30	47.4	[47]
OPC	34	22	15	9	5	85.3	
Geopolymer	38	-	34	28	27	28.9	[57]
OPC	11	-	Cracked	Cracked	Cracked	100	

Table 8. Compressive strength of the geopolymer/cement wasteforms after freeze–thaw cycles test.

Wasteforms	Compressive Strength (MPa)		Loss (%)	Ref.
	Before Freeze–Thaw Test	After Freeze–Thaw Test		
Geopolymer	57	55	3.5	[47]
OPC	34	29	14.7	
Geopolymer	38	34	10.5	[57]
OPC	11	9	18.2	

5. Perspectives on the Sustainability of Geopolymers

The sustainability of any practice is based on its environmental impacts, economical performance, and social acceptance. Ensuring acceptable impacts and performance will enhance the social acceptance of that practice. As indicated in the introduction, the reduced environmental impacts of geopolymers compared to conventional cements and the use of industrial wastes as base materials for geopolymer fabrication boosted their applications as civil engineering materials. The sustainability assessments of the geopolymer concrete prepared from industrial wastes were addressed by identifying the factors that affect their compressive strength as a performance and durability measure [149–151]. In this respect, the effect of the preparation conditions of the geopolymer concrete on its compressive strength as a durability measure was assessed [149]. The results of that study revealed that the compressive strength of the geopolymer concrete prepared with a high content of calcium fly ash increased with increases in the molarity of the sodium activator, the activator-to-binder ratio, and the curing temperature, while it decreased with the increase in coarse aggregate content. In another study, the effect of the incorporation of corncob ash in a GGBFS-based geopolymer on compressive strength was assessed [150]. The study concluded that the environmental impacts, in terms of the transport impact, global warming potential, global temperature potential, embodied energy, sustainability index, and economic index of the studied geopolymer are less than that of conventional concrete.

The use of industrial waste as base material in the preparation of a geopolymer matrix for the immobilization of hazardous wastes were stated to reduce the greenhouse gas emissions, energy requirements, and disposal costs of industrial wastes [86]. The study reported that the annual hazardous waste generation in India is 9.44 million tons, which requires 0.899 million tons of cement to stabilize. The use of an optimized geopolymer for different hazardous wastes was recommended to reduce the conventional cement requirements.

In Sections 3 and 4, the performance measures of different geopolymer materials were presented; these measures provide indicators on the potential sustainability of these materials in radioactive waste management. In addition, the limited large-scale industrial applications of these materials in the immobilization of radioactive wastes, i.e., SIAL, are providing additional measures of the potential sustainability of these materials. Nevertheless, the limited practical applications either in using geopolymers as sorbent or as immobilization matrices are not enough to generate a track record that enable the workers in this field to design specified durability tests that are based on geopolymer characteristics and performance. Additionally, the determinations of the sustainability indices for these applications are lacking. Moreover, the planning for radioactive waste management should be conducted in an integrated way that encompass the entire life cycle of the management practice. In this respect, life-cycle assessment for geopolymers in this field is also lacking.

6. Conclusions

Over the years, geopolymers have gained much attention in different fields, including radioactive waste management. The scientific efforts were focused on assessing the potential applications of these materials, as removal agents for radio-contaminants from liquid waste, and as immobilization matrices for different radioactive waste streams. This trend was supported by the need to reduce the material footprints in radioactive-waste-management activities and the promising performance of geopolymers in the water-and-wastewater-management field. Based on the literature reviewed in this work, the following gaps are identified in the sustainable performance of geopolymers in radioactive waste management:

- The application of a geopolymer in the pre-treatment of aqueous radioactive waste effluent was not addressed. This application is supported by the chemical stability of these materials in slightly acidic and alkaline solutions and its high buffering capacity, which allow an acceptable pH regulation performance.
- Geopolymer applications in membrane separation were not addressed in radioactive waste management. These applications are supported by the mechanical stability of these materials that are preserved even for porous geopolymers. This allows the application of geopolymer as a substrate or active layer in the membrane. Moreover, advanced trends in the literature were directed to assess this potential application in water and wastewater treatment and have provided knowledge that can be transfered to the radioactive-waste-management field.
- The ability of the amorphous geopolymer matrix to entrap metals and oxides can be used as a basis to test these materials for their potential application in photocatalytic degradation. This application, if proven, can be very useful to treat aqueous radioactive wastes that contain organic decontamination residues.
- As mentioned here, numerous batch studies were dictated to assess the promising application of geopolymers in radio-contaminant removal. These studies covered several types of geopolymer base materials, either single or blends, and targeted the removal of cations and anions of concern. Only a few studies have addressed the column operation and the reusability and regeneration ability of these materials, and there is still a need to investigate these aspects in depth and to have a clear understanding of the factors that affect them.
- The durability tests and standards were developed based on the long-term track record of the vulnerable characteristics of Portland cements. Despite geopolymers have been applied in certain countries for the immobilization of radioactive wastes, there is as yet no similar record to allow the adaptation of specific durability tests and standards for geopolymers.

- The life-cycle assessments for geopolymers used in radioactive waste management either as sorbent or as an immobilization matrix are lacking in the literature.

Even though most of the studies have yet to be implemented for actual application, they nevertheless serve as invaluable input for the further development of geopolymers in the field of radioactive waste management.

Author Contributions: Conceptualization, E.P., T.F.C. and R.O.A.R.; Resources, all the authors. All authors have read and agreed to the published version of the manuscript.

Funding: This research received no external funding.

Institutional Review Board Statement: Not applicable.

Informed Consent Statement: Not applicable.

Data Availability Statement: Not applicable.

Acknowledgments: The authors acknowledge the comments by anonymous reviewers, which helped to improve the review.

Conflicts of Interest: The authors declare no conflict of interest.

References

1. Kim, H.J.; Kim, J.S.; Kim, K.P.; Kwon, J.H.; Kim, J.C.; Jeong, J.Y.; Kim, Y.K. Radiological impact assessment of radioactive effluents emitted from nuclear power plants in Korea and China under normal operation. *Prog. Nucl. Energy* **2022**, *153*, 104464. [CrossRef]
2. Abdel Rahman, R.O.; Kozak, M.W.; Hung, Y.-T. Radioactive pollution and control. In *Handbook of Environment and Waste Management*; Hung, Y.-T., Wang, L.K., Shammas, N.K., Eds.; World Scientific Publishing Co.: Singapore, 2014; pp. 949–1027. [CrossRef]
3. Ministry of Foreign Affairs, Report on the Discharge Record and the Seawater Monitoring Results at Fukushima Daiichi Nuclear Power Station during January, Information (18:00), March 1. 2022. Available online: https://www.meti.gc.jp/english/earthquake/nuclear/decommissioning/pdf/sd202203.pdf (accessed on 2 January 2023).
4. IAEA. *Status and Trends in Spent Fuel and Radioactive Waste Management, Nuclear Energy Series, NW-T-1.14*; International Atomic Energy Agency: Vienna, Austria, 2022.
5. Gasser, M.S.; Kadry, H.F.; Helal, A.S.; Abdel Rahman, R.O. Optimization and modeling of uranium recovery from acidic aqueous solutions using liquid membrane with Lix-622 as Phenolic-oxime carrier. *Chem. Eng. Res. Des.* **2022**, *180*, 25–37. [CrossRef]
6. Ansari, S.A.; Mohapatra, P.K. Hollow fibre supported liquid membranes for nuclear fuel cycle applications: A review. *Clean. Eng. Technol.* **2021**, *4*, 100138. [CrossRef]
7. Kim, K.W.; Shon, W.J.; Oh, M.K.; Yang, D.; Foster, R.I.; Lee, K.Y. Evaluation of dynamic behavior of coagulation-flocculation using hydrous ferric oxide for removal of radioactive nuclides in wastewater. *Nucl. Eng. Technol.* **2019**, *51*, 738–745. [CrossRef]
8. Zheng, Y.; Qiao, J.; Yuan, J.; Shen, J.; Wang, A.J.; Niu, L. Electrochemical removal of radioactive cesium from nuclear waste using the dendritic copper hexacyanoferrate/carbon nanotube hybrids. *Electrochim. Acta* **2017**, *257*, 172–180. [CrossRef]
9. Chen, R.; Tanaka, H.; Kawamoto, T.; Asai, M.; Fukushima, C.; Kurihara, M.; Watanabe, M.; Arisaka, M.; Nankawa, T. Preparation of afilm of copper hexacyanoferrate nanoparticles for electrochemical removal of cesium from radioactive wastewater. *Electrochem. Commun.* **2012**, *25*, 23–25. [CrossRef]
10. Osmanlioglu, A.E. Decontamination of radioactive wastewater by two-staged chemical precipitation. *Nucl. Eng. Technol.* **2018**, *50*, 886–889. [CrossRef]
11. Su, T.; Han, Z.; Qu, Z.; Chen, Y.; Lin, X.; Zhu, S. Effective recycling of Co and Sr from Co/Sr-bearing wastewater via an integrated Fe coagulation and hematite precipitation approach. *Environ. Res.* **2020**, *187*, 109654. [CrossRef] [PubMed]
12. Garai, M.; Yavuz, C.T. Radioactive strontium removal from seawater by a MOF via two-step ion exchange. *Chem* **2019**, *5*, 750–752. [CrossRef]
13. Gasser, A.M.S.; El Sherif, E.; Mekhamer, H.S.; Abdel Rahman, R.O. Assessment of Cyanex 301 impregnated resin for its potential use to remove cobalt from aqueous solutions. *Environ. Res.* **2020**, *185*, 10940. [CrossRef]
14. Allahkarami, E.; Rezai, B. Removal of cerium from different aqueous solutions using different adsorbents: A review. *Process Saf. Environ. Prot.* **2019**, *124*, 345–362. [CrossRef]
15. Luo, W.; Huang, Q.; Antwi, P.; Guo, B.; Sasaki, K. Synergistic effect of $ClO^{4-}$and Sr^{2+} adsorption on alginate-encapsulated organo-montmorillonite beads: Implication for radionuclide immobilization. *J. Colloid Interface Sci.* **2020**, *560*, 338–343. [CrossRef] [PubMed]
16. Kim, J.; Lee, K.; Seo, B.K.; Hyun, J.H. Effective removal of radioactive cesium from contaminated water by synthesized composite adsorbent and its thermal treatment for enhanced storage stability. *Environ. Res.* **2020**, *191*, 110099. [CrossRef] [PubMed]

17. Alipour, D.; Keshtkar, A.R.; Moosavian, M.A. Adsorption of thorium(IV) from simulated radioactive solutions using anovel electrospun PVA/TiO_2/Zno nanofiber adsorbent functionalized with mercapto groups: Study in single and multi-component systems. *Appl. Surf. Sci.* **2016**, *366*, 19–29. [CrossRef]
18. Hasan, M.N.; Shenashen, M.A.; Hasan, M.M.; Znad, H.; Awual, M.R. Assessing of cesium removal from wastewater using functionalized wood cellulosic adsorbent. *Chemosphere* **2020**, *270*, 128668. [CrossRef]
19. Iwagami, S.; Tsujimura, M.; Onda, Y.; Konuma, R.; Sato, Y.; Sakakibara, K.; Yoschenko, V. Dissolved ^{137}Cs concentrations in stream water and subsurface water in a forested headwater catchment after the Fukushima Dai-Ichi Nuclear Power Plant accident. *J. Hydrol.* **2019**, *573*, 688–696. [CrossRef]
20. Chen, Y.L.; Tong, Y.Y.; Pan, R.W.; Tang, J. The research on adsorption behaviors and mechanisms of geopolymers on Sr^{2+}, CO^{2+} and Cs^{+}. *Adv. Mater. Res.* **2013**, *704*, 313–318. [CrossRef]
21. Tian, Q.; Sasaki, K. Application of fly ash-based geopolymer for removal of cesium, strontium and arsenate from aqueous solutions: Kinetic, equilibrium and mechanism analysis. *Water Sci. Technol.* **2019**, *79*, 2116–2125. [CrossRef]
22. Jang, J.G.; Park, S.M.; Lee, H.K. Cesium and strontium retentions governed by aluminosilicate gel in alkali-activated cements. *Materials* **2017**, *10*, 447. [CrossRef]
23. Tian, Q.; Nakama, S.; Sasaki, K. Immobilization of cesium in fly ash-silica fume based geopolymers with different Si/Al molar ratios. *Sci. Total Environ.* **2019**, *687*, 1127–1137. [CrossRef]
24. Lei, H.; Muhammad, Y.; Wang, K.; Yi, M.; He, C.; Wei, Y.; Fujita, T. Facile fabrication of metakaolin/slag-based zeolite microspheres (M/Szms) geopolymer for the efficient remediation of Cs^{+} and Sr^{2+} from aqueous media. *J. Hazard. Mater.* **2021**, *406*, 124292. [CrossRef] [PubMed]
25. Lee, N.K.; Khalid, H.R.; Lee, H.K. Adsorption characteristics of cesium onto mesoporous geopolymers containing nano-crystalline zeolites. *Micropor. Mesopor. Mater.* **2017**, *242*, 238–244. [CrossRef]
26. Chen, S.; Qi, Y.; Cossa, J.J.; Dos, S.I.D.S. Efficient Removal of radioactive iodide anions from simulated wastewater by HDTMA-geopolymer. *Prog. Nucl. Energy* **2019**, *117*, 103112. [CrossRef]
27. El-Naggar, M.R.; Amin, M. Impact of alkali cations on properties of metakaolin and metakaolin/slag geopolymers: Microstructures in relation to sorption of ^{134}Cs radionuclide. *J. Hazard. Mater.* **2018**, *344*, 913–924. [CrossRef] [PubMed]
28. Petlitckaia, S.; Barr, Y.; Piallat, T.; Grauby, O.; Ferry, D.; Poulesquen, A. Functionalized geopolymer foams for cesium removal from liquid nuclear waste. *J. Clean. Prod.* **2020**, *269*, 122400. [CrossRef]
29. IAEA. *Treatment and Conditioning of Radioactive Organic Liquids*; IAEA-TECDOC-656; IAEA: Vienna, Austria, 1992.
30. IAEA. *Predisposal Management of Organic Radioactive Waste*; Technical Reports Series No. 427; IAEA: Vienna, Austria, 2004.
31. Abdel Rahman, R.O.; Rakhimov, R.Z.; Rakhimova, N.R.; Ojovan, M.I. *Cementitious Materials for Nuclear Waste Immobilization*; Wiley: New York, NY, USA, 2014.
32. Varlakov, A.; Zherebtsov, A. Innovative and conventional materials and designs of nuclear cementitious systems in radioactive-waste management. In *Sustainability of Life Cycle Management for Nuclear Cementation-Based Technologies*; Abdel Rahman, R.O., Ojovan, M.I., Eds.; Elsevier-Woodhead Publishing: Chichester, UK, 2021; pp. 297–338.
33. Abdel Rahman, R.O.; Ojovan, M.I. Toward sustainable cementitious radioactive waste forms: Immobilization of problematic operational wastes. *Sustainability* **2021**, *13*, 11992. [CrossRef]
34. Abdel Rahman, R.O.; Abdel Moamen, O.A.; El-Masry, E.H. Life cycle of polymer nanocomposites matrices in hazardous waste management. In *Handbook of Polymer and Ceramic Nanotechnology*; Hussain, C.M., Thomas, S., Eds.; Springer Nature: Cham, Switzerland, 2021; pp. 1603–1625. [CrossRef]
35. Ahn, J.; Kim, W.S.; Um, W. Development of metakaolin-based geopolymer for solidification of sulfate-rich hybrid sludge waste. *J. Nucl. Mater.* **2019**, *518*, 247–255. [CrossRef]
36. Orlova, A.I.; Ojovan, M.I. Ceramic mineral waste-forms for nuclear waste immobilization. *Materials* **2019**, *12*, 2638. [CrossRef]
37. Fereiduni, E.; Ghasemi, A.; Elbestawi, M. Characterization of composite powder feedstock from powder bed fusion additive manufacturing perspective. *Materials* **2019**, *12*, 3673. [CrossRef]
38. Harker, A.B. Tailored ceramics. In *Radioactive Waste Forms for the Future*; Lutze, W.R.E., Ed.; North-Holland: Amsterdam, The Netherlands, 1988; pp. 335–392.
39. Frolova, A.V.; Vinokurov, S.E.; Gromyak, I.N.; Danilov, S.S. Medium-temperature phosphate glass composite material as a matrix for the immobilization of high-level waste containing volatile radionuclides. *Energies* **2022**, *15*, 7506. [CrossRef]
40. Farid, O.M.; Ojovan, M.I.; Massoud, A.; Abdel Rahman, R.O. An assessment of initial leaching characteristics of alkali-borosilicate-glasses for nuclear waste immobilization. *Materials* **2019**, *12*, 1462. [CrossRef] [PubMed]
41. Lönartz, M.I.; Dohmen, L.; Lenting, C.; Trautmann, C.; Lang, M.; Geisler, T. The effect of heavy ion irradiation on the forward dissolution rate of borosilicate glasses studied in situ and real time by Fluid-Cell Raman Spectroscopy. *Materials* **2019**, *12*, 1480. [CrossRef] [PubMed]
42. Ojovan, M.I.; Lee, W.E.; Kalmykov, S.N. Immobilization of radioactive waste in cement. In *An Introduction to Nuclear Waste Immobilization*, 3rd ed.; Ojovan, M.I., Lee, W.E., Kalmykov, S.N., Eds.; Elsevier: Amsterdam, The Netherlands, 2019; pp. 271–303.
43. Isaacs, M.; Lange, S.; Deissmann, G.; Bosbach, D.; Milodowski, A.E.; Read, D. Retention of technetium-99 by grout and backfill cements: Implications for the safe disposal of radioactive waste. *Appl. Geochem.* **2020**, *116*, 104580. [CrossRef]
44. Bayoumi, T.A.; Saleh, H.M. Characterization of biological waste stabilized by cement during immersion in aqueous media to develop disposal strategies for phytomediated radioactive waste. *Prog. Nucl. Energy* **2018**, *107*, 83–89. [CrossRef]

45. Yu, Q.; Li, S.; Li, H.; Chai, X.; Bi, X.; Liu, J.; Ohnuki, T. Synthesis and characterization of Mn-slag based geopolymer for immobilization of Co. *J. Clean. Prod.* **2019**, *234*, 97–104. [CrossRef]
46. He, P.; Wang, R.; Fu, S.; Wang, M.; Cai, D.; Ma, G.; Wang, M.; Yuan, J.; Yang, Z.; Duan, X.; et al. Safe trapping of cesium into doping-enhanced pollucite structure by geopolymer precursor technique. *J. Hazard. Mater.* **2019**, *367*, 577–588. [CrossRef]
47. Li, Q.; Sun, Z.; Tao, D.; Xu, Y.; Li, P.; Cui, H.; Zhai, J. Immobilization of simulated radionuclide ^{133}Cs+ by fly ash-based geopolymer. *Radiat. Phys. Chem.* **2013**, *262*, 325–331. [CrossRef]
48. Fu, S.; He, P.; Wang, M.; Cui, J.; Wang, M.; Duan, X.; Yang, Z.; Jia, D.; Zhou, Y. Hydrothermal synthesis of pollucite from metakaolin-based geopolymer for hazardous wastes storage. *J. Clean. Prod.* **2020**, *248*, 119240. [CrossRef]
49. He, P.; Cui, J.; Wang, M.; Fu, S.; Yang, H.; Sun, C.; Duan, X.; Yang, Z.; Jia, D.; Zhou, Y. Interplay between storage temperature, medium and leaching kinetics of hazardous wastes in metakaolin-based geopolymer. *J. Hazard. Mater.* **2020**, *384*, 121377. [CrossRef]
50. Lin, W.K.; Chen, H.Y.; Huang, C.P. Performance study of ion exchange resins solidification using metakaolin-based geopolymer binder. *Prog. Nucl. Energy* **2020**, *129*, 103508. [CrossRef]
51. Lee, W.H.; Cheng, T.W.; Ding, Y.C.; Lin, K.L.; Tsao, S.W.; Huang, C.P. Geopolymer technology for the solidification of simulated ion exchange resins with radionuclides. *J. Environ. Manag.* **2019**, *235*, 19–27. [CrossRef] [PubMed]
52. Williams, B.D.; Neeway, J.J.; Snyder, M.M.V.; Bowden, M.E.; Amonette, J.E.; Arey, B.W.; Pierce, E.M.; Brown, C.F.; Qafoku, N.P. Mineral assemblage transformation of a metakaolin-based waste form after geopolymer encapsulation. *J. Nucl. Mater.* **2016**, *473*, 320–332. [CrossRef]
53. Hanzlicek, T.; Steinerova, M.; Straka, P. Radioactive metal isotopes stabilized in a geopolymer matrix: Determination of a leaching extract by a radiotracer method. *J. Am. Ceram. Soc.* **2006**, *89*, 3541–3543. [CrossRef]
54. Chervonnyj, A.D.; Chervonnaya, N.A. Geopolymeric agent for immobilization of radioactive ashes from biomass burning. *Radiochemistry* **2003**, *45*, 182–188. [CrossRef]
55. Liu, X.; Ding, Y.; Lu, X. Immobilization of simulated radionuclide ^{90}Sr by fly ash-slag-metakaolin-based geopolymer. *Nucl. Technol.* **2017**, *198*, 64–69. [CrossRef]
56. Walkley, B.; Ke, X.; Hussein, O.H.; Bernal, S.A.; Provis, J.L. Incorporation of strontium and calcium in geopolymer gels. *J. Hazard. Mater.* **2020**, *382*, 121015. [CrossRef] [PubMed]
57. Xu, Z.; Jiang, Z.; Wu, D.; Peng, X.; Xu, Y.; Li, N.; Qi, Y.; Li, P. Immobilization of strontium-loaded zeolite A by metakaolin-basedgeopolymer. *Ceram. Int.* **2017**, *43*, 4434–4439. [CrossRef]
58. Tian, Q.; Guo, B.; Sasaki, K. Immobilization mechanism of Se oxyanions in geopolymer: Effects of alkaline activators and calcined hydrotalcite additive. *J. Hazard. Mater.* **2020**, *387*, 121994. [CrossRef]
59. Cannes, C.; Rodrigues, D.; Barré, N.; Lambertin, D.; Delpech, S. Reactivity of uranium in geopolymers, confinement matrices proposed to encapsulate Mgzr Waste. *J. Nucl. Mater.* **2019**, *518*, 370–379. [CrossRef]
60. Girke, N.A.; Steinmetz, H.J.; Bukaemsky, A.; Bosbach, D.; Hermannand, E.; Griebel, I. Cementation of Nuclear Graphite using Geopolymers. In Proceedings of the NUWCEM 2011, Avignon, France, 11–14 October 2011.
61. Xu, H.; Gong, W.; Syltebo, L.; Lutze, W.; Pegg, I.L. Duralith geopolymer waste form for Hanford secondary waste: Correlating setting behavior to hydration heat evolution. *J. Hazard. Mater.* **2014**, *278*, 34–39. [CrossRef]
62. Cuccia, V.; Freire, C.B.; Ladeira, A.C.Q. Radwaste oil immobilization in geopolymer after non-destructive treatment. *Prog. Nucl. Energy* **2020**, *122*, 103246. [CrossRef]
63. Ren, B.; Zhao, Y.; Bai, H.; Kang, S.; Zhang, T.; Song, S. Eco-friendly geopolymer prepared from solid wastes: A critical review. *Chemosphere* **2021**, *267*, 128900. [CrossRef]
64. Sercombe, J.; Gwinner, B.; Tiffreau, C.; Simondi-Teisseire, B.; Adenot, F. Modelling of bituminized radioactive waste leaching. Part I: Constitutive equations. *J. Nucl. Mater.* **2006**, *349*, 96–106. [CrossRef]
65. Zakharova, K.P.; Masanov, O.L. Bituminization of liquid radioactive waste. Safety assessment and operational experience. *Atomic Energy* **2000**, *89*, 135–139. [CrossRef]
66. Ikladious, N.E.; Ghattasv, N.K.; Tawfik, M.E. Some Studies on the incorporation of simulate radioactive waste into polymer matrix. *Radioact. Waste Manag. Nucl. Fuel Cycle* **1993**, *17*, 119–137.
67. Özdemir, T.; Usanmaz, A. Monte Carlo simulations of radioactive waste embedded into polymer. *Rad. Phys. Chem.* **2009**, *78*, 800–805. [CrossRef]
68. Davidovits, J. *Geopolymer Chemistry and Applications*, 4th ed.; Geopolymer Institute: Saint-Quentin, France, 2015.
69. Zakka, W.P.I.; Lim, N.H.A.S.; Khun, M.C. A scientometric review of geopolymer concrete. *J. Clean. Prod.* **2021**, *280*, 124353. [CrossRef]
70. Wang, S.; Li, H.; Zou, S.; Zhang, G. Experimental research on afeasible rice husk/geopolymer foam building insulation material. *Energy Build.* **2020**, *226*, 110358. [CrossRef]
71. Jiang, C.; Wang, A.; Bao, X.; Ni, T.; Ling, J. A review on geopolymer in potential coating application: Materials, preparation and basic properties. *J. Build. Eng.* **2020**, *32*, 101734. [CrossRef]
72. Villaquirán-Caicedo, M.A.; Mejía de Gutiérrez, R. Synthesis of ceramic materials from ecofriendly geopolymer precursors. *Mater. Lett.* **2018**, *230*, 300–304. [CrossRef]
73. Lahoti, M.; Tan, K.H.; Yang, E.H. A critical review of geopolymer properties for structural fire-resistance applications. *Constr. Build. Mater.* **2019**, *221*, 514–526. [CrossRef]

74. Zhang, Y.J.; Han, Z.C.; He, P.Y.; Chen, H. Geopolymer-based catalysts for cost-effective environmental governance: A review based on source control and end-of-pipe treatment. *J. Clean. Prod.* **2020**, *263*, 121556. [CrossRef]
75. Łach, M.; Mierzwiński, D.; Korniejenko, K.; Mikuła, J.; Hebda, M. Geopolymers as a material suitable for immobilization of fly ash from municipal waste incineration plants. *J. Air Waste Manag. Assoc.* **2018**, *68*, 1190–1197. [CrossRef] [PubMed]
76. Tochetto, G.A.; Simão, L.; de Oliveira, D.; Hotza, D.; Immich, A.P.S. Porous geopolymers as dye adsorbents: Review and perspectives. *J. Clean. Prod.* **2022**, *374*, 133982. [CrossRef]
77. Freire, A.L.; José, H.J.; Moreira, R.D.F.P.M. Potential applications for geopolymers in carbon capture and storage. *Int. J. Greenh. Gas Control.* **2022**, *118*, 103687. [CrossRef]
78. Luhar, I.; Luhar, S.; Abdullah, M.M.A.B.; Razak, R.A.; Vizureanu, P.; Sandu, A.V.; Matasaru, P.D. A state-of-the-art review on innovative geopolymer composites designed for water and wastewater treatment. *Materials* **2021**, *14*, 7456. [CrossRef] [PubMed]
79. Sadiq, M.; Naveed, A.; Arif, M.; Hassan, S.; Afridi, S.; Asif, M.; Sultana, S.; Amin, N.; Younas, M.; Khan, M.N.; et al. Geopolymerization: A promising technique for membrane synthesis. *Mater. Res. Express* **2021**, *8*, 112002. [CrossRef]
80. Della Rocca, D.G.; Peralta, R.M.; Peralta, R.A.; Rodríguez-Castellón, E.; de Fatima Peralta Muniz Moreira, R. Adding value to aluminosilicate solid wastes to produce adsorbents, catalysts and filtration membranes for water and wastewater treatment. *J. Mater. Sci.* **2021**, *56*, 1039–1063. [CrossRef]
81. Novais, R.M.; Pullar, R.C.; Labrincha, J.A. Geopolymer foams: An overview of recent advancements. *Prog. Mater. Sci.* **2020**, *109*, 100621. [CrossRef]
82. Wu, Y.; Lu, B.; Bai, T.; Wang, H.; Du, F.; Zhang, Y.; Cai, L.; Jiang, C.; Wang, W. Geopolymer, green alkali activated cementitious material: Synthesis, applications and challenges. *Constr. Build. Mater.* **2019**, *224*, 930–949. [CrossRef]
83. Ji, Z.; Pei, Y. Bibliographic and visualized analysis of geopolymer research and its application in heavy metal immobilization: A review. *J. Environ. Manag.* **2019**, *231*, 256–267. [CrossRef] [PubMed]
84. El-eswed, B.I. Chemical evaluation of immobilization of wastes containing Pb, Cd, Cu and Zn in alkali-activated materials: A critical review. *J. Environ. Chem. Eng.* **2020**, *8*, 104194. [CrossRef]
85. Taki, K.; Mukherjee, S.; Patel, A.K.; Kumar, M. Reappraisal review on geopolymer: A new era of aluminosilicate binder for metal immobilization. *Environ. Nanotechnol. Monit. Manag.* **2020**, *14*, 100345. [CrossRef]
86. Singh, R.; Budarayavalasa, S. Solidification and stabilization of hazardous wastes using geopolymers as sustainable binders. *J. Mater. Cycles Waste Manag.* **2021**, *23*, 1699–1725. [CrossRef]
87. Tian, Q.; Bai, Y.; Pan, Y.; Chen, C.; Yao, S.; Sasaki, K.; Zhang, H. Application of geopolymer in stabilization/solidification of hazardous pollutants: A review. *Molecules* **2022**, *27*, 4570. [CrossRef] [PubMed]
88. Liang, K.; Wang, X.Q.; Chow, C.L.; Lau, D. A review of geopolymer and its adsorption capacity with molecular insights: A promising adsorbent of heavy metal ions. *J. Environ. Manag.* **2022**, *322*, 116066. [CrossRef]
89. Li, J.; Chen, L.; Wang, J. Solidification of radioactive wastes by cement-based materials. *Prog. Nucl. Energy* **2021**, *141*, 103957. [CrossRef]
90. Reeb, C.; Pierlot, C.; Davy, C.; Lambertin, D. Incorporation of organic liquids into geopolymer materials—A review of processing, properties and applications. *Ceram. Int.* **2021**, *47*, 7369–7385. [CrossRef]
91. Provis, J.L.; van Deventer, J.S.J. Geopolymerisation kinetics. 2. Reaction kinetic modelling. *Chem. Eng. Sci.* **2007**, *62*, 2318–2329. [CrossRef]
92. Luhar, I.; Luhar, S. A comprehensive review on fly ash-based geopolymer. *J. Compos. Sci.* **2022**, *6*, 219. [CrossRef]
93. Davidovits, J. *Geopolymer Chemistry and Applications*; Geopolymer Institute: Saint-Quentin, France, 2008.
94. Weng, L.; Sagoe-Crentsil, K. Dissolution processes, hydrolysis and condensation reactions during geopolymer synthesis: Part I—Low Si/Al ratio systems. *J. Mater. Sci.* **2007**, *42*, 2997–3006. [CrossRef]
95. Abdel Rahman, R.O.; El-Kamash, A.M.; Hung, Y.-T. Applications of nano-zeolite in wastewater treatment: An overview. *Water* **2022**, *14*, 137. [CrossRef]
96. Blackford, M.G.; Hanna, J.V.; Pike, K.J.; Vance, E.R.; Perera, D.S. Transmission electron microscopy and nuclear magnetic resonance studies of geopolymers for radioactive waste immobilization. *J. Am. Ceram. Soc.* **2007**, *90*, 1193–1199. [CrossRef]
97. Duxson, P.; Provis, J.L.; Lukey, G.C.; Mallicoat, S.W.; Kriven, W.M.; Van Deventer, J.S. Understanding the relationship between geopolymer composition, microstructure and mechanical properties. *Colloids Surf. A Physicochem. Eng. Asp.* **2005**, *269*, 47–58. [CrossRef]
98. Tian, Q.; Sasaki, K. Application of fly ash-based materials for stabilization/solidification of cesium and strontium. *Environ. Sci. Pollut. Res.* **2019**, *26*, 23542–23554. [CrossRef] [PubMed]
99. Youssef, M.A.; El-Naggar, M.R.; Ahmed, I.M.; Attallah, M.F. Batch kinetics of ^{134}Cs and $^{152+154}$Eu radionuclides onto poly-condensed feldspar and perlite based sorbents. *J. Hazard. Mater.* **2021**, *403*, 123945. [CrossRef]
100. Xiang, Y.; Hou, L.; Liu, J.; Li, J.; Lu, Z.; Niu, Y. Adsorption and enrichment of simulated ^{137}Cs in geopolymer foams. *J. Environ. Chem. Eng.* **2021**, *9*, 105733. [CrossRef]
101. Rooses, A.; Steins, P.; Dannoux-Papin, A.; Lambertin, D.; Poulesquen, A.; Frizon, F. Encapsulation of Mg–Zr alloy in metakaolin-based geopolymer. *Appl. Clay Sci.* **2013**, *73*, 86–92. [CrossRef]
102. Leay, L.; Potts, A.; Donoclift, T. Geopolymers from fly ash and their gamma irradiation. *Mater. Lett.* **2018**, *227*, 240–242. [CrossRef]
103. Lambertin, D.; Boher, C.; Dannoux-Papin, A.; Galliez, K.; Rooses, A.; Frizon, F. Influence of gamma ray irradiation on metakaolin based sodium geopolymer. *J. Nucl. Mater.* **2013**, *443*, 311–315. [CrossRef]

104. Chartier, D.; Sanchez-Canet, J.; Bessette, L.; Esnouf, S.; Renault, J.P. Influence of formulation parameters of cement based materials towards gas production under gamma irradiation. *J. Nucl. Mater.* **2018**, *511*, 183–190. [CrossRef]
105. Kumagai, Y.; Kimura, A.; Taguchi, M.; Nagaishi, R.; Yamagishi, I.; Kimura, T. Hydrogen production in gamma radiolysis of the mixture of mordenite and seawater. *J. Nucl. Sci. Technol.* **2013**, *50*, 130–138. [CrossRef]
106. Deng, N.; An, H.; Cui, H.; Pan, Y.; Wang, B.; Mao, L.; Zhai, J. Effects of gamma-ray irradiation on leaching of simulated $^{133}Cs+$ radionuclides from geopolymer wasteforms. *J. Nucl. Mater.* **2015**, *459*, 270–275. [CrossRef]
107. Ojovan, M.I. Radioactive waste characterization and selection of processing technologies. In *Handbook of Advanced Radioactive Waste Conditioning Technologies*; Woodhead Publishing Limited Chichester: Chichester, UK, 2011; pp. 1–18.
108. Ojovan, M.I.; Lee, W.E. Introduction to immobilization. In *An Introduction to Nuclear Waste Immobilization*; Elsevier: Amsterdam, The Netherlands, 1989; pp. 1–8. ISBN 9780080444628.
109. El-Naggar, M.R. Applicability of alkali activated slag-seeded Egyptian Sinai kaolin for the immobilization of ^{60}Co radionuclide. *J. Nucl. Mater.* **2014**, *447*, 15–21. [CrossRef]
110. Tian, Q.; Sasaki, K. Anovel composite of layered double hydroxide/geopolymer for co-immobilization of Cs^+ and SeO_4^{2-} from aqueous solution. *Sci. Total Environ.* **2019**, *695*, 133799. [CrossRef]
111. Skorina, T. Ion exchange in amorphous alkali-activated aluminosilicates: Potassium based geopolymers. *Appl. Clay Sci.* **2014**, *87*, 205–211. [CrossRef]
112. Abdel Rahman, R.O.; Ojovan, M.I. (Eds.) Techniques to test cementitious systems through their life cycles. In *Sustainability of Life Cycle Management for Nuclear Cementation-Based Technologies*; Woodhead Publishing: Chichester, UK, 2021; pp. 407–430. [CrossRef]
113. Varlakov, A.; Zherebtsov, A.; Ojovan, M.I.; Petrov, V. Long-term irradiation effects in cementitious systems. In *Sustainability of Life Cycle Management for Nuclear Cementation-Based Technologies*; Woodhead Publishing: Chichester, UK, 2021; pp. 161–180.
114. Mast, B.; Gerardy, I.; Pontikes, Y.; Schroeyers, W.; Reniers, B.; Samyn, P.; Gryglewicz, G.; Vandoren, B.; Schreurs, S. The effect of gamma radiation on the mechanical and microstructural properties of Fe-rich inorganic polymers. *J. Nucl. Mater.* **2019**, *521*, 126–136. [CrossRef]
115. Yeoh, M.L.; Ukritnukun, S.; Rawal, A.; Davies, J.; Kang, B.J.; Burrough, K.; Aly, Z.; Dayal, P.; Vance, E.R.; Gregg, D.J. Mechanistic impacts of long-term gamma irradiation on physicochemical, structural, and mechanical stabilities of radiation-responsive geopolymer pastes. *J. Hazard. Mater.* **2021**, *407*, 124805. [CrossRef] [PubMed]
116. *ANSI-ANS-16.1*; Measurement of the Leachability of Solidified Low-Level Radioactive Waste by a Short-Term Test Procedure. American Nuclear Society: La Grange Park, IL, USA, 2003.
117. Ji, Z.; Pei, Y. Immobilization efficiency and mechanism of metal cations (Cd^{2+}, Pb^{2+} and Zn^{2+}) and anions (AsO_4^{3-} and $Cr_2O_7^{2-}$) in waste-based geopolymer. *J. Hazard. Mater.* **2020**, *384*, 121290. [CrossRef] [PubMed]
118. Melo, L.G.A.; Periera, R.A.; Pires, E.F.C.; Darwish, F.A.I.; Silva, F.J. Physicochemical characterization of pulverized phyllite rock for geopolymer resin synthesis. *Mater. Res.* **2017**, *20*, 236–243. [CrossRef]
119. Farid, O.M.; Abdel Rahman, R.O. Preliminary assessment of modified borosilicate glasses for chromium and ruthenium immobilization. *Mater. Chem. Phys.* **2017**, *186*, 462–469. [CrossRef]
120. Kozai, N.; Sato, J.; Osugi, T.; Shimoyama, I.; Sekine, Y.; Sakamoto, F.; Ohnuki, T. Sewage sludge ash contaminated with radiocesium: Solidification with alkaline-reacted metakaolinite (geopolymer) and Portland cement. *J. Hazard. Mater.* **2021**, *416*, 125965. [CrossRef] [PubMed]
121. Abdel Rahman, R.O.; Metwally, S.S.; El-Kamash, A.M. Life Cycle of Ion Exchangers in Nuclear Industry: Application and Management of Spent Exchangers. In *Handbook of Ecomaterials*; Martínez, L., Kharissova, O., Kharisov, B., Eds.; Springer: Cham, Switzerland, 2019; Volume 5, pp. 3709–3732. [CrossRef]
122. Jang, J.G.; Lee, N.K.; Lee, H.K. Fresh and hardened properties of alkali-activated fly ash/slag pastes with superplasticizers. *Constr. Build. Mater.* **2014**, *50*, 169–176. [CrossRef]
123. Lee, N.K.; Lee, H.K. Reactivity and reaction products of alkali-activated, fly ash/slag paste. *Constr. Build. Mater.* **2015**, *81*, 303–312. [CrossRef]
124. Phillip, E.; Khoo, K.S.; Yusof, M.A.W.; Rahman, R.A. Mechanistic insights into the dynamics of radionuclides retention in evolved POFA-OPC and OPC barriers in radioactive waste disposal. *Chem. Eng. J.* **2022**, *437*, 135423. [CrossRef]
125. Pan, Y.; Bai, Y.; Chen, C.; Yao, S.; Tian, Q.; Zhang, H. Effect of calcination temperature on geopolymer for the adsorption of cesium. *Mater. Lett.* **2023**, *330*, 133355. [CrossRef]
126. Ma, S.; Yang, H.; Fu, S.; He, P.; Duan, X.; Yang, Z.; Jia, D.; Colombo, P.; Zhou, Y. Additive manufacturing of geopolymers with hierarchical porosity for highly efficient removal of Cs+. *J. Hazard. Mater.* **2023**, *443*, 130161. [CrossRef] [PubMed]
127. El-Eswed, B.I. *Solidification Versus Adsorption for Immobilization of Pollutants in Geopolymeric Materials: A Review*; Intechopen: London, UK, 2018.
128. Abdel Rahman, R.O.; Zaki, A.A. Comparative analysis of nuclear waste solidification performance models: Spent ion exchanger-cement based wasteforms. *Process Saf. Environ. Prot.* **2020**, *136*, 115–125. [CrossRef]
129. Shi, C.; Fernández-Jiménez, A. Stabilization/solidification of hazardous and radioactive waste with alkali-activated cements. *J. Hazard. Mater.* **2006**, *137*, 1656–1663. [CrossRef] [PubMed]
130. Khalil, M.Y.; Merz, E. Immobilization of intermediate-level wastes in geopolymers. *J. Nucl. Mater.* **1994**, *211*, 141–148. [CrossRef]
131. El-Eswed, B.I.; Aldagag, O.M.; Khalili, F.I. Efficiency and mechanism of stabilization/solidification of Pb(II), Cd(II), Cu(II), Th(IV) and U(VI) in metakaolin based geopolymers. *Appl. Clay Sci.* **2017**, *140*, 148–156. [CrossRef]

132. Perera, D.S.; Blackford, M.G.; Vance, E.R.; Hanna, J.V.; Finnie, K.S.; Nicholson, C.L. Geopolymers for the immobilization of radioactive waste. *Mat. Res. Soc. Symp. Proc.* **2004**, *824*, 432–437. [CrossRef]
133. Jia, L.; He, P.; Jia, D.; Fu, S.; Wang, M.; Wang, M.; Duan, X.; Yang, Z.; Zhou, Y. Immobilization behavior of Sr in geopolymer and its ceramic product. *J. Am. Ceram. Soc.* **2019**, *103*, 1372–1384. [CrossRef]
134. Ke, X.; Bernal, S.A.; Sato, T.; Provis, J.L. Alkali aluminosilicate geopolymers as binders to encapsulate strontium-selective titanate ion-exchangers. *Dalton Trans.* **2019**, *48*, 12116–12126. [CrossRef]
135. Kuenzel, C.; Cisneros, J.F.; Neville, T.P.; Vandeperre, L.J.; Simons, S.J.R.; Bensted, J.; Cheeseman, C.R. Encapsulation of Cs/Sr contaminated clinoptilolite in geopolymers produced from metakaolin. *J. Nucl. Mater.* **2015**, *466*, 94–99. [CrossRef]
136. Peng, X.; Xu, Y.; Xu, Z.; Wu, D.; Li, D. Effect of simulated radionuclide strontium on geopolymerization process. *Procedia Environ. Sci.* **2016**, *31*, 325–329. [CrossRef]
137. Ofer-Rozovsky, E.; Arbel Haddad, M.; Bar-Nes, G.; Borojovitz, E.J.C.; Binyamini, A.; Nikolski, A.; Katz, A. Cesium immobilization in nitrate-bearing metakaolin-based geopolymers. *J. Nucl. Mater.* **2019**, *514*, 247–254. [CrossRef]
138. Cantarel, V.; Motooka, T.; Yamagishi, I. Geopolymers and their potential applications in the nuclear waste management field—A bibliographical study. *JAEA Rev.* **2017**, *14*.
139. Jing, Z.; Hao, W.; He, X.; Fan, J.; Zhang, Y.; Miao, J.; Jin, F. A novel hydrothermal method to convert incineration ash into pollucite for the immobilization of asimulant radioactive cesium. *J. Hazard Mater.* **2016**, *306*, 220–229. [CrossRef]
140. Al-Mashqbeh, A.; Abuali, S.; El-Eswed, B.; Khalili, F.I. Immobilization of toxic inorganic anions ($Cr_2O_7^{2-}$, MnO_4^- and $Fe(CN)_6^{3-}$) in metakaolin based geopolymers: A preliminary study. *Ceram. Int.* **2018**, *44*, 5613–5620. [CrossRef]
141. Lichvar, P.; Rozloznik, M.; Sekely, S. Behaviour of aluminosilicate inorganic matrix SIAL during and after solidification of radioactive sludge and radioactive spent resins and their mixtures. *Amec Nucl. Slovak.* **2010**. Available online: https://www-pub.iaea.org/MTCD/Publications/PDF/TE-1701_add-CD/PDF/Slovakia.pdf (accessed on 3 January 2022).
142. Jang, J.G.; Park, S.M.; Lee, H.K. Physical barrier effect of geopolymeric waste form on diffusivity of cesium and strontium. *J. Hazard. Mater.* **2016**, *318*, 339–346. [CrossRef] [PubMed]
143. Wang, G.; Wang, X.; Chai, X.; Liu, J.; Deng, N. Adsorption of Uranium (VI) from aqueous solution on calcined and acid-activated kaolin. *J. Appl. Clay Sci.* **2010**, *47*, 448–451. [CrossRef]
144. Künzel, C. Metakaolin Based Geopolymers to Encapsulate Nuclear Waste. Ph.D. Thesis, Imperial College, London, UK, 2013.
145. Jiang, Z.; Xu, Z.H.; Shuai, Q.; Li, P.; Xu, Y.H. Thermal stability of geopolymer—Sr contaminated zeolite A blends. *Key Eng. Mater.* **2017**, *727*, 1089–1097. [CrossRef]
146. Boškovi, I.V.; Nenadovi, S.S.; Kljajevi, L.M.; Vukanac, I.S.; Stankovi, N.G.; Lukovi, J.M.; Vukčević, M. Radiological and physico-chemical properties of red mud based geopolymers. *Nucl. Technol. Rad. Prot.* **2018**, *33*, 188–194. [CrossRef]
147. Abdel Rahman, R.O.; Ojovan, M.I. (Eds.) Sustainability of cementitious structures, systems, and components (SSC's): Long-term environmental stressors. In *Sustainability of Life Cycle Management for Nuclear Cementation-Based Technologies*; Elsevier-Woodhead Publishing: Chichester, UK, 2021; pp. 181–232.
148. De Oliveira, L.B.; de Azevedo, A.R.; Marvila, M.T.; Pereira, E.C.; Fediuk, R.; Vieira, C.M.F. Durability of geopolymers with industrial waste. *Case Stud. Constr. Mater.* **2022**, *16*, e00839. [CrossRef]
149. Kishore, Y.S.N.; Nadimpalli, S.G.D.; Potnuru, A.K.; Vemuri, J.; Khan, M.A. Statistical analysis of sustainable geopolymer concrete. *Mater. Today Proc.* **2022**, *61*, 212–223. [CrossRef]
150. Oyebisi, S.; Olutoge, F.; Kathirvel, P.; Oyaotuderekumor, I.; Lawanson, D.; Nwani, J.; Ede, A.; Kaze, R. Sustainability assessment of geopolymer concrete synthesized by slag and corncob ash. *Case Stud. Constr. Mater.* **2022**, *17*, e01665. [CrossRef]
151. Mishra, J.; Nanda, B.; Patro, S.K.; Krishna, R.S. Sustainable Fly Ash Based Geopolymer Binders: A Review on Compressive Strength and Microstructure Properties. *Sustainability* **2022**, *14*, 15062. [CrossRef]

sustainability

MDPI

Review

Recent Advances in Alternative Cementitious Materials for Nuclear Waste Immobilization: A Review

Nailia Rakhimova

Department of Building Materials, Institute of Building Technology, Engineering and Ecology Systems, Kazan State University of Architecture and Engineering, 420043 Kazan, Russia; nailia683@gmail.com; Tel.: +7-906-323-4529

Abstract: Since the emergence of the problem of nuclear waste conditioning, cementation has become an important and developing part of the waste management system, owing to its simplicity and versatility. The continued development of the cementation technique is driven by the improvement and expansion of cementitious materials that are suitable and efficient for nuclear waste solidification. Advances in cement theory and technology have significantly impacted improvements in nuclear waste cementation technology, the quality of fresh and hardened waste forms, waste loading rates, and the reliability and sustainability of the nuclear industry. Modern mineral matrices for nuclear waste immobilization are a broad class of materials with diverse chemical–mineralogical compositions, high encapsulation capacities, and technological and engineering performance. These matrices include not only traditional Portland cement, but also non-Portland clinker inorganic binders. This review focuses on recent trends and achievements in the development of calcium aluminate, calcium sulfoaluminate, phosphate, magnesium silicate, and alkali-activated cements as cementitious matrices for nuclear waste stabilization/solidification.

Keywords: cement; nuclear waste; solidification

Citation: Rakhimova, N. Recent Advances in Alternative Cementitious Materials for Nuclear Waste Immobilization: A Review. *Sustainability* **2023**, *15*, 689. https://doi.org/10.3390/su15010689

Academic Editors: Michael I. Ojovan, Vladislav A. Petrov and Sergey V. Yudintsev

Received: 4 December 2022
Revised: 19 December 2022
Accepted: 20 December 2022
Published: 30 December 2022

1. Introduction

Nuclear energy, which is characterized by a low carbon footprint, high power density, and ability to generate electricity quickly, holds a stable position among other forms of energy and is currently considered one of the most viable forms of base-load electrical generation for the next 50–100 years [1–3]. Recent statistics indicate that nuclear energy accounts for 10% of global power production and is anticipated to rise to 1/3 of world power by 2060 [4]. However, because one of the challenging deficiencies of nuclear power is the generation of radioactive waste (RW), further sustainable development of the nuclear industry must be accompanied by the consistent development of an RW management system.

According to the definition given by the International Atomic Energy Agency basic safety standard "Radiation protection and safety of radiation sources: International basic safety standards" [5], radioactive waste is material that contains, or is contaminated with, radionuclides at activity concentrations greater than clearance levels, as established by the regulatory body, for which, for legal and regulatory purposes, no further use is foreseen. The terms "radioactive waste" and "nuclear waste" are generically used as synonyms in the context of safety and waste management. Nuclear waste management is a complex system and comprises all administrative and operational activities involved in the handling, pre-treatment, treatment, conditioning, transport, storage, and disposal of RW [6]. The continuous improvement of each stage makes an important contribution to improving the efficiency and reliability of the entire system [7–16].

Extensive experience has been accumulated in the field of environmentally benign RW handling, and a wide range of approaches, methods, and materials for the treatment and conditioning of diverse RW have been developed and adopted over the history of the generation and use of nuclear power. Because of the simplicity and versatility of solidification/stabilization of RW by cementitious materials, as well as long-term testing and experience gained by practical usage, this technique is one of the main and most widely used methods for converting various low- and intermediate-level radioactive wastes (LILW) into a safe form [17–19].

RW cementation remains a developing area in the RW management system. Progress in this field has also been supported by significant achievements in materials science, particularly in the chemistry of inorganic binders, as well as in the processing of a wide range of materials. Advances in cement theory and technology significantly impact improvements in RW cementation technology, the quality of fresh and hardened wasteforms, RW loading, and the reliability and sustainability of the nuclear industry. Portland cement (PC) has been the only cementitious material for LILW immobilization for a long time. The consistent implementation of supplementary cementitious materials and chemical additives for PC and PC concretes in recent decades and achievements in this area led to the increasing application of blended and modified PC-based systems in RW cementation [20–22]. It turned out the introduction of pozzolanic and chemical modifiers into PC remarkably improved not only the technological and engineering performance of waste forms but that it could also be used to control the reaction products assemblage and structure of hardened materials, action mechanism of immobilized contaminants, and mineral matrix. Consequently, PC in combination with mineral and chemical modifiers is now normally used and adopted into the practice of RW cementation.

In the past decades, non-Portland clinker alternative binders have emerged as another progressively developing direction. Non-Portland clinker alternative binders are promising for sustainable development of the cement industry and for advancing RW solidification technology [23–32]. Alternative or so-called non-traditional cements form a large group of binders, significantly differing from PC and from each other in terms of the composition and type of the raw materials, composition of the reaction products, the mechanism of formation of the hardened cement pastes, research experience, adoption, and practical application. Most alternative cements are special cements developed in attempts to eliminate the ecological and technical disadvantages of PC and/or develop special binders for special non-building applications. Certain alternative binders presented in Table 1 have become promising for the partial replacement of PC for RW solidification.

Table 1. Alternative cements and their characteristics.

Cementitious Material (Abbreviation)	Reaction Products Providing RW Immobilization (Reaction Process Mechanism)	Specific Properties	Adoption Experience
Calcium aluminate cement (CAC)	$2CaO{\cdot}Al_2O_3{\cdot}8H_2O$, $Al(OH)_3$ (dissolution-precipitation)	fast hardening, high strength, low permeability, high freeze-thaw, corrosion resistance	France [33]
Calcium sulfoaluminate cement (CSAC)	$3CaO{\cdot}Al_2O_3{\cdot}3CaSO_4{\cdot}12H_2O$, $3CaO{\cdot}Al_2O_3{\cdot}3CaSO_4{\cdot}32H_2O$, $Al(OH)_3$ (dissolution-precipitation)		
Magnesium silicate cement (M-S-H cement)	M-S-H gel (dissolution-precipitation)		
Phosphate cements	Magnesium phosphate cement (MPC)	Fast setting, high early strength, adhesive properties, low water demand and drying shrinkage, high temperature and chemical resistance	Russian Federation, USA [34–38]
	Calcium phosphate cement (CPC)		
Alkali-activated cement	Alkali-activated slag cement (AASC)	Fast setting, high strength, low porosity, and high temperature and chemical resistance	Ukraine [39]
	Geopolymer (GP)		Australia, Czech Republic, Slovak Republic, France, USA [40–45]

Years of experience in research and application have demonstrated the effectiveness of alternative binders for the immobilization of toxic materials and RW [23–32]. The reactive phases obtained with alternative cements have diverse compositions compared with those obtained with PC, where the former are characterized by lower solubility and better ion exchange properties, different pH, hardened pastes demonstrate faster hardening, lower permeability, and durability. Moreover, due to the complexity of the starting materials, the cement flexibility is impacted by a greater number of contributing factors, enabling versatile design of the cementitious wasteforms, control of the composition of the reaction phases, and the achievement of desirable performance. Therefore, alternative binders have expanded the possibilities and perspectives for the cementation of toxic materials and RW from the following aspects:

- In some cases, higher efficiency for both physical and chemical immobilization of heavy metals and radionuclides;
- Widening the acceptance of wastes that can be treated and conditioned by cementation;
- Optimizing waste cementation technology in cases of problematic waste components, providing faster curing of cementitious wasteforms, and eliminating the need for waste pre-treatment;
- Enabling the use of alternative binders as adsorbents and chemical additives.

The scientific and practical interest in the use of cementitious materials as a whole and the research and adoption of alternative cements, in particular, has only increased in recent years. This study reviews recent trends and achievements in the development of calcium aluminate, calcium sulfoaluminate, phosphate, magnesium silicate, and alkali-activated cements as cementitious matrices for RW stabilization/solidification.

2. Alternative Cements as Cementitious Materials for RW Immobilization

2.1. Calcium Aluminate (CAC) and Calcium Sulfoaluminate Cements (CSAC)

The production of CAC and CSAC cements, similar to PC, is based on the heat treatment of a prepared mixture of natural rocks, including limestone, bauxite, and gypsum (in the case of CSAC), leading to the formation of hydration hardening minerals. The mineral composition of the resultant clinker of CAC and CSAC cements differs from that of PC, and heat treatment of the former two cements is carried out at lower temperatures.

2.1.1. CACs

CAC consists mainly of monocalcium aluminate ($CaO{\cdot}Al_2O_3$ (CA)), with some secondary minerals, such as $CaO{\cdot}2Al_2O_3$ (CA_2) and $12CaO{\cdot}7Al_2O_3$ ($C_{12}A_7$) [46,47]. At temperatures of 22–30 °C in the presence of water, CA is gradually converted into dicalcium aluminate hydrate ($2CaO{\cdot}Al_2O_3{\cdot}8H_2O$) in the form of lamellar crystals of a hexagonal system through a dissolution–precipitation mechanism. Simultaneously, gel-like aluminum hydroxide, $Al(OH)_3$, with sorptive properties is formed.

CAC is an effective matrix for immobilizing RW as it undergoes fast hardening, and the hardened CAC pastes have high strength, low permeability, and high freeze–thaw and corrosion resistance. The sorptive ability of $Al(OH)_3$ and ion-exchange ability of ettringite (formed with the introduction of lime and calcium sulphate) [48,49] provide the chemical binding of many contaminants by CAC-based matrices. It is necessary to add to the listed advantages the applicability of CAC in respect of tolerance with the waste components retarding the setting and hardening of PC. The feasibility of CAC and blended CAC for cementation of waste containing Cr, Cd, Pb, Zn, Mg, Sn, Cs, liquid borates, radioiodide, etc., has been demonstrated [50–54].

The chemical binding of hazardous and radioactive contaminants by CACs can be improved by introducing various mineral admixtures in which reaction products with ion-exchange properties are formed. It is feasible to include up to 50% calcium sulfate, in the form of gypsum or anhydrite [48] or 5–10% slaked or non-slaked lime and limestone [49], into CAC-based cement matrices. These additives form calcium sulfoaluminate and provide siliceous mineral materials for the formation of zeolitic phases [52], thereby improving the technological and physical–mechanical properties of CACs.

CACs are used in France for (non-radioactive) hazardous waste encapsulation [33].

Recently, CAC has been studied for the solidification and stabilization of ion-exchange resins (IERs) [55,56], as well as Sr and Cl ions [57,58]. Kononenko et al. [55,56] reported that CAC introduced with ^{137}Cs sorbent (modified diatomite) can be incorporated with 22–25% more of a mixture of ion-exchange resins (IERs) (Na^+, NO^{3-}) and 50–83% of a (Na^+, $B_4O_7{}^{2-}$) mixture, as compared with PC-based matrices. In order to prevent the decomposition of $CaO{\cdot}1.64Al_2O_3$ under the action of $B_4O_7{}^{2-}$ and preclude the accompanying decrease in the strength of the wasteforms, the authors proposed suppressing the reactivity of $B_4O_7{}^{2-}$ ions by treating IERs with alkaline earth metal (Ca, Sr, Ba) nitrates, resulting in the formation of insoluble alkaline earth metal tetraborates.

2.1.2. CSAC

Regardless of the merits of PC, one drawback is the shrinkage of PC-based materials. Attempts to create shrinkage-free cement have led to the development of CSACs. These special cements can be shrinkage compensating, expansive, and self-stressing, and are used for various purposes. The invention and manufacture of CSACs were introduced between the 1960s–1970s of the last century. Currently, CACS are produced in industrial volumes. Since 2004, 1.2–1.3 million tons of CACs has been produced globally each year [59].

The main mineral in CSAC, comprising 30–70%, is tetracalcium trialuminate sulfate $C_4A_3\overline{S}$ ($4CaO{\cdot}Al_2O_3{\cdot}SO_3$) ye'elimite, also known as Klein's compound [60]. The second most important mineral in sulfoaluminate–belite cements is belite C_2S–$2CaO{\cdot}SiO_2$ [61,62].

CSA clinker is ground simultaneously with 25% gypsum for the purpose of regulating the setting, strength, strength development, and soundness. Introducing other admixtures, such as PC and limestone, is also possible and effective [30,63,64]. During the interaction of calcium sulfoaluminate with water, calcium monosulfoaluminate hydrate (AFm), and aluminum hydroxide are formed as follows:

$$C_4A_3\overline{S} + 18H \rightarrow C_3A{\cdot}C\overline{S}{\cdot}12H + 2AH_3$$

The AFm phase belongs to the lamellar double hydroxide family. Its crystal structure is composed of positively charged main layers of $[Ca_2Al(OH)_6]^+$ and negatively charged interlayers of $[1/2\,SO_4{\cdot}nH_2O]^-$ [65].

In the presence of gypsum, the interaction of $C_4A_3\overline{S}$ with water is accelerated. In addition to amorphous aluminum hydroxide, ettringite ($3CaO{\cdot}Al_2O_3{\cdot}3CaSO_4{\cdot}32H_2O$ or $C_3A{\cdot}3C\overline{S}{\cdot}32H$) is formed, where the molar ratio of $C_4A_3\overline{S}$ to $C\overline{S}$ is no less than 1:2. This process occurs via the following reaction:

$$C_4A_3\overline{S} + 2\,C\overline{S}H_2 + 34H \rightarrow C_3A{\cdot}3C\overline{S}{\cdot}32H + 2AH_3$$

A mixture of ettringite ($C_3A{\cdot}3C\overline{S}{\cdot}32H$) and $C_3A{\cdot}C\overline{S}{\cdot}12H$ may also be formed, with the full consumption of gypsum in the reaction.

Ettringite is composed of positively charged $[Ca_3Al(OH)_6]^{3+}$ columns and negatively charged channels of $[3/2SO_4{\cdot}nH_2O]^{3-}$. The structural flexibility of AFm and ettringite in terms of ion exchange provides CSACs with the ability to chemically bind many elements of both anionic and cationic nature [66–68] (Figures 1 and 2).

The intensive hydration of $C_4A_3\overline{S}$ and the binding of free water enable fast consolidation of the structure, short setting, and accelerated strength development. Ettringite formation is accompanied by a volume increase in the solid phase [49]. The hydration of C_2S causes prolonged strength development and facilitates the relaxation of pressure in the crystallization of ettringite. Strätlingite (C_2ASH_8), C-S-H, CAH_{10}, or siliceous hydrogarnets can also be formed depending on the clinker composition, presence, and type of supplementary cementitious materials [30].

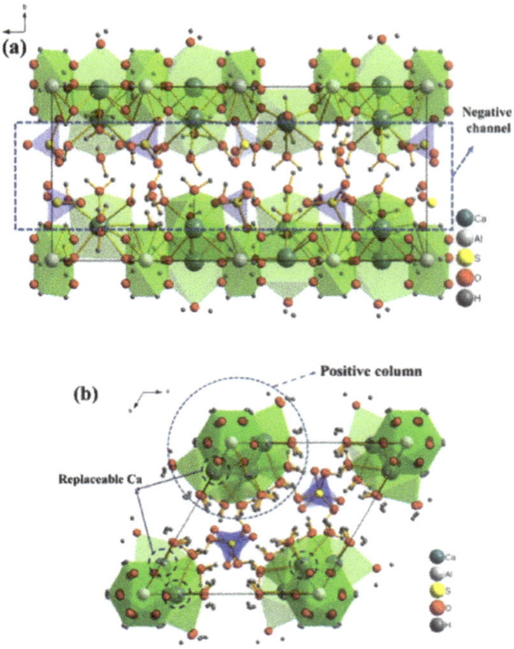

Figure 1. Structure of ettringite: positive columns $[Ca_3Al(OH)_6]^{3+}$ (green), negative channel $[3/2SO_4{\cdot}nH_2O]^{3-}$ (blue), and replaceable Ca (green black); (**a**) c view, (**b**) b view [68].

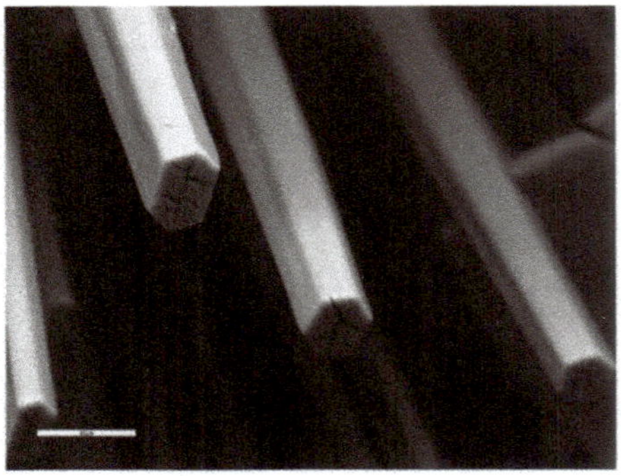

Figure 2. Scanning electron microscope image of ettringite crystals. Bar = 20 μm [69].

Numerous studies [30,70–74] have shown the following features of CSACs as matrices for the immobilization of toxic substances and RW:

- Accommodate heavy metals (Cr, Pb, Zn, Cd, etc.) and IERs;
- Enable the immobilization of waste that is problematic for immobilization by PC, such as those containing Al and U and wastes that produce hydrogen by interacting with cement and radioactive sludge with a high content of sulfate and borate ions;
- Allow for the precipitation of radionuclides as hydroxides (for example, $Sr(OH)_2$) due to the lower pH of hydrated CSACs while potentially decreasing the corrosion reactions of some encapsulated metals, such as Al;
- Shorten the waste cementation process owing to the high rate of hydration, and avoid pre-treatment, enabling the solidification of wastes containing components that make setting and hardening of PC-based systems difficult, for example, B, Zn, and waste.

Studies on the mechanism of solidification of borates by CSACs depending on the presence and content of gypsum and the pH have been continued by Champenois et al. [72], Chen et al. [75], and Cau-Dit-Coumes et al. [70]. Chen et al. [75] observed a dense amorphous ulexite layer with a foil-like morphology and a thickness of approximately 100 nm (Figure 3). This layer fully covered the surface of CSA clinker particles three days after mixing with 0.5 M borate solution at pH < 7, which strongly impeded the dissolution of ye'elimite. Champenois et al. [72] studied the hydration of CSAC incorporated with 0 and 10% gypsum in the presence of 1 mol/L borate ions at pH 11 and revealed that the retardation reaction of the fresh paste increased with the gypsum percentage and was correlated with the content of ulexite ($NaCaB_5O_9{\cdot}8H_2O$). The gypsum content affected the pH of the cementitious system and, consequently, the amount of ulexite formed. Cau-Dit-Coumes et al. [70] investigated the combined influence of lithium hydroxide (as an accelerator) in CSAC and sodium borate on the hydration of CSACs containing 0 or 10% gypsum. The simultaneous presence of borates and lithium led to the superimposition of acceleration and retardation effects. In the gypsum-free system, lithium promoted precipitation of the borated AFm phase. Authors believe that lithium salts can counteract the retardation caused by sodium borate. The results presented by Xu et al. [68] and Guo et al. [76] contribute to the understanding of the mechanism of immobilization of contaminant simulants in ettringite. Xu et al. [68] reported that hardened CACs due to binding capacity of ettringite and $Al(OH)_3$ along with the dense physical structure of hardened paste were better in the leaching performance of Cs^+ and Sr^{2+}, in comparison with the PC-based cementitious matrix (Figure 4). The authors proposed two superimposed mechanisms of Cs^+ leaching: (i) a first-order reaction between the surface of the radwaste matrix and the leachant, (ii) diffusion of Cs^+ through the waste matrix, (iii) release of loosely bound Cs^+; (i) diffusion/dissolution of Cs^+ and (ii) release of loosely bound Cs^+. The authors described Sr^{2+} leaching using a combination model, including the dissolution and diffusion of Sr^{2+} and the release of loosely bound Sr^{2+} in the wasteform. Guo et al. [64] revealed differences in the interaction mechanisms of I^-, IO^{3-}, and ettringite. The authors observed minimal I^- incorporation into ettringite (0.05%), whereas IO^{3-} demonstrated high affinity for ettringite through anion substitution for $SO_4{}^{2-}$ (96%). Substituting IO^{3-} for $SO_4{}^{2-}$ was energetically favorable (−0.41 eV), whereas an unfavorable substitution energy of 4.21 eV was observed for I^- substitution.

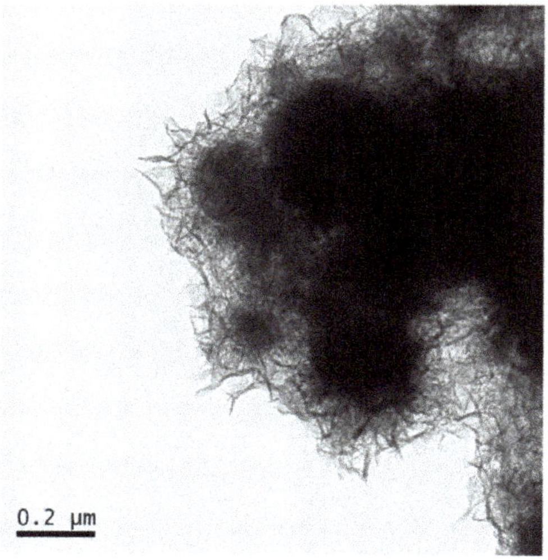

Figure 3. TEM image of powder CSA clinker after 3 days hydration with borate solution [75].

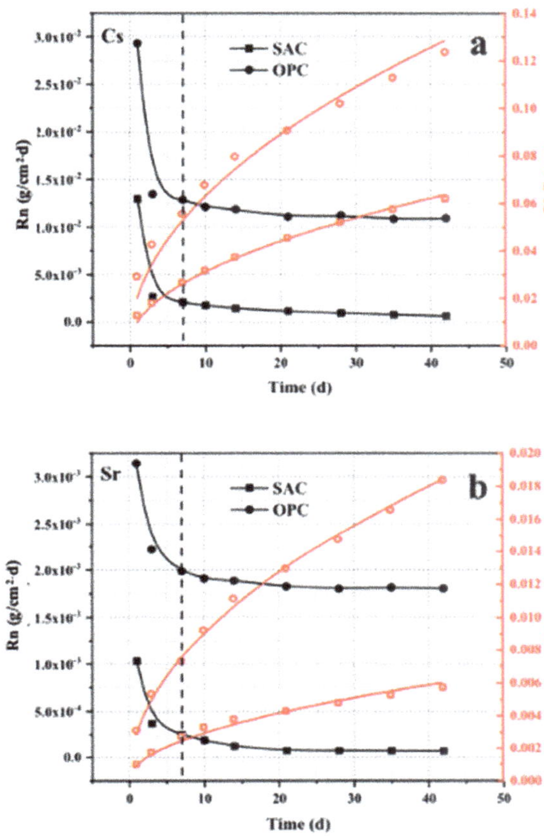

Figure 4. Leaching rate (Rn) and cumulative leaching fraction (CLF) fitting curves for Cs^+ (**a**) and Sr^{2+} (**b**) [68].

Xu et al. [77] improved the performance of cementitious wasteforms based on CSAC and IERs by incorporation of MK in order to increase the resistance of IERs to prolonged water immersion. Cemented by optimal composition of 40 wt.% spent resin, 55.8 wt.% sulfoaluminate cement, 2.2 wt.% MK, and 2 wt.% water reducer, the resin loading in wasteforms was as high as 64% and the compressive strength of hardened wasteforms was 13.7 MPa. It is supposed that MK as an Al source promotes the formation of ettringite, thereby improving the stability of the solidified IERs in acidic environments or during frequent freezing-thawing. Moreover, a greater amount of ettringite provided the retention of Cs(I), with a 42 d leaching rate of 2.3×10^{-4} cm/d.

2.2. Phosphate Cements

The production of phosphate binders is based on the synthesis of phosphate compounds using acid-base reactions of solids of a basic nature (CuO, FeO, ZnO, CaO, MgO, etc.) and highly reactive liquid activators comprising phosphate anions. Activators, such as aqueous phosphoric acid (mainly orthophosphoric acid H_3PO_4) and acid phosphate salt solutions, can be used. These include solutions of KH_2PO_4, $NH_4H_2PO_4$, and $CaHPO_4$. In addition to fundamental differences in the mechanism of the formation of the hardened paste, phosphate binders have a wider chemical and structural composition than other types of binders, as proven by the evaluation of a number of fundamental chemical characteristics of phosphate compounds.

Orthophosphoric acid is polybasic and has three stages of ionization, enabling multi-dimensional stereometric chemical binding and the formation of numerous connection options with varying degrees of substitution (mono-, di-, and trisubstituted salts). An important source of strength formation is also the structural characteristics of phosphoric acid and phosphates with a branched network of hydrogen bonds. Finally, orthophosphoric acid and its derivatives have a high predisposition for association with functional groups, polycondensation, and complexation [78].

The powder part of phosphate cements influences the binding properties of the systems "oxide-phosphoric acid" and the ionic potential of cations in the oxide. The conditions for activating the binding properties of the oxide-orthophosphoric acid systems are listed in Table 2.

Table 2. Conditions of exhibiting binding properties of oxide-orthophosphoric acid systems [47].

Oxide	Electron Work Function, eV	Ionic Potential of Cation, z/r	Conditions of Exhibiting Binding Properties
SiO_2, TiO_2, Al_2O_3, ZrO_2, MnO_2, Cr_2O_3, Co_2O_3, SnO_2	>4.5	5.0	Intensification of acid-base interactions required
Fe_2O_3, Mn_2O_3, NiO, CoO, FeO, CuO	3.3–4.3	2.5–4.4	Hardening in normal conditions
Nd_2O_3, La_2O_3, MgO, ZnO, CdO	2.5–3.3	2.0–3.0	Passivation of acid-base interactions required
CaO, SrO, BaO, PbO	<2.0	1.4–2.0	Emergency measures of passivation of acid-base interactions required

Magnesium (MPCs) and calcium phosphate cements (CPCs) are the most promising for RW cementation purposes. These cements have been widely studied and have already obtained practical adoption.

2.2.1. MPCs

The raw materials for MPCs are orthophosphoric acid (or $(NH_4)_2HPO_4$ (diammonium hydrogen phosphate) and MgO, which is a product of the thermal treatment of magnesite $MgCO_3$. The main reaction product of these interactions is struvite ($NH_4MgPO_4{\cdot}6H_2O$) produced by the reaction $MgO + (NH_4)2HPO_4 + 5H_2O \rightarrow NH_4MgPO_4{\cdot}6H_2O + NH_3$, which determines the setting and hardening of this type of phosphate cement. Mixing MgO with KH_2PO_4 results in the formation of struvite-K.

The performance of MPCs is mainly controlled by the magnesium to phosphorus (M/P) ratio and the water/solid (W/S) ratio [79].

MPCs combine the high ion-exchange capacity of struvite, near-neutral pH, and high physical–mechanical properties of hardened paste, such as quick setting, high early strength, adhesive properties, low water demand and drying shrinkage, high temperature, and chemical resistance [34,80,81]. The struvite structure is able to take on many elements (Figure 5) [34], including monovalent cations (NH^{4+}, K^+, Rb^+, Cs^+, Tl^+), divalent cations (Mg^{2+}, Ni^{2+}, Zn^{2+}, Co^{2+}, Cd^{2+}, Cr^{3+}, Mn^{2+}, VO^{2+}), and trivalent oxyanions ($PO_4{}^{3-}$ and $AsO_4{}^{3-}$) [34] (Figures 5 and 6).

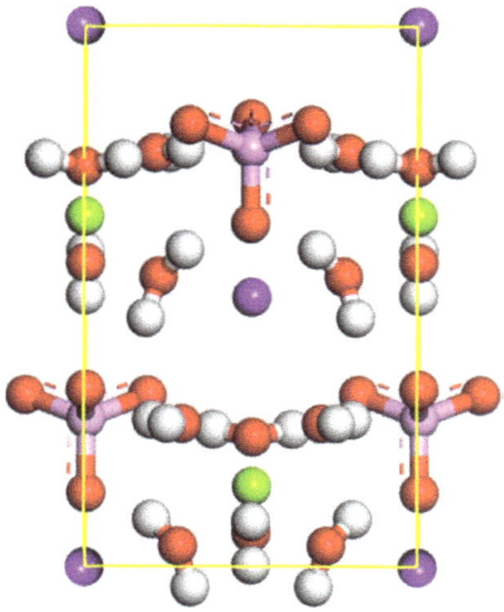

Figure 5. Crystal structure of struvite [82].

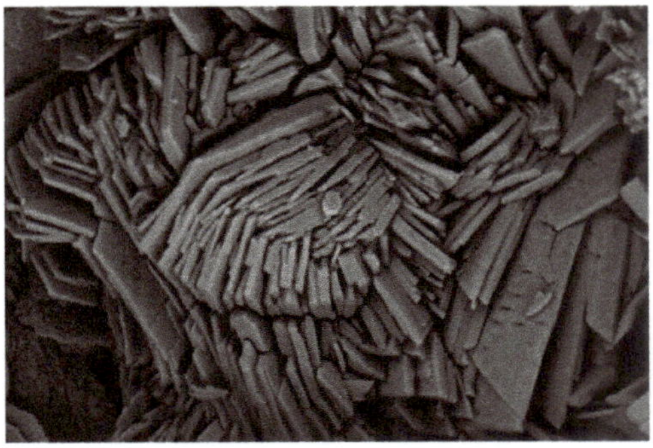

Figure 6. Micrographs of the MPC sample (M/P = 4/1, w/s = 0 14) [83].

Thus, solidification can be realized for a wide range of wastes:

- Chloride, nitrate, and radioactive nitrate-, nitrite-, and sulfate-containing solutions [84–87];
- Cs- and plutonium-contaminated ash [35,36,83,88,89];
- Reactive metals (Al, Mg, and U), which may corrode at high pH [90–95].

In addition, the H_2 radiolytic yield of MPC-based materials is 2–3 times less than that of PC-based materials because most of the mixing water participates in K-struvite formation. Gamma irradiation at a dose of 10 MGy has no notable effect on the mechanical performance and mineralogy of MPC mortars [96]. Bykov et al. [97] proposed a model of cement radiolysis in MPCs-based materials. The comparison of the radiation stability of PC and MPCs under γ-irradiation up to the absorbed dose of 100 MGy showed that the radiation-induced chemical decomposition of the materials in water was accompanied by the evolution of hydrogen. The evolution was retarded as the dose increased, where the limiting gas concentrations (~2.6 and ~0.7 L kg^{-1} for PC and MPC, respectively) were reached at ~20 MGy, demonstrating that oxygen was entrapped by the constituents of the materials.

With the purpose to reduce the risk of radiolytic gas generation, as well as the corrosion of reactive metals in studies [98,99], MPCs were incorporated with fly ash (FA) and blast furnace slag (BFS). Gardner et al. [100] studied the behavior of blended MPCs at elevated temperatures to determine how waste packages behave when exposed to fire The purpose of research was formulated after fire and subsequent radionuclide release at the Waste Isolation Pilot Plant (WIPP) in the USA in 2014 [101]. Hardened pastes were exposed to a range of temperatures between 400 and 1200 °C to study the high-temperature behavior of FA/MKPC and GBFS/MKPC. At 400 °C, the dehydration of struvite-K ($MgKPO_4{\cdot}6H_2O$) was observed, leading to the loss of long-range crystallographic order. In the blended FA/MPC and GBFS/MPC binders exposed to temperatures of 1000 and 1200 °C, the formation of potassium aluminosilicate minerals (leucite and kalsilite), among other crystalline phases (hematite, spinel, and forsterite), was detected. The authors concluded that although the reactive phases assemblage and microstructure of the FA/MPC and GBFS/MPC binders were considerably altered at high temperatures, the binders formed stable products while retaining physical stability, with no evidence of spalling/cracking.

The possibility of solidifying borate-and nitrate-containing wastes was recently investigated by Lahalle et al. [102], Kononenko et al. [103], and Tao et al. [104]. Lahalle et al. [102] proposed a mechanism of retardation of MPC in the presence of borates. $B(OH)_3$ slows down the formation of hydrates in two ways: (i) by stabilizing in solution the cations that outbalance the negative charges of the polyborates formed at pH > 6 and (ii) through the precipitation of an amorphous mineral containing borate and orthophosphate. The first process proceeds in both diluted suspensions and pastes, whereas the second is specific to pastes. Kononenko et al. [103] used struvite-K ($KMgPO_4{\cdot}6H_2O$) and struvite ($NH_4MgPO_4{\cdot}6H_2O$)-based phosphate binders as a matrix for the solidification of liquid wastes. Authors simulated evaporator bottoms for a pressurized water reactor nuclear power plant (PWR NPP) with the following composition: $NaNO_3$—236.6 g dm^{-3}; H_3BO_3—168.2 g dm^{-3}; NaOH—189.6 g dm^{-3}; total salt content—509 g dm^{-3} (37.3 wt%); pH—11.8; solution density—1.364 g cm^{-3}. The borates promoted struvite synthesis. The designed matrices contained up to 17–17.5 wt% salts, which was 1.7–2.5 times greater than that of the PC-based matrices. The volume of the struvite-based matrix was 1.6 times larger than the volume of the liquid waste from which it was obtained. With a Cs-selective nickel-potassium ferrocyanide sorbent or 10–20% MgO in excess of the reaction stoichiometry, the average rate of ^{137}Cs leaching from the cementitious wasteform was less than 10^{-3} $g{\cdot}cm^{-2}{\cdot}d^{-1}$, with a mechanical strength over 5 MPa. Tao et al. [104] reported that incorporating simulated high-nitrate waste into MPCs changed the crystallization degree of struvite-K, where the microstructure changed from dense, plate-like, and prismatic crystals into loose, cluster-like crystals when the amount of nitrates exceeded 5%. Incorporating simulated high-nitrate waste into MPCs also retarded the hydration of the MKPC specimens and increased their porosity.

Vinokurov et al. [88] in 2009 studied MPCs as matrices for the solidification of simulated liquid alkaline high-level wastes containing actinides, as well as fission and corrosion products. These studies were continued by Lai et al. [81] who investigated the rapid immobilization of Cs and Sr in wastes from the PUREX process. The compressive strength of cemented wasteforms incorporating up to 50 wt% waste was 4.2 MPa and 13.2 MPa at an M/P ratio of 1 after 3 h and 1 d, respectively. The leaching rates of Sr^{2+} and Cs^{+} from the cemented forms were less than 10^{-7} $g/m^2/d$ and 10^{-4} $g/m^2/d$, respectively. Zhenyu et al. [105] combined the benefits of ceramics and MPC materials. Particulate solidified ceramic forms with a composition of $Ca_{0.8}Ce_{0.1}TiSiO_5$ were first prepared by heating at 1300 °C for 2 h and then introduced into an MPC-based matrix. The obtained solidified forms demonstrated excellent mechanical properties, high-temperature stability, soaking resistance, and freeze–thaw resistance. The compressive strength of the samples decreased with increasing ceramic content, reaching 27.8 MPa with 50 wt% loading of the ceramic. However, the leaching rate of the simulated nucleus was found to be 1.86×10^{-7} cm/d, which was less than that of the solidified ceramic form.

Pyo et al. [106] stated that radioactive concrete waste generated during the decommissioning of nuclear power plants can be effectively solidified using MPCs. The replacement of MPCs with 50% concrete waste even increased the compressive strength from 41 to 56 MPa. Moreover, the compressive strength remained >45 MPa after thermal-cycling and water-immersion tests. The leaching indices of Cs, Co, and Sr, analyzed according to the ANS 16.1 procedure, were 11.45, 17.63, and 15.66, respectively.

However, based on a comprehensive analysis of solid waste-based MPCs, Zhang et al. [107] pointed out that the promotion of long-term, dynamic, and multi-dimensional research on MPC is an urgent task for the solid waste treatment of MPC.

MKPCs have also been described as "chemically bonded phosphate ceramics (CBPCs)" or by the trade name "Ceramicrete" and have been extensively developed and tested in the United States and Russia for conditioning various challenging nuclear wastes, including plutonium-contaminated ash, heavy metal and radium wastes, and ^{99}Tc-bearing wastes (using $SnCl_2$ as a reductant), as well as liquid Hanford vitrification wastes and Mayak salt wastes [34–38].

2.2.2. CPCs

CPCs comprise the calcium phosphates of diverse compositions or their blends with calcium salts (sulfate, carbonate, hydroxide, aluminate, calcium, etc.), magnesium orthophosphates, strontium, etc. [108–110]. The diverse combinations of calcium and phosphorus oxides (in the presence or absence of water) give a sufficiently large variety of different calcium phosphates; therefore, a wide range of raw materials is available for CPC production. The solubility in water, binding properties, and pH of the calcium phosphate cement are substantially influenced by the Ca/P ratio.

Hardened CPCs consist of stoichiometric or calcium-deficient hydroxyapatite. Their formation results from two reactions. A classic example of the first type of reaction is based on acid–base interactions; for example, the reaction of basic tetracalcium phosphate and acidic anhydrous dicalcium phosphate in an aqueous medium, leading to the formation of poorly crystallized hydroxyapatite (HA) [111]:

$$2Ca_4(PO_4)_2O + 2CaHPO_4 \rightarrow Ca_{10}(PO_4)_6(OH)_2$$

The second type of reaction involves the hydrolysis of metastable ortophosphate in an aqueous medium [110].

Hydroxyapatite resembles the structure of zeolites, characterized by presence of channels with diameters of 2.5 Å and 3–4,5 Å. Hydroxyapatite provides structural flexibility in the ion exchange with contaminant ions, often containing trivalent lanthanides and actinides, which can be replaced by Ca [22]. Hydroxyapatite also has low solubility, being 3–4 times less soluble than C-S-H and portlandite [30].

Recently, an efficient method for the consolidation of cobalt (Co(II))-adsorbed calcium hydroxyapatite was studied to design a simplified route for the decontamination of the coolant system of nuclear power plants and for the direct immobilization of the spent adsorbent [112]. Calcium hydroxyapatite nanopowder, produced by a wet precipitation method, was used as an adsorbent, resulting in a 94% removal of a Co(II) surrogate from simulated cooling water. The relative density after cold sintering was >97%; the obtained materials had a high compressive strength of 175 MPa. The normalized leaching rate of Co(II) was measured, as per the ASTM-C1285 standard, and found to be 2.5×10^{-5} g/m^2/d. The ANSI/ANS-16.1 test procedure was used to analyze the leachability of the sintered matrices, where the measured leaching index was 6.5.

2.3. Magnesium Silicate Hydrate Cements (M-S-H Cements)

M-S-H cements are based on the interaction of MgO or $Mg(OH)_2$ with amorphous silica, resulting in the formation of a M-S-H binder gel. This type of mineral matrix is relatively new among other cementitious materials for RW solidification but has already received scientific attention. Walling et al. [113] stated that Magnox sludge waste (a significant UK nuclear sector waste stream), consisting mainly of $Mg(OH)_2$, can be used as a primary constituent of M-S-H cement-based wasteforms, in combination with silica fume and an inorganic phosphate dispersant. Feasibility studies for the immobilization of Cs^+/Sr^{2+} and Al by low-pH M-S-H cement have also been performed [114–116].

2.4. Alkali-Activated Materials

Invented more than 70 years ago, chemical, and particularly alkali, activation of glassy aluminosilicates—which is an approach for the non-fired or low-temperature production of inorganic binders from various natural and technogenic starting materials—has gained an ever-increasing appeal from the standpoints of theoretical research and industrial implementation, including the stabilization/solidification technique. This is largely due to the possibility of alkali-activated cements (AACs) achieving desirable properties, such as high fluidity, enhanced chemical resistance in aggressive environments, enhanced chemical tolerance to problematic and complex waste streams, potentially high waste loadings, and resilience against security of supply issues [18].

The general mechanism for the formation of hardened paste through alkali activation of glassy aluminosilicates consists of three different stages: (i) the destruction of aluminosilicate glass in an alkali medium, rupture of Si-O-Si and Al-O-Si bonds, and coagulation of transitional species, (ii) coagulation–condensation, and (iii) condensation–crystallization of calcium or sodium aluminosilicate hydrogel as a major reaction product [117,118]. The distinguishing feature of AAC is the greater number of influencing factors than those in PC-based systems. Generally, the formation process, structure, and properties of AACs depend on many factors, including the following: (1) precursor factors, such as the shape and size of the particles, crystal/vitreous phase ratio, and chemical composition (e.g., reactive SiO_2/Al_2O_3 and CaO content); (2) alkali activator factors, such as type (MOH, $M_2O{\cdot}rSiO_2$ (SiO_2/Na_2O), and $NaAlO_2$), molarity, pH, and addition methods (e.g., dry form and solution form); as well as (3) processing factors, such as grinding and mixing methods and the curing regime (e.g., temperature, humidity, and time) [28,119–124]. By varying the controlling parameters, it is possible to design AAC-based cementitious materials with pre-determined reaction products and physical performance, enabling efficient RW encapsulation, as well as the sequestration of specific contaminants and wastes. As alkali activation allows the use of precursors with a wide range of chemical compositions in terms of the percentage of reactive Ca, Si, and Al, the reactive phases assemblage of hardened AACs varies widely. Fast setting, high strength, low porosity, and high chemical and heat resistance are typical for the appropriate formulations of AACs. The range of potential starting materials has changed and expanded continuously throughout the history of AACs [28,119,125–128]. As regards the sources of AACs used as matrices for the immobilization of RW; granulated BFS, FA, and MK; and their combinations, they are now the basic precursors, whereas sodium and potassium hydroxides and silicates are normally used as alkali reactants.

Historically, studies in the field of AACs as cementitious matrices for RW began in the early 1990s, with initial studies on BFS-based AACs [39,129]. The alkali activation of high-Ca precursors, including BFS, results in the formation of a tobermorite-like aluminum-substituted calcium silicate hydrogel C-(A)-S-H [130,131].

Many studies [18,19,23,24,26,28–32,132–134] have demonstrated the efficiency of BFS-based AACs for the solidification of wastes containing heavy metals (such as Zn^{2+}, Pb^{2+}, Cd^{2+} and Cr^{6+}, Hg^{2+}, etc.) and radionuclides (Cs^{+}, Sr^{2+}), as well as sodium-borate-containing liquid wastes, ion-exchange resins, etc.

Studies on the mechanisms of action of the BFS-based AACs with Cs^{+} and Sr^{+} have been conducted in the last few years by several researchers [135–137]. Vandevenne et al. [135] evaluated the mechanism of immobilization of Cs^{+} and Sr^{+} (0.5–2% wt%) by 6 M NaOH-activated BFS-based AAC. The authors reported that Cs^{+} was almost fully incorporated into the mineral matrix, whereas Sr^{2+} mainly precipitated as $Sr(OH)_2$ throughout the AAC-hardened paste. Huang et al. [136] reported that the addition of sodium hexametaphosphate to sodium silicate/sodium hydroxide-activated BFS1 paste enhanced the chemical binding of Sr^{2+} ions via hydroxyapatite formation and Sr^{2+} substitution. Microwave irradiation further increased the mechanical performance of the hardened pastes and inhibited the leaching of Sr^{2+} ions from the matrices by strengthening hydration reactions and Sr^{2+} encapsulation. According to Komljenovic et al. [137], introducing 2% and 5%

Cs into sodium silicate BFS-paste increased the early strength of the hardened paste, with no noticeable effect on the composition of the binder gel.

Since the end of the 70s of the last century, increasing attention has been paid to a subclass of AACs now termed "geopolymers", which are based on low-Ca or Ca-free precursors, such as class F FA and MK. The major reaction product of alkali-activated FA or MK is a three-dimensional, cross-linked, and structurally disordered sodium aluminosilicate hydrate gel, N-A-S-H (Figure 7). The binding gel comprises Si and Al in tetrahedral coordination, connected by oxygen atoms in a pseudo-zeolitic framework structure. Si exists in Q^4(mAl) environments ($1 \leq m \leq 4$, depending on the Al/Si ratio of the gel). The negative charge arising from Al^{3+} in tetrahedral (four-fold) coordination is charge-balanced by the alkali cations provided by the activating solution, commonly Na or K. The secondary reaction products are zeolites, such as hydrosodalite, zeolite P, Na-chabazite, zeolite Y, and faujasite [118] (Figures 7 and 8). Such a reaction product assemblage is favorable for the chemical binding of many contaminants, providing high immobilization potential, accompanied by high physical–mechanical performance of the hardened wasteforms. Thus, the geopolymers have higher efficiency than AACs based on high-Ca precursors, fueling increasing studies in this field started by Davidovits et al. [40,41] at the end of 1990s. Thus far, the binding efficiency of geopolymers for 37 elements, including Sr, Cs, Pb, Cr, and Zn, has been proven [24,28,138,139]. However, the encapsulation of reactive metals and oils by geopolymers and the radiolysis of water in the binder gel under gamma irradiation require further investigation and new approaches [7,140–144]. It is worth noting that geopolymers have been intensively and increasingly studied in recent decades for RW solidification (Figure 9) [107].

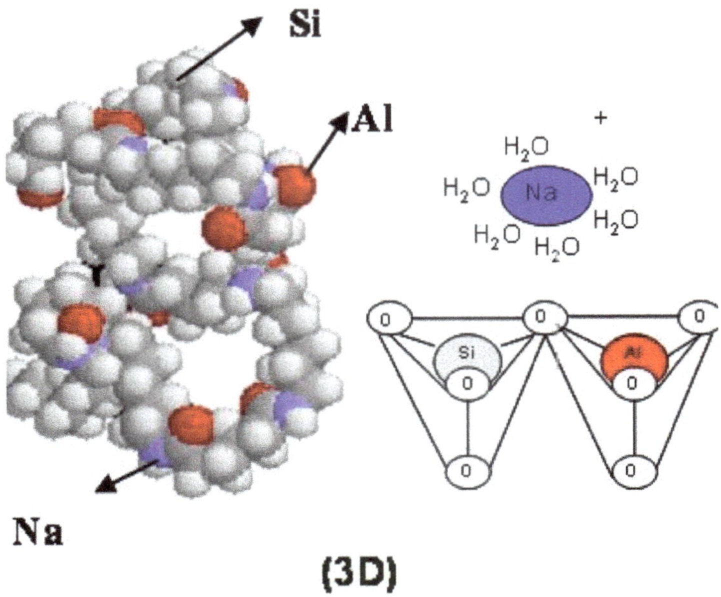

Figure 7. N-A-S-H gel structure [118].

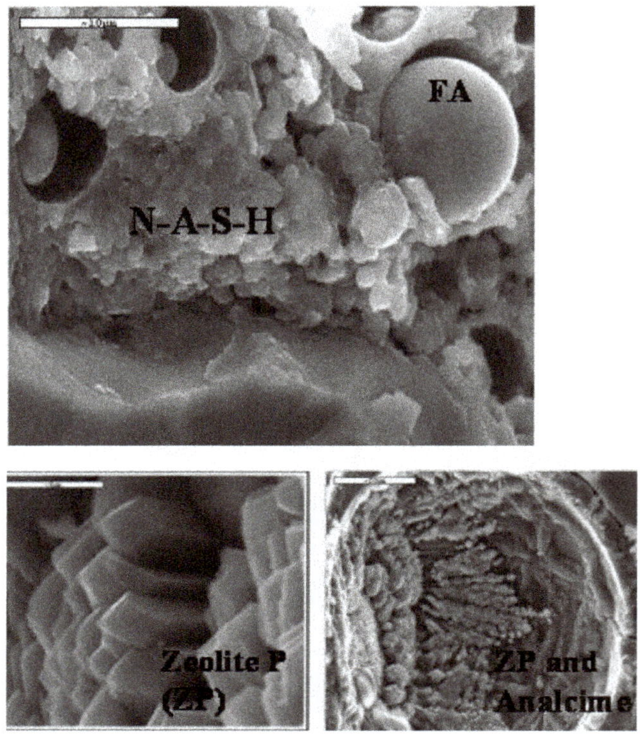

Figure 8. SEM micrographs of N-A-S-H and zeolites [118].

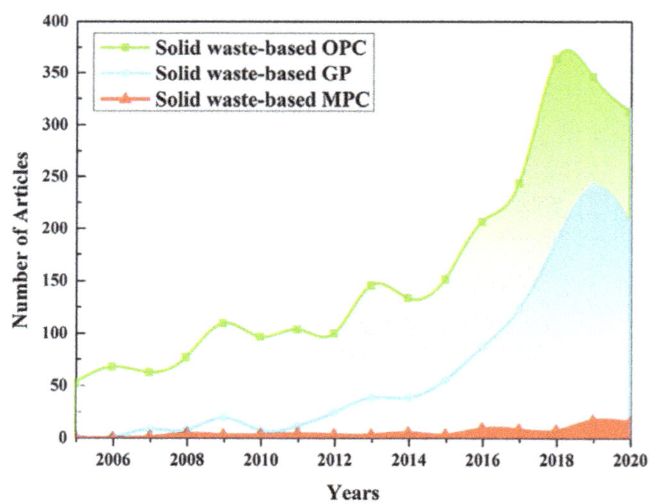

Figure 9. Number of OPC-, GP-, and MPC-related references in the past 15 years [107].

The immobilization of Cs^+ and Sr^+ remains the subject of ongoing research. The results presented by Walkley et al. [139] contribute to further understanding the mechanism of Sr and Ca immobilization in MK-based geopolymers. The incorporation of alkaline earth cations resulted in a minor decrease in the Si/Al ratio of the (N,K)-A-S-H gel; no other changes were found for pastes hardened at 20 °C; however, in those cured at 80 °C, the incorporation of Sr appeared to promote the formation of zeolite A over the faujasite zeolite phases. According to El Alouani et al. [145], the kinetics of Cs^+ adsorption by AACs followed pseudo-first-order and pseudo-second-order kinetic models, indicating that both the physisorption and chemisorption mechanisms controlled the adsorption process. Many studies have stated different ways for effective chemical immobilization of Cs and Sr by geopolymers, such as: (i) in the form of clinoptilolite incorporated with Cs and Sr, Sr-loaded zeolite A [146,147], (ii) Sr-loaded titanate spent adsorbents [148,149], (ii) Cs waste as an activating solution [150–152]; (iii) Cs and Sr hydroxides [153], and (iv) sewage sludge ash contaminated with radiocesium [154]. Lin et al. [155] reported that MK-based AAC safely and effectively solidified IERs (up to12 wt%) for immobilizing both Cs^+ and Sr^{2+}. Tan et al. [156] also found that the MK-based AAC binder exhibited better leaching resistance than the PC binder in deionized water, solution of H_2SO_4, $MgSO_4$, and acetic acid buffer. The compressive strength of MK-based AAC declined to a lesser extent after freeze–thaw cycles and high-temperature tests than that of PC.

Based on the presented data, Arbel-Haddad et al. [157] concluded that designing GP formulations to provide a higher amount of zeolite F is reasonable for the production of matrices to immobilize Cs, because Cs is mainly bound by zeolite F, rather than by other reaction phases of low-Si MK geopolymers.

Curing at elevated temperatures and high-temperature sintering of the hardened products, leading to the formation of different crystalline phases in AAC production, is one approach for improving Cs^+ immobilization. Fu et al. [158] observed the formation of analcime and pollucite in Na- and Cs-rich MK-based systems cured at 170 °C under hydrothermal conditions.

Chaerun et al. [159] found that K-based MK-GP incorporated with a chabazite adsorbent was more effective than Na-based AAM or PC for the immobilization of Cs. The crystallographically disordered nature of K-AAM and its pH were the main contributors to K-ion migration and the structural change of aluminosilicate rings in chabazite, thus resulting in the formation of K-type chabazite with amorphous properties similar to those of K-AAM. Due to the similar ionic radii and retention selectivities of Cs and K, both can be confined during the reconstruction of the aluminosilicate and are crystallized into pollucite during the AAM fabrication process. Jain et al. [160] investigated the effect of the Cs content on the reaction products and pore structure of FA-based geopolymers. A higher Cs loading ($\geq$8 wt.%) facilitated in situ pollucite crystallization within the FA-GP matrix (cured at 90 °C for 7 d) and significantly enhanced Cs immobilization (leachability index of 11.5–14.5).

Ahn et al. [161] investigated MK-based geopolymers for the solidification of sulfate-rich HyBRID sludge waste, consisting of cristobalite (SiO_2) and barite ($BaSO_4$) as major components. The K-based geopolymer had a higher mechanical strength (up to 14.3 MPa) than the Na-based geopolymer, and could also solidify more HyBRID sludge waste, thereby increasing the waste loading to 53.8 wt%. The pure geopolymer with the HyBRID sludge waste exhibited good mechanical stability at a Si/Al ratio of 1.8. However, the highest compressive strength was achieved for the geopolymer prepared with 40 wt.% HyBRID sludge waste at a Si:Al ratio of 1.6. Authors attributed these differences to the consumption of water and additional Si sources.

Kim et al. [162] investigated the potential of simulated borate waste (sodium tetraborate decahydrate ($Na_2B_4O_7{\cdot}10H_2O$), sodium nitrate ($NaNO_3$), potassium nitrate (KNO_3), calcium nitrate ($Ca(NO_3)_2{\cdot}4H_2O$), zinc nitrate ($Zn(NO_3)_2{\cdot}6H_2O$), and magnesium nitrate ($Mg(NO_3)_2{\cdot}6H_2O$)) as raw materials for producing MK-based geopolymers, all of which had higher compressive strengths than the PC-based cementitious wasteforms. The K-geopolymers (40 MPa) had a higher 7 d compressive strength than the Na-geopolymers (24 MPa). However, the compressive strength tended to increase in proportion to the Si/(Al+B) ratio (1.3–1.5), irrespective of the type of alkaline cation. This variation was attributed to the viscosity of the activator used for geopolymer formation, the atomic size of the alkaline cations, and the increase in the Si content. However, as shown in another study [163], the addition of borax increased the reactivity and geopolymer polycondensation.

He et al. [164] compared the temperature-dependence (from 25 °C to 60 °C) and environment-dependence of the kinetics of Sr^{2+} and Cs^+ leaching during long-term leaching tests and reported a low leaching rate and relatively high mechanical properties of Na geopolymers. A high storage temperature and salt medium accelerated the leaching of Sr^{2+} and Cs^+ from the matrix by enhancing the driving force of the leaching process and the corrosion effect. Compared with the change in temperature, the leaching of Sr^{2+} and Cs^+ is more sensitive to changes in the leaching medium, indicating that the corrosion effect of the salt medium plays a more important role in accelerating the leaching of radioactive elements and degradation of the immobilizing matrix.

The performance of AACs as cementitious materials can be effectively manipulated by using mixed precursors. Combining Ca-rich and Ca-free starting materials in alkali-activated systems produced chemically and structurally different binder gels, such as calcium-containing (C-(A)-S-H) and calcium-free (N-A-S-H) [165,166] gels, as well as mixed (C-N-(A)-S-H) gels. C-N-(A)-S-H represents mechanically strong gels consisting of crystalline tobermorite-like and amorphous cross-linked products with a relatively high content of silica in Q^1, Q^2, and Q^3 sites, which leads to the densification of the binder–gel microstructure [167–169]. Densification positively affects the physical–mechanical properties and immobilizing properties of the matrixes [170–178].

AACs based on both high-Ca and Ca-free precursors demonstrated good effects in the solidification of IERs with loadings up to 60% (by volume) (Table 3).

Table 3. The formulations and properties of AACs incorporated with IERs.

Precursor	Alkali Reactant	Ion-Exchange Resins	Details	Ref.
Ground granulated BFS—100% (wt%)	$Na_2SiO_3{\cdot}9H_2O$ (NSH_9)+ sodium hydroxide. (NaOH) (5–7% by Na_2O)	Loading of cationic borate IERs 35% by volume (pH 8.5–10.5)	28 d compressive strength up to 7.3 MPa	[179]
Ground granulated BFS—31–48% (wt%), wollastonite—6–8% (wt%)	9–12 M NaOH (SiO_2/Na_2O = 0.8 and SiO_2/Al_2O_3 = 50)	Loading of wet IERs 45%, dry IERs 22% (wt))	28 d compressive strength up to 22 MPa. (SiO_2/Na_2O = 0.8 and SiO_2/Al_2O_3 = 50)	[180]
MK 90%, Ground granulated BFS—31–48% (wt%)	Sodium silicate/sodium hydroxide 58% (wt%)	Loading of wet IERs 12% (wt)	28 d compressive strength 13.9 MPa	[155]
FA 56% (wt%)	$NaOH/Na_2SiO_3$ 28% (wt%)	10% (wt)	28 d compressive strength 13.9 MPa	[181]
Ground granulated BFS—35–40% (wt%), bentonite 2.5–8%, $Ca(OH)_2$—4–5%, or OPC—8–10%	$Na_2SiO_3{\cdot}5H_2O$—6% or Na_2CO_3—1.5–3%	Loading of cationic and anionic IERs 60% by volume	28 d compressive strength up to 18 MPa	[182]

In Czech Republic, the waste conditioning using geopolymers has been carried out at the Dukovany NPP by the external supplier AMEC Nuclear Slovakia. By the end of 2012, the Dukovany NPP Units 1 and 2 extracted, conditioned, and disposed of more than 200 t of sludge and IX resins [42].

In Slovak Republic, AllDeco Ltd has developed a proprietary geopolymer matrix (called SIAL) for embedding various intermediate-level wastes resulting from Slovak power reactors [163]. Some of the materials encapsulated in geopolymer matrices are bottom sludges from the long-term storage of spent nuclear fuel elements, sludges from the sedimentation tank of a reactor, and several other sludges. Some of the sludges are formed from an emulsified mixture of organic compounds from the cooling media and contain a large amount of calcium and magnesium hydrocarbonates. The activity of the ^{137}Cs in the sludges is ~105–108 Bq/L. Once these sludges were solidified in the geopolymer matrix and placed in 60 L drums, the surface dosage on the drums was 10–20 mGy/h. The D value for the 137Cs in samples taken from the drum was >8 for the ANSI 16.1 test and the compressive strength was 25 MPa. About 20 wt% (on a dry basis) of waste was encapsulated. Organic ion exchange resins on their own and in mixtures of sludges were also encapsulated in geopolymer matrices. It was possible to encapsulate ~20 wt% (on a dry basis) for geopolymers compared to 10 wt% for OPC. These were placed in 200 L drums. The dosages on the drum surfaces were 130–600 μGy/h and the D value (leachability index) for ^{137}Cs was >9 on cut samples from the drums. All the drums used were made from stainless steel. The SIAL matrix (geopolymer) has been accepted by the Slovak Nuclear Authority (UJDSR) and the Czech Nuclear Authority (SUJB) for placement in their respective repositories. AllDeco Ltd emplaced these drums in the Slovak repository in 2003.

In Australia, ANSTO geopolymers derived from metakaolin and alkaline silicate solutions with nominal Na/Al and Si/Al molar ratios of 1 and 2 were studied for the stabilization of ^{137}Cs and ^{90}Sr with Cs inhabited the amorphous phase, whereas Sr was incorporated only partly, being preferentially partitioned to crystalline $SrCO_3$ [43,44].

In the USA, geopolymers with Si to Al ratios of 1 to 1 and 2 to 1 were investigated for the stabilization of hazardous Resource Conservation and Recovery Act (RCRA) metals, such as Ni, Se, Ba, Hg, Cd, Cr, and Pb [44]. Special geopolymer formulations, marketed under the name DuraLith, have been patented for stabilization of 129I and 99Tc at Hanford Waste Treatment Plant [44]. The DuraLith geopolymer is composed of three components: an activator, a binder, and an enhancer [45]. DuraLith is an alkali-activated geopolymer waste form developed by the Vitreous State Laboratory at The Catholic University of America for encapsulating liquid radioactive waste. A DuraLith waste form developed for treating Hanford secondary waste liquids is prepared by the alkali-activation of a mixture of ground blast furnace slag and metakaolin with sand used as a filler material. The DuraLith geopolymers demonstrates compressive strength above 27 MPa, and ANSI/ANS 16.1 Leachability indexes for Tc as high as 9. Savannah River Site has used the FA-based geopolymer. The Class F FA resulted from the burning of harder, older anthracite and bituminous coal and is pozzolanic in nature, containing less than 7% CaO. Adding a chemical activator, such as sodium silicate (water glass) to a Class F ash can form a geopolymer. The wasteform geopolymer recipe contained in wt.%: waste granules 47.4, Class F ash 12.8, Na_2O $2SiO_2$ 44.1, NaOH (50 wt%) 12.5, and water 8.1.

3. Conclusions

Modern mineral matrices for nuclear waste immobilization now include a wide class of cementitious materials with various chemical–mineralogical compositions, high encapsulation capacities, and technological and engineering performance, comprising not only traditional Portland cements but also non-Portland clinker inorganic binders. Consistent development of research in this field has extended the theoretical basis, potential, and versatility of cementation technology. Based on the review and analysis of trends and achievements in the immobilization of nuclear wastes using CACs, CSACs, M-S-H, phosphate, and AACs, the following conclusions can be drawn:

1. Scientific and practical interest in the use of alternative cementitious materials for nuclear waste treatment and conditioning has only increased in recent years.
2. The appropriate formulations of alternative cements combine the high ion-exchange capacity and high physical–mechanical properties of hardened pastes, such as quick setting, high strength, high temperature, and chemical resistance. The design of cemen-

titious materials with "targeted" reactive phase assemblage and excellent physical–mechanical performance is achievable by: (i) varying a range of factors that govern the properties of binder systems; (ii) introducing supplementary cementitious materials into CACs, CSACs, and phosphate cements.

3. Radionuclides (Cs, Sr, etc.), borate- and nitrate-containing wastes, oils, IERs, solid wastes, etc., remain the subject of a great number of studies, most of which highlight the superiority of alternative cements as solidifying matrices compared to PC in terms of effective physical encapsulation and chemical binding of RW, the waste loading rate, and durability of the cementitious wasteforms.
4. New insights into the mechanism of action of Cs, Sr, B, I, etc., as well as the resultant reaction products, have been proposed by several researchers. Presented results proved high efficiency of both chemical binding and physical encapsulation capacity of alternative cementitious materials.
5. The cementation of RW as a "raw material" for cements is a new perspective trend of cementation technique, demonstrating good results. Thus, Cs waste as an activating solution and simulated borate waste were used in recent studies for producing geopolymers.
6. Several studies have demonstrated that phosphate cements and AACs are effective for high-level waste immobilization. MPCs provided good results in rapid immobilization of Cs and Sr in wastes from the PUREX process.

Further studies in the development of alternative-cements-based mineral matrices for the solidification of various types of RW, as well as research into reaction mechanisms of matrices and contaminants, long-term immobilization and durable performance of the wasteforms, the behavior of the cemented wastes in emergency situations, new analytical techniques, and predictive computational modeling of cementitious wasteforms, will contribute to the further sustainable development of RW management systems.

Funding: This research received no external funding.

Institutional Review Board Statement: Not applicable.

Informed Consent Statement: Not applicable.

Data Availability Statement: Not applicable.

Conflicts of Interest: The authors declare no conflict of interest.

References

1. Sadiq, M.; Wen, F.; Dagestani, A.A. Environmental footprint impacts of nuclear energy consumption: The role of environmental technology and globalization in ten largest ecological footprint countries. *Nucl. Eng. Technol.* **2022**, *54*, 3672–3681. [CrossRef]
2. Pioro, I.; Duffey, R.B.; Pioro, R. Overview of current status and future trends of nuclear power industry of the world. In *Fundamental Issues Critical to the Success of Nuclear Projects*; Boucau, J., Ed.; Elsevier: Amsterdam, The Netherlands, 2022; pp. 23–68.
3. Sadiq, M.; Wen, F.; Bashir, M.F.; Amin, A. Does nuclear energy consumption contribute to human development? Modeling the effects of public debt and trade globalization in an OECD heterogeneous panel. *J. Clean. Prod.* **2022**, *375*, 133965. [CrossRef]
4. Christoforidis, T.; Katrakilidis, C.; Karakotsios, A.; Dimitriadis, D. The dynamic links between nuclear energy and sustainable economic growth. Do institutions matter? *Prog. Nucl. Energy.* **2021**, *139*, 103866. [CrossRef]
5. IAEA. *Radiation Protection and Safety of Radiation Sources: International Basic Safety Standards*; IAEA Safety Standards Series GSR Part 3; IAEA: Vienna, Austria, 2014.
6. Ojovan, M.I.; Steinmetz, H.J. Approaches to Disposal of Nuclear Waste. *Energies* **2022**, *15*, 7804. [CrossRef]
7. Stanisz, P.; Oettingen, M.; Cetnar, J. Development of a Trajectory Period Folding Method for Burnup Calculations. *Energies* **2022**, *15*, 2245. [CrossRef]
8. Oettingen, M. The Application of Radiochemical Measurements of PWR Spent Fuel for the Validation of Burnup Codes. *Energies* **2022**, *15*, 3041. [CrossRef]
9. Bell, M.Z.; Macfarlane, A. "Fixing" the nuclear waste problem? The new political economy of spent fuel management in the United States. *Energ. Res. Soc. Sci.* **2022**, *91*, 102728. [CrossRef]
10. Lappi, P.; Lintunen, J. From cradle to grave? On optimal nuclear waste disposal. *Energy Econ.* **2021**, *103*, 105556. [CrossRef]
11. Kurniawan, T.A.; Othman, M.H.D.; Singh, D.; Avtar, R.; Hwang, G.H.; Setiadi, T.; Lo, W.-h. Technological solutions for long-term storage of partially used nuclear waste: A critical review. *Annals Nucl. Energy* **2022**, *166*, 108736. [CrossRef]

12. Esser, M.; Borremans, A.; Dubgorn, A.; Shaban, A. Nuclear Waste Transportation: Quality Assurance and Control. *Transp. Res. Proc.* **2021**, *54*, 871–882. [CrossRef]
13. Petković, D. Modeling of Information System for Nuclear Waste Management. In *Encyclopedia of Renewable and Sustainable Materials*; Hashmi, S., Choudhury, I.A., Eds.; Elsevier: Amsterdam, The Netherlands, 2020; pp. 490–496.
14. Barton, N. A review of mechanical over-closure and thermal over-closure of rock joints: Potential consequences for coupled modelling of nuclear waste disposal and geothermal energy development. *Tunnel. Undergr. Space Techn.* **2020**, *99*, 103379. [CrossRef]
15. Sanders, M.C.; Sanders, C.E. (Eds.) *International nuclear waste management framework, In Nuclear Waste Management Strategies*; Academic Press: Cambridge, MA, USA, 2020; pp. 31–45.
16. Lin, Z.; Dagnelie, R.V.H.; Blin, V.; Vinsot, A.; Lettry, Y.; Hautefeuille, B.; Blanchin, M.-G. Design of a robust and compact gamma sensor: Assessment of uranium detection in underground COx mudstone for nuclear waste management. *Nucl. Instr. Meth. Phys. Res. Sec. A: Accel. Spectr. Det. Assoc. Equipm.* **2019**, *938*, 14–19. [CrossRef]
17. Ojovan, M. *Handbook of Advanced Radioactive Waste Conditioning Technologies*; Woodhead Publishing: Sawston, UK, 2011; pp. 1–15.
18. El Khessaimi, Y.; Taha, Y.; Elghali, A.; Mabroum, S.; Hakkou, R.; Benzaazoua, M. Green and low-carbon cement for stabilization/solidification. In *Low Carbon Stabilization and Solidification of Hazardous Wastes*; Elsevier: Amsterdam, The Netherlands, 2022; pp. 15–30.
19. Abdel Rahman, R.O.; Rakhimov, R.Z.; Rakhimova, N.R.; Ojovan, M.I. *Cementitious Materials for Nuclear Waste Immobilization*; Wiley: New York, NY, USA, 2014; ISBN 9781118512005.
20. Aitcin, P.C. Science and Technology of Concrete Admixtures. In *Supplementary Cementitious Materials and Blended Cements*; Aitcin, P.C., Flatt, R.J., Eds.; Woodhead Publishing: Sawston, UK, 2016.
21. Juenger, M.C.G.; Snellings, R.; Bernal, S.A. Supplementary cementitious materials: New sources, characterization, and performance insights. *Cem. Concr. Res.* **2019**, *122*, 257–273. [CrossRef]
22. Glasser, F. Application Inorganic Cements to the Conditioning and Immobilisation of Radioactive Wastes. In *Handbook of Advanced Radioactive Waste Conditioning Technologies*; Ojovan., M., Ed.; Woodhead Publishing: Sawston, UK, 2011; pp. 67–134.
23. Shi, C.; Fernández Jiménez, A.; Palomo, A. New cements for the 21st century: The pursuit of an allternative to Portland cement. *Cem. Concr. Res.* **2011**, *41*, 750–763. [CrossRef]
24. Zhu, Y.; Zheng, Z.; Deng, Y.; Shi, C.; Zhang, Z. Advances in immobilization of radionuclide wastes by alkali activated cement and related materials. *Cem. Concr. Comp.* **2022**, *126*, 104377. [CrossRef]
25. Li, J.; Chen, L.; Wang, J. Solidification of radioactive wastes by cement-based materials. *Progr. Nucl. Energy.* **2021**, *141*, 103957. [CrossRef]
26. Abdel Rahman, R.O.; Ojovan, M.I. Toward Sustainable Cementitious Radioactive Waste Forms: Immobilization of Problematic Operational Wastes. *Sustainability* **2021**, *13*, 11992. [CrossRef]
27. Abdel Rahman, R.O.; Ojovan, M.I. Introduction to the Nuclear Industry Sustainability. In *Sustainability of Life Cycle Management for Nuclear Cementation-Based Technologies*; Abdel Rahman, R.O., Ojovan, M.I., Eds.; Elsevier-Woodhead Publishing: Chichester, UK, 2021; pp. 3–47.
28. Provis, J.L.; Deventer, J.S.J. *Geopolymers, Structure, Processing, Properties and Industrial Applications*; Woodhead Publishing Limited: Sawston, UK, 2009; pp. 401–416.
29. Vu, T.H.; Gowripalan, N. Mechanism of heavy metal immobilisation using geopolymerization techniques. *J. Adv. Concr. Techn.* **2018**, *16*, 124–135. [CrossRef]
30. Cau-dit-Coumes, C. Alternative Binders to Ordinary Portland Cement for Radwaste Solidification and Stabilization. In *Cement-Based Materials for Nuclear Waste Storage*; Bart, F., Cau-dit-Coumes, C., Frizon, F., Lorente, S., Eds.; Springer: Berlin/Heidelberg, Germany, 2012; pp. 171–191.
31. Kearney, S.; Yorkshire, A.; Geddes, D.A.; Hanein, T.; Nelson, S.; Provis, J.; Walkley, B. Cement-based stabilization/solidification of radioactive waste. In *Low Carbon Stabilization and Solidification of Hazardous Wastes*; Tsang, D.C.W., Wang, L., Eds.; Elsevier: Amsterdam, The Netherlands, 2022; pp. 407–431.
32. Moneghan, D.; Shkoukani, G.; Eapen, J.; Bourham, M.; Mustafa, W.; Kilani, R. Corrosion resistance and mechanical properties of Armakap cement for nuclear applications. *Nucl. Eng. Des.* **2021**, *374*, 111049. [CrossRef]
33. Berger, S.; Fryda, H. Calcium aluminate cements for nuclear wastes conditioning: Literature review and new approaches. In Proceedings of the 1st International Symposium on Cement-Based Materials for Nuclear Wastes, Avignon, France, 11–14 October 2011.
34. Walling, S.A.; Provis, J.L. Magnesia-based cements: A journey of 150 years, and cements for the future? *Chem. Rev.* **2016**, *116*, 41704204. [CrossRef]
35. Wagh, A.S.; Strain, R.; Jeong, S.Y.; Reed, D.; Krause, T.; Singh, D. Stabilization of rocky flats Pu-contaminated ash within chemically bonded phosphate ceramics. *J. Nucl. Mater.* **1999**, *265*, 295–307. [CrossRef]
36. Wagh, A.S.; Sayenko, S.Y.; Shkuropatenko, V.A.; Tarasov, R.V.; Dykiy, M.P.; Svitlychniy, Y.O.; Virych, V.D.; Ulybkina, E.A. Experimental study on cesium immobilization in struvite structures. *J. Hazard. Mater.* **2016**, *302*, 241–249. [CrossRef] [PubMed]
37. Wagh, A.S.; Singh, D.; Jeong, S.Y.; Strain, R.V. Ceramicrete stabilization of low-level mixed wastes, a complete story. In Proceedings of the 18th Annual DOE Low-Level Radioactive Waste Management Conference, Salt Lake City, UT, USA, 16 January 1998.

38. Wagh, A.S.; Jeong, S.Y.; Singh, D.; Strain, R.; No, H.; Wescott, J. *Stabilization of Contaminated Soil and Wastewater with Chemically Bonded Phosphate Ceramics*; USDOE Office of Environmental Restoration and Waste Management: Washington, DC, USA, 1997.
39. Krivenko, P.; Skurchinskaya, J.; Lavrinenko, L.; Starkov, O.; Konovalov, E. Physico-chemical bases of radioactive wastes immobilization in a mineral-like solidified stone. In Proceedings of the First International Conference on Alkaline Cements and Concretes, Kiev, Ukraine, 11–14 October 1994; pp. 1095–1106.
40. Comrie, D.C.; Davidovits, J. Long term durability of hazardous toxic and nuclear waste disposals. In Proceedings of the Geopolymer'88, 1st European Conference on Soft Mineralurge, ComPiegne, France, 1–3 June 1988; pp. 125–134.
41. Kunze, C.; Kiessig, G. Solidification of various radioactive residues by geopolymere with special emphasis on long-term stability. In Proceedings of the Geopolymer'99, 2nd International Conference, Saint-Quentin, France, 30 June–2 July 1999.
42. IAEA. *Use of Benchmarking System for Operational Waste from WWER Reactors*; IAEA TECDOC-1815; IAEA: Vienna, Austria, 2017.
43. Vance, E.R.; Perera, D.S. Development of Geopolymers for Nuclear Waste Immobilization. In *Handbook of Advanced Radioactive Waste Conditioning Technologies*; Woodhead: Oxford, UK, 2011; Chapter 8; p. 207.
44. Jantzen, C.M.; Lee, W.E.; Ojovan, M.I. Radioactive Waste (RAW) Conditioning, Immobilisation, and Encapsulation Processes and Technologies: Overview and Advances. In *Radioactive Waste Management and Contaminated Site Clean-Up: Processes, Technologies and International Experience*; Woodhead: Cambridge, UK, 2013; Chapter 6; pp. 171–272.
45. Mattigod, S.V.; Westsik, J.H., Jr. *Secondary Waste Form Down-Selection Data Package—DuraLith*; PNNL-Pacific Northwest National Laboratory: Richland, WA, USA, 2011.
46. Volgensky, A.V. *Mineral Binders*; Stroiizdat: Moscow, Russia, 1979; pp. 102–136.
47. Kouznetsova, T.V.; Sychev, M.M.; Osokin, A.P.; Korneev, V.I.; Sudakas, L.G. *Special Cements*; Stroiizdat: St. Petersburg, Russia, 1997; pp. 57–69.
48. Amathieu, L.; Estienne, F. Impact of the conditions of ettringite formation on the performance of products based on CAC + CS + OPC. In Proceedings of the 15th Ibausil: International Conference on Building Materials, Weimar, Germany, 24–27 September 2003.
49. Amathieu, L.; Kwasny-Echterhagen, R. In-depth investigation of ettringite formation in high performance self levelling compounds made with calcium aluminate. In Proceedings of the 16th Ibausil: International Conference on Building Materials, Weimar, Germany, 20–23 September 2006.
50. Fryda, H.; Vetter, G.; Ollitrault-Fichet, R.; Boch, P.; Capmas, A. Fomation of chabazite in mixes of calcium aluminate cement and silica fume used for cesium immobilization. *Adv. Cem. Res.* **1996**, *29*, 29–39. [CrossRef]
51. Goni, S.; Guerrero, A. Stability of calcium aluminate cement matrices mixed with borate solution. In Proceedings of the Conference of Calcium Aluminate Cement, Edinburgh, UK, 16–19 July 2001.
52. Madrid, J.; Macias, A.; Fernandez, E. Characterisation of cadmium and lead in calcium aluminate cement matrices. In Proceedings of the Conference of Calcium Aluminate Cement, Edinburgh, UK, 16–19 July 2001.
53. Toyohara, M.; Kaneko, M.; Mitsutska, N.; Fujihara, H.; Saito, N.; Murase, T. Contribution to understanding iodine sorption mechanism onto mixed solid alumina cement and calcium compounds. *J. Nucl. Sci. Technol.* **2002**, *39*, 950–956. [CrossRef]
54. Nithya, R.; Barathan, S.; Gopalarishan, M.; Sivakumar, G.; Gobinath, B. The hydration of heavy metal salt admixtured high alumina cement—A X-ray diffraction study. *Appl. Phys. Res.* **2009**, *2*, 19–24. [CrossRef]
55. Kononenko, O.A.; Milyutin, V.V. Solidification of boron-containing ion-exchange resins with aluminosilicate binder. *Radiochem* **2019**, *61*, 607–611. [CrossRef]
56. Kononenko, O.A.; Milyutin, V.V.; Nekrasova, N.A. Composite Binders for Solidification of Spent Ion-Exchange Resins. *Atom. Energy* **2019**, *125*, 257–261. [CrossRef]
57. Irisawa, K.; Namiki, M.; Taniguchi, T.; Kinoshita, H. Solidification and stabilization of strontium and chloride ions in thermally treated calcium aluminate cement modified with or without sodium polyphosphate. *Cem. Concr. Res.* **2022**, *156*, 106758. [CrossRef]
58. Takahatake, Y.; Watanabe, S.; Irisawa, K.; Shiwaku, H.; Watanabe, M. Structural characterization by X-ray analytical techniques of calcium aluminate cement modified with sodium polyphosphate containing cesium chloride. *J. Nucl. Mater.* **2021**, *556*, 153170. [CrossRef]
59. Wang, J. Hydration Mechanism of Cements Based on low-CO_2 Clinkers Containing Belite, Ye'elimite and Calcium Alumino-ferrite. Ph.D. Thesis, Université des Sciences et Technologie de Lille-Lille I, Lille, France, 2010.
60. Klein, A. Calcium Aluminosulfate and Expansive Cements Containing Same. U.S. Patent 3,155,526, 3 November 1964.
61. Odler, I. *Special Inorganic Cements*; Taylor & Francis Group: London, UK, 2000.
62. Juenger, M.C.G.; Winnefeld, F.; Provis, J.L.; Ideker, J.H. Advances in alternative cementitious binders. *Cem. Concr. Res.* **2011**, *41*, 1232–1243. [CrossRef]
63. Paglia, C.S.B.; Wombacher, F.J.; Böhni, H.K. Hydration, strength, and microstructural development of high early-strength (C4A3SP) activated burnt oil shale based cement system. *ACI Mater. J.* **2001**, *98*, 379–385.
64. Marchi, M.; Costa, U. Influence of the Calcium Sulphate and W/C Ratio on the Hydration of Calcium Sulfoaluminate Cement. In Proceedings of the XIII International Congress on the Chemistry of Cement, Madrid, Spain, 3–8 July 2011; p. 191.
65. Albino, V.; Cioffi, R.; Marroccoli, M.; Santoro, L. Potential application of ettringite generating systems for hazardous waste stabilization. *J. Hazard. Mater.* **1996**, *51*, 241–252. [CrossRef]
66. Peysson, S.; Péra, J.; Chabannet, M. Immobilization of heavy metals by calcium sulfoaluminate cement. *Cem. Concr. Res.* **2005**, *35*, 2261–2270. [CrossRef]

67. Zhou, Q.; Milestone, N.B.; Hayes, M. An alternative to Portland cement for waste encapsulation—the calcium sulfoaluminate cement system. *J. Hazard. Mater.* **2006**, *136*, 120–129. [CrossRef]
68. Xu, X.; Bi, H.; Yu, Y.; Fu, X.; Wang, S.; Liu, Y.; Hou, P.; Cheng, X. Low leaching characteristics and encapsulation mechanism of Cs^{+} and Sr^{2+} from SAC matrix with radioactive IER. *J. Nucl. Mater.* **2021**, *544*, 152701. [CrossRef]
69. Cody, A.M.; Lee, H.; Cody, R.D.; Spry, P.G. The effects of chemical environment on the nucleation, growth, and stability of ettringite $[Ca_3Al(OH)_6]2(SO_4)3{\cdot}26H_2O$. *Cem. Concr. Res.* **2004**, *34*, 869–881. [CrossRef]
70. Cau-Dit-Coumes, C.; Dhoury, M.; Champenois, J.-B.; Mercier, C.; Damidot, D. Combined effects of lithium and borate ions on the hydration of calcium sulfoaluminate cement. *Cem. Concr. Res.* **2017**, *97*, 50–60. [CrossRef]
71. Champenois, J.-B.; Mesbah, A.; Cau-Dit-Coumes, C.; Renaudin, G.; Leroux, F.; Mercier, C.; Revel, B.; Damidot, D. Crystal structures of Boro-AFm and Boro-AFt phases. *Cem. Concr. Res.* **2012**, *42*, 1362–1370. [CrossRef]
72. Champenois, J.-B.; Dhoury, M.; Cau-Dit-Coumes, C.; Mercier, C.; Revel, B.; Le Bescop, P. Influence of sodium borate on the early age hydration of calcium sulfoaluminate cement. *Cem. Concr. Res.* **2015**, *70*, 83–93. [CrossRef]
73. Wu, K.; Shi, H.; Guo, X. Utilization of municipal solid waste incineration fly ash for sulfoaluminate cement clinker production. *Waste Manage.* **2011**, *31*, 2001–2008. [CrossRef]
74. Sun, Q.; Li, J.; Wang, J. Solidification of borate radioactive resins using sulfoaluminate cement blending with zeolite. *Nucl. Eng. Des.* **2011**, *241*, 5308–5315. [CrossRef]
75. Chen, W.; Ling, X.; Li, Q.; Yuan, B.; Li, B.; Ma, H. Experimental evidence on formation of ulexite in sulfoaluminate cement paste mixed with high concentration borate solution and its retarding effects. *Constr. Build. Mater.* **2019**, *215*, 777–785. [CrossRef]
76. Guo, B.; Xiong, Y.; Chen, W.; Saslow, S.A.; Kozai, N.; Ohnuki, T.; Dabo, I.; Sasaki, K. Spectroscopic and first-principles investigations of iodine species incorporation into ettringite: Implications for iodine migration in cement waste forms. *J. Hazard. Mater.* **2019**, *389*, 121880. [CrossRef] [PubMed]
77. Xu, J.; Wang, M.; Li, C.; Han, M.; Wang, Q.; Sun, Q. Metakaolin-Reinforced Sulfoaluminate-Cement-Solidified Wasteforms of Spent Radioactive Resins—Optimization by a Mixture Design. *Coatings* **2022**, *12*, 1466. [CrossRef]
78. Golynko-Volfson, S.L.; Sychev, M.M.; Sudakas, L.G.; Skoblo, L.I. *The Chemical Basis of the Technology and Application of Phosphate Coatings and Ligaments*; Chemistry Publisher: Moscow, Russia, 1968.
79. Xu, B.; Winnefeld, F.; Kaufmann, J.; Lothenbach, B. Influence of magnesium-to phosphate ratio and water-to-cement ratio on hydration and properties of magnesium potassium phosphate cements. *Cem. Concr. Res.* **2019**, *123*, 105781. [CrossRef]
80. Yang, Q.; Zhu, B.; Zhang, S.; Wu, X. Properties and applications of magnesia–phosphate cement mortar for rapid repair of concrete. *Cem. Conc. Res.* **2000**, *30*, 1807–1813. [CrossRef]
81. Lai, Z.; Wang, H.; Hu, Y.; Yan, T.; Lu, Z.; Lv, S.; Zhang, H. Rapid solidification of highly loaded high-level liquid wastes with magnesium phosphate cement. *Ceram. Int.* **2019**, *45*, 5050–5057.
82. Hou, D.; Yan, H.; Zhang, J.; Wang, P.; Li, Z. Experimental and computational investigation of magnesium phosphate cement mortar. *Constr. Build. Mater.* **2016**, *112*, 331–342. [CrossRef]
83. Lai, Z.; Lai, X.; Shi, J.; Lu, Z. Effect of Zn^{2+} on the early hydration behavior of potassium phosphate based magnesium phosphate cement. *Constr. Build. Mater.* **2016**, *129*, 70–78. [CrossRef]
84. Kulikova, S.A.; Belova, K.Y.; Tyupina, E.A.; Vinokurov, S.E. Conditioning of spent electrolyte surrogate LiCl-KCl-CsCl using magnesium potassium phosphate compound. *Energ.* **2020**, *13*, 1963. [CrossRef]
85. Vinokurov, S.E.; Kulikova, S.A.; Myasoedov, B.F. Magnesium potassium phosphate compound for immobilization of radioactive waste containing actinide and rare earth elements. *Materials* **2018**, *11*, 976. [CrossRef] [PubMed]
86. Vinokurov, S.E.; Kulikova, S.A.; Krupskaya, V.V.; Tyupina, E.A. Effect of characteristics of magnesium oxide powder on composition and strength of magnesium potassium phosphate compound for solidifying radioactive waste. *Russ. J. Appl. Chem.* **2019**, *92*, 490–497. [CrossRef]
87. Shkuropatenko, V.A. High-level wastes immobilization in ceramic and hydrated phosphate matrix. *East Eur. J. Phys.* **2016**, *3*, 49–60.
88. Vinokurov, S.E.; Kulyako, Y.M.; Slyuntchev, O.M.; Rovny, S.I.; Myasoedov, B.F. Low-temperature immobilization of actinides and other components of high-level waste in magnesium potassium phosphate matrices. *J. Nucl. Mater.* **2009**, *385*, 189–192. [CrossRef]
89. Sayenko, S.Y.; Shkuropatenko, V.A.; Pylypenko, O.V.; Karsim, S.O.; Zykova, A.V.; Kutnii, D.V.; Wagh, A.S. Radioactive waste immobilization of Hanford sludge in magnesium potassium phosphate ceramic forms. *Progr. Nucl. En.* **2022**, *152*, 104315. [CrossRef]
90. Sharp, J.H.; Milestone, N.B.; Hill, J.; Miller, E.W. Cementitious systems for encapsulation of intermediate level waste. In Proceedings of the 9th International Conference on Radioactive Waste Management and Environmental Remediation, Oxford, UK, 21–25 September 2003.
91. Milestone, N.B. Reactions in cement encapsulated nuclear wastes: Need for toolbox of different cement types. *Adv. Appl. Ceram.* **2006**, *105*, 13–20. [CrossRef]
92. Hayes, M.; Godfrey, I.H. Development of the use of alternative cements for the treatment of intermediate level waste. In Proceedings of the Waste Management Symposia, Tucson, AZ, USA, 1 March 2007; pp. 1–14.
93. Covill, A.; Hyatt, N.C.; Hill, J.; Collier, N.C. Development of magnesium phosphate cements for encapsulation of radioactive waste. *Adv. Appl. Ceram.* **2011**, *110*, 151–156. [CrossRef]

94. Cau Dit Coumes, C.; Lambertin, D.; Lahalle, H.; Antonucci, P.; Cannes, C.; Delpech, S. Selection of a mineral binder with potentialities for the stabilization/solidification of aluminum metal. *J. Nucl. Mater.* **2014**, *453*, 31–40. [CrossRef]
95. De Campos, M.; Davy, C.A.; Djelal, N.; Rivenet, M.; Garcia, J. Development of a stoichiometric magnesium potassium phosphate cement (MKPC) for the immobilization of powdered minerals. *Cem. Concr. Res.* **2021**, *142*, 106346. [CrossRef]
96. Chartier, D.; Sanchez-Canet, J.; Antonucci, P.; Esnouf, S.; Renault, J.P.; Farcy, O.; Lambertin, D.; Parraud, S.; Lamotte, H.; Coumes, C.C.D. Behaviour of magnesium phosphate cement-based materials under gamma and alpha irradiation. *J. Nucl. Mater.* **2020**, *541*, 152411. [CrossRef]
97. Bykov, G.L.; Abkhalimov, E.V.; Ershov, V.; Ershov, B.G. Radiolysis of Portland cement and magnesium phosphate cement: Effect of the content and state of water on the Physicochemical properties and the mechanism and kinetics of hydrogen formation. *Rad. Phys. Chem.* **2022**, *190*, 109822. [CrossRef]
98. Gardner, L.J.; Bernal, S.A.; Walling, S.A.; Corkhill, C.L.; Provis, J.L.; Hyatt, N.C. Characterisation of magnesium potassium phosphate cements blended with fly ash and ground granulated blast furnace slag. *Cem. Concr. Res.* **2015**, *74*, 78–87. [CrossRef]
99. Gardner, L.J.; Corkhill, C.L.; Walling, S.A.; Vigor, J.E.; Murray, C.A.; Tang, C.C.; Provis, J.L.; Hyatt, N.C. Early age hydration and application of blended magnesium potassium phosphate cements for reduced corrosion of reactive metals. *Cem. Concr. Res.* **2021**, *143*, 106375. [CrossRef]
100. Gardner, L.J.; Walling, S.A.; Corkhill, C.L.; Bernal, S.A.; Lejeune, V.; Stennett, M.C.; Provis, J.L.; Hyatt, N.C. Temperature transformation of blended magnesium potassium phosphate cement binders. *Cem. Concr. Res.* **2021**, *141*, 106332. [CrossRef]
101. U.S. Department of Energy. *Waste Isolation Pilot Plant Recovery Plan, Revision 0*; U.S. Department of Energy: Washington, DC, USA, 2014.
102. Lahalle, H.; Coumes, C.C.D.; Mercier, C.; Lambertin, D.; Cannes, C.; Delpech, S.; Gauffinet, S. Influence of the w/c ratio on the hydration process of a magnesium phosphate cement and on its retardation by boric acid. *Cem. Concr. Res.* **2018**, *109*, 159–174. [CrossRef]
103. Kononenko, O.A.; Milyutin, V.V.; Makarenkov, V.I.; Kozlitin, E.A. Immobilization of NPP evaporator bottom high salt-bearing liquid radioactive waste into struvite-based phosphate matrices. *J. Hazard. Mater.* **2021**, *416*, 125902. [CrossRef] [PubMed]
104. Tao, Y.; Zhenyu, L.; Yuanyuan, W.; Xin, H.; Jie, W.; Zhongyuan, L.; Shuzhen, L.; Feng, L.; Xiaoling, F.; Haibin, Z. Hydration process and microstructure of magnesium potassium phosphate cement with nitrate solution. *Sci. Total Environ.* **2020**, *703*, 134686. [CrossRef] [PubMed]
105. Zhenyu, L.; Yang, H.; Tao, Y.; Xin, H.; Zhongyuan, L.; Shuzhen, L.; Haibin, Z. Immobilization of solidified ceramic forms with magnesium phosphate cement. *Ceram. Int.* **2019**, *45*, 13164–13170. [CrossRef]
106. Pyo, J.-Y.; Um, W.; Heo, J. Magnesium potassium phosphate cements to immobilize radioactive concrete wastes generated by decommissioning of nuclear power plants. *Nucl. Eng. Techn.* **2021**, *53*, 2261–2267. [CrossRef]
107. Zhang, Q.; Cao, X.; Ma, R.; Sun, S.; Fang, L.; Lin, J.; Luo, J. Solid waste-based magnesium phosphate cements: Preparation, performance and solidification/stabilization mechanism. *Constr. Build. Mater.* **2021**, *297*, 123761. [CrossRef]
108. Culliton, A.; De Barra, E. Calcium Phosphate Cements. *Encycl. Mater.Techn. Ceram. Glass.* **2021**, *3*, 624–645.
109. Laniesse, P.; Coumes, C.C.D.; Le Saout, G.; Mesbah, A. Understanding the setting and hardening process of wollastonite-based brushite cement. Part 1: Influence of the Ca/P ratio and H_3PO_4 concentration of the mixing solution. *Cem. Concr. Res.* **2020**, *134*, 106094. [CrossRef]
110. Dorozhkin, S.V. Calcium Orthophosphate Cements and Concretes. *Materials* **2009**, *2*, 221–291. [CrossRef]
111. Brown, W.E.; Chow, L.C. Dental Restorative Cement Pastes. U.S. Patent 4,518,430, 21 May 1985.
112. Ul Hassan, M.; Iqbal, S.; Yun, J.-I.; Jin Ryu, H. Immobilization of radioactive corrosion products by cold sintering of pure hydroxyapatite. *J. Hazard. Mater.* **2019**, *374*, 228–237. [CrossRef] [PubMed]
113. Walling, S.A.; Kinoshita, H.; Bernal, S.A.; Collier, N.C.; Provis, J.L. Structure and properties of binder gels formed in the system $Mg(OH)_2$-SiO_2-H_2O for immobilisation of Magnox sludge. *Dalton Trans.* **2015**, *44*, 8126–8137. [CrossRef] [PubMed]
114. Zhang, T.; Li, T.; Zou, J.; Li, Y.; Zhi, S.; Jia, Y.; Cheeseman, C.R. Immobilization of radionuclide ^{133}Cs by magnesium silicate hydrate cement. *Materials* **2020**, *13*, 146. [CrossRef]
115. Zhang, T.; Zou, J.; Li, Y.; Jia, Y.; Cheeseman, C.R. Stabilization/solidification of strontium using magnesium silicate hydrate cement. *Processes* **2020**, *8*, 163. [CrossRef]
116. Bernard, E.; Lothenbach, B.; Cau-Dit-Coumes, C.; Pochard, I.; Rentsch, D. Aluminum incorporation into magnesium silicate hydrate (M-S-H). *Cem. Concr. Res.* **2020**, *128*, 105931. [CrossRef]
117. Glukhovsky, V. Ancient, Modern and Future Concretes. In Proceedings of the First International Conference of Alkaline Cements and Concretes, Kiev, Ukraine, 11–14 October 1994; VIPOL Stock Company: Kiev, Ukraine, 1994.
118. Garcia-Lodeiro, I.; Maltseva, O.; Palomo, Á.; Fernandez-Jimenez, A. Hybrid alkaline cements. Part I: Fundamentals. *Roman. J. Mater.* **2012**, *42*, 330–335.
119. Pacheco-Torgal, F.; Labrincha, J.; Leonelli, C.; Palomo, A.; Chindaprasit, P. *Handbook of Alkali-Activated Cements, Mortars and Concretes*; Elsevier: Amsterdam, The Netherlands, 2014.
120. Rakhimova, N.R.; Rakhimov, R.Z. A review on alkali-activated slag cements incorporated with supplementary materials. *J. Sustain. Cem. Based Mater.* **2014**, *3*, 61–74. [CrossRef]
121. Rakhimova, N.R.; Rakhimov, R.Z. Reaction products, structure and properties of alkali-activated metakaolin cements incorporated with supplementary materials: A review. *J. Mater. Res. Technol.* **2019**, *8*, 1522–1531. [CrossRef]

122. Rakhimova, N.R.; Rakhimov, R.Z. Literature review of advances in materials used in development of alkali-activated mortars, concretes, and composites. *J. Mater. Civ. Eng.* **2019**, *31*, 03119002. [CrossRef]
123. Lahoti, M.; Wong, K.K.; Yang, E.-H.; Tan, K.H. Effects of Si/Al molar ratio on strength endurance and volume stability of metakaolin geopolymers subject to elevated temperature. *Ceram. Int.* **2018**, *44*, 5726–5734. [CrossRef]
124. Latella, B.A.; Perera, D.S.; Durce, D.; Mehrtens, E.G.; Davis, J. Mechanical properties of metakaolin-based geopolymers with molar ratios of Si = Al ≈ 2 and Na = Al ≈ 1. *J. Mater. Sci.* **2008**, *43*, 2693–2699. [CrossRef]
125. Bernal, S.A.; Rodríguez, E.D.; Kirchheim, A.P.; Provis, J.L. Management and valorisation of wastes through use in producing alkali-activated cement materials. *J. Chem. Techn. Biotechn.* **2016**, *91*, 2365–2388. [CrossRef]
126. Mehta, A.; Siddique, R. An overview of geopolymers derived from industrial by-products. *Constr. Build. Mater.* **2016**, *127*, 183–198. [CrossRef]
127. Rakhimova, N.R. A review of calcined clays and ceramic wastes as sources for alkali-activated materials. *Geosyst. Eng.* **2020**, *23*, 287–298. [CrossRef]
128. Rakhimova, N.R. Calcium and/or magnesium carbonate and carbonate-bearing rocks in the development of alkali-activated cements—A review. *Constr. Build. Mater.* **2022**, *325*, 126742. [CrossRef]
129. Khalil, M.; Merz, E. Immobilization of intermediate-level wastes in geopolymers. *J. Nucl. Mater.* **1994**, *211*, 141–148. [CrossRef]
130. Garcia-Lodeiro, I.; Palomo, A.; Fernández-Jiménez, A.; Macphee, D.E. Compatibility studies between N-A-S-H and C-A-S-H gels. Study in the ternary diagram Na_2O-CaO-Al_2O_3-SiO_2-H_2O. *Cem. Concr. Res.* **2011**, *41*, 923–931. [CrossRef]
131. Ben Haha, M.; Le Saout, G.; Winnefeld, F.; Lothenbach, B. Influence of activator type on hydration kinetics, hydrate assemblage and microstructural development of alkali activated blast-furnace slags. *Cem. Concr. Res.* **2011**, *41*, 301–310. [CrossRef]
132. Blyth, A.; Eiben, C.A.; Scherer, G.W.; White, C.E. Impact of activator chemistry on permeability of alkali-activated slags. *J. Am. Ceram. Soc.* **2017**, *100*, 4848–4859. [CrossRef]
133. Shi, C.; Krivenko, P.V.; Roy, D.M. *Alkali-Activated Cements and Concretes*; Taylor & Francis: Abingdon, UK, 2006.
134. Rakhimova, N.R.; Rakhimov, R.Z.; Morozov, V.P.; Potapova, L.I.; Osin, Y.N. Mechanism of solidification of simulated borate liquid wastes with sodium silicate activated slag cements. *J. Clean. Prod.* **2017**, *149*, 60–69. [CrossRef]
135. Vandevenne, N.; Iacobescu, R.I.; Pontikes, Y.; Carleer, R.; Thijssen, E.; Gijbels, K.; Schreurs, S.; Schroeyers, W. Incorporating Cs and Sr into blast furnace slag inorganic polymers and their effect on matrix properties. *J. Nucl. Mater.* **2018**, *503*, 1–12. [CrossRef]
136. Huang, T.; Song, D.; Yin, L.-X.; Zhang, S.-W.; Liu, L.-F.; Zhou, L. Microwave irradiation assisted sodium hexametaphosphate modification on the alkali-activated blast furnace slag for enhancing immobilization of strontium. *Chemosphere* **2020**, *241*, 125069. [CrossRef]
137. Komljenovic, M.; Tanasijevic, G.; Dzunuzovic, N.; Provis, J. Immobilization of cesium with alkali-activated blast furnace slag. *J. Hazard Mater.* **2020**, *388*, 121765. [CrossRef]
138. Muhammad, F.; Huang, X.; Li, S.; Xia, M.; Zhang, M.; Liu, Q.; Hassan, M.A.S.; Jiao, B.; Yu, L.; Li, D. Strength evaluation by using polycarboxylate superplasticizer and solidification efficiency of Cr^{6+}, Pb^{2+} and Cd^{2+} in composite based geopolymer. *J. Clean. Prod.* **2018**, *188*, 807–815. [CrossRef]
139. Walkley, B.; Ke, X.; Hussein, O.H.; Bernal, S.A.; Provis, J.L. Incorporation of strontium and calcium in geopolymer gels. *J. Hazard. Mater.* **2020**, *382*, 121015. [CrossRef]
140. Chupin, F.; Dannoux-Papin, A.; Ngono Ravache, Y.; d'Espinose de Lacaillerie, J.-B. Water content and porosity effect on hydrogen radiolytic yields of geopolymers. *J. Nucl. Mater.* **2017**, *494*, 138–146. [CrossRef]
141. Bai, C.; Ni, T.; Wang, Q.; Li, H.; Colombo, P. Porosity, mechanical and insulating properties of geopolymer foams using vegetable oil as the stabilizing agent. *J. Eur. Ceram. Soc.* **2018**, *38*, 799–805. [CrossRef]
142. Barneoud-Chapelier, A.; Causse, J.; Poulesquen, A. Synthesis of geopolymer emulsions. *Mater. Lett.* **2020**, *276*, 128138. [CrossRef]
143. Davy, C.A.; Hauss, G.; Planel, B.; Lambertin, D. 3D structure of oil droplets in hardened geopolymer emulsions. *J. Am. Ceram. Soc.* **2019**, *102*, 949–954. [CrossRef]
144. Planel, B.; Davy, C.A.; Adler, P.M.; Hauss, G.; Bertin, M.; Cantarel, V.; Lambertin, D. Water permeability of geopolymers emulsified with oil. *Cem. Concr. Res.* **2020**, *135*, 106108. [CrossRef]
145. El Alouani, M.; Saufi, H.; Moutaoukil, G.; Alehyen, S.; Nematollahi, B.; Belmaghraoui, W.; Taibi, M.h. Application of geopolymers for treatment of water contaminated with organic and inorganic pollutants: State-of-the-art review. *J. Environ. Chem. Eng.* **2021**, *9*, 105095. [CrossRef]
146. Kuenzel, C.; Cisneros, J.F.; Neville, T.P.; Vandeperre, L.J.; Simons, S.J.R.; Bensted, J.; Cheeseman, C.R. Encapsulation of Cs/Sr contaminated clinoptilolite in geopolymers produced from metakaolin. *J. Nucl. Mater.* **2015**, *466*, 94–99. [CrossRef]
147. Xu, Z.; Jiang, Z.; Wu, D.; Peng, X.; Xu, Y.; Li, N.; Qi, Y.; Li, P. Immobilization of strontium-loaded zeolite A by metakaolin based-geopolymer. *Ceram. Int.* **2017**, *43*, 4434–4439. [CrossRef]
148. Soonthornwiphat, N.; Kobayashi, Y.; Toda, K.; Kuroda, K.; Islam, C.R.; Otake, T.; Provis, J.L.; Sato, T. Encapsulation of Sr-loaded titanate spent adsorbents in potassium aluminosilicate geopolymer. *J. Nucl. Sci. Technol.* **2020**, *57*, 1181–1188. [CrossRef]
149. Ke, X.; Bernal, S.A.; Sato, T.; Provis, J.L. Alkali aluminosilicate geopolymers as binders to encapsulate strontium-selective titanate ion-exchangers. *Dalton Trans.* **2019**, *48*, 12116–12126. [CrossRef]
150. Provis, J.L.; van Deventer, J.S.J. *Alkali Activated Materials: State-of-the-Art Report*; RILEM TC 224-AAM; Springer: Dordrecht, The Netherlands, 2014.

151. Bernal, S.A.; van Deventer, J.S.J.; Provis, J.L. What happens to 5 year old metakaolin geopolymers? In *Calcined Clays for Sustainable Concrete*; Springer: Dordrecht, Germany, 2015; pp. 315–321.
152. Berger, S.; Frizon, F.; Joussot-Dubien, C. Formulation of caesium based and caesium containing geopolymers. *Adv. Appl. Ceram.* **2009**, *108*, 412–417. [CrossRef]
153. Provis, J.L.; Walls, P.A.; van Deventer, J.S.J. Geopolymerisation kinetics: Effects of Cs and Sr salts. *Chem. Eng. Sci.* **2008**, *63*, 4480–4489. [CrossRef]
154. Kozai, N.; Sato, J.; Osugi, T.; Shimoyama, I.; Sekine, Y.; Sakamoto, F.; Ohnuki, T. Sewage sludge ash contaminated with radiocesium: Solidification with alkaline-reacted metakaolinite (geopolymer) and Portland cement. *J. Hazard. Mater.* **2021**, *416*, 125965. [CrossRef] [PubMed]
155. Lin, W.; Chen, H.; Huang, C. Performance study of ion exchange resins solidification using metakaolin-based geopolymer binder. *Prog. Nucl. Energy* **2020**, *129*, 103508. [CrossRef]
156. Tan, Q.; Li, N.; Xu, Z.; Chen, X.; Peng, X.; Shuai, Q.; Yao, Z. Comparative performance of cement and metakaolin based-geopolymer blocks for strontium immobilization. *J. Ceram. Soc. Jpn.* **2019**, *127*, 44–49. [CrossRef]
157. Arbel-Haddad, M.; Harnik, Y.; Schlosser, Y.; Goldbourt, A. Cesium immobilization in metakaolin-based geopolymers elucidated by 133Cs solid state NMR spectroscopy. *J. Nucl. Mater.* **2022**, *562*, 153570. [CrossRef]
158. Fu, S.; He, P.; Wang, M.; Cui, J.; Wang, M.; Duan, X.; Yang, Z.; Jia, D.; Zhou, Y. Hydrothermal synthesis of pollucite from metakaolin-based geopolymer for hazardous wastes storage. *J. Clean. Prod.* **2020**, *248*, 119240. [CrossRef]
159. Chaerun, N.; Soonthornwiphat, K.; Toda, K.; Niu, R.; Kikuchi, T.; Kuroda, K.; Niu, X.; Kikuchi, R.; Otake, T.; Elakneswaran, Y.; et al. Retention Mechanism of Cesium in Chabazite Embedded into Metakaolin-Based Alkali Activated Materials. *J. Hazard. Mater.* **2022**, *440*, 129732. [CrossRef]
160. Jain, S.; Banthia, N.; Troczynski, T. Leaching of immobilized cesium from NaOH-activated fly ash-based geopolymers, Leaching of immobilized cesium from NaOH-activated fly ash-based geopolymers. *Cem. Concr. Comp.* **2022**, *133*, 104679. [CrossRef]
161. Ahn, J.; Kim, W.-S.; Um, W. Development of metakaolin-based geopolymer for solidification of sulfate-rich HyBRID sludge waste. *J. Nucl. Mater.* **2019**, *518*, 247–255. [CrossRef]
162. Kim, B.; Lee, J.; Kang, J.; Um, W. Development of geopolymer waste form for immobilization of radioactive borate waste. *J. Hazard. Mater.* **2021**, *419*, 126402. [CrossRef]
163. Shastri, D.V.; Annamalai, K.; Rajamane, N.P.; Arunchalam, K.D. Structural and physico-mechanical investigations of $Na_2B_4O_7$ geopolymer for γ radiation attenuating applications. *Ceram. Int.* **2022**, *48*, 3449.
164. He, P.; Cui, J.; Wang, M.; Fu, S.; Yang, H.; Sun, C.; Duan, X.; Yang, Z.; Jia, D.; Zhou, Y. Interplay Between Storage Temperature, Medium and Leaching Kinetics of Hazardous Wastes in Metakaolin-based Geopolymer. *J. Hazard. Mater.* **2019**, *384*, 121377. [CrossRef] [PubMed]
165. Buchwald, A.; Tatarin, R.; Stephan, D. Reaction progress of alkaline-activated metakaolin-ground granulated blast furnace slag blends. *J. Mater. Sci.* **2009**, *44*, 5609–5617. [CrossRef]
166. Yip, C.K.; Lukey, G.C.; van Deventer, J.S.J. The coexistence of geopolymeric gel and calcium silicate hydrate of alkaline activation. *Cem. Concr. Res.* **2005**, *35*, 1688–1697. [CrossRef]
167. Burciaga-Díaz, O.; Escalante-García, J.I. Comparative performance of alkali activated slag/metakaolin cement pastes exposed to high temperatures. *Cem. Concr. Compos.* **2017**, *84*, 157–166. [CrossRef]
168. Myers, R.J.; Bernal, S.A.; San Nicolas, R.; Provis, J.L. Generalized structural description of calcium-sodium aluminosilicate hydrate gel: The cross-linked substituted tobermorite model. *Langmuir* **2013**, *29*, 5294–5306. [CrossRef]
169. Paradal, X.; Brunet, F.; Charpentier, T.; Pochard, I.; Nonat, A. ^{27}Al and ^{29}Si solid-state NMR characterization of calcium-aluminosilicate hydrate. *Inorg. Chem.* **2012**, *5*, 1827–1836. [CrossRef]
170. Van Jaarsveld, J.G.S.; Van Deventer, J.S.J.; Lukey, G.C. A comparative study of kaolinite versus metakaolinite in fly ash based geopolymers containing immobilized metals. *Chem. Eng. Commun.* **2004**, *191*, 531–549. [CrossRef]
171. Jirasit, F.; Rüscher, C.H.; Lohaus, L. A study on the substantial improvement of fly ash-based geopolymeric cement with the addition of metakaolin. In Proceedings of the International Conference on Pozzolan, Concrete and Geopolymer, Khon Kaen, Thailand, 4–25 May 2006; pp. 247–261.
172. Qian, G.; Li, Y.; Yi, F.; Shi, R. Improvement of metakaolin on radioactive Sr and Cs immobilization of alkali-activated slag matrix. *J. Hazard. Mater.* **2002**, *B92*, 289.
173. Lloyd, R.R. The Durability of Inorganic Polymer Cements. Ph.D. Thesis, Department of Chemical and Biomolecular Engineering, University of Melbourne, Melbourne, Australia, 2008.
174. Bao, Y.; Grutzeck, M.W.; Jantzen, C.M. Preparation and properties of hydroceramic waste forms made with simulated hanford low-activity waste. *J. Am. Ceram. Soc.* **2005**, *88*, 3287–3302. [CrossRef]
175. Xu, H.; Gong, W.; Syltebo, L.; Lutze, W.; Pegg, I.L. DuraLith geopolymer waste form for Hanford secondary waste: Correlating setting behavior to hydration heat evolution. *J. Hazard. Mater.* **2014**, *278*, 34–39. [CrossRef]
176. Williams, B.D.; Neeway, J.J.; Snyder, M.M.V.; Bowden, M.E.; Amonette, J.E.; Arey, B.W.; Pierce, E.M.; Brown, C.F.; Qafoku, N.P. Mineral assemblage transformation of a metakaolin-based waste form after geopolymer encapsulation. *J. Nucl. Mater.* **2016**, *473*, 320–332. [CrossRef]

177. Lei, H.; Muhammad, Y.; Wang, K.; Yi, M.; He, C.; Wei, Y.; Fujita, T. Facile fabrication of metakaolin/slag-based zeolite microspheres (M/SZMs) geopolymer for the efficient remediation of Cs^{+} and Sr^{2+} from aqueous media. *J. Hazard Mater.* **2021**, *406*, 124292. [CrossRef] [PubMed]
178. El-Naggar, M.; Amin, M. Impact of alkali cations on properties of metakaolin and metakaolin/slag geopolymers: Microstructures in relation to sorption of ^{134}Cs radionuclide. *J. Hazard Mater.* **2018**, *344*, 913–924. [CrossRef] [PubMed]
179. Rakhimova, N.R.; Rakhimov, R.Z.; Lutskin, Y.S.; Morozov, V.P.; Osin, Y.N. Solidification of borate ion-exchange resins by alkali-activated slag cements. *Roman. J. Mater.* **2018**, *48*, 177–184.
180. Lee, W.-H.; Cheng, T.-W.; Ding, Y.-C.; Lin, K.-L.; Tsao, S.-W.; Huang, C.-P. Geopolymer technology for the solidification of simulated ion exchange resins with radionuclides. *J. Environ. Manag.* **2019**, *235*, 19–27. [CrossRef] [PubMed]
181. Khairudin, N.W.A.; Yasir, M.S.; Majid, A.A.; Wahab, M.A. Immobilisation of spent ion exchange resin from Puspati TRIGA reactor using ash-based geopolymer. *Malays. J. Anal. Sci.* **2015**, *19*, 472–480.
182. Kryvenko, P.; Petropavlovskyi, O.; Cao, H.; Weng, L.; Kovalchuk, O. Applicability of alkali-activated cement for immobilization of low-level radioactive waste in ion-exchange resins. *East. Eur. J. Enterp. Techn.* **2015**, *1*, 79. [CrossRef]

sustainability

MDPI

Review

Toward Sustainable Cementitious Radioactive Waste Forms: Immobilization of Problematic Operational Wastes

Rehab O. Abdel Rahman [1,*] and Michael I. Ojovan [2,3]

1. Hot Laboratory Center, Atomic Energy Authority of Egypt, Cairo 13759, Egypt
2. Department of Materials, South Kensington Campus, Imperial College London, Exhibition Road, London SW7 2AZ, UK; m.ojovan@imperial.ac.uk
3. Institute of Geology of Ore Deposits, Petrography, Mineralogy and Geochemistry of Russian Academy of Sciences (IGEM RAS), 119017 Moscow, Russia

* Correspondence: alaarehab@yahoo.com; Tel.: +20-10-6140-4462

Abstract: Developing effective radioactive waste management practices is essential for ensuring the sustainability of the nuclear industry. The immobilization of radioactive wastes is one of the main activities conducted during the management of these wastes; it aims to produce a durable waste form that has sustainable performance over long periods of time. In this work, the challenges that face the design of durable cementitious waste forms are addressed for problematic operational wastes. In this respect, the problematic characteristics of evaporator concentrates, spent ion exchangers, and organic liquid wastes are overviewed, and the factors that affect the durability of their cementitious waste forms are identified. A summary of potential conventional and innovative cementitious matrices is presented by reviewing the cementation practices in national programs and recent research devoted to developing durable matrices. Finally, a guide to optimize the mix design of these waste forms was proposed that includes the selection of the testing procedure, factors that affect the waste form performance, and the optimization technique. This guide was presented with special focus on leaching tests, which are a means to test the stabilization performance of nuclear waste forms.

Keywords: radioactive waste; evaporator concentrates; spent ion exchangers; organic liquid wastes; conventional cement; innovative cement; mix design optimization; leaching tests

Citation: Abdel Rahman, R.O.; Ojovan, M.I. Toward Sustainable Cementitious Radioactive Waste Forms: Immobilization of Problematic Operational Wastes. *Sustainability* **2021**, *13*, 11992. https://doi.org/10.3390/su132111992

Academic Editor: Silvia Fiore

Received: 27 September 2021
Accepted: 26 October 2021
Published: 29 October 2021

1. Introduction

The nuclear industry is supporting the sustainability of human life, as the continuous development of nuclear energy aims to secure sustainable, clean, and affordable energy and clean water. Moreover, the advance in isotopic techniques aims to improve the agricultural sector toward zero hunger, support the medical sector for good health and well-being, enhance industrial innovation and infrastructure, and support efforts to protect life below water and on land [1]. In general, the operation of the energy production sector results in large extraction, processing, storing, and transportation of raw materials, as well as processed fuel utilization and management of the generated wastes and emissions. The sustainable development indicators for this sector comprise different economic, social, institutional, and environmental dimensions. In particular, the environmental indicators focus on resources, wastes and emissions (e.g., amounts and characteristics of the used fuel and generated wastes and emissions), and land impacts [1–4]. Subsequently, the amount of the generated solid wastes due to the production of unit energy (g/kW·h) is used as an environmental indicator on the sustainability of certain energy sectors [1,3,4]. The International Atomic Energy Agency (IAEA) compared this indicator for different energy production industries for both upstream and operational processes, and the results are summarized in Figure 1 [4]. It is clear from this figure that the upstream operations of renewable energy sources generate very small amounts of solid wastes per unit energy (300–0.98 g/kW·h) compared to that of the coal industry (3000 g/kW·h) (Figure 1a), where most of the generated solid wastes from the nuclear energy sector are non-radioactive and non-hazardous wastes (0.97g/kW·h). The same results are shown for the operational processes of these industries, where lignite coal generates the highest amount of solid waste per unit energy. The nuclear industry generates total operational wastes comparable to that generated from natural gas. Only 5.1% of these wastes have a radiological hazard: where low and intermediate level radioactive wastes (LILW)represent 4.52%, and the rest are high level wastes (HLW), spent fuels, and hazardous wastes (Figure 1b).

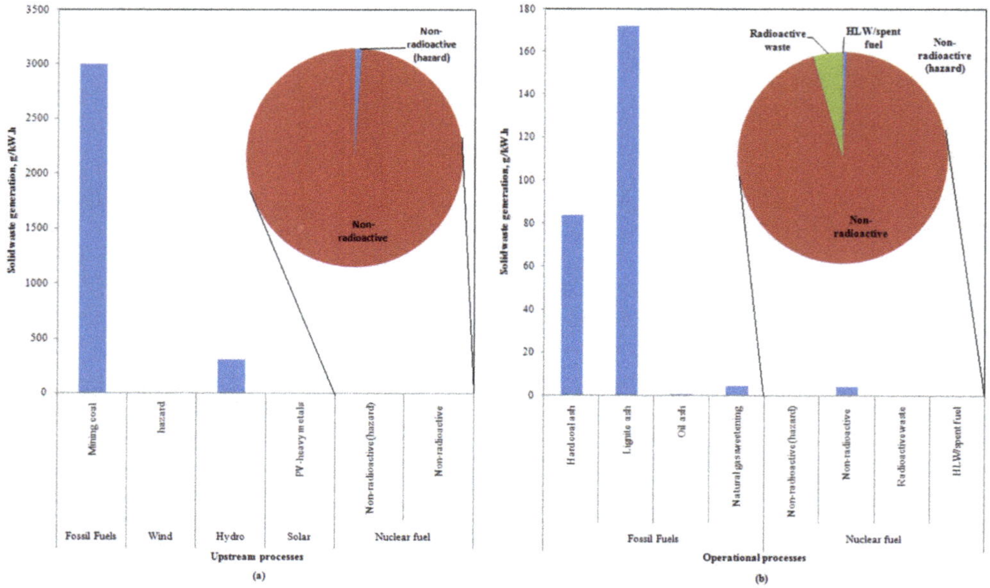

Figure 1. Amount of solid wastes generation per unit energy generation in different energy industries: (**a**) upstream processes; (**b**) operational processes [4].

Although the nuclear energy industry generates small amounts of operational wastes, the analysis of its sustainability considers emissions and radioactive waste generation and management as key sustainability factors [4,5]. This is attributed to the radiological hazards of these wastes and emissions that can last for long times, e.g., several decades for short-lived LILW, according to the 10 half-lives rule of thumb. Subsequently, operational and research efforts are directed at minimizing the waste generation and enhancing the waste management practices, to eliminate the long-term impacts of these wastes on human health and the environment. The management of radioactive wastes typically consists of pre-disposal and disposal activities; the first is sub-categorized into pre-treatment, treatment, and conditioning activities, as illustrated in Figure 2 [6]. Pre-treatment and treatment activities aim to reduce the volume of generated wastes to enhance the safety and/or reduce the costs of the subsequent management activities [7]. The treated wastes are directed to immobilization then packaging, to fabricate a waste package that is able to withstand handling and transport conditions and is durable under storage and disposal conditions (Figure 2). The immobilized waste form should be designed and fabricated in a way that ensures its sustainable performance for centuries in the case of LILW and for millennia in the case of HLW. It should be noted that during the early era of the nuclear industry, LILW were packaged and disposed of without immobilization. Based on the lessons learned from early disposal practices, the waste immobilization activity was introduced. Since then, a large number of research groups focused their work on this topic: among these groups are the Atomic Energy of Canada Limited's (AECL) Chalk River Laboratories, Comisión Nacional de Energia Atómica (CNEA) in Argentina, the Scientific and Industrial Association (RADON) in the Russian Federation, Empresa Nacional de Residuos Radioactivos S.A. (ENRESA) in Spain, the Brookhaven National Laboratory in USA, etc. Detailed information on early immobilization and disposal practices can be found elsewhere [8–18]. An analysis of the bibliometric data in the Scopus database was conducted in order to gain insights into the scientific research interests in the field of radioactive waste management in general and, in particular, on the topics related to radioactive waste forms. The analysis conducted in July 2021 showed 57,887 documents in the database using the keywords "Radioactive AND wastes". The search was refined using more keywords with the operator "AND", and the results are summarized in Figure 3a–c. It is noted that research related to radioactive waste forms represent nearly 10% of the work that addressed radioactive waste (Figure 3a). The factors that affect its performances represent about 8.72% of the research conducted in this field (Figure 3b).

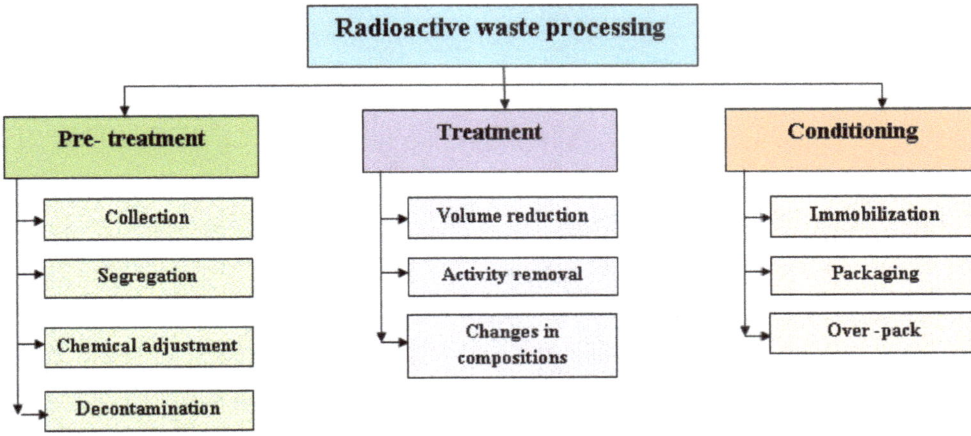

Figure 2. Pre-disposal activities in a radioactive waste management system (copyrighted R.O. Abdel Rahman, et al., *Improving the Performance of Engineering Barriers in Radioactive Waste Disposal Facilities: Role of Nano-Materials*. In: Kharissova O., Martínez L., Kharisov B. (eds) Handbook of Nanomaterials and Nanocomposites for Energy and Environmental Applications. Springer, Cham, 1183–1200 [6]).

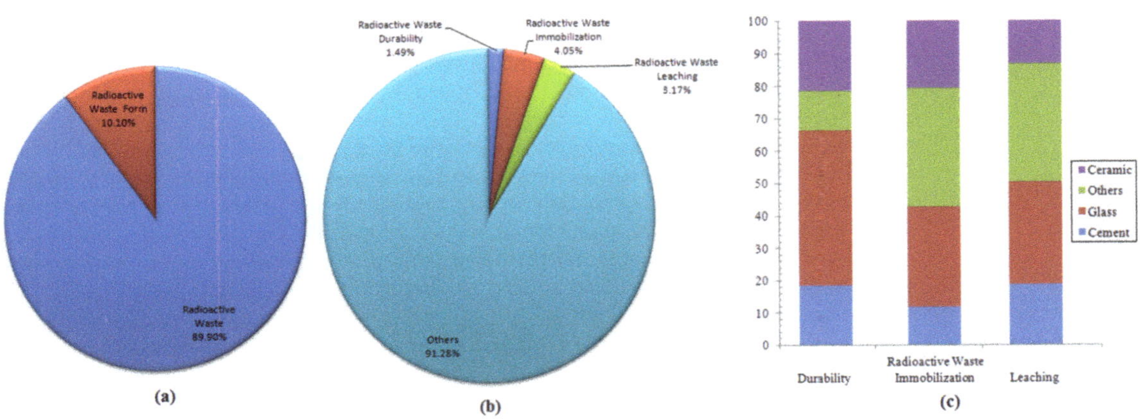

Figure 3. Bibliometric data in the field of radioactive wastes, the contribution of research related to (**a**) waste forms, (**b**) radioactive waste immobilization, durability, and leaching. (**c**) relative contributions of immobilization media in researches related to radioactive waste immobilization, durability and leaching.

Cements, ceramics, glasses, and polymers are widely used as immobilizing matrices for radioactive wastes; they contain and confine several types of radio-contaminants. The sub-categorization of the refined bibliometric data for research directed at investigating radioactive waste immobilization, durability, and leaching is illustrated in Figure 3c. These data show that research focused on solving challenges in the application of cementitious, glass, and ceramic wastes. Figure 4 shows a summary of the annual distribution of the research conducted in the field of radioactive waste for glass and cement materials. It is clear that the research related to the latter increased over the last 10 years compared to that for the first (Figure 4a). The annual distribution of the published glass research in the database is slightly higher than that of the cement research (Figure 4b).

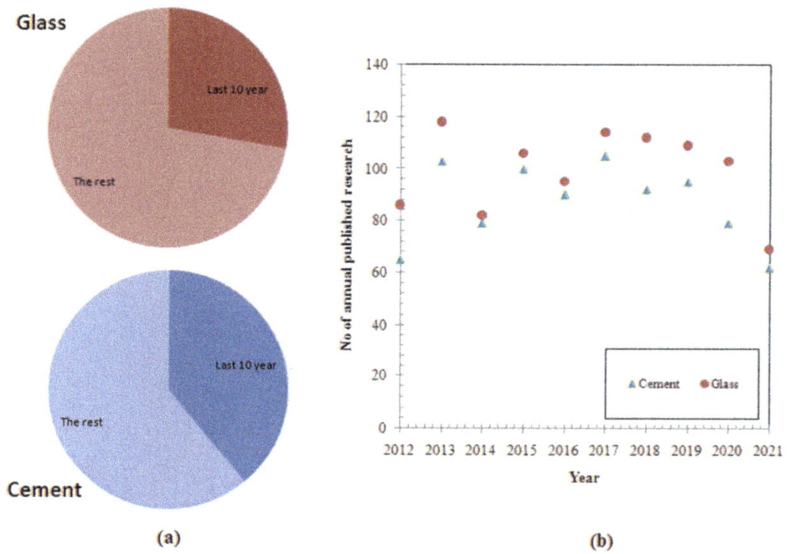

Figure 4. Summary of the progress in the annual published research in the radioactive waste field: (**a**) contribution of the published research in the last decade; (**b**) number of annual published research in the last decade.

The immobilization of radioactive wastes in cementitious matrices is achieved via two simultaneous processes, i.e., solidification and stabilization. The first aims to enhance the mechanical performance of the produced waste form, and the second aims to reduce the mobility of radio-contaminants [19]. In this paper, the focus is on challenges that face the design of sustainable cementitious waste forms for problematic operational wastes. In general, problematic wastes are wastes that require special considerations during the selection of management options, owing to their specific characteristics [20]. Spent ion exchangers, borate- and sulfate-containing wastes, organic liquid wastes, and decontamination solutions containing detergent, organic, and/or acidic solutions are operational wastes that are considered problematic during the design of the cementitious waste forms [21–23]. If directly immobilized, some components in these streams interfere in the hydration reactions of conventional Ordinary Portland Cements (OPC), leading to undesired micro- and macro-changes in the produced waste forms that affect their solidification and stabilization performances. To counteract these changes, the mix design of the cementitious immobilization matrices is modified to include additives or use of innovative cements. The main aim of this work is to summarize the key problematic characteristics of these waste streams, assess the factors that affect the durability of the produced cementitious waste forms, and identify the approaches used in designing these forms.

2. Cementitious Matrices to Immobilize Operational Wastes

Cementitious matrices are widely applied as immobilization media to host several types of radioactive wastes, and this application is supported by their [23–25]:

- Compatibility with different waste streams.
- Capability to activate different immobilization mechanisms.
- Chemical, thermal, and radiological stabilities.
- Low operational cost, ease of operation, and ambient temperature operation.
- Ability to maintain the high pH environment in the disposal facility.

These matrices are categorized into conventional and innovative matrices. The first matrix is composed of a conventional hydraulic binder, i.e., Ordinary Portland Cement (OPC) with or without additives, whereas the second includes calcium aluminate cements (CAC), calcium sulfo-aluminate cements (CSAC), magnesium phosphate cements (MPC), and alkali-activated cements (AAC) [23,26,27]. The design of cementitious waste forms is a crucial process, aiming toward the achievement of a durable form that possesses reliable long-term performance under storage and disposal conditions. During the design phase, quality requirements on raw materials and the chemical and physical characteristics of the waste stream should be considered [26–29]. In addition, the factors that affect the operation of the cementation process, e.g., the setting time and flow-ability of the paste and the radiological dose on the surface of the waste form, should be taken into account. Finally, the nature of the storage or disposal concept and the required safety functions of the waste forms are key aspects in identifying a proper mix design. In general, the compressive strength, leaching resistance, and permeability are used as performance indicators during the design of the cementitious waste forms to assess solidification, stabilization, and hydraulic performances [30]. The design of the immobilization process includes the following steps:

- Identification of the waste characteristics, e.g., components that influence the hydration reactions.
- Preliminary selection of the immobilization matrices components.
- Optimization of the mix-design, e.g., determination of the waste loading percentage, cement to additive ration, cement to water ratio, etc.

2.1. Immobilization of Evaporator Concentrates

Evaporation is a conventional wastewater treatment technology that aims to reduce the volume of the generated aqueous wastes; this technology is reliable and efficient with decontamination factors in the range 10^4–10^6. Evaporators are usually used in the combined aqueous waste treatment process in nuclear power plants and in centralized waste management facilities. As a result of the evaporation process, the aqueous waste splits into two streams: the first is of large a volume and very low radioactivity content that could be discharged from the regulatory control, i.e., condensate, and the second is a small volume stream that contains high radioactivity, i.e., concentrates or bottom residues. Concentrates are usually classified as LILW, which are directed to post-treatment then immobilization or just directed to immobilization, based on the selected immobilization matrix type.

2.1.1. General Characteristics of Concentrates and Their Problematic Nature

Concentrates are high salt solutions with typical specific activity concentrations in the range 10^6–10^8 Bq/L [31]. Major contaminants in the evaporator concentrates are fission products such as ^{137}Cs, ^{60}Co, ^{54}Mn, borates ($NaBO_2$ and $Na_2B_4O_7$), nitrates ($NaNO_3$, KNO_3), hydroxides (NaOH and KOH), and some organic compounds, i.e., oxalates [31–33]. In general, alkali and alkaline fission products, e.g., ^{85}Sr, ^{137}Cs, are characterized by their respective high solubility in the cementitious matrices [34,35]. This behavior affects only the stabilization performance of the waste form. Borates, phosphates, sulfates, nitrates, fluorides, and organic compounds are known for their interference in the hydration reactions, leading to them either retarding or accelerating the hydration process [23]. If the content of these components is high enough, they can affect drastically the solidification, stabilization, and hydraulic performances, e.g., the presence of sulfates at considerable amounts reduces the setting time and was reported to cause false rapid setting and increase the expansion and cracking of cement paste [21]. Radioactive borate wastes are produced in pressurized water reactors (PWR). The typical characteristics of the evaporator concentrates, produced from deep and normal evaporators installed in nuclear power plants (NPP) with pressurized reactors of the water–water energetic reactor (WWER) type, are listed in Table 1 [36]. Despite the fact that there are clear variations in the composition of the concentrate in terms of chemical composition and specific radioactivity, borate represents a major component in this stream, which limits the waste loading to a very low wt.% [37].

Table 1. Typical characteristics of different evaporator concentrates in WWER.

Characteristics	Salt Cake	NPPs Jaslovske	
		V-1 NPP	V-2 NPP
pH	-	11.3–13.3	11.2–13.2
Density, kg/m^3	1500–1800	1414	-
Specific radioactivity, Bq/kg	<3.7 × 10^7	<1.2 × 10^7	<3.39 × 10^5
Dry salt content, Kg/L	1.3–1.5	0.15–0.397	0.15–0.321
Chemical composition, g/L			
Na^+	180–220	42–100	25.3–75
K^+	30–55	9.4–21.5	9.1–26
H_3BO_3	280–350	73–160	64–117
Cl^-	2–5	0.9–1.8	1.1–1.8
NO_3^-	200–350	5–29.6	-
SO_4^{-2}	10–15	14.5–24.8	-
Organic content	20–30	14.5–34.8	0.6–5.6

In particular, borates are known for their effects on the solidification process, i.e., strength reduction and the operability of the process, i.e., retarding effects. The effect of borate on the hydration of different cementitious phases had been studied over the past three decades. These studies indicated that [21,23,27,38–42]:

1. During tri-calcium silicate (C_3S) hydration, borate will consume some of the formed portlandite to form amorphous or poor crystalline calcium borate hydrates (CBHs).
2. The clinker grains will be partially or fully covered with an impermeable CBHs layer, which in turn suppresses the hydration reactions.
3. The solubility of the CBH is affected by the changes in the pore solution pH and is precipitated as crystallized phases at pH > 12 and the hydration is resumed.

A recent study that explored the immobilization of 5.26% B_2O_3 in OPC indicated that the heat release during the early hydration stage is considerably affected by the presence of B_2O_3 [39]. Within the first hour of hydration, B_2O_3 dissolution leads to a sensible increase in the released heat (0–a in Figure 5a). Then, the heat release is suppressed (a–d in Figure 5a). XRD patterns for OPC and OPC-B_2O_3 samples at 28 days showed that the formation of ettringite and portlandite formation are hindered in the borated sample. The presence of unhydrated phases, e.g., C_3S and C_2S, evidences the suppression of the hydration process in that sample and the formation of CaB_2O_4. $4H_2O$ refers to a reaction between the formed portlandite at an early time with the borate. These changes were associated with a reduction in the compressive strength from 46.4 to 6.6 MPa at 28 days [39]. Thus, the mix design of specified evaporator concentrates should account for the effect of borate on the hydration of the cementitious matrix. In addition, the durability of the waste from under long-term exposure to environmental conditions and the radiation field should be accounted for during the design [43,44].

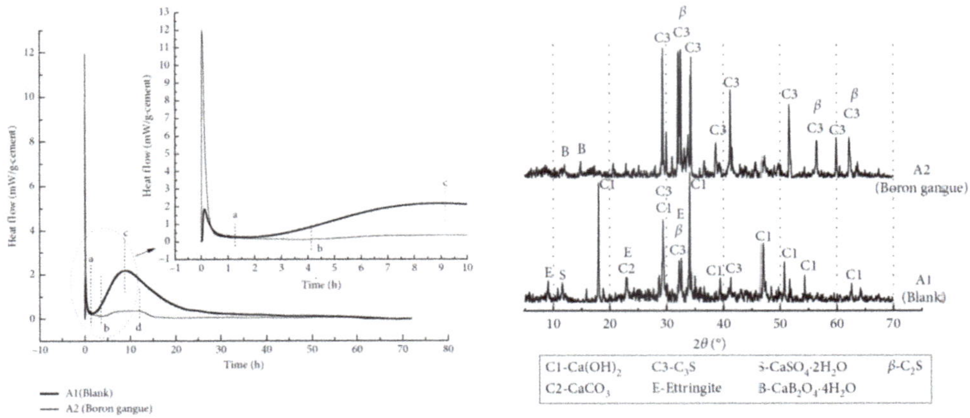

(a) (b)

Figure 5. Effect of B_2O_3 on the hydration of ordinary Portland cement 42.5: (**a**) heat flow; (**b**) XRD patterns for 28-day hydrated cement (copyrighted Q. Zhao, J. Tu, W. Han, X. Wang, Y. Chen, Hydration Properties of Portland Cement Paste with Boron Gangue, Advances in Materials Science and Engineering (2020) 7194654, 9 pages, https://doi.org/10.1155/2020/7194654).

2.1.2. Potential Cementitious Immobilization Matrices for Concentrates

Both conventional and innovative cementitious waste matrices have been evaluated and practiced, to immobilize evaporator concentrates. The conventional cementitious waste form relies on the use of OPC and one or more additives: typical additives include alkali and alkaline compounds, waste minerals, and clay minerals [21,23,27,36,45–55]. Table 2 summarizes the additives widely used for evaporator concentrates immobilization and their effects on the hydration process and the stabilization of the radio-contaminants [23].

Table 2. Potential additives used in designing an evaporator concentrate waste form.

Category	Compound	Effect on Hydration	Effect on The Stabilization
Alkali and alkaline compounds	LiOH NaOH Na_2SiO_3	Accelerators to enhance the super-saturation of the liquid phase during early hydration stage	-
	$Ca(OH)_2$	Calcium sources and pH buffer	Stabilize amphoteric elements
	$CaCl_2$	Accelerator with several acceleration mechanisms	-
Waste	BFS *	Pozzolanic materials	Reducing agent to enhance metal precipitation Bind alkali and alkaline metals
	Fly ash		Enhance the contaminant sorption and ion-exchange
Clay mineral	Benotnite Zeolite Vermiculite Diatomite	-	Enhance the contaminant sorption and ion-exchange leading to reduced leaching
	Siliceousmineral	Reducing C/S ratio in CSH	-

* BFS = blast furnace slag.

Blends of OPC and non-gypsum cements were investigated for their potential use to solidify-simulated evaporator concentrates [40,56,57]. In a comparative study on the effect of irradiation on a specific binder matrix designed to host evaporator concentrates marked as "NP", which was composed of 50%OPC and 50% non-gypsum cements, and

OPC and NP loaded with simulated evaporator concentrates (NP-C) [56], it was found that the presence of the dissolved salts e.g., nitrates, affects the performance of the latter. The compressive strengths of the samples were enhanced after irradiation, yet surface cracks were detected in NP and NP-C and were attributed to radiolysis drying [57]. Finally, the effect of wet–dry cycles on the durability of NP and 50% non-gypsum cement with 50% metakaolin (NM), with and without 5% zeolite additives in the presence of pure fine silica and calcium sulfo-aluminate cements (CSAC) containing simulated concentrates, was investigated [40]. All the non-gypsum cement matrices showed enhanced solidification performance in terms of compressive strength and flexural strength after exposure to three wet–dry cycles. Despite CSAC showing superior compressive strength and water absorption, its exposure to the cycles led to its distortion [40].

Calcium aluminate cements (CAC), calcium sulfo-aluminate cements (CSAC), magnesium phosphate cements (MPC), alkali-activated cements (AAC), and their blends with OPC were studied to immobilize borate solutions. On the short term, these materials were found to counteract the retarding effects of boron as follows [58–71]:

- CSAC loaded with borate have higher early strength compared to OPC. This result is highly dependent on the gypsum content in the CSAC and/or the sodium content in the mix design [58,59].
- Optimized AAC based on the fly ash and slag reduced the diffusion of boron in the fly ash-based matrices 100 times less than that in OPC.
- AAC based on using NaOH activator and slag cement was found to counteract the retarding effect of boron if the activator is properly optimized. The use of 7% NaOH in AAC yielded a compressive strength >49.7 MPa at 9–10% sodium borate loading [64].
- Geopolymers were investigated to immobilize borate wastes, where AAC based on metakaolin and silica fume activated with KOH was found efficient in immobilizing 9–14% borate wastes with a 7-day compressive strength in the range 19–40 MPa.
- Geopolymer, known as DuraLith, has been investigated as alternative for conventional cement-based matrix to immobilize the secondary wastes from Hanford waste treatment plant in USA. The secondary wastes include evaporator concentrates and melter scrub solutions. DuraLith is AAC based on mixture of ground BFS and metakaolin with sand used as a filler material. It demonstrates compressive strength above 27 MPa and American National Standards Institute/American Nuclear Society (ANSI/ANS) 16.1 leachability indexes for Tc as high as 9 [69–71].
- CSAC supplemented with $Ca(OH)_2$, zeolite, accelerator, and Dura fiber was found to stabilize boron and have a compressive strength 13.9 MPa and alkali and alkaline elements leaching rates in the range 10^{-5}–10^{-7} cm/d.
- The use of OPC-CAC and OPC-CSAC was found to be effective in stabilizing the boron in the Al_2O_3–Fe_2O_3-mono sulfate (AFm) and tri-sulfate (AFt) phases [65].

2.2. Immobilization of Spent Ion Exchangers

Ion exchangers/sorbents are widely used to remove/separate various radionuclides; it is applied for effluents of low salt content, compared to evaporators. In particular, ion exchangers are used in nuclear reactors to control system chemistry, minimize corrosion or degradation of system components, remove radioactive contaminants, and clean and decontaminate aqueous streams. Exchangers are widely used to treat the primary coolant, wet storage waters, steam generator blow down, boric acid, and condensate in boiling water reactors (BWR). This technique has efficient performance in removing and separating a wide range of elements. In addition, there are large choices of exchangers/sorbents available at the commercial scale, and its operation is known to be of low cost and easy. Yet, the operation is sensitive to the presence of organic contaminants, high salt content, colloidal/pseudo colloidal radionuclides and the extent of nonionic or non-exchangeable species, and requires regeneration to allow the reuse of the exchangers/sorbents [23,30,72,73]. There are several classifications for exchangers that are used to facilitate their life cycle management. Examples of these classifications and the types of materials used as ion exchangers and their applications in the nuclear industry are illustrated in Table 3 [23,72,74]. Another useful classification for the ion exchangers is based on the type and strength of the function group, which includes strong acidic, weak acidic, strong basic, and weak basic. At the end of their operational life, the exchangers are treated before immobilization or directly immobilized in suitable immobilization media [23,30,72].

Table 3. Examples of the classification of ion-exchangers [23,72,74].

Classification		Examples	Application in Nuclear Industry	
			Exchanger	Target Stream
Chemical nature	Inorganic	Zeolite	Clinoptilolite	SIXEP, Sellafield (remove Cs, Sr)
	Organic	Phenol formaldehyde	Lewatit DN KR	Pond water (remove Cs)
	Hybrid	Mg-Fehydrotalcite loaded with Cyanex 272		-
		Poly-acrylamide-based Ce(IV) phosphate		
Physical form	Bead	Phenol formaldehyde	Lewatit DN KR	Pond water
		Divinylbenzene	IRN 77	AGR/Sizewell B pond water (remove borate)
	Hydrogel	Potassium copper-hexacyanoferrate embedded 3D-interconnected porous hydrogel		-
		Modified silica hydrogel $C_{16}H_{35}O_3P$		-
Target contaminant	Cation	Zeolite	AW500 (synthetic)	Pond water in Magnox Sites (remove Cs)
			Clinoptilolite (natural)	SIXEP, Sellafield, (remove Cs, Sr)
		Divinylbenzene	IRN 77/77L Lewatit S100	AGR/Sizewell B pond water (treatment).
	Anion	Divinylbenzene	IRN 78/78L Lewatit MP 62	

2.2.1. General Characteristics of Spent Exchangers and Their Problematic Nature

Spent ion exchangers as operational wastes are most often classified as LILW; the major contaminants in these exchangers are fission products, e.g., Cs, Sr, corrosion products, borates, nitrates, and alkali components. As mentioned above, the presence of some of these contaminants can interfere with the hydration reactions. Alkali and alkaline contaminants have high solubility in the high pH environment. In addition, the stability of the spent exchanger can affect the performance of the produced waste form. In general, inorganic exchangers are known for their better thermal and radiation stabilities compared with the organic exchangers, and the chemical stability of both types is dependent on the solution pH [75]. Despite the immobilization of the exchangers in the cements generally proceed via encapsulation, limited reactions occur in the transition zone between the encapsulated exchangers and the hydrated cement. These reactions occur on the microstructure scale, but depending on their extents, they can affect the characteristics of the waste form widely. To avoid these effects, as well as potential mechanical destruction of the matrix resulting from the swelling of ion exchangers, the waste loading into the cement is limited to a small value [23,30]. In this respect, the following characteristics are important to be evaluated during the design of a cementitious waste form for spent ion exchangers/sorbents [23,30,72,75–82]:

1. Buffering effects of spent inorganic exchanger, which is dependent on the degree of depletion of the exchanger/sorbent. Unloaded exchangers can buffer the pH to fixed value over a wide range of solution pH, this behavior is depending on the composition of the exchanger, e.g., magnetic zeolite composite can buffer the pH in the range 3–9 to pH = 4, and hydrotalcite-ferrocyanide composite buffers the solution in the same pH range to approximately neutral values [79,80]. If acidic buffer effect

is noted for the spent exchanger, there will be a need to condition the stream before direct immobilization.

2. pH of the spent organic resin slurry, which is also dependent on the degree of the depletion of the resins, where unloaded cation and anion organic resins buffer the solution to acidic and basic media, respectively. Fully depleted resin slurry has neutral pH. Reported study on actual spent resins slurry indicated that the solution pH is nearly 5 which is acceptable for direct immobilization without prior conditioning [77].
3. Water content in the exchangers, as the water content in dump resin is typically in the range of 30% and in saturated slurry is nearly 40%. There is a need to optimize the water used in the mix-design to avoid the generation of large volume of bleeding water.
4. Stability of the spent ion exchangers in high pH environment; for inorganic exchanger, phosphates and oxyhydrates of non-ferrous metals based inorganic exchangers and zeolites are respectively stable at basic pH with minimum stability at pH = 3. Transition metal ferro-cyanides exchangers are unstable at pH = 10, yet the use of modified exchangers can enhance the sorption capacity and the stability in alkaline media [79].
5. Ability of the inorganic exchanger to react with cement; inorganic exchangers have a potential to interfere in the hydration reaction of the cementitious matrix. The extent of this interference is highly dependent on the reactivity of the spent exchanger and its particle size. In particular:
   - Zeolite and alumina were reported to interfere in the early age hydration reaction, leading to enhanced formation of aluminosilicate phases, i.e., C3AH6, AFT, AFM.
   - Cation exchanger can remove calcium and aluminum ions during the early age of hydration, leading to slow formation of C-S-H of low calcium to silica ration.
   - Some inorganic exchangers were reported to have a pozzolanic activity, which is more influential during the late hydration stages.
6. Ability of the spent organic exchanger to react with cement; the ability of the exchanged contaminant to interfere with the hydration reaction is well known for organic exchangers, where:
   - Resins used to treat borate streams are suspected to retard hydration as a result of boron release from the exchangers.
   - Resins containing salts of short chain water soluble organic acids, e.g., acetic, formic, picolinic acids, can interfere with the hydration reactions. Examples of these resins are Purolite A200 and A400.
   - Production of bleeding water should be investigated during the immobilization of anion exchange resins.
   - Gel type resins, e.g., Dowex 50w-x8 (H^+ form) and Dowex 11 (NO_3^- form), have lower rigidity than beads and can lead to larger structural changes [81].
   - Resins swelling can lead to varying effects on the solidification performance in terms of compressive strength. This behavior is more prominent in cationic exchangers.

The swelling behavior of the spent organic exchangers and the composition of the transition zone were investigated [83–86]. In a simple cementitious system, Amberlite IR120H (Na^+ form) in Alite, a transient zone of small dimension was reported to occur just after setting, due to the decrease in the osmotic pressure of the external solution. Gaps were detected around the beads that separate them from the hydration products; the thickness of the gaps was reported to be dependent on the relative humidity of the SEM chamber [86]. The main internal pressure in the spent organic resin cementitious matrices is supposed to be attributed to the osmotic swelling of the resin, where cracking was proposed to occur when the generated tensile stresses exceed the maximum tensile strength of the waste forms [83]. A detailed investigation on the swelling behavior of the ion exchange resins in Alite was conducted to understand the chemical and mechanical behavior of a cementitious waste matrix containing C-S-H and portlandite only [84,85]. The studies used Amberlite IR-120 (Ca^{+2} form) with a mean diameter of 0.5 mm and effective capacity of 5.03 mEq/g of dry resin. Figure 6 illustrates the SEM micrograph with elemental mapping for the studied conditions [84]. The system did not show any cracks, which employs that the stresses during solidification did not affect the matrix and the portlandite precipitated around the bead (Figure 6A). Samples immersed in NaCl at 1 mol/L for three weeks did not show a massive portlandite precipitate, nor mechanical degradation (Figure 6B). This implies that the internal pressure due to the ion exchange processes is not sufficient to form cracks in the composite. Samples immersed in KOH showed shrinkage in the bead associated with the formation of a portlandite shell around the bead, without significant effect on the mechanical integrity (Figure 6C). Figure 6D shows the immersed sample in NaOH solution, and obvious cracks and bead swelling are noted and attributed to the combination of the bead swelling and portlandite precipitation. The effects of re-immersion in other alkali solutions, namely KOH, NaCl, and NaOH, are illustrated in Figure 6E–G, respectively. For these samples, enhanced precipitation of portlandite was noted and cracks appeared for samples (F) and (G), which entail that the composition of the interstitial pore water plays a critical role in determining the mechanical integrity of the sample, where excess formation of portlandite around the bead leads to a sensible increase in the internal pressure.

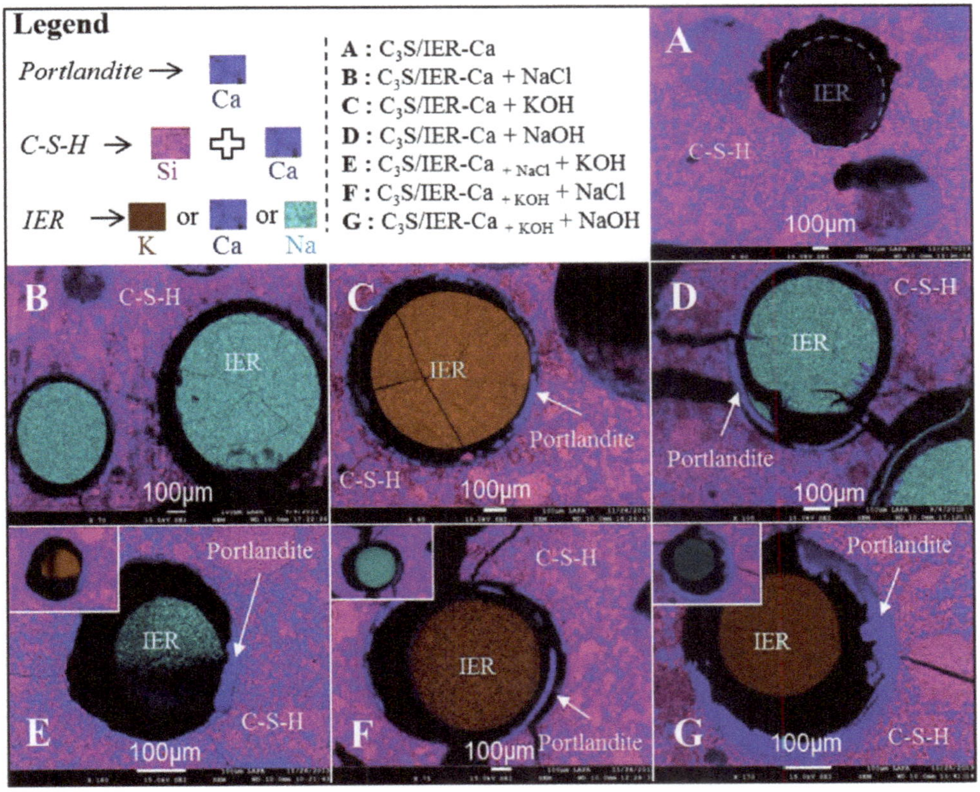

Figure 6. Elemental mapping of hydrated Alite/AmberLite IR-120 (Ca form) matrix immersed in different solutions (copyrighted from M. Neji, B. Bary, P. Le Bescop, N. Burlion, Swelling behavior of ion exchange resins incorporated in tri-calcium silicate cement matrix: I. Chemical analysis, J. Nucl. Mater. 467 (2015) 544–556).

2.2.2. Potential Cementitious Immobilization Matrices for Spent Ion Exchangers

Direct immobilization of spent exchangers, both organic and inorganic, in cementitious-based materials has been practiced in different countries for more than six decades [73,87]. This practice is based either on mixing evaporator concentrates, precipitation sludge, and the spent exchangers using in drum mix technology or immobilizing the spent exchanger only in the cementitious matrix [23]. Examples of these applications are at Beloyarsk, Kola, Smolensk NPPs in the Russian Federation. Various conventional cementitious matrices were tested, including OPC, OPC with additives, and mixture of BFS/OPC and OPC/polymer. Below are some findings from these tests for the direct immobilization of this problematic operational waste [21,23,30,52,73–78,81,82,87–98]:

- OPC matrices host efficiently 10% spent exchangers without any additives. As the waste loading increases, the compressive strength of the produced waste form is reduced within the first 28 days. This behavior is attributed to the reduction of the OPC content. For inorganic exchangers, in addition to the previously mentioned reason, the formation of low strength hydration phases and reduction of the Ca/Si ratio could be responsible for this reduction. If the inorganic exchanger has a pozzolanic effect, changes in this behavior could be noted depending on the reactivity of the exchanger.
- OPC-additive matrices: additives are used increase the waste loading up to 20 and 35% for beads and gel type resins, respectively. These matrices showed improved leaching resistance and enhanced workability, depending on the nature of the additive and exchangers. In particular, the following findings are drawn:
    1. OPC/natural pozzolanic materials: the addition of 1–2% natural zeolite was found to enhance the solidification of hybrid exchanger, e.g., transition metal oxide–PAN, in OPC matrix and increase the waste loading to 30% wet exchangers. Condensed silica fume/cement (II/V) was found to counteract the release of sulfate and sulfonic acid from irradiated strong acidic exchangers.
    2. OPC/clay, e.g., vermiculite, bentonite, red clay, sand, optimum matrices, have enhanced stabilization performance towards Cs, where these clays provide active sites to capture Cs into the matrices.
- BFS/OPC matrices enable low hydration heat reactions, denser immobilization matrix, and better stabilization of alkali contaminants. An optimum formulation based on the type of the exchanger and the contaminants could be used to increase the waste loading safely:
    1. The use of BFS/OPC with additives was also investigated to host polystyrene organic resin. The use of micro silica in that system led to achieving a waste loading 36% with acceptable Cs and Co leaching behavior.
    2. BFS/OPC matrix was found to enhance the stability of the phenol formaldehyde resins in the immobilization matrix by mean of the pozzolanic effect.
    3. OPC/polymer: this type of matrices issued to densify the structure, subsequently the solidification and stabilization performance are enhanced.

On another hand, novel cements were investigated for their application in the immobilization of spent exchangers. Zeolites (an inorganic exchanger) immobilization in CSA was found to be affected by the reaction of this exchanger with CSA between 28–90 days, leading to higher Cs leaching rates. This result was attributed to the distortion of the crystalline structure of zeolite and reduced C-S-H formation [74]. A ratio of 1:1 mixed bed styrene-based resins was successfully immobilized in CSA with a 20% zeolite additive at 42% loading percentage [99]. Geopolymers were extensively studied and applied to address the problems associated with the immobilization of spent exchangers. The Slovak Republic and Czech Republic have licensed this application and used it on an industrial scale [26,27,100,101]. At Dukovany NPP in Slovakia, 195 tones of spent resins and 4.5 tones of sludge were immobilized using a SIAL-geopolymer. The compressive strength of waste forms was in the range 12–35 MPa, where organic resins or mixtures of sludge and resins were encapsulated. The highest waste loading for the resin was 20% (on dry basis) with stainless steel drum surface dose rates in the range 130–600 μGy/h, and the D value (leachability index) for ^{137}Cs was >9 on cut samples from the drums. Researchs in this area were directed to investigate simple and complex immobilization matrices; the first was composed of a single precursor and alkali reactant, and the latter was composed of two or three precursors, an alkaline solution, additive, and/or admixture [101–108]. Table 4 summarizes the waste loading and solidification performance for the studied matrices. The investigations of the solidification and radiological stability indicated the presence of a transition zone around the cationic borate resin, immobilized in geopolymer loaded and radiolysis degradation for the resins [101,108].

Table 4. Summary of spent exchanger immobilization in different geopolymer matrices.

AAC–MATRIX Components, wt.%			Exchanger Loading, %	Solidification at 28 Days, MPa
Precursor	Alkali Reactant	Additive/Admixture		
GGBFS [a]	$Na_2SiO_3 \cdot 9H_2O$ + NaOH		Cationic borate resin 35% (pH 8.5–10.5)	4.7–7.3
Fly ash (56%)	Na_2SiO_3 20% NaOH 8%	Super-plasticizer 6%	10%	6.1
MK [b] (29.4%)	Water glass 23.5	H_2O 17.6	Zeolite-29.4	37.6
MK	Alkaline silicate solution	-	Resin 10	48
GGBFS (32.7–40.8%) OPC (8.4–9.9%)	$Na_2SiO_3 \cdot 5H_2O$–5.6–6.7%	Bentonite 2.8–8.2%, $Ca(OH)_2$–4.9–5.6%	43.2–45.1	6.4–12.6 (at 14 day)
GGBFS (34.2–40.2%) OPC (8.8%)	Na_2CO_3–1.6–3%	Bentonite 2.5–5.2%, $Ca(OH)_2$4–4.9% H_2O 5.9–11.1	41.9–45	8.4–13.2 (at 11 day)
MK+ Feldspar MK+ Feldspar + BFS	Alkaline silicate solution		Resin 8–12%	42 48
MK (41.9–33.5%) BFS (4.2–8.4%)	Sodium silicate/NaOH 58.1%		12%	13.63

[a] GGBFS = ground granulated blast furnace slag. [b] Metakaolin = MK.

2.3. Immobilization of Organic Liquid Operational Wastes

Organic materials are used during the operation of nuclear power plants; they are classified based on their physical nature as solid and liquid materials. At the end of their operational life, most of these materials become radiologically contaminated and should be treated as radioactive wastes. Examples of organic solid wastes include contaminated clothing, plastic sheets and bags, rubber gloves, mats, shoe covers, and paper wipes. Examples of organic liquid wastes include contaminated lubricating oils and hydraulic fluids from reactor operation, scintillation liquids from radio-analytical laboratories, and miscellaneous solvents/diluents generated from decontamination activities and nuclear fuel processing [21,33,109–111]. These wastes could be treated then immobilized or directly immobilized, and there are several available techniques to treat these wastes, including destructive technologies, either chemical or thermal, and non-destructive technologies, i.e., physical [23,112,113]. Organic solid wastes could be physically treated using compaction and then directed to immobilization in cement, where they are encapsulated without potential interference in the hydration reactions.

2.3.1. General Characteristics of Organic Liquid Wastes and Their Problematic Nature

In general, the characteristics of organic liquid wastes are dependent on the chemical composition of the organic agents, the nature of the applications, and the extent of hydrolysis and radiolysis of the organic components. The common radiological characteristics of these wastes are as follow [109–114]:

- Oils are generated in small volume compared to aqueous wastes and include lubricating oils, hydraulic fluids, and vacuum pump oils. Oils from reactor operations are classified as LLW, where they are contaminated with relatively small amounts of beta and gamma emitting radionuclides, e.g., ^{137}Cs, ^{134}Cs, ^{58}Co, ^{60}Co, and ^{65}Zn. Some lubricating oils in hot cells are contaminated with alpha emitting radionuclides. In PHWR, spent vacuum pump oils are classified as ILW.

- Scintillation liquids are used in radiochemical analysis of low energy alpha and beta emitters. They include both non-aqueous and non-polar solvents. The radioactivity content in these wastes is about 350 MBq/m^3,
- Miscellaneous solvents are used in the decontamination activities, including organic acids and solvent, e.g., toluene, carbon tetrachloride, acetone, alcohols, and trichloroethane. Their activity contents are <200 MBq/m^3. In addition, solvents used in the solvent extraction processes, widely used in nuclear fuel reprocessing, belong to this class. The latter is generated in respectively high volumes.

Depending on the chemical structure of these wastes, they can be volatile, flammable, and toxic. Safe management of these wastes includes either treatment or direct immobilization [109–117]. Direct immobilization of organic liquid wastes in a cement matrix is challenged by [23,74,118,119]:

- The interference of the organic components with the hydration reactions of the cementitious matrix leads to retardation in the setting of the cementitious paste and creation of porous matrix, where the organic liquid components cover the anhydrous cement grains and prevent them from reaction with water. This behavior is more notable for polar solvents.
- The retention of the organic components is mainly dependent on physical entrapment, i.e., encapsulated in cavities, not the chemical bonds with cement, which will make them vulnerable to leaching.
- Organic liquid wastes are susceptible to radiolysis, thermal and microbial degradation, which is associated with gas releases that can initiate cracking.

These challenges limit the full-scale practice of organic liquid wastes immobilization into conventional cementitious matrices to a 10–12% loading.

2.3.2. Cementitious Immobilization Matrices for Organic Liquid Wastes

Several additives and admixtures were proposed to increase the organic liquid waste loading into conventional cementitious waste matrices. These include clays, silica- or calcium-based additives, and emulsifiers. Three techniques were practiced for producing acceptable cementitious waste forms, which are pre-emulsification, direct mixing, and pre-impregnation [21,23,113,119,120]:

1. Pre-emulsification is a two-step technique, in which a stable emulsion is prepared by mixing the organic wastes with water in the presence or absence of emulsifier. In the second step, the emulsion is added to the mixed blend of OPC and the additive. This method was applied in USA and reported to be sensitive to changes in the waste composition. Pre-emulsification was also used in Cernavoda nuclear power plant in Romania to prepare cementitious waste forms for the immobilization of spent scintillation liquid contaminated with tritium [119,120]. In that practice, the oil and water are emulsified using high shear mixer then added to the cement.
2. Direct mixing is a single step technique, in which all the waste matrix components, e.g., OPC, additive, waste, water, and emulsifier if needed, are mixed until achieving a homogenous paste. The waste loading is limited in this case.
3. Pre-impregnation is also a two-step technique, in which the organic liquid is impregnated onto suitable sorbent, then the impregnated sorbent is mixed with the OPC and water to produce homogenous waste form paste. The increase in the waste form volume, due to the use of sorbent, is counterweighed with the low generated volumes of these wastes.

Figure 7 illustrates different mix designs for typical applications of pre-emulsification and pre-impregnation techniques in the immobilization of spent oil in conventional cementitious matrices. Data presented reveal that the pre-impregnation technique is sensitive to the type of used additive, where variable (15–56%) organic liquid waste loadings were achieved.

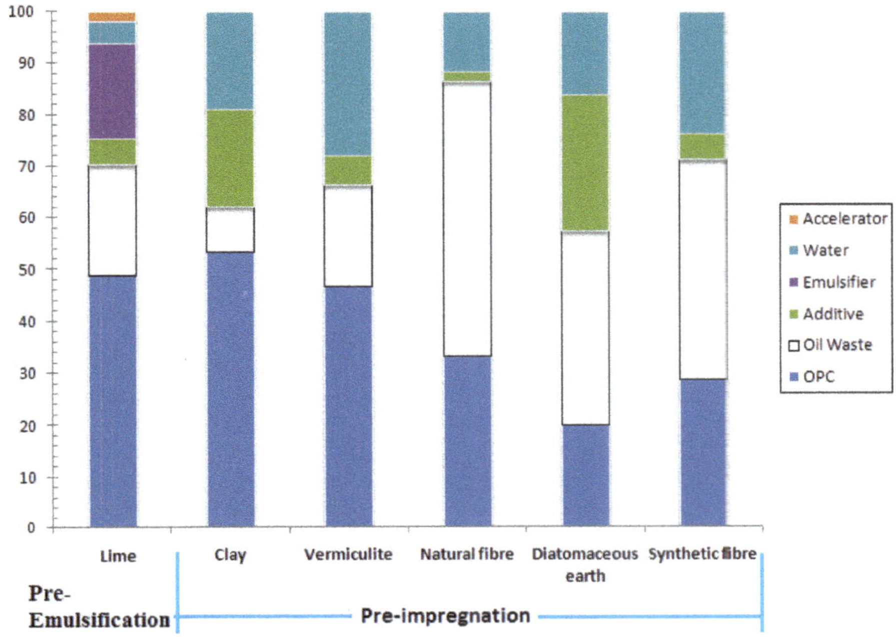

Figure 7. Typical composition of a conventional cementitious waste immobilization matrix for spent oil.

Research efforts are continuing to find suitable design mixes for improving the immobilization of organic liquid wastes, and among these efforts are [118,120–123]:

- Pre-emulsification technique was investigated to immobilize alpha contaminated lubricating oil in conventional cementitious matrix. Stable emulsion was prepared by mixing the oil and NaOH solution. Then, silica fumes were added to form solubilized silicate. Finally, the cement was added. This waste form was reported to enable the accommodation of 20% oil in cement and has acceptable alpha, beta, and gamma stabilization performances.
- Thermal expanded graphite (TEG) and bentonite clay were tested to immobilize TBP and spent oil contaminated with Cs and Sr. Pre-impregnation technique was followed, where TEG was impregnated with the organic liquid wastes then mixed with OPC, bentonite, and water. The effect of water immersion, freeze-thaw cycles, irradiation up to 250 KGy on the solidification performance, and Cs and Sr stabilizations were investigated. 25% TBP and 20% spent oil were successfully immobilized in the matrix at w/c ration 0.3. This matrix was recommended to have 42 days curing prior to transportation.
- Natural clay was used as additive to immobilize spent scintillation liquid following direct mixing technique. The liquid waste was dispersed into OPC-3% clay composite at 0.3 w/c ration then mixed to achieve homogenous paste. The matrix showed acceptable resistance to freeze-thaw and good stabilization performance under flooding conditions.

A study was performed to compare the performance of conventional and innovative cementitious forms in immobilizing organic liquid wastes [124]. In that study, lime, zeolite, and emulsifier were proposed to immobilize simulated TBP from the PUREX process in the conventional OPC matrix and innovative CSA matrix. The results indicated that at 28 days, the OPC matrix has a higher compressive strength and is more durable under freezing and irradiation conditions. Under flooding conditions, the compressive strength of CSA increased over that of OPC. The study concluded that CSA has a better stabilization performance than OPC.

A recent review summarized the progress of the incorporation of organic liquid in geopolymers, classifying the incorporation process into three classes similar to those used in conventional cementitious matrices [125]. Examples on the application of these classes in recent research to immobilize different organic liquid wastes include [126–128]:

- Pre-emulsification: the immobilization of simulated TBP from PUREX in phosphate-acid base geopolymer was investigated. The matrix, loaded with 18% waste, was reported to have compressive strength equals 59.19 MPa. Exposure to flooding and freeze conditions led to a reduction in the strength by 10 and 25%, respectively [126].
- Pre-impregnation: simulated radwaste oils immobilization in metakaolin-based geopolymers was studied. Bentonite and commercial polymer N910 were tested for oil pre-impregnation, then the matrix was prepared using Metakaolin and mixture of Na_2SiO_3 and NaOH. The analysis showed that the pre-impregnation of the oil using bentonite with 2% commercial polymer meets the Brazilian requirements on the stabilization performance [127].
- Pre-impregnation: the immobilization of simulated Lix84 loaded with copper in geopolymer was investigated. The metakaolin was impregnated with the extractant. Then, the binder was activated by adding the alkaline solution. 8% waste loading was achieved with acceptable stabilization performance [128].

3. Optimizing the Mix Design

The optimization of the mix design for any waste form is guided by available quality requirements on nuclear cement, paste operability characteristics, and the waste acceptance criteria (WAC) for the storage or disposal facility, as indicated in Section 2. Based on the national legislation, quality requirements on nuclear cements can be applied to the cement and additives used in producing the cementitious waste form. In this case, quality requirements on the raw materials and the produced waste form should be followed. Compliance with these requirements is usually ensured by conducting a standardized index experiment, e.g., grading and density [129,130]. The operability characteristics are determined in view of available cementation technologies in the facility/country. There are two classes of mixing technologies that are widely used in nuclear waste cementation, namely in drum mixing and grout batch mixing [21,23]. The features of each and available technologies are addressed elsewhere [21,23]. The mix design should be optimized so the paste possesses the following characteristics:

- Adequate flow-ability to allow the cementitious paste to flow freely from the mixer either in-drum or batch.
- Sufficient liquid content to minimize bleeding water.
- Setting time is longer than the process cycle, e.g., mixing, pouring, and mixer dripping, if any.

WAC are derived based on the safety case of a facility and include certain performance measures, e.g., limits on release rates, compressive strength, permeability, surface dose, and others [23]. These performances are usually evaluated based on the expected (natural) evolution scenario of the facility combined with hydraulic failure [131]. Within this context, accelerated aging or accidental procedures are applied to the waste form, then its performance is investigated [130]. These procedures can include repeated freeze–thaw cycles, wet–dry cycles, flooding, and dropping. Then, the performance measures are evaluated, e.g., the stabilization performance is evaluated using a suitable leaching test, and the solidification performance is evaluated by measuring the compressive strength, etc. During the optimization process, there are several options that could be considered during the design of the testing program. These include selection of the testing procedure, the factors that affect the waste form performance, and the optimization technique. Figure 8a–c illustrates the classification of the leaching tests, the factors considered during the design of these tests, and the optimization options.

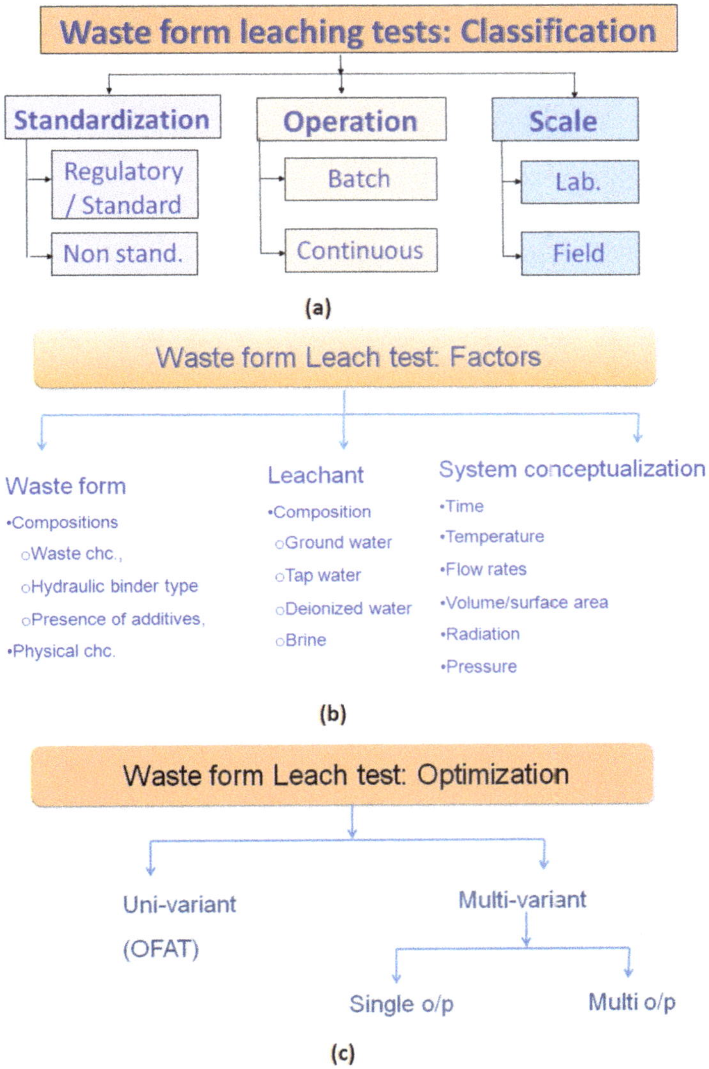

Figure 8. Waste form leaching test (**a**) classifications, (**b**) factors that affect the performance, and (**c**) optimization technique.

3.1. Leaching Test Procedure

Leaching procedures are conducted by contacting the waste form and a leachant (typically either deionized or natural water). Then, samples of the leachant are drawn and analyzed to determine the release of elements of interest, e.g., contaminants or structural elements. The features, advantages, and limitations of each test class, illustrated in Figure 8a, are listed in Table 5 [23,26,130,132]. Leaching tests are not only used to design the waste form composition but are used to support the quality control program and provide inputs for the disposal assessment as well. Leaching can be accelerated by using granular particles, de-mineralized water, and applying high flow rates, temperature, or pressure. The selection of the leaching test procedure should be guided by the aims of conducting the tests, stage of the design phase, and WAC requirements.

Table 5. Features, advantages, and limitations of different leaching test classes.

Class	Feature	Advantage	Limitation
		Standardization	
Standardized	Well-defined and controlled conditions, e.g., ASTM C 1308 and ANSI/ANS-16.1	Allows inter-comparison of the results of different waste forms from different places	Usually provide conservative estimates of radio-contaminants release
Non-standardized	Flexible condition, e.g., large scale field experiments	Allow mimicking actual concern about specified waste form	Results cannot be generalized easily
		Mode of Operation	
Batch	Static and semi-dynamic standardized and non-standardized leaching tests are used The semi-dynamic tests include total renewal of the leachate at pre-determined periods, e.g., ANS16.1 Static leach tests do not include renewal orcirculation of the leachant, e.g., MCC	Mimics slow flow conditions Simplicity of operation Small scale Robust and simple models are available to analyze the experimental data	Relatively slow compared with continuous models
Continuous	Flow through measure the release under conservative conditions, e.g., ASTM extraction column	Accelerated tests Simulate conservative conditions compared to that in Batch experiment	The modeling aspects are more complicated than those used for analyzing the batch experimental data
		Scale of application	
Lab	Both standardized and non-standardized tests are used	Simpler and cheaper compared to field tests Complete control of the experiment conditions	Need further investigations to up scale the results
Field	Non-standardized tests	Simulate the performance under true site-specific conditions	Hard to achieve complete control

3.2. Selection of the Waste Simulants

There are three options applied to investigating the potential use of a specified mix design and simulating the interactions of the radioactive wastes with the cementitious matrix. The selection of the most appropriate option is dependent on the problem formulation, and the options can be as follows:

1. Use stable solutions containing only potential radio-contaminants, e.g., Cs, Sr, I, etc.
2. Use simulated waste stream, e.g., simulated stable salt solutions of evaporator concentrate, precipitate, or loaded resin with stable isotopes.
3. Use simulated waste stream loaded with radioactive isotope.

The use of a stable isotope stimulant is justified by the similar chemical performance of the radioactive and stable isotopes, so the stabilization and solidification results will not be affected. Moreover, using stable simulants eliminates the radiological doses to workers. A recent review summarized the used solutions in studying the stable solutions for assessing the performance of nuclear waste forms and indicated that for, Cs, Sr, and Co, the most commonly used solutions for cementitious waste form testing are chlorides and nitrates [133]. In depth analysis of the contaminant speciation during the early hydration stage is important to evaluate the binding mechanisms of these contaminants [91,92]. The speciation results of Cs, Sr, and Co chloride solutions indicated that the solutions contain 100% Cs^+, a mixture of (Sr^{+2}, and $SrCl^+$), and a mixture of (Co^{+2}, $CoOH^+$, $Co(OH)_2aq$, $Co(OH)_3^-$) [91,92,134]. Cl^- is the only available species for Cs and Co solutions, and a mixture of Cl^- and $SrCl^+$ species exists in Sr solutions. The precise proportion of available species for stabilization/binding is pH dependent. Upon mixing these solutions with OPC-based materials, the unhydrated cement grains start to dissolute, leading to an increase in the solution pH. A reported study indicated that Cs^+ is the only detected species in the mixing solution independent on the presence of the OPC, additive, or another contaminant [91]. Using simulated evaporator concentrates and precipitates will complicate the speciation study for divalent and higher valency contaminants, depending on the composition of the solutions. An analysis of the interfacial water composition during the initial leaching of borosilicate glass indicated that a small portion of Sr will react with borate [131].

3.3. Selection of Additive Type

As indicated in Section 2, additives are used to improve the operability, solidification, and stabilization performances of conventional and innovative cementitious waste forms via several mechanisms. In a recent study, the general effects of the additives on the solidification performance, based on their shape, were concluded as follows [135]:

- Fine particle fillers, e.g., slag, fly ash, and silica fume can improve the pore structure and increase the compressive strength.
- Fiber additives are used to improve the tensile strength.

Recently, nano-materials have been tested as additives for cementitious waste forms [136,137]. The practical implementation still requires a lot of investigation [6,135,138]. During the selection of the additive, a prior precise evaluation of the problematic nature of the waste stream and required WAC should be conducted.

3.4. Solid and Liquid Contents Proportion

As the proportion of the mix design for both conventional and innovative cementitious waste forms highly affect the form performance, there is a need to identify the optimum proportion of the mix design, taking into consideration the following [21,23,30,56,93,139]:

- For liquid wastes, e.g., evaporator concentrates, saline wastes, increasing the waste loading will be associated with the formation of porous materials and subsequently reduce the compressive strength and increase the permeability.
- For matrices that contain additives of considerable water sorbtivity, the determination of the water content is crucial, where using low water content will reduce the degree of hydration and subsequently reduced the solidification and stabilization performance and affect the flow-ability of the paste.
- For wet solid wastes, e.g., spent ion exchangers, the water content should be carefully adjusted to minimize the bleeding water.
- For matrices that contain additives, the reactivity of the additives should be evaluated, and the amount of the additives should be optimized to ensure the formation of sufficient hydration phases to achieve the required stabilization performance.

3.5. Selection of Optimization Techniques

Optimization techniques are divided into uni-variant, i.e., one factor at a time (OFAT), and multivariate techniques (MVT) (Figure 8c). The first is conducted by fixing all the studied factors at a fixed value and changing only one factor at a time, whereas in the second, all the factors are changed simultaneously. Table 6 lists the features, advantages, and limitations of both techniques [80]. Due to the complexity of the design process, where there are large numbers of factors and performance measures that should be considered, the optimization process usually proceeds via three steps toward the final durable waste form, as follow:

1. Identifying the mix design: in this step, the precise waste loading and liquid and solid proportions are identified. Both optimization techniques can be used in this step.
2. Test the robustness of the mix design: in this step, the ability of the optimum mix design to accommodate wastes with fluctuated chemical composition is tested. In this

case, multi-variant techniques are preferred, as they provide an easy way to identify the main interacting factors and enable setting controls on them.

3. Validation for industrial scale application: in this step, the performance of industrial scale waste forms istested and usually OFAT is adopted.

Table 6. Comparison between the optimization techniques.

Technique	Feature	Advantage	Limitation
OFAT	Evaluate isolated effects of the studied factors on a single performance measure Empirical, mechanistic, and black box models are used to analyze the data	Allow the determination of mechanisms, interpolate and extrapolate the process performance	Does not allow the determination of the effect of interaction between the factors that affect the performance
MVT	Evaluate the effects of the studied factors' variability and their interactions on single and multi-performance measures	Identify the main influencing factors Provide insights into the system reliability	Does not allow the determination of the mechanism

4. Conclusions

Cementitious materials are widely applied as immobilization media to host radioactive wastes, including problematic waste streams for which routine technologies need modifications to account for their specific parameters. Sustainability considerations during the design of cementitious matrices are important, and here we have analyzed the characteristics of problematic operational waste streams resulting from nuclear energy utilization, such as evaporator concentrates, spent ion exchangers, and organic liquid wastes. The components that interfere with the hydration reaction of cements are the key characteristics required for assessing the factors that affect the production of cementitious waste forms. A guide for optimizing the mix design of cementitious waste forms was developed that considers the features, advantages, and disadvantages of different leaching test categories, options to simulate the wastes, additives, water and solid components, and the adopted optimization techniques. Over the past decades, cementitious matrices proved their ability to produce sustainable waste forms for containing problematic operational wastes. A lot of efforts are still needed to optimize the mix design of these forms, and research in these areas can include the investigation of the use of wastes and nano-materials as additives to reduce the amount of cement used and improve the operational and long-term performance of these matrices. Investigations of sustainable and green solutions to increase the waste loading and maintain the long-term durability of the produced waste form are of great importance.

Author Contributions: Conceptualization, R.O.A.R.; resources, M.I.O. and R.O.A.R.; writing—original draft preparation, R.O.A.R. and M.I.O.; writing—review and editing, R.O.A.R. and M.I.O. All authors have read and agreed to the published version of the manuscript.

Funding: This research received no external funding.

Institutional Review Board Statement: Not applicable.

Informed Consent Statement: Not applicable.

Data Availability Statement: Not applicable.

Acknowledgments: The authors acknowledge the critical comments by anonymous reviewers, which helped to improve the review.

Conflicts of Interest: The authors declare no conflict of interest.

References

1. Abdel Rahman, R.O.; Ojovan, M.I. Introduction to the nuclear industry sustainability. In *Sustainability of Life Cycle Management for Nuclear Cementation-Based Technologies*; Abdel Rahman, R.O., Ojovan, M.I., Eds.; Elsevier-Woodhead Publishing: Chichester, UK, 2021; pp. 3–47. [CrossRef]
2. Kroger, W. Measuring the sustainability of energy systems. *NEA News* **2001**, *19*, 21–24.
3. OECD. *Nuclear Energy in a Sustainable Development Perspective*; Nuclear Energy Agency: Paris, France, 2000.
4. IAEA. *Nuclear Power and Sustainable Development*; International Atomic Energy Agency: Vienna, Austria, 2016.
5. Pearce, J.M. Limitations of Nuclear Power as a Sustainable Energy Source. *Sustainability* **2012**, *4*, 1173–1187. [CrossRef]
6. Abdel Rahman, R.O.; Metwally, S.S.; El-Kamash, A.M. Improving the Performance of Engineering Barriers in Radioactive Waste Disposal Facilities: Role of Nano-Materials. In *Handbook of Nanomaterials and Nanocomposites for Energy and Environmental Applications*; Kharissova, O., Martínez, L., Kharisov, B., Eds.; Springer: Cham, Switzerland, 2021; pp. 1183–1200. [CrossRef]
7. IAEA. *IAEA Safety Glossary, Terminology Used in Nuclear Safety and Radiation Protection*, 2018th ed.; International Atomic Energy Agency: Vienna, Austria, 2019.
8. Conner, J.R.; Hoeffner, S.L. The History of Stabilization/Solidification Technology. *Crit. Rev. Environ. Sci. Technol.* **1998**, *28*, 325–396. [CrossRef]
9. Heacock, H.W. *Alternative Nuclear Waste Solidification Processes*; University of Arizona: Tucson, AZ, USA, 1975; pp 177–206.
10. Mullarkey, T.B.; Jentz, T.L.; Connelly, J.M.; Kane, J.P. *A Survey and Evaluation of Handling and Disposing of Solid Low-Level Nuclear Fuel Cycle Wastes*; Atomic Industrial Forum Inc.: Washington, DC, USA, 1976.
11. NRC. *Solidification of High-Level Radioactive Wastes*; U.S. National Regulatory Center National Academy of Engineering: Washington, DC, USA, 1979.
12. Schneider, K.J. Solidification of Radioactive Wastes. *Chem. Eng. Prog.* **1970**, *66*, 35.
13. Conner, J.R.; Hoeffner, S.L. A Critical Review of Stabilization/Solidification Technology. *Crit. Rev. Environ. Sci. Technol.* **1988**, *28*, 397–462. [CrossRef]
14. Abdel Rahman, R.O.; Guskov, A.; Kozak, M.W.; Hung, Y.T. Recent Evaluation of Early Radioactive Disposal Practice. In *Handbook of Environmental Engineering*; Wang, L., Wang, M.H., Hung, Y.T., Shammas, N., Eds.; Natural Resources and Control Processes; Springer: Cham, Switzerland, 2016; pp. 371–400.
15. Cote, P.; Gilliam, M. *Environmental Aspects of Stabilization and Solidification of Hazardous and Radioactive Wastes*; STP 1033; ASTM: Philadelphia, PA, USA, 1989.
16. Gilliam, T.M.; Wiles, C.C. *Stabilization and Solidification of Hazardous, Radioactive, and Mixed Wastes*; STP 1123; ASTM: Philadelphia, PA, USA, 1992; Volume 2.
17. Laguna, W. Radioactive Waste Disposal by Hydraulic Fracturing. In *Purdue Univ. Eng. Bull. Ext. Ser.*; **1970**, *137*. Available online: https://www.osti.gov/biblio/5980790 (accessed on 27 October 2021).
18. Conner, J.R. Ultimate Disposal of Liquid Wastes by Chemical Fixation. In Proceedings of the 29th Annual Purdue Industrial Waste Conference, West Lafayette, IN, USA, 7–9 May 1974.
19. Batchelor, B. Overview of waste stabilization with cement. *Waste Manag.* **2006**, *26*, 689–698. [CrossRef] [PubMed]
20. IAEA. *Management of Problematic Waste and Material Generated during the Decommissioning of Nuclear Facilities*; TRS 441; International Atomic Energy Agency: Vienna, Austria, 2006.
21. Varlakov, A.; Zherebtsov, A. Innovative and conventional materials and designs of nuclear cementitious systems in radioactive waste management. In *Sustainability of Life Cycle Management for Nuclear Cementation-Based Technologies*; Abdel Rahman, R.O., Ojovan, M.I., Eds.; Elsevier-Woodhead Publishing: Chichester, UK, 2021; pp. 297–338. [CrossRef]
22. IAEA. *Improved Cement Solidification of Low and Intermediate Level Radioactive Waste*; TRS. 350; International Atomic Energy Agency: Vienna, Austria, 1993.
23. Abdel Rahman, R.O.; Rakhimov, R.Z.; Rakhimova, N.R.; Ojovan, M.I. *Cementitious Materials for Nuclear Waste Immobilization*; Wiley: New York, NY, USA, 2014; ISBN 9781118512005. [CrossRef]
24. Heath, T.; Schofield, J.; Shelton, A. Understanding cementitious backfill interactions with groundwater components. *J. Appl. Geochem.* **2020**, *113*, 104495. [CrossRef]
25. Stefan, L.; Varet, T.; Lamouroux, C. Cementation of Nuclear Wastes: From Past to Future in a Sustainable Approach. In Proceedings of the WM2018 Conference, Phoenix, AZ, USA, 18–22 March 2018.
26. Drace, Z.; Mele, I.; Ojovan, M.I.; Abdel Rahman, R.O. An overview of research activities on cementitious materials for radioactive waste management. *Mater. Res. Soc. Symp. Proc.* **2012**, *1475*, 253–264. [CrossRef]
27. IAEA. *Behaviour of Cementitious Materials in Long-Term Storage and Disposal of Radioactive Waste*; IAEA-TECDOC-1701; International Atomic Energy Agency: Vienna, Austria, 2013.
28. International Atomic Energy Agency (IAEA). *Requirements and Methods for Low and Intermediate Level Waste Package Acceptability*; IAEA TECDOC-864; International Atomic Energy Agency: Vienna, Austria, 1996.
29. IAEA. *Characterization of Radioactive Waste Forms and Packages*; TRS-383; International Atomic Energy Agency: Vienna, Austria, 1997.
30. Abdel Rahman, R.O.; Zaki, A.A. Comparative analysis of nuclear waste solidification performance models: Spent ion exchanger-cement based wasteforms. *Process Saf. Environ. Prot.* **2020**, *136*, 115–125. [CrossRef]

31. Dmitriev, S.A.; Lifanov, F.A.; Savkin, A.E.; Lashchenov, C.M. Handling of the Bottom Residues of a Nuclear Power Plant. *At. Energy* **2000**, *89*, 884–889. [CrossRef]
32. Kononenko, O.A.; Milyutin, V.V.; Makarenkov, V.I.; Kozlitin, E.A. Immobilization of NPP Evaporator Bottom High Salt-Bearing Liquid Radioactive Waste into Struvite-Based Phosphate Matrices. *J. Hazard. Mater.* **2021**, *416*, 125902. [CrossRef]
33. Pátzay, G.; Weiser, L.; Feil, F.; Patek, G. Analysis and Selective Treatment of Radioactive Waste Waters and Sludges. In *Waste Water—Evaluation and Management*; Einschlag, F.S.G., Ed.; InTech: Rijeka, Croatia, 2011; ISBN 9789533072333. Available online: https://www.intechopen.com/chapters/14576 (accessed on 27 October 2021).
34. Matsuzuru, H.; Moriyama, N.; Wadachi, Y.; Ito, A. Leaching behavior of Cesium-137 in cement–waste composites. *Health Phys.* **1977**, *32*, 529–534. [CrossRef]
35. Abdel Rahman, R.O.; Ojovan, M.I. Recent Trends in the Evaluation of Cementitious Material in Radioactive Waste Disposal. In *Handbook of Environmental Engineering*; Wang, L., Wang, M.H., Hung, Y.T., Shammas, N., Eds.; Natural Resources and Control Processes; Springer: Cham, Switherland, 2016; Volume 17, pp. 401–448. [CrossRef]
36. Anton, S.; Milan, Z. Borate compound content reduction in liquid radioactive waste from nuclear power plants with VVER reactor. In Proceedings of the International Conference Nuclear Energy in Central Europe 2000, Golf Hotel, Bled, Slovenia, 11–14 September 2000.
37. Kononenko, O.A.; Gelis, V.M.; Milyutin, V.V. Incorporation of bottoms from nuclear power plants into a matrix based on portland cement and silicic additives. *At. Energy* **2011**, *109*, 278–284. [CrossRef]
38. Li, B.; Ling, X.; Liu, X.; Li, Q.; Chen, W. Hydration of Portland cement in solutions containing a high concentration of borate ions: Effects ofLiOH. *Cem. Concr. Compos.* **2019**, *102*, 94–104. [CrossRef]
39. Zhao, Q.; Tu, J.; Han, W.; Wang, X.; Chen, Y. Hydration Properties of Portland Cement Paste with Boron Gangue. *Adv. Mater. Sci. Eng.* **2020**, *2020*, 7194654. [CrossRef]
40. Zatloukalova, J.; Zatloukal, J.; Hraníček, J.; Kolář, K.; Konvalinka, P. Study on the properties of cement composites for immobilization of evaporator concentrates. *Prog. Nucl. Energy* **2021**, *140*, 103919. [CrossRef]
41. Wang, Y.; Huang, B.; Mao, Z.; Deng, M.; Cao, H. Effect of a Boric Acid Corrosive Environment on the Microstructure and Properties of Concrete. *Materials* **2020**, *13*, 5036. [CrossRef] [PubMed]
42. Csetenyi, L.J.; Glasser, F.P. Borate retardation of cement set and phase relations in the system Na_2O-CaO-B_2O_3-H_2O. *Adv. Cem. Res.* **1995**, *7*, 13–19. [CrossRef]
43. Abdel Rahman, R.O.; Ojovan, M.I. Sustainability of cementitious structures, systems, and components (SSC's): Long-term environmental stressors. In *Sustainability of Life Cycle Management for Nuclear Cementation-Based Technologies*; Abdel Rahman, R.O., Ojovan, M.I., Eds.; Elsevier-Woodhead Publishing: Chichester, UK, 2021; pp. 181–232. [CrossRef]
44. Varlakov, A.; Zherebtsov, A.; Ojovan, M.I.; Petrov, V. Long-term irradiation effectsin cementitious systems. In *Sustainability of Life Cycle Management for Nuclear Cementation-Based Technologies*; Abdel Rahman, R.O., Ojovan, M.I., Eds.; Elsevier-Woodhead Publishing: Chichester, UK, 2021; pp. 161–180. [CrossRef]
45. Goni, S.; Guerrero, A.; Lorenzo, M.P. Efficiency of fly ash belite cement and zeolite matreices for immobilizing cesium. *J. Hazard. Mater.* **2006**, *B 137*, 1608–1617. [CrossRef]
46. Sun, Q.; Li, J.; Wang, J. Effect of borate concentration on solidification of radioactive wastes by different cements. *Nucl. Eng. Des.* **2011**, *241*, 4341–4345. [CrossRef]
47. Guerrero, A.; Goñi, S. Efficiency of a blast furnace slag cement for immobilizing simulated borate radioactive liquid waste. *J. Waste Manag.* **2002**, *22*, 831–836. [CrossRef]
48. Sinha, P.K.; Shanmugamani, A.G.; Renganathan, K.; Muthiah, R. Fixation of radioactive chemical sludge in a matrix containing cement and additives. *Ann. Nucl. Energy* **2009**, *36*, 620–625. [CrossRef]
49. Reddy, D.A.; Khandelwal, S.K.; Muthiah, R.; Shanmugamani, A.G.; Paul, B.; Rao, S.V.S.; Sinha, P.K. Conditioning of sludge produced through chemical treatment of radioactive liquid waste; operating experiences. *Ann. Nucl. Energy* **2010**, *37*, 934–941. [CrossRef]
50. Osmanlioglu, A.E. Immobilization of radioactive borate liquid waste using natural diatomite. *Desalin. Water Treat.* **2016**, *57*, 15146–15153. [CrossRef]
51. Polyakov, A.S.; Masanov, O.L.; Zakharova, K.P. Cementing of radioactive salt concentrates. *At. Energy* **1994**, *77*, 468–470. [CrossRef]
52. El-Kamash, A.M.; El-Naggar, M.R.; El-Dessouky, M.I. Immobilization of cesium and strontium radionuclides in zeolite–cement blends. *J. Hazard. Mater.* **2006**, *136*, 310–316. [CrossRef]
53. Varlakov, A.P. Development of a unified technological process for cementing liquid radioactive wastes. *At. Energy* **2010**, *109*, 16–21. [CrossRef]
54. Poluektova, G.B.; Smirnov, Y.V.; Sokolova, I.D. *Processing and Disposal of Radioactive Waste from Nuclear Industry Companies in Foreign Countries. Central Research and Development Institute for Management, Economics and Information of the Ministry of Nuclear Industry of the Russian Federation, Moscow*; CNIIatominform: Moscow, Russia, 1990.
55. Plecas, I.; Dimovic, S. Immobilization of Cs and Co in concert matrix. *Ann. Nucl. Energy* **2003**, *30*, 1899–1903. [CrossRef]
56. Zatloukalová, J.; Marty, V.D.; Zatloukal, J.; Kolár, K.; Guillot, W.; Konvalinka, P. Investigation of radiolysis in cement pastes immobilizing simulatedevaporatorconcrentrates. *Ann. Nucl. Energy* **2021**, *151*, 107901. [CrossRef]

57. Zatloukalova, J.; Marty, V.D.; Zatloukal, J.; Kolar, K.; Hlavac, Z.; Konvalinka, P. Mechanical properties of irradiated cement pastes for immobilization of evaporator concentrates. *Prog. Nucl. Energy* **2020**, *127*, 103437. [CrossRef]
58. Champenois, J.B.; Coumes, C.C.D.; Poulesquen, A.; Le Bescop, B.; Damidot, D. Conditioning highly concentrated borate solutions with calcium sulfoaluminate cement. In *Cement-Based Materials for Nuclear Waste Storage*; Bart, F., Coumes, C.C.D., Frizon, F., Lorente, S., Eds.; Springer: New York, NY, USA, 2013; pp. 203–214.
59. Champenois, J.B.; Dhoury, M.; Coumes, C.C.D.; Mercier, C.; Revel, B.; Le Bescop, P.; Damidot, D. Influence of sodium borate on the early age hydration of calcium sulfoaluminate cement. *Cem. Concr. Res.* **2015**, *70*, 83–93. [CrossRef]
60. Qina, S.; Wang, J. Cementation of radioactive borate liquid waste produced in pressurized water reactors. *Nucl. Eng. Des.* **2010**, *240*, 3660–3664.
61. Hall, D.A. The effect of retarders on the microstructure and mechanical properties of magnesiaephosphate cement mortar. *Cem. Concr. Res.* **2001**, *31*, 455–465. [CrossRef]
62. Yang, J. Effect of borax on hydration and hardening properties of magnesium and potassium phosphate cement pastes. *J. Wuhan. Univ. Techn. Mater. Sc. Ed.* **2010**, *25*, 613–618. [CrossRef]
63. Palomo, A.; De la Fuente, J.I. Alkali-activated cementitious materials: Alternative matrices for the immobilisation of hazardous wastes, part I. Stabilisation of boron. *Cem. Concr. Res.* **2003**, *33*, 281–288. [CrossRef]
64. Rakhimova, N.R.; Rakhimov, R.Z.; Morozov, V.P.; Potapova, L.I.; Osin, Y.N. Mechanism of solidification of simulated borate liquid wastes with sodium silicate activated slag cements. *J. Clean. Prod.* **2017**, *149*, 60–69. [CrossRef]
65. Champenois, J.B.; Mesbah, A.; Coumes, C.C.D.; Renaudin, G.; Leroux, F.; Mercier, C.; Revel, B.; Damidot, D. Crystal structures of Boro-AFm and sBoro-AFt phases. *Cement Concr. Res.* **2012**, *42*, 1362–1370. [CrossRef]
66. Coumes, C.C.D.; Courtois, S.; Peysson, S.; Ambroise, J.; Pera, J. Calcium sulfoaluminate cement blended with OPC A promising binder to encapsulate low-level radioactive slurries of complex chemistry. *Cem. Concr. Res.* **2009**, *39*, 740–747. [CrossRef]
67. Hernandez, S.; Guerrero, A.; Goni, S. Leaching of borate waste cement matrices: Pore solution and solid phase characterization. *Adv. Cem. Res.* **2000**, *12*, 1–8. [CrossRef]
68. Kwan, B.; Lee, J.; Kang, J.; Um, W. Development of geopolymer waste form for immobilization of radioactive borate waste. *J. Hazard. Mater.* **2021**, *419*, 126402.
69. Jantzen, C.M.; Lee, W.E.; Ojovan, M.I. Radioactive waste (RAW) conditioning, immobilisation, and encapsulation processes and technologies: Overview and advances. Chapter 6. In *Radioactive Waste Management and Contaminated Site Clean-Up: Processes, Technologies and International Experience*; Woodhead: Cambridge, UK, 2013; pp. 171–272.
70. Pierce, E.M.; Um, W.; Cantrell, K.J.; Valenta, M.M.; Westsik, J.H.; Serne, R.J., Jr.; Parker, K.E. *Secondary Waste Form Screening Test Results—Cast Stone and Alkali Alumino-Silicate Geopolymer*; PNNL-19505; PNNL: Washington, WA, USA, 2011.
71. Mattigod, S.V.; Westsik, J.H. *Secondary Waste Form Down-Selection Data Package—DuraLith*; PNNL-20718; Pacific Northwest National Laboratory: Richland, WA, USA, 2011.
72. Abdel Rahman, R.O.; Metwally, S.S.; El-Kamash, A.M. Life Cycle of Ion Exchangers in Nuclear Industry: Application and Management of Spent Exchangers. In *Handbook of Ecomaterials*; Martínez, L., Kharissova, O., Kharisov, B., Eds.; Springer: Cham, Switzerland, 2019; pp. 3709–3732. [CrossRef]
73. IAEA. *Applications of Ion Exchange Processes for the Treatment of Radioactive Waste and Management of Spent Ion Exchangers*; TRS-408; International Atomic Energy Agency: Vienna, Austria, 2002.
74. Utton, C.; Godfrey, I.H. *Review of Stability of Cemented Grouted Ion-Exchange Materials, Sludges and Flocs*; National Nuclear Laboratory: Warrington, UK, 2010.
75. Ojovan, M.I.; Lee, W.E. *An Introduction to Nuclear Waste Immobilisation*; Elsevier: Amsterdam, The Netherlands, 2014.
76. Arnold, G.; Fuhrmann, M.; Colombo, P.; Kalb, P.; Doty, R.; Kemeny, P.; Franz, E.M.; Neilson, R.M.; Reilly, S., Jr. *Solidification of Ion Exchange-Resin Wastes, Brookhaven National Laboratory*; Brookhaven National Laboratory Upton: New York, NY, USA, 1982.
77. Faiz, Z.; Bouih, A.; Fakhi, S.; Laissaoui, A.; Hannache, H.; Idrissi, A. Improvement of conditions for the radioactive ion exchange resin immobilization in the cement Portland. *J. Mater. Environ. Sci.* **2015**, *6*, 289–296.
78. Laili, Z.; Yasir, M.S.; Abdul Wahab, M.; Mahmud, N.A.; Zainal Abidin, N. Evaluation of the compressive strength of cement-spent resins matrix mixed with biochar. *Malays. J. Anal. Sci.* **2015**, *19*, 565–573.
79. Gasser, M.S.; Mekhamer, H.S.; Abdel Rahman, R.O. Optimization of the utilization of Mg/Fe hydrotalcite like Compounds in the removal of Sr(II) from aqueous solution. *J. Environ. Chem. Eng.* **2016**, *4*, 4619–4630. [CrossRef]
80. Abdel Rahman, R.O.; Abdel Moamen, O.A.; Abdelmonem, N.; Ismaill, M. Optimizing the removal of strontium and cesium ions from binary solutions on magnetic nano-zeolite using response surface methodology (RSM) and artificial neural network (ANN). *Environ. Res.* **2019**, *173*, 397–410. [CrossRef] [PubMed]
81. Howard, C.G.; Jolliffe, C.B.; Lee, D.J. *Immobilization of Ion-Exchange Resins in Cement, Contract No FI1W-0006-UK, Commission of the European Communities, Directorate-General Telecommunications*; Information Industries and Innovation L-2920; Commision of the European Communities: Luxembourg, 1991.
82. Veazey, G.W.; Ames, R.L. *Cement Waste-Form Development for Ion-Exchange Resins at the Rocky Flats Plant*; LA-13226-MS; Los Alamos National Laboratory: Los Alamaos, NM, USA, 1997.
83. Matsuda, M.; Nishi, T.; Chino, K.; Kikuchi, M. Solidification of Spent Ion Exchange Resin Using New Cementitious Material, (I) Swelling Pressure of Ion Exchange Resin. *J. Nucl. Sci. Technol.* **1992**, *29*, 883–889. [CrossRef]

84. Neji, M.; Bary, B.; Le Bescop, P.; Burlion, N. Swelling behavior of ion exchange resins incorporated in tri-calcium silicate cement matrix: I. Chemical analysis. *J. Nucl. Mater.* **2015**, *467*, 544–556. [CrossRef]
85. Neji, M.; Bary, B.; Le Bescop, P.; Burlion, N. Swelling behavior of ion exchange resins incorporated in tri-calcium silicate cement matrix: II. Mechanical analysis. *J. Nucl. Mater.* **2015**, *467*, 863–875. [CrossRef]
86. Lafond, E.; Coumes, C.C.D.; Gauffinet, S.; Chartier, D.; Le Bescop, P.; Stefan, L.; Nonat, A. Investigation of the swelling behavior of cationic exchange resins saturated with Na+ ions in a C3S paste. *Cem. Concr. Res.* **2015**, *69*, 61–71. [CrossRef]
87. IAEA. *Treatment and Conditioning of Spent Ion Exchange Resins from Research Reactors, Precipitation Sludges and Other Radioactive Concentrates*; IAEA-TECDOC-689; International Atomic Energy Agency: Vienna, Austria, 1993.
88. Abdel Rahman, R.O.; Zaki, A.A. Comparative study of leaching conceptual models: Cs leaching from different ILW cement based matrices. *Chem. Eng. J.* **2011**, *173*, 722–736. [CrossRef]
89. Li, J.; Wang, J. Advances in cement solidification technology for waste radioactive ion exchange resins: A review. *J. Hazard. Mater.* **2006**, *B135*, 443–448. [CrossRef]
90. Osmanlioglu, A.E. Immobilization of radioactive waste by cementation with purified kaolin clay. *Waste Manag.* **2015**, *2*, 481–483. [CrossRef]
91. Abdel Rahman, R.O.; Zein, D.H.; Abo Shadi, H. Cesium binding and leaching from single and binary contaminant cement-bentonite matrices. *Chem. Eng. J.* **2014**, *245*, 276–287. [CrossRef]
92. Abdel Rahman, R.O.; Zein, D.H.; Abo Shadi, H. Assessment of strontium immobilization in cement–bentonite matrices. *Chem. Eng. J.* **2013**, *228*, 772–780. [CrossRef]
93. Phillip, E.; Khoo, K.S.; Yusof, M.A.W.; Abdel Rahman, R.O. Assessment of POFA-cementitious material as backfill barrier in DSRS borehole disposal: ^{226}Ra confinement. *J. Environ. Manage.* **2021**, *28*, 111703. [CrossRef]
94. Sebesta, F.; John, J.; Motl, A. *Development of Composite Ion Exchangers and Their Use in Treatment of Liquid Radioactive Wastes, Waste Treatment and Technologies Involving Inorganic Sorbents*; Final Report of a Co-Ordinated Research Programme 1992–1996 IAEA-TECDOC-947; IAEA: Vienna, Austria, 1997; pp. 79–105.
95. Nurokhim, L.T. *Immobilization Of 137Cs on Cement-Zeolite Composites, Waste Treatment and Technologies Involving Inorganic Sorbents*; Final Report of a Co-Ordinated Research Programme 1992–1996 IAEA-TECDOC-947; IAEA: Vienna, Austria, 1997; pp. 153–163.
96. Plecas, I.; Pavlovic, R.; Pavlovic, S. Leaching behaviour of 60Co and 137Cs from spent ion exchange resins in cement-bentonite clay matrix. *J. Nucl. Mater.* **2004**, *327*, 171–174. [CrossRef]
97. Osmanlioglu, A.E. Progress in cementation of reactor resins. *Biol. Sci.* **2007**, *49*, 20–26. [CrossRef]
98. Plecas, I.; Dimovic, S. Influence of natural sorbents on the immobilization ofspent ion exchange resins in cement. *J. Radioanal. Nucl. Chem.* **2006**, *269*, 181–185. [CrossRef]
99. Junfeng, L.; Gong, Z.W.; Jianlong, W. Solidification of low level-radioactive resins in ASC zeolite blends. *Nucl. Eng. Des.* **2005**, *235*, 817–821. [CrossRef]
100. IAEA. *Use of Benchmarking System for Operational Waste from WWER Reactors*; TECDOC-1815; IAEA: Vienna, Austria, 2017.
101. Lichvar, P.; Rozloznik, M.; Sekely, S. Behaviour of Aluminosilicate Inorganic Matrix Sial during and after Solidification of Radioactive Sludge and Radioactive Spent Resins and Their Mixtures; Amec Nuclear Slovakia. 2010. Available online: https://www-pub.iaea.org/MTCD/Publications/PDF/TE-1701_add-CD/PDF/Slovakia.pdf (accessed on 27 October 2021).
102. Rakhimova, N.R.; Rakhimov, R.Z.; Lutskin, Y.S.; Morozov, V.P.; Osin, Y.N. Solidification of borate ion-exchange resins by alkali—activated slag cements. *Rom. J. Mater.* **2018**, *48*, 177–184.
103. Lee, W.H.; Cheng, T.W.; Ding, Y.C.; Lin, K.L.; Tsao, S.W.; Huang, C.P. Geopolymer technology for the solidification of simulated ion exchange resins with radionuclides. *J. Environ. Manag.* **2019**, *235*, 19–27. [CrossRef] [PubMed]
104. Lin, W.H.; Chen, H.Y.; Huang, C.P. Performance study of ion exchange resins solidification using metakaolin-based geopolymer binder. *Prog. Nucl. Energy* **2020**, *129*, 103508. [CrossRef]
105. Khairudin, N.W.A.; Yasir, M.S.; Majid, A.A.; Wahab, M.A. Immobilisation of spent ion exchange resin from PuspatiTRIGA reactor using ash-based geopolymer. *Malays. J. Anal. Sci.* **2015**, *19*, 472–480.
106. Kryvenko, P.; Petropavlovskyi, O.; Cao, H.; Weng, L.; Kovalchuk, O. Applicability of alkali-activated cement for immobilization of low-level radioactive waste in ion-exchange resins. *East. Eur. J. Enterp. Technol.* **2015**, *6*, 40–45. [CrossRef]
107. Xu, Z.; Jiang, Z.; Wu, D.; Peng, X.; Xu, Y.; Li, N.; Qi, Y.; Li, P. Immobilization of Strontium-Loaded Zeolite A by Metakaolin Based-Geopolymer. *Ceram. Int.* **2017**, *43*, 4434–4439. [CrossRef]
108. El-Naggar, M.R.; El-Masry, E.H.; El-Sadek, A.A. Assessment of individual andmixed alkali activated binders for solidification of a nuclear grade organic resin loaded with 134Cs, 60Co and 152+154 Eu radionuclides. *J. Hazard. Mater.* **2019**, *375*, 149–160. [CrossRef]
109. Abdel Rahman, R.O.; Kozak, M.W.; Hung, Y.-T. Radioactive pollution and control, Ch (16). In *Handbook of Environment and Waste Management*; Hung, Y.T., Wang, L.K., Shammas, N.K., Eds.; World Scientific Publishing Co.: Singapore, 2014; pp. 949–1027.
110. Abdel Rahman, R.O.; Ibrahium, H.A.; Hung, Y.-T. Liquid radioactive wastes treatment: A Review. *Water* **2011**, *3*, 551–565. [CrossRef]
111. Romanovsky, V.; Pokhitonov, Y.; Kelley, D. Prospects on Immobilization of Liquid Organic Radwastes with the Use of High Technology Polymers—14065. In Proceedings of the WM2014 Conference, Phoenix, AZ, USA, 2–6 March 2014.
112. IAEA. *Design and Operation of Radioactive Waste Incineration Facilities*; Safety Series No. 921; IAEA: Vienna, Austria, 1992.
113. IAEA. *Treatment and Conditioning of Radioactive Organic Liquids*; TECDOC 656; IAEA: Vienna, Austria, 1992.

114. IAEA. *Reference Design for a Centralized Waste Processing and Storage Facility*; IAEATECDOC-776; IAEA: Vienna, Austria, 1994.
115. IAEA. *Handling and Processing of Radioactive Waste from Nuclear Applications*; Technical Reports Series No. 402; IAEA: Vienna, Austria, 2001.
116. IAEA. *Combined Methods for Liquid Radioactive Waste Treatment*; IAEA-TECDOC 1336; IAEA: Vienna, Austria, 2003.
117. IAEA. *Predisposal Management of Organic Radioactive Waste*; Technical Reports Series No. 427; IAEA: Vienna, Austria, 2004.
118. Manohar, S.; Sasidharan, N.S.; Wattal, P.K.; Shah, N.J.; Chander, M.; Bansal, N.K. Immobilization of Alpha Contaminated Lubricantingoul in Cement Matrix, Government of India. BARC/2000/E/036; Bhabha Atomic Energy Commission: Mumbai, India, 2000.
119. Teoreau, I.; Deneanu, N.; Dulama, M. Matrix Materials for the conditioning organic radioactive wastes. *Romn J Mater.* **2010**, *40*, 112–121.
120. Dianu, M.; Podină, C. The safety of environment in final disposal of ultima gold scintillation liquid cocktail used for determination of the radioactive content in various samples at cernavoda nuclear power plant. *Rev. Roum Chim.* **2007**, *52*, 509–519.
121. Nikitin, A.V.; Kondakova, Y.V.; Sazonov, A.B. Combined Matrixes for Solidification of Organic Radioactive Liquid Wastes Containing Cs-137 and Sr-90. *Inorg. Mater. Appl. Res.* **2017**, *8*, 681–690. [CrossRef]
122. Eskander, S.B.; Abdel Aziz, S.M.; El-Didamony, H.; Sayed, M.I. Immobilization of low and intermediate level of organic radioactive wastes in cement matrices. *J. Hazard. Mater.* **2011**, *190*, 969–979. [CrossRef]
123. El-Didamony, H.; Bayoumi, T.A.; Sayed, M.I. Evaluation of the Properties of Cemented Liquid Scintillator Wastes under Flooding Scenario in Various Aqueous. *Int. Sch. Res. Netw.* **2012**, *2012*, 373795. [CrossRef]
124. Zhang, W.; Li, J.; Wang, J. Solidification of spent radioactive organic solvent by sulfoaluminate and Portland cements. *J. Nucl. Sci. Technol.* **2015**, *52*, 1362–1368. [CrossRef]
125. Reeb, C.; Pierlot, C.; Davy, C.; Lambertin, D. Incorporation of organic liquids into geopolymer materials—A review of processing, properties and applications. *Ceram. Int.* **2021**, *47*, 7369–7385. [CrossRef]
126. Dong, T.; Xie, S.; Wang, J.; Zhao, G.; Song, Q. Solidification and stabilization of spent TBP/OK organic liquids in a phosphate acid-based geopolymer. *Sci. Technol. Nucl. Install.* **2020**, *2020*, 8094205. [CrossRef]
127. Cuccia, V.; Freire, C.B.; Ladeira, A.C.Q. Radwaste oil immobilization in geopolymer after non-destructive treatment. *Prog. Nucl. Energy* **2020**, *122*, 103246. [CrossRef]
128. El-Naggar, M.R.; El-Sherief, E.A.; Mekhemar, H.S. Performance of geopolymers for direct immobilization of solvent extraction liquids: Metakaolin/LIX-84 formulations. *J. Hazard. Mater.* **2018**, *360*, 670–680. [CrossRef] [PubMed]
129. Al-Tabbaa, A.; Perera, A.S.R. Stabilisation/solidification treatment and remediation part IV: Testing & performance criteria. In Proceedings of the International Conference on Stabilisation/Solidification Treatment and Remediation, Cambridge, UK, 12–13 April 2005; pp. 415–435.
130. Abdel Rahman, R.O.; Ojovan, M.I. Leaching Tests and Modelling of Cementitious Waste forms Corrosion. *Innov. Corros. Mater. Sci.* **2014**, *4*, 90–95. [CrossRef]
131. Farid, O.M.; Ojovan, M.I.; Abdel Rahman, R.O. Evolution of cations speciation during the initial leaching stage of alkali-borosilicate-glasses. *MRS Adv.* **2020**, *5*, 185–193. [CrossRef]
132. Abdel Rahman, R.O.; Ojovan, M.I. Techniques to test cementitious systems through their life cycles. In *Sustainability of Life Cycle Management for Nuclear Cementation-Based Technologies*; Abdel Rahman, R.O., Ojovan, M.I., Eds.; Woodhead Publishing: Chichester, UK, 2021; pp. 407–430. [CrossRef]
133. Li, J.; Xu, D.; Wang, W.; Wang, X.; Mao, Y.; Zhang, C.; Jiang, W.; Wu, C. Review on Selection and Experiment Method of Commonly Studied Simulated Radionuclides in Researches of Nuclear Waste Solidification. *Sci. Technol. Nucl. Install.* **2020**, *2020*, 3287320. [CrossRef]
134. Gasser, M.S.; El Sherif, E.; Mekhamer, H.S.; Abdel Rahman, R.O. Assessment of Cyanex 301 impregnated resin for its potential use to remove cobalt from aqueous solutions. *Environ. Res.* **2020**, *185*, 109402. [CrossRef] [PubMed]
135. Li, J.; Chen, L.; Wang, J. Solidification of radioactive wastes by cement-based materials. *Prog. Nucl. Energy* **2021**, *141*, 103957. [CrossRef]
136. Cao, B.; Fan, S.; Tan, X.; Li, M.; Hu, Y. Cementitious materials modified with hematite nanoparticles for enhanced cement hydration and uranium immobilization. *Environ. Sci. Nano* **2017**, *4*, 1670–1681. [CrossRef]
137. Fan, S.; Cao, B.; Deng, N.; Hu, Y.; Li, M. Effects of ferrihydrite nanoparticle incorporation in cementitious materials on radioactive waste immobilization. *J. Hazard. Mater.* **2019**, *379*, 120570. [CrossRef]
138. Abdel Rahman, R.O.; Ojovan, M.I. Application of nano-materials in radioactive waste management. In *Environmental Science and Engineering. Industrial Processes & Nanotechnology Ch (15)*; Tian, C.Z., Bhola, R.G., Govil, J.N., Eds.; Studium Press LLC: New York, NY, USA, 2017; Volume10, pp. 361–378, ISBN 101626990980.
139. Abdel Rahman, R.O.; Zaki, A.A. Assessment of the leaching characteristics of incineration ashes in cement matrix. *Chem. Eng. J.* **2009**, *155*, 698–708. [CrossRef]

sustainability

MDPI

Review

Glass Crystalline Materials as Advanced Nuclear Wasteforms

Michael I. Ojovan *, **Vladislav A. Petrov** and **Sergey V. Yudintsev**

Institute of Geology of Ore Deposits, Petrography, Mineralogy and Geochemistry of Russian Academy of Sciences (IGEM RAS), 119017 Moscow, Russia; vlad243@igem.ru (V.A.P.); syud@igem.ru (S.V.Y.)
* Correspondence: m.i.ojovan@gmail.com

Abstract: Glass crystalline materials (GCM) are of increasing interest as advanced nuclear wasteforms combining the advantages of vitreous and crystalline matrices. The GCM are versatile wasteforms envisaged for a wider use to immobilise various types of both radioactive and chemically hazardous wastes. They can be produced either via low temperature sintering using precursors composed of glass frit, oxides, and crystalline phases or through conventional melting aiming to produce first a parent glass, which is then crystallised by a controlled thermal schedule to obtain target crystalline phases within the GCM. Utilization of GCM is highlighted as a perspective wasteform for immobilization of partitioned radionuclide streams.

Keywords: high level radioactive waste (HLW); radionuclide partitioning; immobilisation; melting; sintering; crystalline matrix; vitreous matrix; glass crystalline materials (GCM)

Citation: Ojovan, M.I.; Petrov, V.A.; Yudintsev, S.V. Glass Crystalline Materials as Advanced Nuclear Wasteforms. *Sustainability* **2021**, *13*, 4117. https://doi.org/10.3390/su13084117

Academic Editor: Changhyun Roh

Received: 1 March 2021
Accepted: 6 April 2021
Published: 7 April 2021

1. Introduction

The spent nuclear fuel (SNF) reprocessing generates radioactive wastes, including high level radioactive waste (HLW), which is industrially immobilised in the Na-Al-P glass in Russia and A-B-Si glass elsewhere, where A stands for alkaline elements. By 2013, there were about 30,000 accumulated tonnes of vitrified HLW overall in the world [1,2]. Considering the processing rates of vitrification facilities [2–6], the current mass of vitrified HLW can be estimated at about 35,000–36,000 tonnes, of which almost 80% are A-B-Si, and the rest are Na-Al-P glasses. Vitrification of HLW is, however, not the optimal method of immobilisation due to the relative low radionuclide loading of glasses and their susceptibility to crystallisation, which can begin immediately after the melt pouring into canisters due to the residual heat of the melt [6,7]. Partitioning of HLW radionuclides onto groups can provide a better solution for their immobilisation by incorporation within crystalline lattice of silicates, titanates, zirconates, and phosphates. Their natural analogues are minerals zircon, britholite, pyrochlore, zirconolite, murataite, perovskite, monazite, and garnet [8–16]. Fission products (Cs, Sr) can be isolated in both crystalline phases such as hollandite, pollucite, perovskite, langbeinite, and glasses. Glass crystalline materials (GCM) with the same mineral-like phases are optimal for wastes of complex composition. The major component of GCM may be either the crystalline phases with the glass acting as a binding agent or alternatively the vitreous phase may be the major component, with crystalline particles dispersed in the glass matrix [17–21].

The aim of this paper is to highlight the effectiveness of GCM as a universal nuclear wasteform which overcomes limitations of classical vitrification processes related to the low solubility in glasses of some nuclear waste components such as minor actinides (MA) and enhances the waste loading while improving the durability and long-term stability of the wasteform. The review focuses on universality of GCM wasteforms for immobilisation of HLW radionuclides utilizing the fractionation (partitioning) approach when the most dangerous components are immobilized in most durable and most stable crystals whereas the vitreous phase accommodates less dangerous, shorter-lived radionuclides. The vitrification technology currently accepted as baseline for HLW immobilization can be accordingly modified to produce GCM wasteforms.

2. HLW from SNF Reprocessing

The composition of SNF depends on the type of reactor, initial fuel used, the depth of its burnup, and storage time after discharge from reactor [22,23]. HLW radionuclides are fission products, actinides, and activated elements of fuel assemblies. The main fission products are rare earth elements (REE), Zr, Mo, Tc, Ru, Pd, Cs, Sr, Rh, Te, Xe, Kr, and I, with REE accounting for up to 25 wt.%, platinoids—16%, Zr—15%, Mo—12%, Cs—6% [24]. After three years of storage of SNF with a burnup of 33 GW·day (enrichment of 3.5% ^{235}U) it contains in kg/tonne of SNF [19]: Alkalis (Cs, Rb)—3, alkaline earths (Sr, Ba)—2.4, rare earths (mainly light of Ce group)—10.2, transition 4d metals (Mo, Zr, Tc)—7.7, platinoids (Ru, Rh, Pd)—3.9, 0.5 kg Se and Te, 0.2 kg I and Br, 0.1 kg Ag, Cd, Sn, and Sb. These quantities of fission products are determined by their bimodal distribution (yield) which depends on the atomic mass, where the maxima are located in the ranges of 85–105 (Rb, Sr, Zr, Mo, Tc, platinoids, Ag) and 130–150 (Te, I, Xe, Cs, Ba, REE). With the burnup increase the content of fission products in SNF and thus in HLW increases too. Over time, the heat release of SNF decreases and within a few hundred years the main role in heat generation transfers from the short-lived fission products to long-lived actinides.

The SNF reprocessing generates from 1 to 13 m^3 of liquid HLW per 1 tonne of SNF [23], although after evaporation its volume is reduced to 250 L [24]. Specific features of the SNF composition are inherited by the HLW; in addition, it contains technological impurities [21,23,24]. In the dry residue of HLW there are [19,25]: 19% of transition metals (Mo, Zr, Tc), 18% of rare earths (La, Ce, Pr, Nd, Sm), 10% alkaline (Gs, Rb) and alkaline earth (Sr, Ba) elements, 7% of platinoids (Ru, Rh, Pd), 2% of minor actinides (Np, Am, Cm), up to 1% Se and Te, and another 43 wt.% belong for Fe, Ni, Cr, Na, P, and less significant elements. Calcined HLW contains (mol.%) [26,27]: Fission products—26.4 REE, 13.2 Zr, 12.2 Mo, 7.6 Ru, 7.0 Cs, 4.1 Pd, 3.5 Sr, 3.5 Ba, 1.3 Rb; 9% of other fission and corrosion products, mainly Tc, Rh, Te, I, Ni, and Cr; actinides: 1.4 (U + Th) and 0.2 (Am + Cm + Pu + Np), impurities—6.4 Fe, 3.2 P, 1.0 Na. According to [28], the HLW calcine after 10 years of storage consists of, wt.%: 6 Cs_2O, 6 TcO_2, 3 SrO, and 6 minor actinides (MA): Np, Am, and Cm, 4 BaO, 10 RuO_2, 15 REE_2O_3, 6 PdO, 15 ZrO_2, 2 Rh_2O_3, 15 MoO_3.

3. HLW Vitrification

Industrial vitrification of liquid HLW from SNF reprocessing is done for more than 50 years. It started in France in 1968–1973 [29] with 12 tonnes of radioactive glass produced by vitrifying 25 m^3 of waste. Large-scale vitrification of HLW is done in France since 1978 using A-B-Si glass whereas Russia uses Na-Al-P glass composition since 1987 [2,4,19,21,30–34]. A drawback of glasses is the limited content of nuclear waste which is typically from 3 to 5 wt.% for Na-Al-P glasses and from 5 to 20 wt.% for A-B-Si glasses. This results in a relatively large volume of vitreous product to be disposed of which entails high costs for the construction of underground disposal facilities. Therefore, approximately 1.5–1.8 tonnes of Na-Al-P glass is produced at immobilization of HLW derived at reprocessing of 1 tonne of SNF, while for A-B-Si glasses this value is significantly lower (about 0.4 tonnes only). Low content of radionuclides in the Na-Al-P glass is caused by its lower heat resistance [35]. The temperature of the vitrified waste block is requested to be at least 100 °C lower than the glass transition temperature (T_g) of glass to avoid crystallisation effects [19]. The temperature of the glass block depends on the intensity of radiogenic heat generation. The maximum design HLW radioactivity for Na-Al-P glass is 2500 Ci/L, with practical values being from 60 to 600 Ci/L [24,35]. The intensity of heat generation of vitrified HLW decreases with time from 18 kW/m^3 after 5 years to 6 kW/m^3 after 25 years, and to 1 kW/m^3 after 100 years of storage. Within the first 150 years, the main heat source of HLW is due to the decay of fission products with a dominance of ^{137}Cs and ^{90}Sr, and then due to the MA decay [22,36]. Regulatorily, the heat emission of vitrified HLW in the interim storage should not exceed 5 kW/m^3 [24] and 2 kW/m^3 on disposal [37]. The last document establishes the requirement for the temperatures up to 450 °C. Glasses are fundamentally metastable and depending on temperature can crystallize over time at elevated temperatures. Cooling of the glass melt on its draining into the canisters should be fast enough to form a uniform glass. This is achieved by quenching the melt at high cooling rates about 500 °C/h, whereas at cooling rates of 50 °C/h or less, crystalline phases appear [38]. Crystallisation of glass can occur after the melt is drained into canisters due to residual heat [6,7]. Often T_g is taken as the temperature at which the melt viscosity is 10^{12} Pa·s (10^{13} Poise) [39] although the glass transition is a quasi-equilibrium second order phase transformation, and the glassy state of matter differs topologically from that of molten state [40]. The Kauzmann empirical rule $T_g = (2/3)\cdot T_m$ ($T_{g,m}$ values are given in the Kelvin's scale) implies that the T_g increases with the increase of melting temperature (T_m). Heating of glass above T_g accelerates devitrification due to the decrease in glass viscosity and the increase of diffusion rate of elements in it. Na-Al-P glasses have T_g about 400 °C and the crystallization rate maximum is at ~500 °C, whereas A-B-Si glasses with T_g about 550 °C exhibit maximum crystallisation rates at about 650 °C [38,39]. Partial glass devitrification occurred when cooling rates about 30–50 °C/h close to that of HLW vitrification facilities were used [6]. On melt pouring from the EP-500 Joule-heated ceramic melter, the temperature of the container with vitrified HLW decreases within 17 h from 700 to 500 °C [7], that is, the cooling rate was just about 10–15 °C/h. The question therefore remains open about the state of vitrified HLW stored at "Mayak" PA since 1987 as it can be a relative homogeneous glass, although glass with crystalline phases or even a fully crystallized material are also possible variants. A partially crystallized glass in containers with HLW seems more likely and such wasteform is prone to an accelerated corrosion in the presence of water vapours [41,42].

4. GCM with Mineral-Like Phases

4.1. Importance of Novel Matrices

Much attention is paid to the modernization of existing glass matrices, as well as the search for new types of wasteforms, for example, for HLW radionuclide fractions. Partitioning of HLW and incorporation of the most dangerous long-lived actinides and fission products in a compact and capacious matrix will improve the use of underground repository space. This will reduce the need for the construction of new storage facilities and lead to savings in finances and time for their search and construction. Potential novel matrices for HLW immobilisation are crystalline and GCM which have been studied since the 1970s, almost simultaneously with research of glasses [32,34,43,44].

4.2. Crystaline Matrices

The best known crystalline wasteforms are the Synroc polyphase ceramic and the monophase NZP matrix. In the first, artificial phases of minerals with the structure of perovskite, zirconolite, pyrochlore, and hollandite serve as carriers of radionuclides and the HLW elements are distributed between them in accordance with the radius and charge of cations. The structure of the NZP matrix of the composition $NaZr_2(PO_4)_3$ as a natural analogue of the mineral kosnarite is formed by a three-dimensional network of PO_4^{3-} octahedra connected by vertices to ZrO_6 octahedra, and large Na^+ cations occupy voids. The HLW components can enter three positions of the structure: Alkalis, alkaline earths instead of sodium; REE and actinides are in the Zr position, hexavalent Mo replaces phosphorus, etc. Usually, there is an additional phase—REE phosphate with a monazite structure. Kosnarite, $KZr_2(PO_4)_3$ is a natural analogue of the NZP matrix, although unlike other phases (pyrochlore, zirconolite, brannerite, monazite, etc.), it does not contain radioactive elements such as U and Th [32]. The waste loading of Synroc and NZP ceramics is about 20 wt.%. Crystalline phases for immobilization of waste have been overviewed in many publications such as [15,16,32,45].

4.3. GCM as an Universal Nuclear Wasteform

GCM are thermodynamically more stable materials compared with homogeneous glasses. Indeed, the free Gibbs energy G_{GCM} of a GCM containing the volume fraction φ of crystalline phase will be:

$$G_{GCM} = \varphi G_C + (1 - \varphi)G_G,$$

where the free Gibbs energy of crystalline phase G_C is lower than the free Gibbs energy of glass G_G. Thus, the driving force of crystallization of GCM toward a most stable fully crystalline material (for which φ = 1) will be smaller.

GCM containing both crystalline and glassy phases are optimal for radioactive waste of complex composition [8,17,19,21,32,38,46–52]. Compared to homogeneous glassy materials, GCMs can incorporate larger amounts of waste elements and they can be produced using lower processing temperatures than those of conventional melting. Indeed, hazardous and nuclear waste constituents can be immobilised both by direct chemical incorporation into the glass structure in a classical vitrification approach and by the physical encapsulation of the waste in a glass matrix, forming a GCM consisting of both vitreous and crystalline phases (Figure 1).

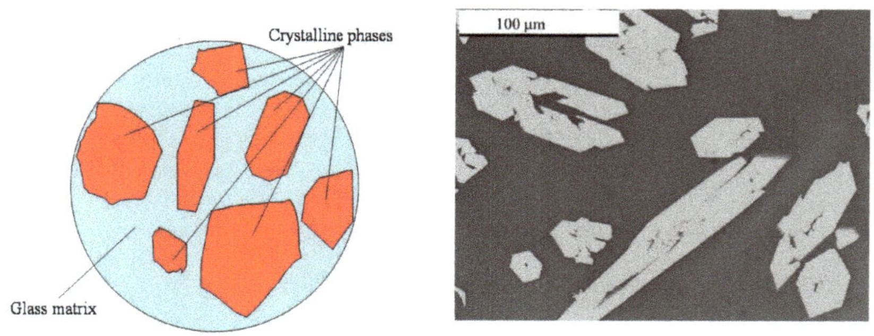

(a) (b)

Figure 1. Schematic of a Glass crystalline material (GCM) nuclear wasteform (**a**) and SEM/BSE image (**b**) of a real specimen loaded with rare earth elements (REE)-imitator of actinides. Dark areas are composed of glass and light grains are crystals of britholite.

4.4. Examples of GCM Nuclear Wasteforms

GCM include: Glass-ceramics where a glassy wasteform is crystallised using a separate heat treatment schedule, materials in which a refractory crystalline phase is encapsulated in glass, such as hot pressed lead silicate or sodium borosilicate glass matrix encapsulating up to 30 vol.% of $La_2Zr_2O_7$ or $Ga_2Zr_2O_7$ pyrochlore-structure crystals to immobilize actinides [53–57], borosilicate glasses for encapsulation of TRISO-UO_2 particles from pebble bed modular reactor [58], materials in which spent clinoptilolite from aqueous waste reprocessing is immobilised by pressureless sintering [59]; U/Mo-containing materials immobilized in a GCM termed U-Mo, glass formed by cold crucible melting (which partially crystallise on cooling) [60]; yellow phase containing up to 15 vol.% of sulphates, chlorides, and molybdates [61]; and materials to immobilise ashes from incineration of solid radioactive wastes [62]. An example of a recently developed GCM system obtained via sintering is the GCM designed for spent clinoptilolite waste arising from low-level aqueous waste treatment facilities [52,62] where a 2 h pressureless sintering at 750 °C was used. Waste loading ranging from 1:1 up to 1:10 of glass to clinoptilolite volume ratios corresponding to 37–88 mass% were analysed. Water durability of the GCMs assessed by 7 day leaching tests in deionised water at 40 °C based on ASTM C1220-98 standard remained below 6.35×10^{-6} g/(cm^2 day) in a GCM with 73 mass% waste loading. The sintering process typically takes place at lower temperatures compared to the classical melting route and is of great interest for the immobilization of highly volatile radionuclides such as $^{134,135,137}Cs$. Crystalline phases within a glassy matrix of GCM can be zirconolite, pyrochlore, brannerite, britholite within a matrix made of a A-B-Si glass, or monazite, or NZP in a Na-Al-P glass. In silicate glass ceramics, radionuclides are incorporated both in the phases of silicates (britholite), titanates, and zirconates (pyrochlore, zirconolite), and molybdates (powellite), and in phosphate glass ceramics—into phosphates with the structure of monazite, kosnarite, or langbeinite [63–71]. Table 1 outlines various waste types that have been immobilised into GCM both at laboratory and pilot-scale facilities [72,73].

4.5. GCM Phase Assemblage

Generally, host glasses used to develop GCM should have the ability to incorporate various waste constituents, be processed at relatively low processing temperatures, as well as be chemically durable and radiation resistant. High silica-content glasses including borosilicates, alumina silicates, alumina borosilicates, soda-lime silicates, lead silicates, lead borosilicates, phosphates, copper phosphates, silver phosphates, lead-iron phosphates, and soda alumina-phosphate glasses have been examined as host matrices for GCM production [17,18,72,73]. Several processing routes have been examined to produce GCM including controlled crystallisation of a glass, powder methods and sintering, petrurgical method, and sol-gel precursor glass, which give flexibility in the choice of the optimal immobilisation technology for the target waste as outlined in Table 2.

GCM can be used to immobilise long-lived radionuclides such as MA by incorporating them into the durable crystalline phases, whereas the short-lived radionuclides can readily be accommodated in the less durable glass. Acceptable durability will result if species of concern are locked into the crystal phases that are encapsulated in a durable glass matrix. The processing, compositions, phase assemblages, and microstructures of GCM can be tailored to achieve the necessary properties for improved performance of the wasteform.

4.6. Technology Readiness Level for Industrial-Scale Advanced Wasteforms Fabrication

When choosing a matrix, its hydraulic stability, its capacity for radionuclides, the ability to include other elements of waste, and the availability of industrial production technology play a key role. GCM samples were already obtained in laboratory conditions by viscous- and solid-phase sintering at normal and high pressure; high-speed electro-pulse sintering (SPS); microwave heating or SHS (self-propagating high-temperature synthesis); melting in a resistance furnace or in a "cold" induction heating crucible (CCIM). On an industrial scale, inductance melting in a "hot" or "cold" crucible is already used for the manufacture of A-B-Si glasses to immobilise HLW and intermediate level radioactive waste (ILW) in France and in Russia. The production of highly radioactive wasteforms weighing 10–100 kg corresponds to the value of technology readiness level (TRL) equal to 7, and in the case of their production in an amount of more than 1 tonne—as 8 and 9 [74]. The readiness of the manufacture of radioactive wasteforms via melting or sintering is estimated at TRL values from 7 to 9. These methods are already used in real production, or at semi-industrial and pilot plants. The capabilities of CCIM have been demonstrated in laboratory conditions to obtain blocks of crystalline matrices to 18 kg at TRL 6 [75,76]. The current state of the separation of HLW and methods for obtaining wasteforms with fractionated waste (radionuclide partitioning) envisages a relatively quick transition to the TRL equal to 9, which corresponds to the industrial implementation of this approach to handling HLW and acceptable financial costs [8]. Another way is to separate radionuclides for transmutation although this technique is still at an early stage of development with TRL about 3 to 4. A combined approach uses both options: The utilization of cubic ZrO_2 solid solution with fluorite-type structure, Zr-pyrochlore, or Y-Al garnet as a fuel for actinides (Np, Am) with an inert (composed of non-fertile elements) matrix for a single combustion in a reactor and disposal without reprocessing. The wasteforms to be used for the immobilisation of HLW resulting from SNF reprocessing including the fuel are summarized in Table 3 following the reference [8].

Table 1. Waste streams immobilised into GCM wasteforms.

Waste Type	Waste Description	Waste Loading
LILW	High sulphate, molybdate/noble metal content waste. High concentration of Cl.	30 wt%
Intermediate level radioactive waste (ILW)	High content of Mo, PbO, MnO, and simulated zircaloy chips.	20 wt%
High level radioactive waste (HLW)	Liquid waste arising from reprocessing of natural uranium (UO_2).	10 wt%
The same	Calcine containing transuranics, fission products, and transition elements.	60–80 wt%
	Acidic HLW with Al, Zr, Na, nitrate, and fluoride ions.	75 wt%
	From reprocessing light water reactor and fast breeder fuel.	5–35 wt%
	Zeolite occluded salt waste (transuranics, fission products, and halides).	75 wt%
	Fission product and transuranic actinides containing salt immobilized in glass bonded sodalite.	75wt%
	Zeolite occluded with molten LiCl-KCl-NaCl and ~6wt% of fission product chlorides.	50–67 wt%
	Simulated waste with spinel forming components.	45 wt%
	Simulated HLW consisting $\leq$ 35wt% ZrO_2.	30–50 wt%
	Plutonium bearing nuclear legacy waste in pyrochlore phases.	30 vol.%
	Containing both actinides and chlorides.	11 wt%
	Simulated HLW waste chemically immobilized in a mixture of chlorapatite, $Ca(PO_4)Cl$ and spodiosite, $Ca_2(PO_4)Cl$ mineral phase.	65–90 wt%
	Long-lived nuclear simulated waste (actinides).	20 wt%
	$Ag^{129}I$	
	TRISO-UO_2 particles.	16 vol.%
	Actinides surrogate (Ce_2O_3,Nd_2O_3, Eu_2O_3,Gd_2O_3,Yb_2O_3,ThO_2 in highly durable zirconolite ($CaZrTi_2O_7$).	4–6wt.%
	Waste fission product and actinides in titanite, $CaTiSiO_5$.	
	Waste ions distributed in sphene, $CaTiSiO_5$ phase and glass phase.	5 wt%
	Estimated simulation of HLW from reprocessing of nuclear fuels in Japan Atomic Institute.	25 wt%
	Simulated ^{90}Sr HLW partitioned from of high level liquid waste in China immobilized in apatite glass-ceramic.	35 wt%
Ashes	From incineration of plutonium bearing organics.	50 wt%
	From incineration of solid radioactive waste.	15–40 wt%

Table 2. Processing routes of GCM wasteforms.

Route	Description	Process Parameters
Pressureless sintering	Powder mixing, cold pressing. Relatively low temperature (≤1100 K).	Particle size, compacting pressure, temperature and duration of sintering, heating and cooling rate.
Hot Pressing	Pressure and temperature applied during Hot Uniaxial Press (HUP), Hot Isostatic Press (HIP) or Hydrothermal hot-pressing (HHP).	Glass composition, particle size, maximum temperature, pressure, soaking time, heating and cooling rate.
Reaction Sintering	Sintering under isostatic pressure by adding amorphous silica. Chemical reaction between waste components and surrounding glass.	Particle size, HIP temperature, pressure and duration of sintering.
Sintering with aerogels	Porous network of silica is soaked in a solution containing the actinide, then dried and fully sintered.	Mechanical properties, capillary forces, permeability of aerogel, sintering temperature.
Cold crucible melting	Electric currents generated inside waste contained in water-cooled crucible, surrounded by an induction coil.	Operating frequency, input vibrating power, operating temperature, melting duration.
Self-sustaining vitrification	Utilises the energy released during exorthermic chemical reactions.	Composition of initial waste and Powder Metal Fuel.
In situ sintering	Utilises ambient pressure of a disposal environment, its radiation shielding and extended time of storage.	Disposal environment, ambient pressure and temperature.
Controlled crystallization	Additional heat-treatment on to vitrified glass forming glass-ceramic.	Temperature, duration, heating and cooling rate.
Petrurgic method	Crystals nucleate and grow directly upon cooling glass from the melting temperature.	Cooling rate from melt temperature.
Sol-gel followed by sintering	Allows the formation of a very reactive powder in which components are mixed on a molecular scale.	Calcine temperature, milling media, drying temperature, sintering temperature.

Table 3. Possible wasteforms for HLW fractions. GCM—glass crystalline material, is shown in bold italics.

<table>
<tr><th>HLW Fraction</th><th>U + Pu</th><th>Np</th><th>Am</th><th>Cm</th><th>REE</th><th>Cs-Sr</th><th>FP [2]–UR [3]</th></tr>
<tr><td rowspan="7">Methods for managing waste fractions or separate radionuclides</td><td colspan="4">New (fresh) nuclear fuel</td><td>Ceramics</td><td rowspan="5">Glass, Ceramics or GCM</td><td>Alloy</td></tr>
<tr><td colspan="3">Fuel</td><td colspan="2">Ceramics</td><td>Alloy</td></tr>
<tr><td colspan="2">Fuel</td><td colspan="3">Ceramics</td><td>Alloy</td></tr>
<tr><td>Fuel</td><td colspan="4">Ceramics</td><td>Alloy</td></tr>
<tr><td>Fuel</td><td colspan="4">GCM (An/Ln/TM [1])</td><td>Alloy (UR)</td></tr>
<tr><td colspan="2">Fuel</td><td colspan="3">Ceramics</td><td colspan="2">GCM</td></tr>
<tr><td colspan="2">Fuel</td><td colspan="5">GCM</td></tr>
</table>

[1] TM—4d transition metals ([2] FP—fission products: Mo, Zr, Ru, Rh, Pd; corrosion elements—Zr, Fe, Cr, Ni, etc). [3] UR—undissolved SNF residues, composed of Mo-Tc-Ru-Rh-Pd alloy (ε-phase).

Although the review did not discuss the economics of immobilisation technologies used, one can note that the SNF reprocessing with fractionation of HLW (radionuclide partitioning) is more expensive than the standard PUREX technology. However, the GCM immobilization option will be cheaper and safer due to the reduced need for expensive deep geological disposal facilities [77,78] and because the dangerous long-lived isotopes are incorporated into the most durable host phases for industrial applying.

5. Conclusions

HLW derived in the close nuclear fuel cycle is currently immobilised into Na-Al-P and A-B-Si glasses using vitrification technology. Waste immobilisation in GCMs has emerged as a versatile technology enabling reliable immobilisation of complex and varying composition waste streams, including both radioactive and hazardous residues, which are otherwise difficult to immobilise using the traditional vitrification technology. The optimisation of the GCM phase assemblages as a function of waste stream composition is important for achieving simultaneously high nuclear waste loadings and corrosion resistance. Future research may focus on practical aspects of GCM utilisation through one or another technological process using either controlled devitrification of synthesized parent glasses or sintering routes using crystalline and vitreous precursors.

Author Contributions: Conceptualization, S.V.Y. and M.I.O.; methodology, S.V.Y.; investigation, S.V.Y. and M.I.O.; resources, V.A.P.; data curation, S.V.Y. and M.I.O.; writing—original draft preparation, M.I.O.; writing—review and editing, S.V.Y.; supervision, V.A.P.; project administration, V.A.P.; funding acquisition, V.A.P. All authors have read and agreed to the published version of the manuscript.

Funding: The research was carried out within the framework of the state assignment of IGEM RAS, the topic of research work: "Geoecological problems of nuclear power".

Institutional Review Board Statement: Not applicable.

Informed Consent Statement: Not applicable.

Data Availability Statement: Data used to support findings of this study can be made available from the corresponding author upon request.

Acknowledgments: Authors acknowledge critical comments by anonymous reviewers which helped to improve the review.

Conflicts of Interest: The authors declare no conflict of interest.

References

1. Gin, S.; Abdelouas, A.; Criscenti, L.J.; Ebert, W.L.; Ferrand, K.; Geisler, T.; Harrison, M.T.; Inagaki, Y.; Mitsui, S.; Mueller, K.T.; et al. An international initiative on long-term behavior of high-level nuclear waste glass. *Mater. Today* **2013**, *16*, 243–248. [CrossRef]
2. Ojovan, M.I. On Alteration Rate Renewal Stage of Nuclear Waste Glass Corrosion. *MRS Adv.* **2020**, *5*, 111–120. [CrossRef]

3. Luzhetsky, A.V.; Petrov, V.A.; Yudintsev, S.V.; Malkovsky, V.I.; Ojovan, M.I.; Nickolsky, M.S.; Shiryaev, A.A.; Danilov, S.S.; Ostashkina, E.E. Effect of gamma irradiation on structural features and dissolution of nuclear waste Na–Al–P glasses in water. *Sustainability* **2020**, *12*, 4137. [CrossRef]
4. Gin, S.; Jollivet, P.; Tribet, M.; Peuget, S.; Schuller, S. Radionuclides containment in nuclear glasses: An overview. *Radiochim. Acta* **2017**, *105*, 927–959.
5. Jantzen, C.M.; Ojovan, M.I. On selection of matrix (wasteform) material for higher activity nuclear waste immobilisation (Review). *Russ. J. Inorg. Chem.* **2019**, *64*, 1611–1624. [CrossRef]
6. Kozlov, P.V.; Remizov, M.B.; Belanova, E.A.; Vlasova, N.V.; Orlova, V.A.; Martynov, K.V. Modification of the composition of aluminophosphate glasses with HLW simulators to increase their stability. 1. The effect of modifiers on the viscosity and crystallization ability of melts. *Rad. Saf. Issues* **2019**, *1*, 3–15.
7. Stefanovsky, S.V.; Stefanovsky, O.I.; Remizov, M.B.; Kozlov, P.V.; Belanova, E.A.; Makarovsky, R.A.; Myasoedov, B.F. Sodium–aluminum–iron phosphate glasses as legacy high-level waste forms. *Prog. Nucl. Energy* **2017**, *94*, 229–234. [CrossRef]
8. Yudintsev, S.V. Isolation of fractionated waste of nuclear industry. *Radiochemistry* **2021**, in press.
9. Ewing, R.C. The Design and Evaluation of Nuclear-Waste Forms: Clues from Mineralogy. *Canad. Miner.* **2001**, *39*, 697–715. [CrossRef]
10. Omel'yanenko, B.I.; Livshits, T.S.; Yudintsev, S.V.; Nikonov, B.S. Natural and artificial minerals as matrices for immobilization of actinides. *Geol. Ore Depos.* **2007**, *49*, 173–193. [CrossRef]
11. Vance, E.R. Ceramic Waste Forms. In *Comprehensive Nuclear Materials*; Konings, R., Allen, T., Stoller, R., Yamanaka, S., Eds.; Elsevier: Amsterdam, The Netherlands, 2012; Volume 5, Chapter 5.19; pp. 485–503.
12. Stefanovsky, S.V.; Yudintsev, S.V. Titanates, zirconates, aluminates and ferrites as waste forms for actinide immobilization. *Russ. Chem. Rev.* **2016**, *85*, 962–994. [CrossRef]
13. Orlova, A.I.; Koryttseva, A.K.; Loginova, E.E. A family of phosphates of langbeinite structure. Crystal-chemical aspect of radioactive waste immobilization. *Radiochemistry* **2011**, *53*, 51–62. [CrossRef]
14. Orlova, A.I.; Lizin, A.A.; Tomilin, S.V.; Lukinykh, A.N. On the possibility of realization of monazite and langbeinite structural types in complex americium and plutonium phosphates. Synthesis and X-ray diffraction studies. *Radiochemistry* **2011**, *53*, 63–68. [CrossRef]
15. Orlova, A.I.; Ojovan, M.I. Ceramic Mineral Waste-Forms for Nuclear Waste Immobilization. *Materials* **2019**, *12*, 2638. [CrossRef]
16. Burakov, B.E.; Ojovan, M.I.; Lee, W.E. *Crystalline Materials for Actinide Immobilization*; Imperial College Press: London, UK, 2011; 197p.
17. Ojovan, M.I.; Lee, W.E.; Kalmykov, S.N. *An Introduction to Nuclear Waste Immobilisation*, 3rd ed.; Elsevier: Amsterdam, The Netherlands, 2019; 497p.
18. Ojovan, M.I.; Juoi, J.M.; Boccaccini, A.R.; Lee, W.E. Glass Composite Materials for Nuclear and Hazardous Waste Immobilisation. *Mater. Res. Soc. Symp. Proc.* **2008**, *1107*, 245–252. [CrossRef]
19. Caurant, D.; Loiseau, P.; Majérus, O.; Aubin-Cheval-Donnet, V.; Bardez, I.; Quintas, A. *Glasses, Glass-Ceramics and Ceramics for Immobilization of Highly Radioactive Nuclear Wastes*; Nova Science Publishers: New York, NY, USA, 2009; 445p.
20. Ojovan, M.I.; Lee, W.E. *New Developments in Glassy Nuclear Wasteforms*; Nova Science Publishers: New York, NY, USA, 2007; 131p.
21. Donald, I.W. *Waste Immobilization in Glass and Ceramic-Based Hosts*; Wiley: Chichester, UK, 2010; 507p.
22. Fedorov, Y.S.; Zil'berman, B.Y.; Aloi, A.S.; Puzikov, E.A.; Shadrin, A.Y.; Alyapyshev, M.Y. Problems of modernization of spent nuclear fuel extraction processing. *Russ. J. Gen. Chem.* **2011**, *81*, 1932–1948. [CrossRef]
23. Kopyrin, A.A.; Karelin, A.I.; Karelin, V.A. *Technology of Production and Radiochemical Processing of Nuclear Fuel*: Atomizdat Publishing House CJSC: Moscow, Russia, 2006; 576p.
24. Batorshin GSh Kirillov, S.N.; Smirnov, I.V.; Sarychev, G.A.; Tananaev, I.G.; Fedorova, O.V.; Myasoedov, B.F. Integrated separation of valuable components from technogenic radioactive waste as an option to create cost-eective nuclear fuel cycle. *Radiat. Saf. Issues* **2015**, *3*, 30–36.
25. Caurant, D. Spectroscopic investigations on glasses, glass-ceramics and ceramics developed for nuclear waste immobilization. *Opt. Spectrosc.* **2014**, *116*, 667–676. [CrossRef]
26. McCarthy, G.J. High Level Waste Ceramics: Materials Considerations, Process Simulation, and Product Characterization. *Nucl. Technol.* **1977**, *32*, 92–105. [CrossRef]
27. Ringwood, A.E.; Kesson, S.E.; Ware, N.G.; Hibberson, W.; Major, A. Immobilisation of high-level nuclear reactor wastes in SYNROC. *Nature* **1979**, *278*, 219–223. [CrossRef]
28. Stewart, M.W.A.; Vance, E.R.; Moricca, S.A.; Brew, D.R.; Cheung, C.; Eddowes, T.; Bermudez, W. Immobilisation of Higher Activity Wastes from Nuclear Reactor Production of ^{99}Mo. *Sci. Technol. Nucl. Install.* **2013**, *2013*, 926026. [CrossRef]
29. Vernaz, É.; Bruezière, J. History of Nuclear Waste Glass in France. *Proc. Mater. Sci.* **2014**, *7*, 3–9. [CrossRef]
30. Laverov, N.P.; Velichkin, V.I.; Omelyanenko, B.I.; Yudintsev, S.V.; Petrov, V.A.; Bychkov, A.V. *Isolation of Spent Nuclear Materials: Geological and Geochemical Foundations*; IPZ RAS: Moscow, Russia, 2008; 280p.
31. Remizov, M.B.; Kozlov, P.V.; Logunov, M.V.; Koltyshev, V.K.; Korchenkin, K.K. Conceptual and technical solutions on construction at PA "Mayak" plants for vitrification of current and stored liquid HLW. *Radiat. Saf. Issues* **2014**, *3*, 17–25.
32. Lutze, W.; Ewing, R.C. *Radioactive Waste Forms for the Future*; Elsevier: New York, NY, USA, 1988; 778p.

33. Poluektov, P.P.; Sukhanov, L.P.; Matyunin, Y.I. Scientific approaches and technical solutions in the field of liquid high-level waste management. *Russ. Chem. J.* **2005**, *XLIX*, 29–41.
34. Ojovan, M.I. *Handbook of Advanced Radioactive Waste Conditioning Technologies*; Woodhead: Cambridge, UK, 2011; 512p.
35. Vlasova, N.V.; Remizov, M.B.; Kozlov, P.V.; Belanova, E.A. Investigation of Chemical Stability ofAluminum-Phosphate Glasses Simulating Solidified HLW to be Returned to Foreign Contractors. *Radiat. Saf. Issues* **2017**, *3*, 32–37.
36. Carter, J.T.; Luptak, A.J.; Gastelum, J.; Stockman, S.; Miller, A. *Fuel Cycle Potential Waste Inventory for Disposition. Prepared for U.S. Department of Energy*; SRNL, INL, PNNL, SNL: Washington, DC, USA, 2012; 328p.
37. Acceptance criteria for radioactive waste for disposal (NP-093-14). *Nucl. Rad. Saf.* **2015**, *3*, 59–82.
38. Vashman, A.A.; Demin, A.V.; Krylova, N.V.; Kushnikov, V.V.; Matyunin, Y.I.; Poluektov, P.P.; Polyakov, A.S.; Teterin, E.G. *Phosphate Glasses with Radioactive Waste*; CNIIatominform: Moscow, Russia, 1997; p. 172.
39. Remizov, M.B.; Kozlov, P.V.; Borisenko, V.P.; Dementeva, I.I.; Blokhin, P.A.; Samoylov, A.A. Development of an Algorithm for Estimating the Radionuclide Composition of Vitrified HLW of FSUE "PA "Mayak" for the Purpose of Their Safe Disposal. *Radioact. Waste* **2018**, *3*, 87–94.
40. Ojovan, M.I. Glass formation. In *Encyclopedia of Glass Science, Technology, History, and Culture*; Richet, P., Conradt, R., Takada, A., Dyon, J., Eds.; Wiley: Hoboken, NJ, USA, 2021; pp. 249–259. [CrossRef]
41. Yudintsev, S.V.; Pervukhina, A.M.; Mokhov, A.V.; Malkovsky, V.I.; Stefanovsky, S.V. Influence of phosphate glass recrystallization on the stability of a waste matrix to leaching. *Dokl. Earth Sci.* **2017**, *473 Pt 2*, 427–432. [CrossRef]
42. Malkovsky, V.I.; Yudintsev, S.V.; Aleksandrova, E.V. Influence of Na-Al-Fe-P glass alteration in hot non-saturated vapor on leaching of vitrified radioactive wastes in water. *J. Nucl. Mater.* **2018**, *508*, 212–218. [CrossRef]
43. Hayward, P.J. Glass-ceramics. In *Radioactive Waste Forms for the Future*; Lutze, W., Ewing, R.C., Eds.; Elsevier: New York, NY, USA, 1988; pp. 427–494.
44. National Research Council. *Waste Forms Technology and Performance: Final Report*; National Academies Press: Washington, DC, USA, 2011; 308p.
45. Weber, W.J.; Ewing, R.C.; Catlow, C.R.A.; Diaz de la Rubia, T.; Hobbs, L.W.; Kinoshita, C.; Matzke Hj Motta, A.T.; Nastasi, M.; Salje, E.K.H.; Vance, E.R.; et al. Radiation effects in crystalline ceramics for the immobilization of high-level nuclear waste and plutonium. *J. Mater. Res.* **1998**, *13*, 1434–1484. [CrossRef]
46. McCloy, J.S.; Goel, A. Glass-ceramics for nuclear-waste immobilization. *Mater. Res. Soc. Bull.* **2017**, *42*, 233–238. [CrossRef]
47. Donald, I.W. Immobilisation of radioactive and non-radioactive wastes in glass-based systems: An overview. *Glass Technol.* **2007**, *48*, 155–163.
48. Rawling, R.D.; Wu, J.P.; Boccaccini, A.R. Glass-ceramics: Their production from wastes—A Review. *J. Mater. Sci.* **2006**, *41*, 733–761. [CrossRef]
49. Colombo, P.; Brusatin, G.; Bernardo, E.; Scarinci, G. Inertization and reuse of waste materials by vitrification and fabrication of glass-based products. *Curr. Opin. Solid State Mater. Sci.* **2003**, *7*, 225–239. [CrossRef]
50. Lee, W.E.; Ojovan, M.I.; Stennett, M.C.; Hyatt, N.C. Immobilisation of radioactive waste in glasses, glass composite materials and ceramics. *Adv. Appl. Ceram.* **2006**, *105*, 3–12. [CrossRef]
51. Lewis, M.A.; Fischer, D.F. Properties of glass-bonded zeolite monoliths. *Ceram. Trans.* **1994**, *45*, 277–286.
52. Juoi, J.M.; Ojovan, M.I.; Lee, W.E. Microstructure and leaching durability of glass composite wasteforms for spent clinoptilolite immobilisation. *J. Nucl. Mater.* **2008**, *372*, 358–366. [CrossRef]
53. Loiseau, P.; Caurant, D.; Majerus, O.; Baffier, N.; Fillet, C. Crystallization study of (TiO_2, ZrO_2)-rich SiO_2-Al_2O_3-CaO glasses Part I Preparation and characterization of zirconolite-based glass-ceramics. *J. Mater. Sci.* **2003**, *38*, 843–864. [CrossRef]
54. Boccaccini, A.R.; Bernardo, E.; Blain, L.; Boccaccini, D.N. Borosilicate and lead silicate glass matrix composites containing pyrochlore phases for nuclear waste encapsulation. *J. Nucl. Mater.* **2004**, *327*, 148–158. [CrossRef]
55. Digeos, A.A.; Valdez, J.A.; Sikafus, K.E.; Atiq, S.; Grimes, R.W.; Boccaccini, A.R. Glass matrix/pyrochlore phase composites for nuclear wastes encapsulation. *J. Mater. Sci.* **2003**, *38*, 1597–1604. [CrossRef]
56. Boccaccini, A.R.; Berthier, T.; Seglem, S. Encapsulated gadolinium zirconate pyrochlore particles in soda borosilicate glass as novel radioactive waste form. *Ceram. Int.* **2007**, *33*, 1231–1235. [CrossRef]
57. Pace, S.; Cannillo, V.; Wu, J.; Boccaccini, D.N.; Seglem, S.; Boccaccini, A.R. Processing glass-pyrochlore composites for nuclear waste encapsulation. *J. Nucl. Mater.* **2005**, *341*, 12–18. [CrossRef]
58. Abdelouas, A.; Noirault, S.; Grambow, B. Immobilization of inert TRISO-coated fuel in glass for geological disposal. *J. Nucl. Mater.* **2006**, *358*, 1–9. [CrossRef]
59. Henry, N.; Deniard, P.; Jobic, S.; Brec, R.; Fillet, C.; Bart, F.; Grandjean, A.; Pinet, O. Heat treatments versus microstructure in a molybdenum-rich borosilicate. *J. Non Cryst. Solids* **2004**, *333*, 199–205. [CrossRef]
60. Sobolev, I.A.; Ojovan, M.I.; Scherbatova, T.D.; Batyukhnova, O.G. *Glasses for Radioactive Waste*; Energoatomizdat: Moscow, Russia, 1999.
61. Sobolev, I.A.; Dmitriev, S.A.; Lifanov, F.A.; Kobelev, A.P.; Stefanovsky, S.V.; Ojovan, M.I. Vitrification processes for low, intermediate radioactive and mixed wastes. *Glass Technol.* **2005**, *46*, 28–35.
62. Juoi, J.M.; Ojovan, M.I. The effect of waste loading on the microstructure of glass composite waste form immobilising spent clinoptilolite. *Glass Technol.* **2007**, *48*, 124–129.

63. Maddrell, E.; Thornber, S.; Hyatt, N.C. The influence of glass composition on crystalline phase stability in glass-ceramic wasteforms. *J. Nucl. Mater.* **2015**, *456*, 461–466. [CrossRef]
64. Nicoleau, E.; Angeli, F.; Schuller, S.; Charpentier, T.; Jollivet, P.; Moskura, M. Rare-earth silicate crystallization in borosilicate glasses: Effect on structural and chemical durability properties. *J. Non Cryst. Solids* **2016**, *438*, 37–48. [CrossRef]
65. Zhang, Y.; Kong, L.; Karatchevtseva, I.; Aughterson, R.D.; Gregg, D.J.; Triani, G. Development of brannerite glass-ceramics for the immobilization of actinide-rich radioactive wastes. *J. Am. Ceram. Soc.* **2017**, *100*, 4341–4351. [CrossRef]
66. McCloy, J.S.; Marcial, J.; Patil, D.; Saleh, M.; Ahmadzadeh, M.; Chen, H.; Crum, J.V.; Riley, B.J.; Kamat, H.; Bréhault, A.; et al. Glass structure and crystallization in boro-aluminosilicate glasses containing rare earth and transition metal cations: A US-UK collaborative program. *MRS Adv.* **2019**. [CrossRef]
67. Kong, L.; Karatchevtseva, I.; Zhang, Y.; Wei, T. The incorporation of Nd or Ce in $CaZrTi_2O_7$ zirconolite: Ceramic versus glass-ceramic. *J. Nucl. Mater.* **2021**, *543*, 152583. [CrossRef]
68. Stefanovsky, S.V.; Stefanovsky, O.I.; Danilov, S.S.; Kadyko, M.I. Phosphate-based glasses and glass ceramics for immobilization of lanthanides and actinides. *Ceram. Int.* **2019**, *45*, 9331–9338. [CrossRef]
69. He, Y.; Lu, Y.; Zhang, Q. Characterization of monazite glass–ceramics as wasteform for simulated α-HLLW. *J. Nucl. Mater.* **2008**, *376*, 201–206. [CrossRef]
70. Asuvathraman, R.; Kitheri, J.; Madhavan, R.R.; Sudha, R.; Prabhu, K.R.; Govindan Kutty, K.V. A versatile monazite—IPG glass–ceramic waste form with simulated HLW: Synthesis and characterization. *J. Eur. Ceram. Soc.* **2015**, *35*, 4233–4239. [CrossRef]
71. Liu, J.; Wang, F.; Liao, Q.; Zhu, H.; Liu, D.; Zhu, Y. Synthesis and characterization of phosphate-based glass-ceramics for nuclear waste immobilization: Structure, thermal behavior, and chemical stability. *J. Nucl. Mater.* **2019**, *513*, 251–259. [CrossRef]
72. Juoi, J.M.; Ojovan, M.I.; Lee, W.E. Development of glass-composite materials for radioactive waste immobilization. In Proceedings of the Proc. ICEM'05, ICEM-1069, Glasgow, UK, 28–30 October 2005; ASME: Glasgow, UK, 2005.
73. Ojovan, M.I.; Juoi, J.M.; Lee, W.E. Glass Composite Materials for Nuclear Waste Immobilisation. In Proceedings of the Extended Abstracts ICG 2007 Proc. XXIst International Congress on Glass, Strasbourg, France, 2–6 July 2007. U37, 5p.
74. *State-of-the-Art Report on Light Water Reactor Accident-Tolerant Fuels*; NEA: Paris, France, 2018; 367p.
75. Yudintsev, S.V.; Stefanovsky, S.V.; Kalenova, M.Y.; Nikonov, B.S.; Nikol'skii, M.S.; Koshcheev, A.M.; Shchepin, A.S. Matrices for immobilization of the rare earth–actinide waste fraction, synthesized by cold crucible induction melting. *Radiochemistry* **2015**, *57*, 321–333. [CrossRef]
76. Amoroso, J.W.; Marra, J.; Dandeneau, C.S. Cold crucible induction melter test for crystalline ceramic waste form fabrication: A feasibility assessment. *J. Nucl. Mater.* **2017**, *486*, 283–297. [CrossRef]
77. *Advanced Fuel Cycle Cost Basis. Prepared for U.S. Department of Energy Fuel Cycle Options Campaign*; NTRD-FCO-2017-000265; INL: Idaho Falls, ID, USA, 2017.
78. Sorokin, V.T.; Pavlov, D.I. The cost of radwaste disposal: Foreign assessments. *Radioact. Waste* **2019**, *1*, 39–46.

MDPI
St. Alban-Anlage 66
4052 Basel
Switzerland
www.mdpi.com

Sustainability Editorial Office
E-mail: sustainability@mdpi.com
www.mdpi.com/journal/sustainability

www.ingramcontent.com/pod-product-compliance
Lightning Source LLC
Chambersburg PA
CBHW061807210926
51850CB00006B/638
* 9 7 8 3 0 3 6 5 8 6 0 6 9 *